The Correspondence of John Tyndall

VOLUME 12

THE CORRESPONDENCE OF JOHN TYNDALL

VOLUME 12

The Correspondence, March 1871–May 1872

Edited by
Anne DeWitt and Kathleen Sheppard

General Editors
Roland Jackson, Bernard Lightman, and Michael S. Reidy

Associate Editors
Michael D. Barton, Erin Grosjean, and Michael Laurentius

UNIVERSITY OF PITTSBURGH PRESS

Published by the University of Pittsburgh Press, Pittsburgh, Pa., 15260

Manufactured in the United States of America
Printed on acid-free paper
10 9 8 7 6 5 4 3 2 1

Cataloging-in-Publication data is available from the Library of Congress

ISBN 13: 978-0-8229-4689-2
ISBN 10: 0-8229-4689-0

CONTENTS

ACKNOWLEDGMENTS

First and foremost, we want to thank Mallory Stubbs, who worked as the research assistant for our volume. Mallory is the primary author of the biographical register as well as many of the biographical notes; her ingenuity and resourcefulness enabled her to track down many obscure figures. Mallory contributed in many other ways to this volume, and her attention to detail, organizational skills, and technological prowess made the process much smoother and the end result much better than they would have been in her absence.

We are also grateful to the support and assistance rendered by many people involved in the Tyndall Correspondence Project. We especially acknowledge the work of many graduate students and other colleagues in providing the original transcriptions. We would like to thank the general editors, Bernard Lightman, Roland Jackson, and Michael Reidy, for answering questions and providing feedback and guidance along the way. Erin Grosjean was unfailingly helpful and unflaggingly patient with our many emails. We are also grateful to Michael Laurentius and Michael Barton for their assistance. And we would like to thank the editors of other volumes, especially Elizabeth Neswald, Nathan Kapoor, and Matthew Stanley, for their assistance.

Jane Harrison and her colleagues at the Royal Institution helped us with scans and fielded archival questions. Eleanor Coulter helped us to check French-language letters; James Braund did the same with German-language letters. We are also grateful to James for his assistance in identifying German scientific practitioners and their publications. G. Colussi Arthur translated the Italian-language letter and made the English flow much better.

In addition, many colleagues and friends shared their expertise as we annotated the letters. We would like to thank Petra DeWitt, Gabriel Finkelstein, Andrew Goldstone, Natalie Houston, Patrick Leary, Katherine Mannheimer, Jonah Siegal, Sara Murphy, Thomas Tag, and Gregory Vargo.

Special thanks go to Bernard Lightman for pairing us together. We have found each other to be reliable and hardworking colleagues, but more than that, we have each found a true friend.

Finally, Anne would like to thank the staff of the Edgar Early Learning

Center, whose work and expertise afforded her the time and peace of mind necessary to complete this process. Kate would like to thank her son's teachers over the last few years, especially Amanda Brookshire, Krista Cross, and Brittany Sneed, who not only love him but also taught him to read and do math; also, the staff and teachers at Greentree Childcare Center are invaluable for after-school and summer care. Our husbands, Andrew and Dan, were and are understanding when work takes over. Without these important teachers and caregivers neither of us would be able to do our jobs.

LIST OF ABBREVIATIONS

ANB	*American National Biography*
Ascent of John Tyndall	Roland Jackson, *The Ascent of John Tyndall: Victorian Scientist, Mountaineer, and Public Intellectual* (Oxford: University of Oxford Press, 2018)
BAAS	British Association for the Advancement of Science
BL	British Library
Brit. Assoc. Rep.	*Reports of the British Association for the Advancement of Science*
Brock and MacLeod, *Hirst Journals*	William H. Brock and Roy M. MacLeod, eds., *Natural Knowledge in Social Context: The Journals of Thomas Archer Hirst FRS* (London: Mansell microfiche, 1980)
CDSB	*Complete Dictionary of Scientific Biography*, 27 vols. (Detroit, MI: Charles Scribner's Sons, 2008), Gale eBooks
CUL	Cambridge University Library
CUL SC	Cambridge University Library, Stokes Correspondence
Desmond, *Huxley*	Adrian Desmond, *Huxley: From Devil's Disciple to Evolution's High Priest* (Reading, MA: Perseus Books, 1994)
EH	Elton Hall
Endersby, *Imperial Nature*	Jim Endersby, *Imperial Nature: Joseph Hooker and the Practices of Victorian Science* (Chicago: University of Chicago Press, 2008)
Forms of Water	John Tyndall, *The Forms of Water in Clouds and Rivers, Ice and Glaciers* (London: Henry S. King, 1872)
Fragments of Science	John Tyndall, *Fragments of Science for Unscientific People: A Series of Detached Essays, Lectures, and Reviews*, 1st ed. (London: Longmans, Green, 1871)
FRAS	Fellow of the Royal Astronomical Society
FRS	Fellow of the Royal Society
FRSE	Fellow of the Royal Society of Edinburgh
Glaciers of the Alps	John Tyndall, *The Glaciers of the Alps* (London: John Murray, 1860)
IC HP	Imperial College of Science, Technology and Medicine, College Archives, Huxley Papers
IET	Institution of Engineering and Technology

Hours of Exercise in the Alps	John Tyndall, *Hours of Exercise in the Alps*, 1st ed. (London: Longmans, Green, 1871)
Journal	John Tyndall's Journal, 1855–72, typescript (RI MS JT/2/13c)
LT	Louisa Tyndall, John Tyndall's wife
MacLeod, "Ayrton Incident"	Roy MacLeod, "The Ayrton Incident: A Commentary on the Relations of Science and Government in England, 1870–73," in *Science and Values: Patterns of Tradition and Change*, ed. Arnold Thackray and Everett Mendelsohn, eds., 45–78 (New York: Humanities Press, 1974)
Molecular Physics	John Tyndall, *Contributions to Molecular Physics in the Domain of Radiant Heat* (London: Longmans, Green, 1872)
National Archives MT	National Archives, Kew, Ministry of Transport Records
NLS	National Library of Scotland
Phil. Mag.	*Philosophical Magazine*
Phil. Trans.	*Philosophical Transactions of the Royal Society of London*
RI	Royal Institution of Great Britain
RI MS JT	Royal Institution, London, John Tyndall Papers
Roy. Inst. Proc.	*Proceedings of the Royal Institution of Great Britain*
Roy. Soc. Proc.	*Proceedings of the Royal Society of London*
RS	Royal Society
Scrambles amongst the Alps	Edward Whymper, *Scrambles amongst the Alps in the Years 1860–69* (London: John Murray, 1871)
Sound	John Tyndall, *Sound: A Course of Eight Lectures delivered at the Royal Institution of Great Britain* (London: Longmans, Green, 1867)
Tyndall Correspondence, Volume 6	Michael D. Barton, Janet Browne, Ken Corbett, and Norman McMillan, eds., *The Correspondence of John Tyndall*, vol. 6: *The Correspondence, November 1856–February 1859* (Pittsburgh: University of Pittsburgh Press, 2019)
Tyndall Correspondence, Volume 7	Roland Jackson, Nanna Katrine Lüders Kaalund, and Diarmid A. Finnegan, eds., *The Correspondence of John Tyndall*, vol. 7: *The Correspondence, March 1859–May 1862* (Pittsburgh: University of Pittsburgh Press, 2019)
Tyndall Correspondence: Volume 8	Piers J. Hale, Elizabeth Neswald, Nathan N. Kapoor, and Michael D. Barton, eds., *The Correspondence of John Tyndall*, vol. 8, *The Correspondence, June 1862–January 1865* (Pittsburgh: University of Pittsburgh Press, 2020)
Tyndall Correspondence: Volume 10	Michael D. Barton, Ken Corbett, and Roland Jackson, eds., *The Correspondence of John Tyndall*, vol. 10, *The Correspondence, January 1867–December 1868* (Pittsburgh: University of Pittsburgh Press, 2021)

Tyndall Correspondence, Volume 11	Adrian Kirwan and Elizabeth Neswald, eds., *The Correspondence of John Tyndall*, vol. 11, *The Correspondence, January 1869–February 1871* (Pittsburgh: University of Pittsburgh Press, 2022)
UCL	University College, London

Caricature of John Tyndall, *Vanity Fair*, 6 April 1872. 'Men of the Day, No. 43.' Courtesy of the National Portrait Gallery, London.

INTRODUCTION TO VOLUME 12

The twelfth volume of *The Correspondence of John Tyndall* contains 340 letters and covers the fifteen months of Tyndall's life from March 1871 through May 1872. It begins just before the publication of *Fragments of Science* in April and *Hours of Exercise in the Alps* in May. The volume includes a series of small but public disputes about science. Tyndall had a number of visits from friends and dignitaries, and he traveled to Switzerland, Ireland, and the countryside for scrambles. He was dealing with family issues in Ireland, which were troublesome for him. He was busy administering the Royal Institution (RI) and the Royal Society (RS); he was also working as the scientific consultant to Trinity House, which was involved in overseeing lighthouses in England, Scotland, and Ireland.

Unlike other volumes, this volume is not defined as much by one or two major projects or events for Tyndall, but instead includes a number of smaller projects and issues for him personally and professionally. It is therefore possible for us to learn more about Tyndall's life than his work during this time. Starting in late 1870 and early 1871, he began to refuse both social and professional invitations.[1] He realized that he did not need to attend everything, mainly because he was firmly in position as a leading man of science. On the other hand, he became frustrated with the work required for Trinity House and other smaller projects, so he did not have time for all the socializing or research he wanted to do.

Tyndall already had a "substantial" reputation in both Britain and the United States by April of 1871.[2] In contrast to his position earlier in life, Tyndall was no longer working hard to create himself as a man of science; he had become a central figure in the field by the time this volume begins. And yet, although he was well established, Tyndall remained concerned with his image, and we see this manifest in a number of ways.

Fragments of Science and *Hours of Exercise in the Alps*

In April and May 1871, Tyndall published two of his best-selling books, *Fragments of Science for Unscientific People: A Series of Detached Essays, Lectures, and Reviews* and *Hours of Exercise in the Alps*. As early as March, Tyndall was sending copies of *Fragments of Science* and *Hours of Exercise in the Alps* to his friends and colleagues. The first edition of *Fragments of Science* is a compilation of about twenty of Tyndall's shorter articles, ranging from "Thoughts on Prayer and Natural Law" and "Miracles and Special Providences" to "On Radiation" and "Dust and Disease." Some of these articles were already being printed in the United States, and *Fragments of Science* was Tyndall's attempt to circumvent those unauthorized publications.[3] The first edition of one thousand copies quickly sold out in Britain, and Anna Helmholtz translated *Fragments of Science* into German that same year. Eight total editions were published in England, and likely more in America, before his death.

Fragments of Science was generally well received by critics and, of course, by his friends. Mary Egerton wrote to him after receiving her copy, "1,000 0000 00000000 . . . thanks for [the book] . . . I recognize 11 out of the 13 longer articles as old friends."[4] Reviews touted Tyndall's "frankness combined with lucidity" and his "graceful eloquence";[5] on the other hand, they argued that the articles were not necessarily easy reading for everyone. The *Daily News* wrote that Tyndall "has the gift of rendering whatever he writes interesting to and comprehendable by any reader of ordinary intelligence."[6] However, they thought that the title, specifically the part that said the book was "for unscientific people" was "faulty." The *Athenaeum*, a leading weekly that reviewed a wide range of books, offered a glowing account of *Fragments of Science*.[7] They wrote that Tyndall's "rollicking jollity" rendered the book useful to communicating scientific ideas to the public. Tyndall wrote in a style in which his "sentences [were] fearfully and wonderfully made, wholly beyond the pale of syntactical law," but that style was what made the book available to the general public.[8] There were some reviewers who took issue with the fact that articles about miracles were included with articles about the natural world, but *Scientific American* also noted that what they called Tyndall's "materialism" was woven through both types of pieces, bringing some coherence to otherwise unconnected ideas.[9] The materialist arguments and scientific naturalism contained in these chapters also gave some indication of the major debates that were about to consume much of Tyndall's career: the "prayer-gauge debate" in 1872 and the fallout from his 1874 Belfast Address.[10]

Hours of Exercise in the Alps was Tyndall's second book of the year, published in May of 1871. He wrote in the preface that the book was meant to be a sort of supplement to *Fragments of Science*. *Hours of Exercise in the Alps* offered distinct chapters and stories about traveling and studying in the Alps instead of "in the Attic and Laboratory."[11] It was largely a memoir of Tyndall's major expeditions in the Alps, "including his struggles with the Matterhorn," which he termed "assaults," as though the mountain were something to be conquered and beaten into submission.[12] One of the most tragic stories in it was of his friend and guide Johann Joseph Bennen, who died in an avalanche on the Haut de Cry.[13] The chapter was written by Philip Gossett, and it was a well-known story by the time Tyndall published *Hours of Exercise in the Alps* but it still shocked and disappointed some of his readers.[14] The book also included a section called "Notes on Ice and Glaciers, Etc." in which Tyndall presented some of his scientific ideas about glaciers, ice, clouds, and eclipses.[15]

The first print run of one thousand copies sold well in Britain and there were a few other editions that same year. Like *Fragments of Science* it was almost immediately translated into German and was sold on the Continent. It was also published in the United States in 1871 and sold well there. He sent it to a number of friends, most notably Emily Peel and Sarah Faraday, who enjoyed the illustrations and stories; Mary Egerton was particularly troubled about Bennen.[16]

Reviews of *Hours of Exercise in the Alps* were, generally, positive. The *Pall Mall Gazette* called it a "pleasant little volume" that aimed to show readers how much adventure awaited them in the Alps.[17] But the *Gazette* also took issue with Tyndall's main "error of the preface": seeming adherence to Herbert Spencer's recent ideas of cultural evolution, when Tyndall argued that modern humans' enjoyment of the outdoors was a survival from our "more barbarous times."[18] The *Examiner* stated that, taking both of Tyndall's 1871 books together, they illustrated the "savant at work and the savant at play," which is exactly what Tyndall seemed to be going for.[19]

In contrast to their response to *Fragments of Science*, the *Athenaeum* was particularly harsh in its review of *Hours of Exercise in the Alps*.[20] As much as they praised the former for its writing style and candor, they described the latter as a good example of how "a man of reputation can endanger his fame by mere book-making; though on this point friendly critics will prefer to be silent."[21] Further, they argued that this book, whose chapters were "truly such a *disjecta membra*" that it was hard to "make an articulated body out of them," was much more suited to Tyndall's friends, who may have been, according to this review, the only people who would want to read the book.[22]

The *Examiner* and other reviewers enjoyed his explanation of exercise as important to masculinity and muscular strength. The *Saturday Review*, for example, used its review of the book to testify to the importance of "adventure, real, old-fashioned adventure."[23] However, the *Athenaeum* seemed to take Tyndall to task for this. First, the author of this review criticized Tyndall for writing too much about "mere personal and petty occurrences and of expressions of the pride of mountain conquest."[24] Further, they argued that Tyndall did not take the appropriate precautions in his climbing, referring to the fact that "poor Bennen" was "a far cleverer and bolder mountaineer" than Tyndall, and much more fit for climbing.[25] To add insult to injury, the

Athenaeum claimed that there would soon be women conquering the Matterhorn, and that the "Maids of the Matterhorn" would do just as good a job as men do.[26] Why wouldn't they? the *Athenaeum* asked. Taking a swipe at Tyndall, they seemed to argue that "Alpine females" were stronger than many men, "at least, literary men."[27] Other than the fact that Tyndall here espoused the virtues of masculine strength, and how important "adventure, real, old-fashioned adventure" was for men, it is hard to figure out why there was such a strong response from the *Athenaeum* reviewer. On the other hand, this may have been a sarcastic attack on female mountain climbers, written at the height of the age that prized masculine, muscular adventure.[28]

Both *Fragments of Science* and *Hours of Exercise in the Alps*—the "little books," as his correspondents repeatedly called them—appear as topics of conversation throughout this volume, and were important pieces of Tyndall's work in terms of communicating science to the public. His friend Rudolf Clausius wrote to him in June of 1871 that he admired "the ease with which you can produce rigorous scientific works, which open new fields in any branch of physics, and how you then contribute to the diffusion of science through fine popular presentations."[29] Most of the public agreed. By the time he had made his way to the United States in November of 1872, where he was already well known by the public, the American periodical *Old and New* told its readers that reading these two books would be good precursors to prepare for Tyndall's visit.[30] In this volume, Tyndall received his initial invitation for the lecture tour from Joseph Henry of the Smithsonian Institution, demonstrating that he had arrived as a professional man of science both in Britain and in the United States.[31]

Many of his friends were suitably shocked by the amount of danger and risk that Tyndall took, and about which he was so frank in *Hours of Exercise in the Alps*. Throughout his letters to friends in the summer of 1871, we see he liked to share in detail the adventures he had. For Tyndall, experiencing, internalizing, and overcoming this risk was central to becoming an established alpine figure, as well as a Victorian man of science.[32] Risk itself was not masculine; overcoming it was. In fact, most alpinists considered it morally bad to go into dangerous situations without taking care to be safe.[33] Tyndall's earlier conflicts with alpinist Edward Whymper and his bad safety choices become clear in this context.[34] K. Theodore Hoppen has argued that Tyndall's alpinism was important both to his *being* a particular type of masculine Victorian gentleman of science and to his *belonging* in all the clubs.[35] It took Tyndall many years of hard work to be accepted into these social groups. On the Ides of March, Tyndall was elected to "The Club," which was "an exclusive dining club of thirty-four men, including the Archbishop of Canterbury, Gladstone, the Duke of Argyll, and the Earl of Derby." This, his biographer argues, was "the ultimate recognition of Tyndall's social standing."[36] He was only the second man of science to be elected to The Club, after Richard Owen, which marked his ascendance to being an establishment scientific, cultural, and alpine figure.

Man of Science, Man of Leisure, Great Conversationalist

The letters in this volume give the reader insight into Tyndall's comfort with his new position as a central figure in nineteenth-century science. He had been accepted into various clubs and associations, which helped make him confident in that status. In a display of his assurance in the summer of 1871, he skipped the BAAS meeting in Edinburgh. He did so at Thomas Archer Hirst's and Henry Bence Jones's urging for the good of his health, but also because his attendance was not required for him to maintain his stature as a man of science.[37] While his friends missed him, they understood his need to stay in the Alps. He usually did not like to go to the BAAS meetings, as he preferred the "metropolitan science" of London, done at the RI and by members of the RS, as opposed to the science done outside the capital.[38] The term *metropolitan science* is used in two ways here. First, in the context of science being done in a large city, with a thriving community of scholars within conversational distance.[39] Second, Tyndall was part of a group of London-based scientific naturalists known as the "Metropolitans," who stood at odds with the "North British," a group of Scottish physicists and engineers.[40]

Thanks to his newfound status, Tyndall had time to travel a bit more. In 1871 he was able to extend his annual climbing trip to the Swiss Alps, which usually lasted around a month. He left London for Pontresina on July 12 and returned to London on September 4. While he was in Pontresina, he wrote some, but not many, letters back to his friends.[41] Tragedy seemed to mar his trip to Switzerland, in the form of a lost climber who was a local student. Tyndall stayed on for a few extra days to help with the search, but sadly to no avail.

Tyndall continued to help others, like his colleagues and family members in need. At this point in his life he had enough money to support himself (he was still single) and to give money to others. His family in Gorey kept asking for money, and he gave them some help in that way.[42] Tyndall was also on the RS's Scientific Relief Fund Committee, which gave help to scientists and their families who had financial difficulties. They had agreed, upon the request by Jane Barnard, to give money to Tony Petitjean to support his stay in a convalescent home.[43] He helped a number of both French and German colleagues financially and professionally after the Franco-Prussian War, including Henri Regnault and Jules Jamin.

In his socializing, issues of how Tyndall interacted in interpersonal relationships inevitably arose. He was considered "one of the great conversationalists of his age" and described as "a sensitive and careful listener,"[44] but his attitudes toward some, as portrayed in his letters, may negate that. He was loyal to his friends and sometimes to his enemies, and he wanted relational harmony among them. However, Tyndall was a man in the nineteenth century, and he shared the problematic attitudes of many of his contemporaries.[45]

Tyndall would not have been called progressive in his views on gender, race, or class, but he had many friends who were women: Sarah Faraday, Jane Barnard, and Emily Peel, to name a few. He had garnered some admirers as well; in our volume we see Henrietta Huxley and Juliet Pollock paying him affectionate and friendly attention. However, they seem to have had very little patience when Tyndall crossed the line to the blatantly insulting. For example, Emily Peel took Tyndall to task over an exchange about the Baron Rothschild, a friend of Peel's, wishing to join the Alpine Club. Peel asked Tyndall if Rothschild would be able to join the English Alpine Club, given his "distinction, having performed various ascensions & passes of importance, such as the Jungfrau, Finsteraarhorn, Monte Rosa, Bernina Spitz, Weissthor, & I believe the Oberaarjoch."[46] Tyndall responded that John Ball of the Alpine Club would be sure to propose him, but only if he heard from Rothschild himself. Tyndall told Peel that he was "almost ashamed" to tell her that Ball "does not place implicit trust in your account of the Baron's achievements, on the ground that ladies are usually inaccurate!"[47] Peel responded swiftly that she was "greatly amused at the discredit which Mr. Ball throws on my accuracy in general, and the accuracy in particular of my memory."[48] Peel was obviously accustomed to dealing with men who doubted her intelligence simply because she was a woman. Tyndall may have genuinely felt "almost ashamed" to pass on Ball's doubts; he seems to have appreciated and sought out intelligent women as friends. Nevertheless, the exchange illuminates how easy it was to express doubts of women's intelligence; it is difficult to imagine Tyndall writing in the same way to a male friend, especially another scientific practitioner.[49]

The letters in this volume demonstrate that personal relationships within science were an ongoing source of tension for Tyndall. In Tyndall's relationships with his colleagues, who were also his friends, it became clear just how much politicking was necessary in science in the nineteenth century. In May, Tyndall defended the character of John Herschel, who had just died but had been accused of vanity by an anonymous author in the *Daily News*.[50] In June, Tyndall further attempted to bring peace to a longstanding feud between Thomas Henry Huxley and Owen, who had not spoken to each other since the Great Hippocampus debate.[51] But the two men never fully reconciled. In November, Tyndall successfully lobbied for the RS's prestigious Copley Medal to be awarded to German physicist Julius Mayer for his work on heat, opposing William Thomson and others who wanted it to go to Hermann

Helmholtz; this was the second year in a row that Tyndall's candidate had succeeded (James Prescott Joule had received the medal in 1870). Tyndall's advocacy for Mayer brings into view the differences between the North British scientists and the Metropolitan group.[52] It also renewed a decade-old spat with William Carpenter, who publicly took Tyndall to task for failing to acknowledge his (Carpenter's) early recognition of Mayer.[53] These controversies remind us not only of the complicated history of thermodynamics but more broadly of how fraught and politicized any interactions among scientific practitioners could be in this period.

The Ayrton Incident

Tyndall also encountered conflict, or generated it, in two matters that brought him into contact with the British government: his work as a scientific advisor to the official bodies concerned with lighthouses, and his endeavors to assist his friend Joseph Hooker, the director of the Royal Botanic Gardens at Kew.[54] Hooker's involvement with the gardens was of long standing: his father, William Hooker, had greatly expanded the gardens during his directorship, which lasted from 1841 until Joseph succeeded him in 1865. As histories by Roy Macleod and Jim Endersby stress, the gardens were tied up with the Hooker family in a way that was intensely personal.[55]

The Hookers had had cordial relationships with previous governments, whose commissioners of works had essentially allowed them complete jurisdiction over Kew. But when the Liberal party came into power in the elections of 1868, the administration formed by William Gladstone pursued austerity policies that were eagerly embraced by Acton Smee Ayrton, Gladstone's first commissioner of works. Ayrton had little patience for Kew's autonomy. To him, Hooker was merely a civil servant, while the gardens were a gratuitous drain on public finances. In 1869, Ayrton sent Hooker a reprimand about problems in Kew's accounts—a reprimand that turned out to be based on Ayrton's own misreading. Undaunted, Ayrton insisted that a clerk at Kew be appointed on the basis of a civil service exam; Hooker, however, insisted that the winning candidate was not qualified for the position.[56] The situation escalated as Hooker and the Board of Works came into conflict over work Hooker wanted done to the gardens' hothouse heating apparatus. Hooker hired a contractor that he had previously employed, but Douglas Galton, an engineer and FRS whom Ayrton had appointed to the post of director of public works, objected that the contract should be awarded competitively. While Hooker seems to have reached a fragile reconciliation with Galton,[57] the matter only exacerbated his relations with Ayrton. It was at this point that Tyndall became involved: on 26 March 1871, he spoke at a dinner of The Club about Hooker's mistreatment. His account of the situation interested Edward Stanley, Lord Derby, who wrote to Tyndall the next day to offer his assistance, including a willingness to put forward, in the House of Lords, "a motion for the production of papers"—the first step in instigating a parliamentary inquiry into the affair.[58]

Tyndall responded with a lengthy account of Hooker's virtues and wrongs,[59] and the matter unfolded slowly over the next two months. Derby may have had second thoughts about his involvement, or at least his offer to bring the matter to Parliament: he hoped the matter could be settled more informally, and asked Tyndall for assurance of Hooker's wish to go ahead with the call for papers.[60] Our volume ends with the matter in suspense, but at this point Tyndall and Hooker's other allies were beginning to prepare a memorial letter to Gladstone that would appear in *Nature* in the summer.[61] Ultimately, the affair ended ambiguously, with no clear winner or loser, but it certainly strained the relationship between Gladstone and men of science, and created a broader sense that science and government were at odds.[62]

To the Lighthouse

That sense, which according to Frank Turner became widespread at the end of the nineteenth century, was also fed by Tyndall's lighthouse work, the piece of research that receives the most attention (often unwilling) from Tyndall in our volume.[63] In 1867 Tyndall succeeded Michael Faraday as the scientific advisor on lighthouses to the Board of Trade and Trinity House.[64] Since the 1850s, the Board of Trade had overseen the three bodies responsible for British lighthouses: Trinity House, which dealt with English and Welsh lighthouses, the Commissioners of Northern Lights in Scotland, and the Commissioners of Irish Lights. Yet the bureaucratic arrangement was complicated by restrictions on the board's control of lighthouse financing, and by the fact that Trinity House, as the oldest institution of the three, had authority over its Scottish and Irish counterparts, an arrangement that seems to have done little to assuage the competitive nature of the relationship among the trio.[65] Lighthouse matters were fraught because so much was at stake: by 1880, 55 percent of all seagoing trade worldwide was British. And of course, sea travel was essential to an island nation with a vast worldwide empire. But not only was the sea bound up with the nation's wealth and power, it was also a significant threat: each year saw the loss at sea of 1,200 ships and 800 to 1,200 lives.[66]

Tyndall's appointment was contentious from the start, with the Commissioners of Northern Lights insisting that they did not want a standing scientific advisor,[67] and tensions with this body continue through the present volume. The first lighthouse correspondence we see here is an exchange with Cecil Trevor, the assistant secretary to the Harbour Department of the Board of Trade, concerning a report written by James and William Stevenson, both of whom were engineers for the Commissioners of Northern Lights.[68] The

Stevensons objected to Tyndall's conclusions concerning the use of gas, rather than oil, in Irish lighthouses: in 1869, Tyndall had reported favorably about a gas burner designed by the Irish inventor John Wigham in 1865 and adopted by the Commissioners of Irish Lighthouses.[69] This correspondence points to a major issue in the regulation of lighthouses during this period: the emergence of new illuminants that replaced the candles that had been used in the early part of the century.[70] The question of fuel dominated Tyndall's lighthouse work in this period, and in our volume we see several different aspects of it. The issue of gas recurred in the summer of 1871, as he investigated an "annular lens" to be used in combination with gas.[71] This investigation took him to Ireland twice, first in June and then in September, and he wrote two separate reports about his findings, both of which concluded in favor of gas and recommended its adoption in Irish lighthouses.[72]

Meanwhile, in the late summer, Tyndall was beginning to hear about a topic that would consume most of his time in the fall of 1871. Again, the question of what illuminant was best for lighthouses was central, but this time the contest was not over gas versus oil but rather concerned the possibility of fueling lighthouse lamps with paraffin. Also known as mineral oil and kerosene, paraffin is distilled from coal or petroleum; the interest in using it in lighthouses was but one small facet of the British economy's transition to fossil fuels in the nineteenth century.[73] From the middle of the century paraffin had been seen as a potentially valuable lighthouse illuminant, due to its low cost and brilliant light, but it had not been adopted because of its volatility and consequent danger.[74] In 1868 Tyndall tested a paraffin burner designed by Henry Harrison Doty, an American ship captain. Though impressed by the lamp's illuminating power, Tyndall concluded that the difficulties presented by paraffin meant that Doty's design did not warrant further investigation.[75]

Rebuffed by Trinity House, Doty took his design first to the French and then to the Stevensons, through whose intervention the burner was adopted in Scotland in 1870, on conditions that were financially extremely favorable to Doty.[76] In the meantime, experiments on paraffin were conducted by Stevenson Macadam, the consulting chemist to the Commissioners of Northern Lights, and by James Nicholas Douglass, the engineer to Trinity House, both of whom found that paraffin produced a bright light at significantly less cost than the vegetable oil colza. After what seems to have been a somewhat tense series of exchanges, the decision was reached to test Doty's lamp against a burner used by Trinity House and designed by Douglass. Tyndall, as a neutral party, oversaw the tests at the Royal College of Chemistry, a neutral space.[77]

Tyndall's work on this matter began almost immediately after he returned from Switzerland in early September, and it lasted through the middle of November, when he composed a thirteen-thousand-word report comparing the Doty and Douglass lamps.[78] After laying out the history of the lamps and of attempts to employ paraffin in lighthouses, the report turns to the experiments that Tyndall oversaw in the fall of 1871, and which took place from September 27 until October 31. The experimenters included Douglass and Doty as well as William Valentin, senior assistant at the Royal College of Chemistry, and Arthur Ayres, an assistant at Trinity House. Each individual experiment involved fueling the lamps with a different oil, then taking a series of photometric readings to determine their illuminating power. The experimenters also tracked the conditions of the lamps over the course of the experiments, noting when they burned steadily and when they flared, smoked, or forked. While the results were, broadly speaking, in favor of Doty's lamp as brighter and more cost efficient, Tyndall did not endorse Doty's invention wholeheartedly. Instead, in his conclusion he reminded the board of his favorable conclusions concerning gas and dwelled on the problem of volatility and consequent danger inherent in mineral oil.

Tyndall's Scientific Self

The extraordinary length of Tyndall's report on the lighthouse lamp trials is partly due to the complexity of the issues under examination. But it is also an expression of how Tyndall saw himself as a scientific practitioner, and an endeavor to make others see him the same way.[79] Intertwined with the details of experimental procedures and tables summarizing photometric results is a story about Tyndall, Doty, and their relationship to a scientific ideal. That ideal is the mechanical objectivity that Lorraine Daston and Peter Galison describe as central to nineteenth-century scientific practice.[80] Mechanical objectivity produced a morality of self-restraint and hard work, as practitioners sought to emulate the impersonality and tireless labor of the machine. In letter 3555, we see how this ideal governs the experimental setup: Tyndall writes that he had hoped to conduct the trials of the lamps in the absence of the two "principals," that is, Doty and Douglass, the inventors of the lamps under investigation. Douglass agreed to this proceeding. Doty refused, but in revealing terms: he told Tyndall that no one else knew his lamp well enough to operate it, and "I am therefore reluctantly compelled to be my own operator, pledging myself to offer no observations or suggestions whatever, relative to the conduct of the trials, and to maintain a perfectly neutral attitude." Doty here seems to pay allegiance to objectivity; he says, in effect, that he will discipline his own partiality by quashing any "observations or suggestions." Yet we have already been primed to doubt any such statement, given that Tyndall has characterized him as "rather eager as regards visits and suggestions." And as the experiments unfold we see that he fails to fulfill this promise to remain neutral, and instead frequently protests and complains.

The representation of Doty in this letter serves to highlight Tyndall's own adherence to the ideal of objectivity. In the report's conclusion he writes,

"Were the combatants less eager, and Captain Doty less fond of 'protests,' the Report might be condensed to a statement of the results; but as the case stands, it is desirable to show upon the face of the Report that nothing was done or neglected which could justify complaint."[81] The laborious detail of this letter is here accounted for as evidence that will enable the reader to appreciate Tyndall's neutrality and his adherence to mechanical objectivity; the very grammar of the sentence, with its reliance on the passive voice ("might be condensed," "nothing was done or neglected") erases Tyndall as an actor.

Ian Hesketh has shown that mechanical objectivity was essential to Tyndall's creation of himself as a man of science, particularly during his challenging early years.[82] In this volume, Tyndall's commitment to objectivity emerges as a matter of self-presentation, impelled by a wish to make others see him a certain way. We see this not only in letter 3555 but also in his repeated insistence that he receive no money for his planned lecture tour of the United States, and in the detailed self-exculpations that he unfolds in various letters in response to the attack on him by German physicist Johann Zöllner.[83] In both of these instances, it was important to Tyndall to tell people that he was not driven by pecuniary concerns.[84] Tyndall presents Joseph Hooker in similar terms when advocating that Derby take up the latter's cause, stressing that Hooker willingly exchanged half a year's salary for a collection to augment the holdings at Kew; this lack of self-regard exemplifies Tyndall's claim that Hooker's character is "of the very finest fibre, a sense of duty of the very highest cast, which would render the slightest neglect of public interests simply impossible to him."[85] At the same time, Tyndall similarly underlines his own scientific character. Writing, "I am told on all hands that the sin of this transaction lies at the door of Mr Gladstone, who backs up this dog in office," he adds, "I desire no privacy for these sentiments of mine. I should be only too happy of the opportunity of expressing them in the presence of the Prime Minister himself."[86] Men of science, Tyndall is effectively saying, are not driven by financial concerns, nor cowed by political power: they are committed solely to the increase of knowledge and expression of the truth.

This emphasis on objectivity in Tyndall's representation of Hooker's character and his own might also be understood in terms of how Tyndall was trying to portray scientific practitioners as a group. Roy MacLeod has shown that, in the context of lighthouse management, scientific practitioners and engineers clashed as, respectively, "the advocates of 'science'" and "the advocates of 'common sense.'"[87] The former were seen as what might now be called "ivory tower" intellectuals, while engineers were regarded in heroic terms by a public awed by their ability to erect durable structures on perilous and precarious seacoasts.[88] Moreover, engineers were well established professionally in a way that men of science were not. According to Thomas Gieryn, the tensions with engineers that we see in Tyndall's writing should be understood in the larger context of Tyndall's endeavors to define what it meant to be a scientific practitioner.[89]

Thus, even as this volume shows Tyndall to be established as a man of science, it also reveals that that position was not static or fixed, but was rather a continual work in progress, something that Tyndall, and other scientific practitioners, were constantly maintaining and creating. We see that work across the types of letters that we have discussed in this introduction: in those that negotiate with other scientific practitioners, in those that address Tyndall's relationship to the public, and even in those that relate to his personal life. If this volume of letters is not marked by a single notable event or piece of research, its interest lies in the very ordinariness and dailiness of its picture of a Victorian man of science.

Notes

1. See Journal, 1855–72, typescript (RI MS JT/2/13c).
2. *Ascent of John Tyndall*, 273.
3. *Ascent of John Tyndall*, 274.
4. See letter 3419.
5. "Science for the Unscientific," *All the Year Round* 6, no. 133 (17 June 1871): 57–60; "Science for Unscientific People," *Examiner*, no. 3298 (15 April 1871): 391–92.
6. "Current Literature," *Daily News*, 7 August 1871, 2.
7. "Science," *Athenaeum*, no. 2274 (27 May 1871): 657.
8. "Science," *Athenaeum*, 657.
9. "Tyndall's *Fragments of Science for Unscientific People*," *Scientific American* 25, no. 6 (5 August 1871): 80.
10. Michael S. Reidy, "Introduction: John Tyndall, Scientific Naturalism and Modes of Communication," in *The Age of Scientific Naturalism*, ed. Bernard Lightman and Michael S. Reidy (Pittsburgh: University of Pittsburgh Press, 2016), 1–13.
11. Tyndall, "Preface," *Hours of Exercise in the Alps*, v.
12. *Ascent of John Tyndall*, 274; Tyndall, "Preface," *Hours of Exercise in the Alps*, vi.
13. Philip C. Gossett, "Death of Bennen on the Haut de Cry," in Tyndall, *Hours of Exercise in the Alps*, 192–205.
14. See letter 3441; also "Hours of Exercise in the Alps," *Athenaeum*, no. 2275 (3 June 1871): 679–80.
15. "Notes on Ice and Glaciers, Etc.," in Tyndall, *Hours of Exercise in the Alps*, 337–473.
16. See letters 3441, 3445, and 3446.
17. "Hours of Exercise in the Alps," *Pall Mall Gazette*, 30 May 1871, 10.
18. "Hours of Exercise in the Alps," *Pall Mall Gazette*, 10.
19. "Hours of Exercise in the Alps," *Examiner*, no. 3307 (17 June 1871): 612–13.
20. "Hours of Exercise in the Alps," *Athenaeum*, 679–80.
21. "Hours of Exercise in the Alps," *Athenaeum*, 679.
22. "Hours of Exercise in the Alps," *Athenaeum*, 679.
23. "Hours of Scrambling Exercise—Tyndall and Whymper," *Saturday Review* 32, no. 819 (8 July 1871): 59.
24. "Hours of Exercise in the Alps," *Athenaeum*, 679.
25. "Hours of Exercise in the Alps," *Athenaeum*, 680.

26. "Hours of Exercise in the Alps," *Athenaeum*, 679.
27. "Hours of Exercise in the Alps," *Athenaeum*, 679.
28. Michael S. Reidy, "Mountaineering, Masculinity, and the Male Body in Mid-Victorian Britain," *Osiris* 30 (2015): 158–81.
29. Letter 3444.
30. "Professor Tyndall," *Old and New* 4, no. 5 (November 1871): 598.
31. Letter 3504.
32. Joseph E. Taylor, "Victorians," in *Pilgrims of the Vertical: Yosemite Rock Climbers and Nature at Risk* (Cambridge, MA: Harvard University Press, 2010), 15–43.
33. See R. D. Eaton, "In the 'World of Death and Beauty': Risk, Control, and John Tyndall as Alpinist," *Victorian Literature and Culture* 41, no. 1 (2013): 55–73.
34. See letters 3525 and 3531.
35. K. Theodore Hoppen, *The Mid-Victorian Generation: 1846–1886* (London: Clarendon Press, 1998), 489.
36. *Ascent of John Tyndall*, 273.
37. See letter 3503.
38. *Ascent of John Tyndall*, 176 and 266.
39. The geography of science is an important factor in the development of scientific ideas and identities, especially in the nineteenth century. For a specific discussion of the BAAS and siting science, see Charles W. J. Withers, "Scale and the Geographies of Civic Science: Practice and Experience in the Meetings of the British Association for the Advancement of Science in Britain and in Ireland, c. 1845–1900," in *Geographies of Nineteenth-Century Science*, ed. D. N. Livingstone and C. W. J. Withers (Chicago: University of Chicago Press, 2011), 99–122.
40. *Ascent of John Tyndall*, 176. See also Crosbie Smith, *The Science of Energy: A Cultural History of Energy Physics in Victorian Britain* (Chicago: University of Chicago Press, 1998).
41. See letters 3494, 3495, 3497, 3498, 3503, and 3510.
42. See, for example, letters between Tyndall and Elizabeth Steuart, 3379, 3401, 3474, 3491, 3536, 3571, 3572, and 3580.
43. See letters 3408, 3409, 3413, 3414, and 3418.
44. *Ascent of John Tyndall*, 273.
45. For a useful discussion of this topic, see Evelleen Richards, "Good Wives," in *Darwin and the Making of Sexual Selection* (Chicago: University of Chicago Press, 2017), 36–61.
46. Letter 3525.
47. Letter 3531.
48. Letter 3533.
49. In addition, see letter 3613 and *Ascent of John Tyndall*, 286–87.
50. See letters 3424, 3425, 3429, and 3430.
51. See letters 3458, 3461, and 3463, and Desmond, *Huxley*, 292–311.
52. Introduction (pp. xiii–xxvii) and letter 2006, *Tyndall Correspondence*, vol. 8; *Ascent of John Tyndall*, 170–87; this volume, letters 3539, 3541, 3544, and 3546.
53. [William Carpenter], "Grove, Carpenter, &c., on the Correlation of Forces, Physical and Vital," *British and Foreign Medico-Chirurgical Review* 8, no. 15 (1851): 206–37. Carpenter briefly discusses Mayer's publication on the final page of his article. See letters 3573, 3575, 3582, and 3584.
54. The following discussion of the Kew affair draws from two studies: Roy M. MacLeod, "The Ayrton Incident: A Commentary on the Relations of Science and Government in England, 1870–1873," in *Science and Values: Patterns of Tradition and Change*, ed. Arnold Thackray and Everett Mendelsohn (New York: Humanities Press, 1974), 43–78; and Jim Endersby, *Imperial Nature: Joseph Hooker and the Practices of Victorian Science* (Chicago: University of Chicago Press, 2008), 282–93.
55. MacLeod, "Ayrton Incident," 54; Endersby, *Imperial Nature*, 290.

56. Letters 3647, 3692, and 3700.
57. Letter 3630.
58. Letter 3633. In what follows we refer to Stanley as "Derby" in order to maintain consistency with the way that he is discussed in the letters themselves.
59. Letter 3638.
60. Letters 3659, 3667, and 3677.
61. Letters 3686, 3700, 3705, and 3706. See also Charles Lyell, Charles Darwin, George Bentham, Henry Holland, George Burrows, George Busk, Henry Creswicke Rawlinson, James Paget, William Spottiswoode, Thomas Henry Huxley, and John Tyndall, "Letter to W. E. Gladstone," *Nature* 6, no. 141 (11 July 1872): 211–16; this letter will appear in the thirteenth volume of *The Correspondence of John Tyndall*.
62. MacLeod, "Ayrton Incident," 69–70; Endersby, *Imperial Nature*, 293.
63. Frank M. Turner, *Contesting Cultural Authority* (Cambridge, UK: Cambridge University Press, 1993), 208. See also Roy M. MacLeod, "Science and Government in Victorian England: Lighthouse Illumination and the Board of Trade, 1866–1886," *Isis* 60, no. 1 (Spring 1969): 5–38.
64. *Ascent of John Tyndall*, 282–83.
65. MacLeod, "Science and Government," 12.
66. MacLeod, "Science and Government," 7.
67. *Ascent of John Tyndall*, 283.
68. Letters 3400 and 3417.
69. MacLeod, "Science and Government," 15.
70. MacLeod, "Science and Government," 14.
71. Letter 3475.
72. See letters 3475, 3521, and 3526.
73. For one account of this transition, see E. A. Wrigley, *Energy and the Industrial Revolution* (Cambridge, UK: Cambridge University Press, 2010), 26–52.
74. Thomas A. Tag, "The Doty Dilemma: Technological Advancements in Lighthouse Lamps and Subsequent Patent Infringement," U.S. Lighthouse Society *Keeper's Log* 16, no. 4 (Summer 2000): 28–33, on 29.
75. Tag, "Doty Dilemma," 29.
76. Tag, "Doty Dilemma," 30; letter 3555.
77. Letter 3555.
78. Letter 3555; the remainder of this paragraph draws on this letter.
79. For a discussion of how Tyndall's "self-fashioning" changed over the course of his lifetime, see Bernard Lightman, "Fashioning the Victorian Man of Science: Tyndall's Shifting Strategies," *Journal of the Dialectics of Nature* 38, no. 1 (January 2016): 66–79.
80. Lorraine Daston and Peter Galison, *Objectivity* (New York: Zone Books, 2007), 115–90.
81. Letter 3555.
82. Ian Hesketh, "Technologies of the Scientific Self: John Tyndall and His Journal," *Isis* 110, no. 3 (September 2019): 460–82, on 462.
83. For the American lecture tour, see letters 3542, 3639, 3669, and 3670; for the Zöllner affair, see letters 3652 and 3655.
84. Thomas Gieryn notes that Tyndall set scientific practitioners in opposition to engineers by claiming that the latter were animated by a desire to make money, whereas the former pursued knowledge in a disinterested manner (Thomas F. Gieryn, *Cultural Boundaries of Science: Credibility on the Line* [Chicago: University of Chicago Press, 1999], 42, 58). Gieryn also argues that Tyndall distinguished between science and engineering by associating science with systematic experimentation and engineering with rule of thumb procedures (*Cultural Boundaries*, 55); this opposition may also be in play in letter 3555 as Tyndall presents his own systematic work and Doty's resistance to it.

85. Letter 3638.
86. Letter 3638.
87. MacLeod, "Science and Government," 15.
88. MacLeod, "Science and Government," 15.
89. Gieryn, *Cultural Boundaries*, 45.

EDITORIAL PRINCIPLES

Our aim is to include all letters to and from John Tyndall that are currently extant. Some of the letters are from published sources, such as letters written by Tyndall to newspapers or journals, but the majority are either the originals or typescripts produced by Louisa Tyndall from originals. One central editorial principle has been to transcribe, wherever possible, from original letters rather than Louisa Tyndall's typescripts. This is because of Louisa Tyndall's conscious and unconscious editorial interventions, a point discussed in more detail below. We have also aimed to reproduce, as accurately as possible, the text of the letters as they were written. Spelling mistakes that have appeared in the handwritten letters have therefore been preserved; mistakes appearing where only the typescript letter has survived are silently corrected, as in our judgment they are almost always typographical errors. But, in general, nonstandard spellings (e.g., the use of dialect) has been retained even when in a typescript-only letter (though there is the possibility that the spelling is Louisa Tyndall's, not her husband's). Where the transcription is based on a typescript transcription by Louisa Tyndall (or one of her assistants), this will be indicated by "LT Transcript Only."

We have included illustrations that are part of the letter. In cases where it was possible to reprint the original handwritten image, we have done so. In other cases, volume editors have reconstructed drawings because it was not possible to reprint an image from the handwritten letter (due, for example, to deterioration of the original document).

The letters are presented in chronological order of writing. Where more than one letter has the same date, letters from Tyndall (in alphabetical order of addressee) come first, followed by letters to Tyndall (in alphabetical order of writer), with the exception that if the order of writing can be determined, that order takes precedence. Where letters cannot be dated precisely they are placed at the latest likely date of writing. (In some cases if it is possible, but unlikely, that they were written later, notes indicate such possibilities.) To indicate the beginning of a new year, the date of that year has been inserted in a large font before the first letter from that year. Readers should note that although volumes are arranged in chronological order, letters that are discovered from years already covered in published volumes will appear in the final volume of the correspondence.

For non-English-language letters, both the transcription of the original and the English translation are presented. The transcription of the non-English letter appears first, followed by the English translation in a smaller font. Editorial notes are inserted in the English translation. Some foreign letters are in an original hand, while others exist only as transcriptions by Louisa or others who often did not have a full command of the language. In the latter case these letters contain frequent spelling, grammatical, and interpretation errors that do not seem to have been committed by the original author. In these typescript-only cases, and only in these cases, spelling and grammatical errors have been corrected, as have poor interpretations where a better word choice was obvious from the context.

There is a standard format for each letter. The first line lists the author, if it is to Tyndall, or the recipient, if it is from Tyndall; the date; and the letter number. The second line lists any information supplied by the writer in their salutation, such as the place, date, or time of day when the letter was written. (In some cases, though, where the writer put the date at the foot of the letter we have left it there in order to reproduce the text accurately.) The opening salutation, the text of the letter, and closing then follow. Then come the source and editorial notes, the latter indicated in the text by arabic numerals.

Editorial notes are intended to provide the contextualization necessary to understand the contents of each letter. There are basically six types of editorial notes. Many of them are informative notes on persons mentioned in the letters. If a person referred to in a letter is not described in a note, they have been mentioned more than twice in the volume and they will have an entry in the biographical register at the back of the volume. For the rest—those mentioned once or twice—biographical information will appear in the editorial note appended to the letter in which he or she is first mentioned. When there is a second reference to that person a note will refer back to the letter in which he or she was first mentioned. Biographical notes on well-known figures such as Faraday or Huxley stress what that person was doing at the time when the letter was written rather than providing a full overview of the person's life.

A second type of editorial note identifies allusions and quotations. In the case of quotations, where possible, the first edition of the work is cited unless a different edition is mentioned in the letter. If a quote is not in English a translation is given. Glosses on obscure places, abstruse words, and words used in an unfamiliar sense constitute the third type of editorial note. The fourth type provides information about the context of the letter, drawing on such sources as contemporary publications, scholarly sources, Tyndall's personal journals and field notes, and other letters in the *Tyndall Correspondence*. The fifth type of editorial note is cross-references to other letters in the correspondence, referring to the letter number. The last type is information on significant textual changes that Tyndall made to the original that is thought to be relevant to the meaning of the letter.

There are a series of conventions for indicating when editors have given additional information about the text of a letter:

When Louisa Tyndall has made an annotation or marginal insertion that the editors have deemed important enough to retain, we have included her annotation in an editorial footnote.

When editors other than she have made insertions, they are indicated by unitalicized square brackets; for example, [word].

When certain words are ambiguous and we have had to conjecture their meaning, they are indicated by italicized square brackets; for example, *[*words*]*; when we have considered such words illegible the number of illegible words is indicated in full italicization and square brackets; for example, *[3 words illeg]*.

If a word or series of words are not present or have been destroyed—by inkblots, holes, or mold, for example—they are indicated as missing, italicized and in angular brackets; for example, *<3 words missing>*.

In extremely rare cases where it is clear that Louisa herself has intentionally destroyed a few words, they are indicated as excised, italicized and in angular brackets; for example, *<3 words excised>*.

Finally, a brief word about Louisa Tyndall. A John Tyndall correspondence project would be very difficult, perhaps impossible, without the work done by Louisa Tyndall, much of which seems to have been done in order to write a biography of her late husband. This biography was not completed. We have therefore treated Louisa Tyndall as a de facto member of the editorial team. Some of her handwritten comments on the typewritten letters are extremely helpful and thus have been retained as footnotes to the relevant passage. Conversely, many of these handwritten annotations seem to have been inserted to help write her biography—for instance, summaries of passages or interesting quotes. When the editors have deemed these notes of Louisa's to add no new information, they have not been included for publication.

It is also important to recognize that Louisa Tyndall was not always reliable as a transcriber. She often threw out the manuscript letter when she had made her transcription, but some original manuscripts remain, which indicate her editing processes. Like any transcriber, she was sometimes inaccurate, but corrections show that she proofread her transcriptions. However, she also edited John Tyndall's writing to make his letters seem more consistent or grammatically correct. For example, she made punctuation more formal, corrected grammar and, occasionally, corrected quotations. She consistently corrected John Tyndall's usual (deliberate) lower case for "god" to "God." Her typewriter was unable to insert the superscripts that Tyndall used in abbreviations, or the "&c" symbols which stood for an address at the foot of the letter. (Modern typefaces are unable to deal with all the variant scrawls which the "etc" symbol took and they are all formalized here to "&c.") Louisa Tyndall also left out details that she considered too private or too salacious for public consumption. She equated postmark and date of writing and overlooked the time taken to write some letters. When she inferred dates from internal evidence her reasoning was sometimes faulty. Similarly, her identifications of people are sometimes in error. Independent evidence has always been sought for her dates and identifications. Lastly, the patterns of punctuation differ between Louisa Tyndall's typescript and manuscript sources. John Tyndall himself used dashes more often than commas, semicolons, and stops, and he used dashes of many different lengths which are all standardized here to em-dashes. It is often difficult to distinguish between very short dashes and commas or very short dashes and stops. Similarly, capitalization is often ambiguous; distinguishing *S/s*, for "science" for example, is a particular problem. Therefore, scholars for whom punctuation and capitalization are significant will need to consult the manuscripts. Users should also be aware that throughout this project the transcribers and editors have generally worked from scans and photographs rather than the original manuscripts.

NOTE ON MONEY

In the nineteenth century the currency used in England was the pound sterling, usually abbreviated to £ (or, occasionally, to *L*). A pound consisted of 20 shillings (abbreviated *s*) and each shilling contained 12 pennies (*d*). A guinea was 21 shillings; that is, £1 1s. A crown was 5 shillings (5s), and half a crown equaled two shillings and sixpence (2s 6d). The smallest coin was a farthing, one quarter of a penny (¼d).

Several forms of representation were in contemporary use. Sometimes the numbers for pounds, shillings, and pence were separated by dots, forward slashes, or spaces; thus, for example, 3 pounds, 4 shillings, and 5 pennies could be written as £3.4.5, £3/4/5, or £3 4 5. A zero could be represented by a dash; for example, 7 shillings was often written as 7/- or just 7/. A half crown was often written as 2/6.

French currency is also occasionally mentioned in this volume. The franc was the national currency in France from 1795 until 1999, when France adopted the Euro. In the mid-nineteenth century the exchange rate between British and French currencies was approximately 25 francs per £. A centime was a coin worth one-hundredth of a franc.

TIMELINE OF JOHN TYNDALL'S LIFE

Year	Event
c. 1822	John Tyndall born at Leighlinbridge, County Carlow, Ireland
c. 1836	Attends John Conwill's National School in Ballinabranna
1839	Begins employment as civil assistant with Ordnance Survey of Ireland in Carlow
1840	Becomes civil assistant in the Ordnance Survey Office at Youghal, Cork County
1841	First appearance in print (poem in *Carlow Sentinel* under pseudonym "W[alter] S[nooks]")
1842	Joins English Ordnance Survey in Preston
1843	Leads written criticism of Ordnance Survey in *Liverpool Mercury*; dismissed November
1844	Home in Ireland, mostly unemployed, until obtaining a position with the firm of Nevins and Lawton, Surveyors, of Manchester
1845–47	Works for Richard Carter, Surveyor, of Halifax
1845	Meets Thomas Archer Hirst in Halifax
1847	Begins teaching mathematics and surveying at Queenwood College, Hampshire
1848–50	At University of Marburg, working for PhD with Robert Bunsen and Friedrich Stegmann
1850	Returns to England; meets Michael Faraday and William Francis and attends his first British Association meeting, before returning to Marburg
1851	Moves from Marburg to Berlin (April–June); then returns to Queenwood
1852	Elected fellow of the Royal Society
1852	Friendship developed with Thomas Henry Huxley
1853	First lecture at Royal Institution (RI)
1853	Meets Herbert Spencer
1853	Appointed professor of natural philosophy at RI
1853	Awarded Royal Medal for magnetic work; declines when controversy arises over award
1856	Meets Thomas Carlyle
1857	Climbs Mont Blanc for first time with Hirst
1859	Demonstrates existence of greenhouse gases and climatic implications
1859	Joins Government School of Mines as professor of natural philosophy
1860	*The Glaciers of the Alps* published
1860	Meets John Lubbock and his wife, Ellen Lubbock

Year	Event
1861	First ascent of the Weisshorn
1861	Delivers his first Christmas Lectures at RI
1862	*Mountaineering in 1861* published
1862	Begins to champion Mayer's priority in discovery of conservation of energy
1863	*Heat Considered as a Mode of Motion* published
1864	Awarded Royal Society's Rumford Medal
1864	First meeting of X Club
1865	*On Radiation* published
1865	Engages in public controversy over the efficacy of prayer
1866	Succeeds Michael Faraday as scientific adviser to Trinity House
1867	Faraday dies; Tyndall becomes superintendent of the RI
1867	*Sound: A Course of Eight Lectures* published
1868	Successfully climbs Matterhorn
1868	Discovers the cause of light scattering, to be known as "Tyndall Effect"
1868	*Faraday as a Discoverer* published
1868	"Scientific Materialism" lecture at the British Association
1868	Resigns from Royal School of Mines
1869	Joins Metaphysical Society
1870	"On the Scientific Use of the Imagination" lecture at British Association
1870	*Three Scientific Addresses* published
1870	*Researches on Diamagnetism and Magne-Crystallic Action* published
1871	Meets Louis Pasteur for first time while in Paris
1871	*Hours of Exercise in the Alps* published
1871	*Fragments of Science* published (April)
1871	*Light and Electricity* published (May)
1872	Debates over how to measure the efficacy of prayer, aka the "Prayer-Gauge Debate"
1872	*Contributions to Molecular Physics in the Domain of Radiant Heat* published
1872	*The Forms of Water in Clouds and Rivers, Ice and Glaciers* published
1872–73	USA Lecture Tour
1873	*Six Lectures on Light* published
1874	As president of the BAAS, delivers the "Belfast Address"
1876	Marries Louisa Charlotte Hamilton
1877	Develops "Tyndallization" (discontinuous heating and cooling process to destroy heat-resistant spores)
1877	Tyndall and Louisa build summer cottage, Alp Lusgen, at Belalp, northern side of Valais, above Brig
1877	*Fermentation and Its Bearings on Phenomena of Disease* published
1879	*Fragments of Science*, which had gradually expanded, first published in two volumes
1881	*Essays on the Floating Matter of the Air* published
1885	Tyndall and Louisa build Hindhead retreat, Surrey Downs
1887	Resigns from RI
1890	Fight with Gladstone in the *Times* over Irish Home Rule
1892	Hirst dies
1892	*New Fragments* published
1893	Dies at Haslemere, Surrey, accidentally poisoned

TIMELINE OF EVENTS IN JOHN TYNDALL'S LIFE SPECIFIC TO VOLUME 12

Month and year	Biographical details	Notable social, political, and cultural events
March 1871	Hector Tyndale, Tyndall's American cousin, arrives in London to visit Elected to The Club (15th)	Marriage of Princess Louise and Lord Lorne (21st) Paris Commune formally established in France (26th) Royal Albert Hall opens in London (29th) Publication of *Desperate Remedies* (anonymous), Thomas Hardy's first novel Publication of *Erewhon* (anonymous), by Samuel Butler
April 1871	*Fragments of Science* published	US president Ulysses S. Grant signs the Civil Rights Act of 1871 to combat white supremacist movements
May 1871	*Hours of Exercise in the Alps* published	Treaty of Frankfurt signed, officially ending the Franco-Prussian War (10th) Tichborne trial opens (11th) John Herschel dies (11th) and is buried (19th) Paris Commune falls (28th)
June 1871	Travels to Ireland to investigate lighthouses; sends report to Board of Trade Delivers "On Dust and Smoke" at the RI (9th) Notices Thomas Henry Huxley and Richard Owen speaking to each other in person and unsuccessfully encourages the two to put their differences behind them (12th–16th)	Universities Tests Act of 1871 removes restrictions limiting access to Oxford, Cambridge, and Durham Universities to members of the Church of England (18th) Trade Union Act of 1871 legalizes trade unions in the UK (29th) Science Schools at South Kensington open

Month and year	Biographical details	Notable social, political, and cultural events
July 1871	Meets with the emperor of Brazil Begins his travels to Pontresina, Switzerland (12th); on the way sees firsthand the destruction caused by the Franco-Prussian war in Paris and the rest of France Climbs Piz Bernina with Christopher Puller and their guide, Peter Jenni (17th–18th)	Tensions flare between Joseph Hooker and Acton Smee Ayrton over the building of greenhouses at Kew
August 1871	BAAS meeting in Edinburgh begins (2nd); Thomas Archer Hirst and Henry Bence Jones urged Tyndall not to attend for his health Receives an invitation from Joseph Henry of the Smithsonian Institution in the United States to give a series of talks on science	Adolphe Thiers becomes president of the French Republic (31st) Hooker begins complaining to William Gladstone about Ayrton
September 1871	Returns from Switzerland (3rd) Undertakes research for Trinity House on competing designs for lighthouse lamps (through November)	
October 1871	Involved in RS discussions of the awarding of the Copley Medal Appointed a governor of Harrow School Attends the first X Club meeting of the season (19th)	Great Chicago Fire destroys two thousand acres in downtown Chicago, Illinois (8th–10th) Charles Babbage dies (18th) Roderick Murchison dies (22nd) Prince of Wales falls ill of typhoid fever
November 1871	Injures his right calf and is laid up on the couch (19th) Attends first meeting as a governor of Harrow School (29th)	London-Australia telegraph cable arrives in Australia (7th) American correspondent Henry Morton Stanley locates missing explorer David Livingstone in Ujiji (10th)
December 1871	Delivers Christmas Lectures at the RI, on "Ice, Water, Vapour, and Air" Enters a dispute with William Carpenter about Julius Mayer, eventually publishing letters and articles in *Nature* related to the subject	Iwakura Mission begins, leaving Yokohama (23rd) Publication of book I of *Middlemarch*, by George Eliot; the novel appeared in eight books over the course of a year

Month and year	Biographical details	Notable social, political, and cultural events
February 1872	Delivers "On the Identity of Light and Radiant Heat" at the RI (2nd) German edition of *Fragments* published (February or March)	Assassination of Richard Southwell Bourke (1822–72), governor-general of India (8th) Metropolitan Museum of Art opens in New York City (20th) National Day of Thanksgiving for the recovery of the Prince of Wales (27th)
March 1872	Offered presidency of BAAS, which he declines Becomes involved in Hooker's conflict with Ayrton over Kew Gardens	Yellowstone National Park established in the western United States Tichborne case decided against the claimant, Arthur Orton
April 1872	*Vanity Fair* publishes his portrait as "Man of the Day" (6th) Gives weekly lectures at RI on "Heat and Light" (through June)	Third Carlist War begins in Spain
May 1872		*Popular Science Monthly* begins publication in the United States

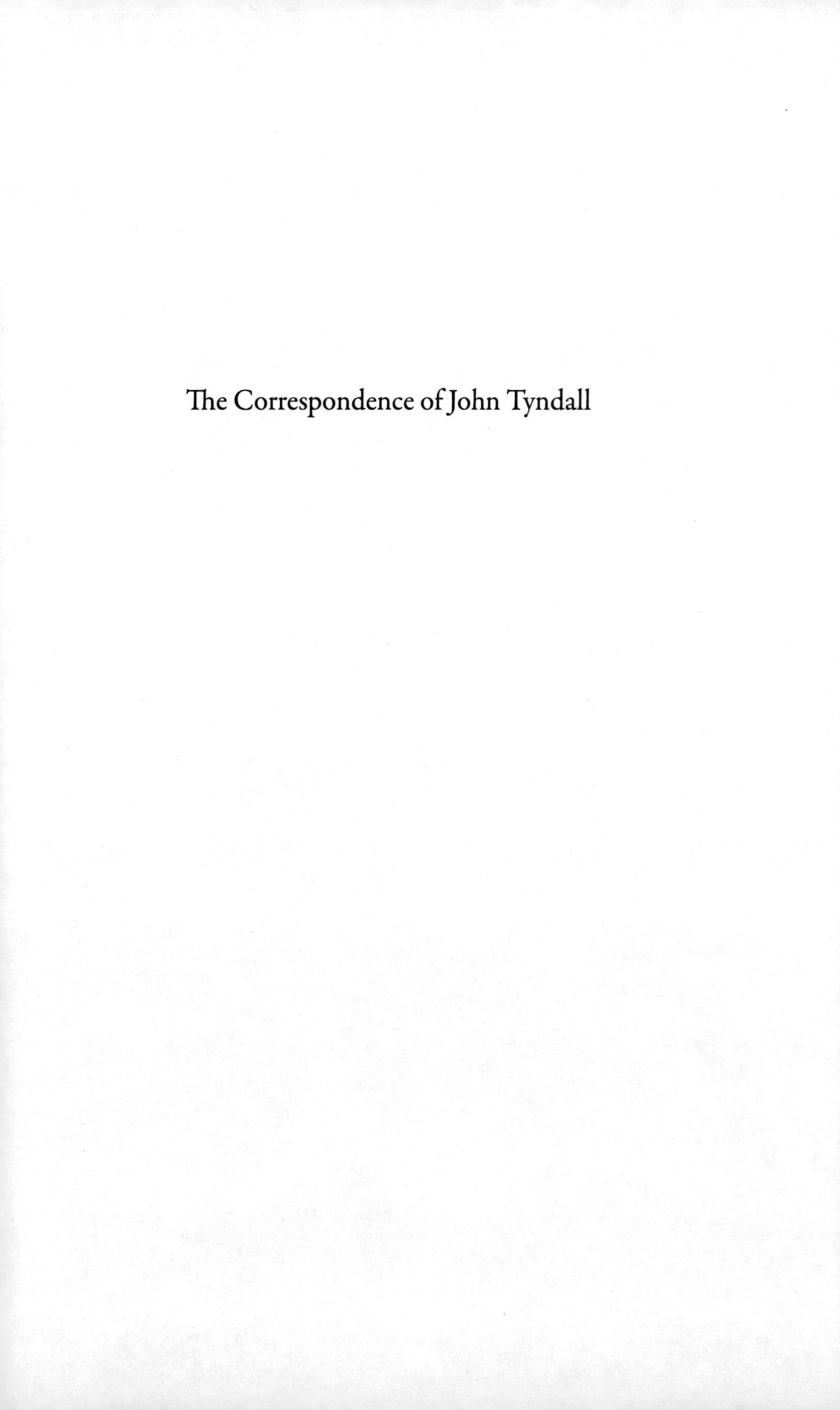

The Correspondence of John Tyndall

1871

From Charles Darwin — 1 March [1871][1] — 3367

Down, | Beckenham, Kent. S.E.[2] | March 1.

My dear Tyndall,

I saw Dr. Ogle,[3] who is a most acute observer, and told him of your suggestion,[4] and he will keep it in mind in reference to the less hairy races of man and the lower animals.

He is very anxious for information on one point, and as it is closely connected with your work,[5] I have thought you would forgive me for troubling you. Does your glycerine &c. respirator deprive odorous substances of their smell—I neglected to ask what sort of substance, but I think from the tenour of his remarks, solid substances, such as camphor, musk, rubbed brass would be most useful to him; but I daresay he would care about volatile oils or any odorous matter.

He is much perplexed with some physiological results, as for instance the relation of colour to the tissues supplied with olfactory nerves, and of differently coloured substances absorbing with different degrees of power odours.[6] A brief note, which I could communicate to Dr. Ogle, would greatly oblige me.

Yours very sincerely | Ch. Darwin.

(I return home to-morrow morning)

RI MS JT/1/TYP/9/2811
LT Typescript Only

---

1. *March 1 [1871]*: LT annotated the typescript with 'probably 1874'. For more of this conversation, see letters 3368, 3369, 3373, 3374 and 3375 in this volume; and letters 3363 and 3365 in *Tyndall Correspondence*, vol. 11.
2. *Down, | Beckenham, Kent. S.E.*: the location of Darwin's home, known as Down House (currently an English Heritage site).
3. *Dr. Ogle*: William Ogle; see letter 3369 for clarification.

4. *your suggestion*: Tyndall's suggestion, concerning his new respirator for firemen, that human noses, with hairs and mucus, work like his respirator as protection against diseases and other particles. See letters 3363 and 3365 in *Tyndall Correspondence*, vol. 11. Tyndall published on this three years later; see J. Tyndall, 'On some recent Experiments with a Fireman's Respirator', *Roy. Soc. Proc.*, 22 (1874), pp. 359–61. See also *Ascent of John Tyndall*, pp. 274–5.
5. *your work*: Tyndall's work on respirators.
6. *power odours*: Ogle had been working on the connection between odorous vapors and pigments. See letter 3373.

To Charles Darwin 1 March [1871][1] 3368

Royal Institution | 1st. March

My dear Darwin,

If Ogle[2] would call here we could arrange together such experiments[3] as he might think suitable for his purpose. I know him and shall be glad to see him.

Yours ever | John Tyndall.

RI MS JT/1/TYP/9/2812
LT Typescript Only

1. *1 March [1871]*: dated by relation to letters 3367 and 3369.
2. *Ogle*: Tyndall may have been referring to John William Ogle (1824–1905), an Oxford physician, but in the next letter Darwin clarified that he was referring to William Ogle, see letter 3369, n. 2.
3. *such experiments*: regarding Tyndall's fireman's respirator; see letter 3367, n. 4.

From Charles Darwin 1 March [1871][1] 3369

Down, | Beckenham, Kent. S.E. | March 1.

My dear Tyndall,

Very sincere thanks. I am sorry to trouble you, but you have mistaken the man, as my Dr. W. Ogle[2] of 34 Clarges St. told me he did <u>not</u> know you. He is an Oxford man, and I have liked much what little I have seen of him. If he may call on you,[3] I do not doubt he would be proud and pleased.

Yours very sincerely | Ch. Darwin.

(off very early tomorrow to home)[4]

RI MS JT/1/TYP/9/2813
LT Typescript Only

1. *March 1 [1871]*: dated by relation to letters 3367 and 3368.
2. *my Dr. W. Ogle*: William Ogle.
3. *If he may call on you*: see letter 3368.
4. *home*: LT left out the word 'home' from her typescript, and the missing word was obtained from the Darwin Correspondence Project's transcription of this letter.

From François-Napoleon -Marie Moigno[1] 2 March 1871 3370

Les Mondes | Revue Hebdomadaire | des Sciences | et de leurs applications | aux Arts & à l'Industrie | par | M. l'abbé MOIGNO | Rédaction | 2, rue d'Erfurth 2.

Paris, le 2 Mars 1871

Mon cher Mr Tyndall,

Permettez moi de vous donner un signe de vie ou de résurrection, après une si longue agonie, un si cruel silence de mort. Vous avez su que j'ai couru un grand danger, un obus est tombé sur mon petit logement que vous connaissez, et y a tout saccagé. Il aurait du me tuer mais j'en ai été quitte pour une très légère blessure.

J'ai reçu la collection du Mecanics Magazine, et j'y ai trouvé deux longs résumés de votre lecture de Liverpool où j'aurais dû être, et de votre dernière conférence de Royal Institution. Je voudrais avoir le texte de ces deux dissertations; et je vous prie de me les envoyer.

J'attends avec une vive impatience les autres publications du Royal Institution; devrai-je vous prier de demander qu'on me les envoie.

Vous le voyez, les libraires sont toujours les mêmes; la seconde édition de la chaleur n'a pas encore paru, et la guerre l'ajourne indéfiniment. Pourquoi faut-il que Mr Giraud m'ait volé, et n'ait pas rempli des engagements envers vous. Nous serions nous aussi à la quatrième édition.

Vous connaissez mes actualités scientifiques; les hésitations de Mr Gauthier-Villars m'ont obligé heureusement à les publier à mes frais; elles ont eu un tel succès que ce même Mr Gauthier-Villars m'en demande aujourd'hui *[quand]* celle qui contient votre programme des leçons sur l'électricité va paraitre. Les notes sur douze leçons d'optique sont traduites et seront bientôt sous-presse, si vous n'y voyez pas d'inconvénient. J'ai reçu en son temps votre beau volume sur le Diamagnétisme; avec quel bonheur je le traduirais si je trouvais un éditeur généreux pour vous.

Je vais reprendre mes Mondes qui sont redemandés comme nécessaires, mais dans de tristes conditions. Les recouvrements de 1870 ont été rendus impossibles par la guerre; les réabonnements de 1871 n'ont pas pu le faire. Ma pauvre caisse est vide. J'aurai besoin qu'on fasse pour moi un peu de propagande; je ne vous demanderai pas de vous abonner, mais je vous prierai de faire abonner les établissements où votre parole est si écoutée: Jermyn Street Museum of practical Geologie; London Institution, etc; etc.

Il est possible que j'aille dans quelques jours faire acte de présence à la soirée du Général Sabine.

Vos conférences s'il vous plait!

Les publications de Royal Institution

Les cours de Mr Odling.

Honteux d'avoir osé vous tant importuner, je suis dans les sentiments du respect le plus profond et le plus affectueux.

Votre humble serviteur et ami | L'abbé F. Moigno

Les Mondes | Weekly Journal | of the Sciences | and their applications | to Arts & Industry | by | Mr. Abbé MOIGNO | Editorial Board | 2, rue d'Erfurth 2.

Paris | 2 March 1871

My dear Mr Tyndall,

Allow me to send you a sign of life or resurrection, following such a long agony, such a cruel deathly silence.[2] You know that I was in great danger, for a shell fell on my modest accommodation, which you know, and devastated everything there.[3] It should have killed me but I got away with no more than a very slight injury.

I have received the collection of the Mechanics Magazine[4] and found in it two long summaries of your lecture in Liverpool[5] where I should have been, and of your last conference in the Royal Institution.[6] I would like to have the text of both these essays; and ask you to send them to me.

I wait with extreme impatience for the other publications from the Royal Institution; must I beg you to have them sent to me.

As you can see, the booksellers have not changed; the second edition of Chaleur[7] has not yet been published, and the war is postponing it indefinitely. Why did Mr Giraud[8] steal from me, and did not fulfil his commitments to you? We would now also be at the fourth edition.

You are aware of my latest scientific news;[9] thankfully, the hesitations of Mr Gauthier-Villars[10] have forced me to publish them at my expense; they were so successful that this same Mr Gauthier-Villars is asking me today *[when]* the one containing your course of the lessons on electricity will be published. The notes on twelve lessons in Optics[11] are translated and will soon go to press, if you have no objections. I received in due time your admirable volume on diamagnetism;[12] with what joy I would translate it were I to find a generous publisher for you.

I will return to my Mondes,[13] which are requested as necessities, but in sad circumstances. The fee collection for 1870 has been prevented by the war; the subscription renewals for 1871 have not been able to do it. My poor coffer is empty. I will need someone to spread some propaganda for me; I will not ask you to subscribe, but will ask you to get the institutions where your word is influential to subscribe: Jermyn Street Museum of Practical Geology, London Institution, etc. etc.

In a few days, I might put in an appearance at General Sabine's reception.[14]

Your conferences please!

The publications of Royal Institution

The lessons by Mr Odling.[15]

Ashamed to have dared to inconvenience you, I am feeling the deepest and the most affectionate regards.

Your humble servant and friend | Abbé F. Moigno

RI MS JT/1/M/120

---

1. *François-Napoleon-Marie Moigno*: Moigno (1804–84) was a science writer and Jesuit priest. He founded the scientific journal *Cosmos* in 1852, and *Les Mondes*, which was a weekly journal of scientific news, in 1863.
2. *such a cruel deathly silence*: Moigno's previous letter to Tyndall was from Paris on 15 February 1871; see letter 3359, *Tyndall Correspondence*, vol. 11.
3. *You learnt . . . there*: The Franco-Prussian War, or War of 1870 (July 1870–January 1871), was fought between the Second French Empire (then the Third French Republic) and the North German Confederation, led by Prussia. Though short, the war devastated thousands of people's homes and killed over a half a million people, most of those civilians. Many of Tyndall's French colleagues lost their homes, labs, families, or lives; see letter 3382, for example. Tyndall was a supporter of the Prussian side, however; see letters 3494 and 3495, for example.
4. *Mechanics Magazine*: *The Mechanics' Magazine*, a weekly periodical started in London in 1823 that spread useful knowledge to the general public about arts and industries. In 1873, the publication merged with a new periodical, *Iron: the Journal of Science, Metals & Manufactures*.
5. *lecture in Liverpool*: At the annual meeting of the BAAS, held in Liverpool from 14–21 September 1870, Tyndall delivered a lecture on 16 September entitled 'Scientific Use of the Imagination', in which he argued that human imagination, bounded by reason, allowed people to figure out new scientific theories (*Brit. Assoc. Rep. 1870*, p. lxxii). This was published as J. Tyndall, *On the Scientific Use of the Imagination* (London: Longmans, Green, and Co., 1870); and together with his 1868 BAAS lecture at Norwich ('Scientific Limit of the Imagination') as J. Tyndall, *Essays on the Use and Limit of the Imagination in Science* (London: Longmans, Green, and Co., 1870). The summary referred to by Moigno is 'The

British Association. On the Scientific Use of the Imagination', *Mechanics' Magazine*, 23 September 1870, pp. 221–2.

6. *last conference . . . Royal Institution*: Tyndall delivered a Friday Evening Discourse at the RI on 20 January 1871, 'On the Colour of Water, and the Scattering of Light in Water and in Air' (*Roy. Inst. Proc.*, 6 (1870–72), pp. 189–99). The summary referred to by Moigno is 'A Lecture by Professor Tyndall', *Mechanics' Magazine*, 27 January 1871, pp. 60–1, 64–5.
7. *the second . . . Chaleur*: The first French translation of Tyndall's *Heat Considered as a Mode of Motion* (London: Longmans, Green, and Co., 1863) appeared as J. Tyndall, *La Chaleur Considérée Comme Un Mode De Mouvement*, trans. F. M. Moigno (Paris: Étienne Giraud, Libraire-Editeur, 1864). The second edition did not appear until 1874, and there would be two more editions (1881, 1887).
8. *Mr Giraud*: Étienne Giraud (fl. 1850–60s), a French publisher and bookseller whose firm was located at 20 rue Saint Sulpice in Paris. He published Moigno's *Les Mondes* (1863), as well as the first French edition of Tyndall's *Heat* (1863), for which see n. 7.
9. *publication*: possibly referring to his book *Saccharimétrie optique, chimique et melassimétrique* (Paris: Les Mondes, 1869), or 'Saccharimetry, the measurement of the concentration of sugar in solution'.
10. *Mr Gauthier-Villars*: Jean-Albert Gauthier-Villars (1828–98), a French engineer and publisher. Born to a family of printers in Lons-le-Saunier, he studied at the École Polytechnique (1848–50), became a telegraph engineer, and served in the military. In 1864 he purchased the Mallet-Bachelier printing house which had published the *Comptes rendus hebdomadaires des séances de l'Académie des sciences*. Under his direction the firm, renamed Gauthier-Villars, specialized in mathematical and scientific books and magazines (H. W. Paul, *From Knowledge to Power: The Rise of the Science Empire in France, 1860–1939* (Cambridge: Cambridge University Press, 1985), p. 252).
11. *notes on twelve lessons in Optics*: Tyndall delivered a course of twelve lectures 'On Experimental Optics' at the RI in January and February of 1864 (*Roy. Inst. Proc.*, 5 (1862–66), p. 157). These were not published in English, and it does not appear that the French translation by Moigno was published.
12. *admirable volume on diamagnetism*: J. Tyndall, *Researches on Diamagnetism and Magnecrystallic Action including the Question of Diamagnetic Polarity* (London: Longmans, Green, and Co., 1870).
13. *Mondes*: *Les Mondes*, the journal of science that Moigno edited. It was active from 1863–73.
14. *General Sabine's reception*: the first of two annual soirées of the RS for 1871, put on by then President of the RS, Edward Sabine, on 11 March ('The Royal Society', *The Lancet*, 18 March 1871, pp. 387–8).
15. *Mr Odling*: William Odling (1829–1921), an English chemist. Odling worked on classifying the elements and refining Dmitri Mendeleev's periodic table of elements. Before becoming Fullerian Professor of Chemistry at the RI in 1868, he was a lecturer in chemistry at St Bartholomew's Hospital Medical School, and from 1872–1912 he was a professor of chemistry at Worcester College, Oxford. Odling was elected FRS in 1859.

To Thomas Archer Hirst 4 March [1871][1] 3371

My dear Tom.

Hector[2] has come! I want you to meet him here on Monday at 6 1/2 P.M.[3] Ever | John | 4th March

RI MS JT/1/T/879

1. *[1871]*: dated by relation to other letters about Hector Tyndale's visit to London.
2. *Hector*: Hector Tyndale.
3. *Monday at 6 1/2 P.M.*: Hirst recorded this dinner in his journal on 6 March 1871 (Brock and MacLeod, *Hirst Journals*, p. 1895).

To Thomas Archer Hirst 7 March 1871 3372

7th March 1871

My dear Tom

It is not right I think to leave Carpenter in this state of delusion.

I met him on the staircase of the Foreign office as I should any other friend. Had you been there I I[1] should have acted substantially as I did, that is, shake hands with you as a friend, make a remark that the moving crowd was like a phantasmagoria,[2] and then relapse into the problem which was uppermost in my mind, and which was this:—"Once in this crowd how am I to get out of it?"

What Carpenter can have to say to my shaking hands with Mr Gladstone[3] I am at a loss to imagine. He held out his hand to me in a kind frank way, I shook it and went on.

But no notion of "snubbing" could have entered Carpenter's brain had it not been prepared to imagine the thing. What were the pre-existing causes?

1. Mrs Carpenter[4] has been kind enough to ask me to dine with her once or twice: I have not been able to accept her invitation.

2. Gwynne Jeffreys[5] asked me to make a proposal to the Government Grant Committee,[6] which would be unjustifiable in any body so scantily informed as I was as to the objects of the proposal.

3. The proposal was then made by Gwynne Jeffreys himself. It was merely a general proposition that a grant should be given to Carpenter. Had it rested there I should have voted for the proposal without enquiry. But Stokes as is duty bound described the experiment;[7] and stated that the result contemplated by Dr. Carpenter was of so certain a character that no demonstration of it was needed. When Sharpey turned and asked me to express my opinion, of course I said what it was my duty to say, namely, that I agreed with Stokes.

Throughout all my conversations with Carpenter I never could see that he had realized the conditions of the problem he has taken in hand. His tub experiment[8] I have made over and over again in substance, but it does not touch the question. Let him take a portion of the ocean equal to his tub in magnitude, and let him prove that the different heating of the northern and southern sides of his tub is able to generate a force of convection competent to overcome the viscosity of the water.[9] This is what he has to do. Every body knows that with a sufficient difference of temperature you must get an exchange of water between the equator and the poles; but the question is are the existing differences competent to produce the existing currents, Carpenter's experiment is worse than useless here; worse I say because it gives a delusive appearance of explanation when no explanation whatever is given.

Yours affectionately | John.

With regard to Dr. Beale[10] he has been persistently attacking me for five or six years, and I never wrote a single unkind sentence regarding him save that one in Imagination[11] where I connected him with a remark of Huxley's in the Lay Sermons.[12] Even that I did not permit to appear in my new volume.[13]

I hope you will be at the Athenaeum[14] tonight. I want that proof.[15]

RI MS JT/1/T/686
RI MS JT/1/HTYP/579

---

1. *I*: Tyndall repeated the word 'I'.
2. *phantasmagoria*: 'the conjuration by preternatural means of a vision, ghost, etc.; an apparition' (*OED*).
3. *Mr Gladstone*: William Gladstone.
4. *Mrs Carpenter*: Louisa Carpenter (née Powell, 1812–87), born in Exeter to the merchant Joseph Powell, was the wife of William Carpenter.
5. *Gwynne Jeffreys*: John Gwyn Jeffreys (1809–85), an English lawyer and conchologist. He retired from law in 1866 to continue his practice of collecting shells. He was elected to the Linnean Society in 1829 and made FRS in 1840. He is best known for his work on Mollusca, and he wrote more than 100 papers on the topic. His collection of Mollusca was purchased by the Smithsonian Institution in 1883.
6. *Government Grant Committee*: The Government Grant Committee of the RS was established in 1849 for the purpose of granting funds to researchers. On the Government Grants, see M. B. Hall, *All Scientists Now: The Royal Society in the Nineteenth Century* (New York: Cambridge University Press, 2002), chapters 5 and 6; and R. M. MacLeod, 'The Royal Society and the Government Grant: Notes on the Administration of Scientific Research, 1849–1914', *The Historical Journal*, 14 (1971), pp. 323–58.
7. *described the experiment*: As Secretary of the RS, this was part of George Stokes' duties. From 1869–71, Carpenter was studying life in the deep sea and the possibility of luminous invertebrates. Based on the results from these expeditions, Carpenter made experiments

about the temperature of the seas at various depths. Clearly, Tyndall disagreed with him. See W. B. Carpenter, 'Lenz's doctrine of ocean circulation', *Nature*, 10 (2 July 1874), pp. 170–1.

8. *His tub experiment*: Carpenter was testing his theory by working on a smaller scale, and he had delivered a Friday Evening Discourse at the RI on the topic on 11 February 1870 (W. B. Carpenter, 'On the Temperature and Animal Life of the Deep Sea', *Roy. Inst. Proc.*, 5 (1870–72), pp. 1–21).
9. *let him prove . . . the water*: see n. 8.
10. *With regard to Dr. Beale*: Lionel Smith Beale (1828–1906), a British physician who was a professor of Pathology at King's College, London, and was elected FRS in 1857. He argued that using microscopy and the scientific method in medicine could help understand disease.
11. *save that one in Imagination*: Tyndall's essay 'Scientific Use of the Imagination' quoted a reference of Huxley's that characterizes Beale as 'one conspicuous member' of 'microscopists, ignorant alike of Philosophy and Biology' (J. Tyndall, 'The Scientific Use of the Imagination', in *Essays on the Use and Limit of the Imagination in Science*, 2nd edn (London: Longmans, Green, and Co., 1870), pp. 13–51, on p. 49).
12. *remark of Huxley's in the Lay Sermons*: T. H. Huxley, 'Prefatory Letter', in *Lay Sermons, Addresses and Reviews* (London: MacMillan, 1870), pp. v–viii. This is the published version of a letter from Huxley to Tyndall on 14 June 1870. The letter is missing and not in *Tyndall Correspondence*, vol. 11.
13. *appear in my new volume*: likely *Fragments of Science.*
14. *Athenaeum*: the Athenaeum Club, a private members' club located at 107 Pall Mall in London, particularly for those with intellectual interests, and accomplishments in art, science, or literature. The club was founded in 1824, at the instigation of John Wilson Croker. Humphry Davy became the first chairman and Michael Faraday the first secretary. Faraday became a member after a few months when he was able to resign the secretaryship. Tyndall became a member in 1860, Thomas Huxley in 1858, Joseph Hooker in 1851 and Thomas Hirst in 1866. On the club, see M. Wheeler, *The Athenaeum: More Than Just Another London Club* (New Haven, CT: Yale University Press, 2020).
15. *I want that proof*: likely referring to a conversation Tyndall and Hirst had that does not appear in writing.

From Charles Darwin 7 March [1871][1] 3373

Down, Beckenham, Kent. S.E. | March 7th

My dear Tyndall,

I wrote about a week ago,[2] explaining that my Dr. Ogle was not the same as yours, and asking whether he might call on you—But the two men may make all the difference to you. As my man has been very obliging to me, I much wish to make him some answer. Forgive me for troubling you. Shall I say that you have not attended to odorous particles?[3] and are sorry that you cannot give him the desired information? Or anything else you may determine on. I know how busy you must be.

Yours very sincerely | Ch. Darwin.

RI MS JT/1/TYP/9/2814
LT Typescript Only

1. *[1871]*: see letter 3367, n. 1.
2. *a week ago*: 1 March 1871; see letters 3367 and 3369. For more on this conversation, see also letters 3368, 3374 and 3375.
3. *odorous particles*: see letter 3367, n. 6.

To Charles Darwin 8 March [1871][1] 3374

Royal Institution. | 8th. March.

My dear Darwin,

I wrote to your friend[2] immediately,[3] saw him, talked to him, and liked him. I thought he would probably tell you all this—otherwise I would have written to you.[4]

The subjects on which he spoke to me have often occupied my attention.

Yours ever | John Tyndall.

RI MS JT/5/16b
LT Typescript Only

1. *[1871]*: see letter 3367, n. 1.
2. *your friend*: William Ogle.
3. *I wrote . . . immediately*: letter missing.
4. *I would have written to you*: see letter 3373.

From Charles Darwin 8 March [1871][1] 3375

Down, Beckenham Kent. S. E. | March 8.

My dear Tyndall,

A million apologies and a million thanks. I have just heard from Dr. Ogle who is much pleased at your kindness.[2]

Ever yours | Ch. Darwin.

RI MS JT/1/TYP/9/2816
LT Typescript Only

1. *[1871]*: year determined by previous letters; see letter 3367, n. 1.
2. *much pleased at your kindness*: see letter 3374.

To Hector Tyndale 10 March [1871][1] 3376

10th March

Dear Hector

Carpenter lectures here tonight at 9[2]—If you would care to hear him pray come at a quarter before 9.

Yours affect[ionate]ly. | John.

RI MS JT/1/T/1466
RI MS JT/1/TYP/4/1686

1. *[1871]*: dated from Carpenter's Friday Evening Discourse at the RI (see n. 2). Letters 3385 and 3395 establish that Tyndale, who traveled more than once to London, was there in March 1871. For additional correspondence relating to Tyndale's trip, see letters 3378, 3380, 3389 and 3393.
2. *Carpenter . . . at 9*: William Carpenter delivered a Friday Evening Discourse at the RI on 10 March 1871, 'On the Latest Scientific Researches in the Mediterranean' (*Roy. Inst. Proc.*, 6 (1870–72), pp. 236–59).

To Alfred Mayer 11 March 1871 3377

Royal Institution of Great Britain | 11th March 1871

My dear Sir

When the exquisite photographs of the Lines of Force[1] reached me I chanced to be putting together for publication an elementary lecture on magnetism.[2] I at once placed one of the photographs in the hands of a wood engraver with a view of rendering my words more acceptable by so beautiful an addition to them.

I hope I have not taken an unwarranted liberty.

Your papers have reached me[3] and I can see in them, even from the superficial glance which up to the present moment I have been able to bestow upon them, firmness of intellectual fibre which is sure to make itself felt, and which it is of exceeding importance to cultivate, in America.

I am glad to be informed that so amiable a philosopher as Prof. Joseph Henry is your scientific father.[4] I wish his son every success

Yours faithfully | John Tyndall

Prof. Mayer | &c. &c.

Tyndall, John (1820–1893); Hyatt and Mayer Collection, C0076, Manuscripts Division, Department of Special Collections, Princeton University Library

---

1. *exquisite photographs of the Lines of Force*: Mayer had developed a method where he could photograph the spectra of magnetic lines of force in the late 1860s (A. G. Mayer and R. S. Woodward, *Biographical Memoir of Alfred Marshall Mayer, 1836–1897* (Washington, DC: National Academy of Sciences, 1916), p. 18). See also J. M. Mayer, 'On a method of fixing, photographing, and exhibiting the magnetic spectra', *American Journal of Science*, 50 (April 1871), pp. 263–6; this article, which included a photograph of lines of force, the same one Tyndall used (see n. 2), was soon republished in England as 'On a Method of Fixing, Photographing, and Exhibiting the Magnetic Spectra', *Phil. Mag.*, 41 (June 1871), pp. 476–80).
2. *lecture on magnetism*: J. Tyndall, 'An Elementary Lecture on Magnetism', in *Fragments of Science*, pp. 373–401. This was from an 1861 lecture to primary school teachers held at the South Kensington Museum on 30 April 1861; Tyndall included a photograph of Mayer's on the page opposite the first page of the lecture.
3. *papers have reached me*: probably recent papers of Mayer's on the topic of magnetism, 'Researches in electro-magnetism', *American Journal of Science*, 50 (1870), pp. 195–212; and 'Abstract of a research in electro-magnetism', *Journal of the Franklin Institute*, 60 (1870), pp. 403–10. Possibly also a draft copy of the article referred to in n. 1, which at this time would have been in press.
4. *Joseph Henry is your scientific father*: see letter 3510.

To Hector Tyndale [12 March 1871][1] 3378

Sunday.

Dear Hector.

I will call upon you in an hour in the hope that you will take a walk with myself and Debus.[2]

Yours affectionately | John.

It would be still better if you w[oul][d] come over here. You breakfast late, this is the reason why I have not asked you to my dejeuner.[3]

RI MS JT/1/T/1481
RI MS JT/1/TYP/4/1688

1. *[12 March 1871]*: date based on the time of Tyndale's travels to London, and that the 12th was a Sunday in March of 1871. For more on this dating, see letter 3376, n. 1; for additional correspondence relating to this trip of Tyndale's, see letters 3380, 3385, 3389, 3393 and 3395.
2. *myself and Debus*: Tyndall, Hirst and Heinrich Debus would often walk on Sundays.
3. *dejeuner*: the morning meal; breakfast (*OED*); for the French, dejeuner means lunch, while the English adapted the word for breakfast.

To Elizabeth Dawson Steuart 14 March [1871][1] 3379

Royal Institution | 14th March.

My dear Mrs Steuart,

Would you kindly send a pound for me to Catharine Washington—I knew her father[2] when a boy.

I will write to you again soon.[3] At the present moment I am awfully pressed and busy.

I have had another application[4] from Caleb Tyndall. There is something wrong in the blood of these people which annuls the sense of dignity and even the sense of shame. His daughter Dorah,[5] to whom I had a short time previously given 50 pounds, had the effrontery to write to me for more.[6] The wife[7] of his son William[8] has also been applying to me.

The only one whose behaviour pleases me is Sarah.[9] She has never troubled me and she appeared a nice modest girl when I saw her.

I would gladly if you thought it would do any permanent good help his two younger sons.[10] They have never troubled me.

Yours ever | John Tyndall.

RI MS JT/1/TYP/10/3393
LT Typescript Only

1. *[1871]*: The year is given by a note LT wrote on the typescript.
2. *her father*: not identified.
3. *I will write to you again soon*: letter 3401.
4. *another application*: letter missing.
5. *Dorah*: Dora, Dorah or Dorothea Tyndall (dates unknown) was Caleb's daughter (see letter 2644, n. 5, *Tyndall Correspondence*, vol. 10).
6. *write to me for more*: letter missing.
7. *the wife*: Frances Tyndall (née Stone, *c.* 1839–89), also called 'Fanny', the wife of William Tyndall. They married in 1860 in Kilkenny.
8. *of his son, William*: William Tyndall (*c.* 1826–1902), Tyndall's cousin and the son of Caleb Tyndall and Dorothea Shirley. He was a civil bill officer.

9. *Sarah*: Sarah Tyndall (1844–1917), Tyndall's cousin and the daughter of Caleb Tyndall and Dorothea Shirley. In 1869 she married David Miller. Letter 2777 in *Tyndall Correspondence*, vol. 10, indicates that Tyndall had committed to providing a 'little marriage portion' for Sarah but had concerns over her suitor's expectation of the money.
10. *his two younger sons*: Caleb's four sons were Caleb (b. 1837), James, William (see n. 7) and John, so it is possible here Tyndall is referring to James and John (see letter 2644, *Tyndall Correspondence*, vol. 10).

To Hector Tyndale 15 March [1871][1] 3380

15th. March.

My dear Hector.

Will you take your chance of meeting Huxley at the Royal Society Club dinner[2] on Thursday at 6 P.M. If you come here at a quarter before 6 I will walk on with you. We dine at Willis's Rooms King St. St. James.[3]

I fear there is no time to arrange a meeting with Huxley here.

I doubt also whether Carlyle is manageable this week.

Yours affectionately | John Tyndall

RI MS JT/1/T/1467
RI MS JT/1/TYP/4/1689

---

1. *[1871]*: for this dating, see letter 3376, n. 1; for additional correspondence relating to this trip of Tyndale's, see letters 3378, 3385, 3389, 3393 and 3395.
2. *Royal Society Club dinner*: Hirst noted that the Council of the RS met this day, then went to the Club later (Brock and MacLeod, *Hirst Journals*, p. 1895). The Royal Society Club was founded in 1743, and the current President of the RS acted as President of this club. A second club, the Philosophical Club, was founded in 1847 by FRS to promote reform within the RS. For both clubs, dinners occurred on the same day as RS meetings. Tyndall was most likely referring to the Philosophical Club. See T. E. Allibone, *The Royal Society and its Dining Clubs* (Oxford; New York: Pergamon Press, 1976); and M. B. Hall, *All Scientists Now: The Royal Society in the Nineteenth Century* (New York: Cambridge University Press, 2002), pp. 4, 82 and 217. See also T. G. Bonney, *Annals of the Philosophical Club of the Royal Society, Written from Its Minute Books* (London: Macmillan and Co., 1919); and A. Geikie, *Annals of the Royal Society Club: The Record of a London Dining-Club in the Eighteenth & Nineteenth Centuries* (London: Macmillan and Co., 1917).
3. *Willis's Rooms King St. St. James*: Almack's, a popular dining, dancing and meeting place in London open to gentlemen and ladies, founded in the mid-eighteenth century, later became a gentlemen's club under the name of Willis's Rooms.

From Henry Reeve[1] 15 March [1871][2] 3381

PRIVY COUNCIL OFFICE | March 15.

Dear Sir,

The Lord Chancellor undertook last night to inform you[3] that we had last night the pleasure of electing you at 'The Club'.[4]

I enclose a list[5] of the members & days of the Dinners of this season.

At your convenience I will request you in my capacity of Treasurer, to pay the sum of £5 to the credit of 'The Club' at Mess[rs]. Cocks & Biddulph[6] in Charing Cross.

There is no regular subscription to the Club. But this sum of £5 is paid on entrance, & we have occasionally a small whip[7] when it is wanted.

Yours sincerely | Henry Reeve | Prof. Tyndall F. R. S.

RI MS JT/2/10/470

---

1. *Henry Reeve*: Henry Reeve (1813–95) was the editor of the *Edinburgh Review*, journalist, translator and diarist. Member of The Club from 1861, and Treasurer from 1867–93, he also recorded details of club membership in his diaries.
2. *[1871]*: letter follows 3383.
3. *to inform you*: see letter 3383.
4. *'The Club'*: 'an exclusive dining club of thirty-four men, including the Archbishop of Canterbury, [Prime Minister William] Gladstone, the Duke of Argyll and the Earl of Derby. [Tyndall] joined with them and others such as George Richmond, Froude and Tennyson, as only the second man of science, after Richard Owen, to be elected' (*Ascent of John Tyndall*, p. 273).
5. *I enclose a list*: enclosure missing.
6. *Messrs. Cocks & Biddulph's*: The banking firm Cocks, Biddulph and Co. was founded in 1757, and located as of 1759 at 43 Charing Cross in London. In 1919 the firm was bought by Bank of Liverpool and Martins, and in 1969 became part of Barclays Bank.
7. *small whip*: a whip-round, or 'a call or appeal to a number of persons for contributions to a sum or fund' (*OED*).

From Henri Regnault[1] 15 March 1871 3382

Mon Cher Mr Tyndall

C'est aujourd'hui 15 Mars que je reçois à Genève la lettre que vous m'avez écrite à Sèvres le 11 Janvier, et par laquelle vous mettez à ma disposition une somme de 1000 fr à distribuer en secours aux habitants malheureux des environs de Paris. Je ne peux que vous féliciter pour cette généreuse intention et je suis convaincu que depuis lors vous avez trouvé un moyen plus efficace pour faire parvenir votre don.

J'ai été expulsé de la Manufacture, avec ce qui me restait de personnel, le 13 Octobre, et amené à Versailles. Je suis resté à Versailles jusqu'au 27 au même mois pour obtenir la protection du Prince Royal de Prusse afin d'éviter la dévastation de notre établissement, et pour ramener à Versailles ce qui pouvait encore être transporté. Le manque d'argent m'a forcé d'aller rejoindre à Genève mes plus jeunes enfants que j'y avais envoyés peu de jours avant l'investissement de Paris. Les savants Genevois, mes amis, m'ont offert l'hospitalité la plus gracieuse.

Depuis lors, tous les malheurs m'ont accablé. Mon second fils, Henri Regnault, âgé de 26 ans et déjà un des premiers peintres de notre temps, a été tué à la fatale sortie du 19 Janvier. C'est un coup de mort pour moi, il était le final directeur pour mon existence déjà si tourmentée.

Les travaux scientifiques ne pourront même pas me faire oublier, par moments, mon malheur, car, malgré les promesses formelles de conservation qui m'avaient été données par le Prince Royal et par le quartier général Prussien, mon laboratoire de recherches de Sèvres, qui contenait toutes mes machines, les appareils qui servaient à mes recherches depuis 30 ans, <u>a été complètement saccagé</u>—les machines, instruments &c &c brisés à coups de marteau et de pavés ....

Enfin, ce qui est encore plus triste pour moi, les cahiers et registres d'expériences qui contenaient les données des expériences non publiées encore, et dont j'annonce la publication prochaine dans mon dernier volume—ont été réunis en un tas avec tous les papiers et brulés en feu de joie—

Et cependant, la défense de faire aucun dégât venant de l'Etat Major Prussien était affichée à la porte de mon laboratoire, ainsi qu'une lettre en Allemand, écrite par moi, par laquelle je priais les troupes allemandes de respecter les instruments et les papiers qui ne se rapportaient qu'à des travaux scientifiques qui intéressent autant l'Allemagne que la France.

Dans l'espoir de conserver le fruit de mes longs travaux, j'avais poussé <u>la bassesse</u> jusqu'à dire que j'étais un des associés de l'Académie de Berlin,

Commandeur de l'Aigle Rouge de Prusse,

Chevalier de l'ordre du mérite civil de Prusse.

C'est probablement cette marque de vanité, la seule que j'ai donnée dans ma vie, car je n'ai jamais porté aucune décoration étrangère—qui les a portés à cet acte de vandalisme.

Je n'ai pas besoin d'ajouter que dans ma maison d'habitation à Sèvres, tous mes meubles ont été brisés ou brulés et que je suis obligé maintenant de me créer une nouvelle installation. Ma bibliothèque seule me reste, parce que j'étais parvenu à la faire transporter au palais de Versailles.

Vous voyez que je suis bien malheureux, et que je termine bien tristement une Carrière, dont on doit au moins honorer les efforts.

Votre très dévoué confrère | V. Regnault | Lancy, ce 15 Mars 1871

Mon adresse est | Campagne Soret | à Lancy, près Genève.

My dear Mr Tyndall,

Today, 15 March, I received in Geneva the letter you wrote me to Sèvres on 11 January,[2] and through which you placed at my disposal a sum of 1000 francs to distribute as relief to the unfortunate residents around Paris. I cannot but congratulate you for this generosity and I am sure that since then you have found a more efficient way to send your donation.

I was evicted from the workshop with my remaining staff on 13 October and was taken to Versailles. I stayed in Versailles until the 27th of the same month in order to gain the Prussian Royal Prince's[3] protection to prevent the devastation of our establishment and to move to Versailles that which could still be transported. Lack of money forced me to go to Geneva to join my youngest children, whom I sent there a few days before the siege of Paris. Genevan scientists, my friends, gave me the most gracious hospitality.

Since then, every misfortune has befallen me. My second son, Henri Regnault,[4] 26 years old and already one of the leading painters of our time, was killed in the fatal sortie of 19 January. This is a death blow for me; he was the final purpose of my already tormented existence.

Even scientific work cannot make me forget, for a time, my misfortune, since despite the formal promises of conservation that I had been given by the Royal Prince and the General Prussian Headquarters, my research laboratory in Sèvres, which had all my machines, the devices that were used in my research for 30 years, has been completely ransacked.[5] Machinery, instruments, etc etc, broken with hammers and stones . . . .

Finally, and even sadder for me, the books and records which contained the data of as yet unpublished experiments, whose forthcoming publication I announced in my last volume—were added to a pile with all the papers and burned in bonfire—

And nevertheless, the interdiction against any damage from the Prussian General Staff was posted on the door of my laboratory, as well as a letter in German, written by me, in which I requested German troops respect the instruments and papers which related only to scientific work that would interest Germany as much as France.

In the hopes of conserving the fruits of my long labours, I sunk to claiming that I was an associate of the Academy of Berlin,[6]

Commander of the Red Eagle of Prussia,

Knight of the Order of Civil Merit of Prussia.

It is probably this sign of vanity, the only one that I have given in my life, because I never wore any foreign decorations—that spurred them to this act of vandalism.

I do not need to add that in my house in Sèvres,[7] all of my furniture was broken or burned, and I am now forced to create a new installation. My library is the only thing that remains to me, because I was able to transport it to the palace of Versailles.

You see that I am miserable, and that I now face the end of my career, whose efforts at least should be honoured.

Your devoted colleague | V. Regnault | Lancy, 15 March 1871

My address is | Campagne Soret | in Lancy, near Geneva.

RI MS JT/1/R/10
RI MS JT/1/TYP/3/1009–10

1. *Henri Regnault*: Henri Victor Regnault (1810–78), a French chemist and physicist who worked with a number of British scientists in this period. He worked on the properties of heat and steam, designed thermometers and calorimeters and was an early proponent of thermodynamics, with Thomson and Rankine (see letter 3387, n. 35). He won the RS's Copley Medal in 1869. His home and lab were destroyed by fighting during the Franco-Prussian War. His son was also killed during the war, and he retired in 1872.
2. *the letter . . . on 11 January*: letter missing from *Tyndall Correspondence*, vol. 11.
3. *Prussian Royal Prince*: Wilhelm I (1797–1888), Kaiser of Prussia from 1861, and Emperor of Germany from the signing of the Treaty of Versailles to end the Franco-Prussian War on 18 January 1871. He was the first head of state of a united Germany. They did not normally use the term King for their heads of state, but instead used a more generic term such as Prince.
4. *Henri Regnault*: Alexandre-Georges-Henri Regnault (1843–71), a French painter, and the son of Henri Victor Regnault. He died in the Second Battle of Buzenval on 19 January 1871.
5. *Even scientific work . . . ransacked*: see n. 1. See also letter 3438.
6. *Academy of Berlin*: the Royal Prussian Academy of Sciences, founded in 1700 in Berlin.
7. *Sèvres*: commune in the southwestern area of Paris.

From William Wood[1] 15 March 1871 3383

31 Great George Street | S.W. | March 15. 1871

My dear Professor Tyndall

As chairman last evening at "The Club"[2] I have the pleasure of informing you that you have the honour of being elected a member of the body[3]

Yours faithfully | Hatherley | Professor Tyndall

RI MS JT 2/10/470–1a

---

1. *William Wood*: William Page Wood, 1st Baron Hatherley (1801–81), a British lawyer and Lord Chancellor from 1868–72 in Gladstone's first ministry.
2. *"The Club"*: see letter 3381, n. 4.
3. *a member of the body*: proof of Tyndall's growing reputation, as membership was only by unanimous election. For further correspondence on this subject, see letters 3381 and 3384. Tyndall was one of two members elected on 14 March 1871; the other was John Emerich Edward Dalberg-Acton, 1st Baron Acton (1834–1902) (*Annals of The Club, 1764–1914* (London: The Club, 1914), pp. 89, 209–10).

To William Wood[1] 16 March 1871 3384

Royal Institution of Great Britain | 16th March 1871.

My dear Lord Hatherley,

I prize highly the honour of the election which you have announced to me;[2] but not less highly do I prize that element of kindness which sweetens honour, and with which this distinction had been accompanied. Thank you much, and believe me

faithfully & gratefully yours[3]

The Lord Chancellor.

RI JT 2/10/470–1b
Transcript Only

---

1. *William Wood*: see letter 3383, n. 1.
2. *the election which you have announced to me*: to 'The Club'; see letters 3381 and 3383.
3. Tyndall's signature was not copied onto this handwritten transcript.

To Hector Tyndale [17 March 1871][1] 3385

Friday

My dear Hector.

Just desire your made to tell my messenger how you feel this morning—Will you be able to receive Tom[2] & myself today? I saw Tom last night—if not forewarned he will be with you between 6 & 6 1/2.

I must leave you two a little before 9.

Ever. John.

RI MS JT/1/T/1479
RI MS JT/1/TYP/4/1773

1. *[17 March 1871]*: date based on the day given and on the information in Hirst's own journals (Brock and MacLeod, *Hirst Journals*, pp. 1895–6). See also letters 3388, 3393 and 3395.
2. *Tom*: Thomas Hirst. See letter 3395.

From Philip Gossett[1] 17 March 1871 3386

Berne: March 17, 1871.

Bennen's body[2] was found with great difficulty the third day after Boissonnet[3] was found. The cord-end had been covered up with snow. The curé d'Ardon[4] informed me that poor Bennen was found eight feet under the snow, in a horizontal position, the head facing the valley of the Lyzerne. His watch had been wrenched from the chain, probably when the cord broke; the chain, however, remained attached to his waistcoat. Three years ago I met one of my Ardon guides; he told me that Bennen's watch had been found by a shepherd seven months after the accident. This shepherd had been one of the party who went up to look for Bennen; during the following summer he had watched the melting of the avalanche. When mounted, the watch obeyed. This reminds me of your fall on the Morteratsch glacier.*

I know you were very much attached to Bennen; the same was the case with him in regard to you. An hour before his death the Matterhorn[5] showed its black head over one of the *aretes*[6] of the Haut de Cry.[7] I asked Bennen whether he thought it would ever be ascended. His answer was a decided "Yes"; but he added, alluding to your last attack on the mountain, "Wir waren fünf; der Professor und ich stimmten für Vorwärts; die drei stimmten dagegen."[8]

There is one circumstance in reference to my fall with the avalanche of the Haut de Cry that I am utterly unable to understand: I mean what physical phenomena took place when the avalanche stopped and froze. It stopped because in its progress downwards the broad couloir down which it was going got narrower, and the mass of snow could not pass. It froze because the successive portions of the body of the avalanche became compressed against the head, which latter had come to a stop. When the layer in which I was stopped, the pressure on my body was enormous—so great, in fact, that I expected I should be crushed flat. This pressure ceased *suddenly*; I know it, for the atrocious pains it was causing ceased *suddenly* too. What happened during that interval?[9]

*See Chapter XIX[10]

Hours of Exercise in the Alps, pp. 204–5
Typed Transcript Only

1. *Philip Gossett*: Philip Charles Gosset (1838–1911), a Swiss-born English engineer, alpinist and glaciologist. He was an urban planner in Switzerland. He went on a number of glacier expeditions and published an article about the ill-fated expedition in 1864 that he related in this letter. See P. C. Gosset, 'Narrative of the Fatal Accident on the Haut-de-Cry, Canton Valais', *Alpine Journal,* 1 (1864), pp. 288–94. This account was made famous in *Hours of Exercise in the Alps*, pp. 192–205. Tyndall introduced this letter in *Hours of Exercise in the Alps*: 'I have been recently favoured with a letter from Mr. Gossett, from which the following is an extract:'.
2. *Bennen's body*: Johann Joseph Bennen (1824–64), a Swiss alpine guide. He had performed a number of ascents with Tyndall, including the first ascent of the Weisshorn in 1861.
3. *Boissonnet*: Louis Boissonnet (1838–64), a Russian-born Swiss engineer and alpine guide.
4. *curé d'Ardon*: the priest of Ardon, a town near the mountain Haut de Cry.
5. *Matterhorn*: mountain that straddles the Swiss and Italian Alps. The first ascent was in 1865, and the team included Edward Whymper (see letters 3462, 3464 and 3470); four of the party of seven were killed on the descent.
6. *aretes*: a sharp mountain ridge (*OED*).
7. *Haut de Cry*: mountain in the Bernese (Swiss) Alps.
8. *"Wir waren fünf; der Professor und ich stimmten für Vorwärts; die drei stimmten dagegen."*: 'We were five; the Professor and I voted forwards; the three voted against it' (German).
9. *interval?*: When Tyndall related the story of the avalanche that took Bennen and Boissonet, he mentioned that Bennen had asked the guides about the danger of an avalanche at their spot, and the guides had said their 'position was perfectly safe' (*Hours of Exercise in the Alps*, p. 195). At the end of this story, Tyndall appended this note onto the original letter: '[Bennen was well acquainted with winter snow; but no man of his temper, and in his position, would place himself in direct opposition to local guides, whose knowledge of the mountain must have been superior to his own.]'.
10. *See Chapter XIX*: This in-text note by Tyndall refers to Chapter 19 (pp. 206–18) in *Hours of Exercise in the Alps* in which he detailed his fall on this glacier near Pontresina, Switzerland, in 1864. See also letter 2149, *Tyndall Correspondence*, vol. 8.

From Alexander Herschel 15–17 March 1871 3387

57 North Hanover Street | Glasgow March 15th '71

Dear Professor Tyndall

I have delayed very long acknowledging the kind present of your book on "Diamagnetism and Magnecrystallic Action"[1] which I received last summer, with its enclosure "From the Author", from the Publishers Mess$^{\underline{rs}}$ Longmans & Co. I have constantly been on the point of writing to thank you for it when the many pressing occupations of this Session[2] have as often prevented me, and I have for the same reason not perused it, as I would like to have done!— Sound, Light, Heat, Astronomy, Steam and Applied Mechanics keep my hands very full of work, to train three pretty numerously attended classes; and they teach me to value more and more the great scope and the admirable clearness of your works! When, in the next Winter's Session I will be lecturing again on "Magnetism and Electricity", I have no doubt that I will owe much to this new boon to Electricians, as I have borrowed in former years from your works on "Sound",[3] and "Heat as a Mode of Motion".[4] These two last books are Prizes for which my class always shows the greatest preference; when, at the end of the half year a present from the President[5] enables me to give a few "rewards of diligence" during the session to the best pupils in an examination; and the stimulus which they derive from them is I am sure very fully shown by the continued popularity of my "Popular Course" of lectures on Physics among the steady and hard working Glasgow Artisans![6]—

The Faculty of Glasgow Physicians wish me to show them at their Annual Conversazione[7] on Wed[nes]d[a]y Evening next some experiments showing the organic nature of dust in the air; and I have tried with the electric lamp to produce the black-clouds above a Bunsen-flame, in its bright beam, very successfully. Before the lecture-day I hope to rehearse the experiments with a stronger beam; and to obtain the result also, if possible with cotton wool.* The Physicians are very much interested in the question of 'Dust and Disease',[8] and perhaps the exhibition of these experiments will help them to form a right opinion of the insidious nature of the particles floating in the air; and I hope that their organic nature, and the power of separating them by cotton-fibres will be well illustrated by them.—I have also lately given two or three lectures on the late eclipse expeditions,[9] in which you were so unfortunately disappointed in seeing the sun's radiations, or whatever the curiously-shaped light of the Corona will, at last be found to be![10]—My father[11] sent me two photographs of the Corona, one of them by Mr Brothers, at Syracuse,[12] and the other taken by the American observers at Xeres,[13] to look at, and to compare together; and they resemble each other so closely, when brought together in a Stereoscope,[14] that there certainly appears to be proof enough of the reality of the object which they both represent!—I knew comparatively little of this interesting result, and of the exact measurement of the bright spectral lines in the Corona[15] by Professor Young and others,[16] when the accompanying lecture was given; and a later lecture, in which I described them, having been read only to the "Chemical Section" of the Glasgow "Philosophical Society",[17] it will not be printed in their Proceedings. Some account of the proceedings of the Sicilian expedition,[18] which Dr Thorpe[19] gave us at the same time, I hope, however to be able to send you very soon. He is Professor Penny's[20] successor in the Chair of Chemistry at the Andersonian Institution, here; and went to Sicily to observe the photographic Intensity of the sun's light during the Eclipse, in continuation of his former researches on the Chemical intensity of direct sun-, and diffuse day-light for Professor Roscoe.[21]

Friday, March 17th;—I find that I can send you a paged copy of Dr Thorpe's and my Papers on the Solar Eclipse[22] which will be placed before the next meeting of the Society on Wednesday next, as the last published number of its "Proceedings"[23] and this contains a short Paper on Helmholtz's "Analysis of the Vowel Sounds",[24] which I wished very much to have sent you in the Proof Sheets, in order that any corrections which you would have been so good as to suggest, might have been made in it; but I have so little time for public lectures during the Session, that the lecture was not written, before; and the Reporter's account of it, given in this Paper, is as he sent it to me to correct, and to supply some missing names, and quotations (as I read them) from your "Lectures on Sound",[25] and from Helmholtz's Work "Die Tonempfindungen, &c."[26]—There is a very slight apparent difference of description between the two passages quoted, which by studying the subject in Professor Helmholtz's book "Die Lehre von den Tonempfindungen" I found resolves itself so nearly into a coincidence, that I hope that the short description of Professor Helmholtz's experiments which this Paper contains embraces really some of the "latter Experiments", which a note (at p. 200) in my 2nd Ed[itio]n copy of the "Lectures on Sound" refers the reader to, in an "Appendix", which seems to have been omitted from the book? Comparing the section on the "artificial imitation of the vowel sounds" in Helmholtz's work with your description of the vowels U O and Ah (the whole being uttered in the pitch of a note about the middle, or upper part of the base-stave) the harmonies reinforced by the tuning-forks which were loudly or softly heard in Prof[esso]r Helmholtz's Exp[erimen]ts are almost identically those named as characteristic of them in your book;

viz for Ū the fundamental note: for O the first octave; and for Ah the higher overtones, being rendered strongly, while the fundamental note, and a few of its other overtones are at the same time rendered more or less distinctly. The vowels A, and E (as written in English) have ambiguous sounds; but the description of their clang or "ring" (my Father's word), assuming the same, or a higher fundamental tone than the last is nearly the same in your "Lecture V" on Sound, and in the second series of artificially produced vowel sounds which Helmholtz describes in his work (at p. 189) as A, "A, E, I.—The experiment of a particular tuning fork resounding very strongly in the cavity of the open mouth, which it is placed before, when the speaker is about to utter the vowel O, as described in an immediately preceding paragraph of the Lecture V, to explain the peculiar quality of that, and of the other vowel sounds shows, I think that you agree with Helmholtz in regarding the characteristic notes of resonance, in the mouth and throat, while speaking or singing any particular vowel (on any note of the scale sufficiently low to give effect to the resonance, in reinforcing some of the overtones of the note sung), as generally fixed in pitch;

so that, for example, if Ah! is sung on the note E (320 Vibr[atio]ns) at the bottom of the treble stave, the overtone of the vocal sound which is reinforced (B in the Ledger-lines; 960 Vibr[atio]ns) will be the 2nd overtone, or the 12th of the base-note. But if Ah! is sung on B in the middle treble-line (480 Vibrations, the overtone B on the ledger-line (960 Vibrations), which is still reinforced, as before, becomes the first overtone of the vocal series, or the first octave above the base-note?—This is a single example which Prof[esso]r Helmholtz gives, to show that the vowel-clang consists in the strengthening of one or more Harmonics of a particular pitch, while the clang of ordinary musical instruments depends on the strengthening of Harmonics of a special type or order in their numerical succession!—I have made a long story of this mightily difficult subject; rather than to let you think that I had not tried to repeat and verify Helmholtz's experiments as far as I could with instruments which I borrowed or prepared for the lecture to the Philosophical Society, and which I had already made several trials with before, to my class.—The 'vowel-flames, and revolving mirror' give I'm afraid too uncertain results to allow the principal harmonies in the vowel-clang to be learned from them; although some facts are very prominently shown, such as the appearance of the octave or first overtone when Oh is sung, and of the second octave (or perhaps the 12th), when Ah is sung without changing the pitch of the note. But I have not studied the shapes of the teeth of the flames very closely, yet; and it is quite evident that the superposition of the same harmonic overtones in various, different phases would give compound, toothed flames of such very different appearances that it would be quite impossible without great labour, and preliminary study to recognize of what harmonics a given toothed flame is composed!—In general the addition together of a series of perfectly cycloidal harmonic oscillations of a stretched string does not, (as Prof[esso]r Helmholtz has shown in the case of a plucked string, p. 93 in his book "Die Lebre der Tonempfindunsen") produce a symmetrically recurring curve in the oscillation of the string!

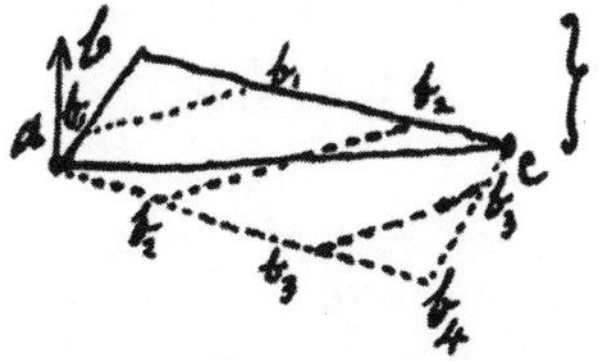

Posit[io]^{n} 1. of such a string, abc plucked upwards at b, and suddenly released from b.—Successive positions of the string after its release (ab_1b_1c; ab_2b_2c; ab_3b_3c; and ab_4c); the same successive positions then recur again in the opposite order from the last position ab_4c to the first, abc from which the same oscillations as the first begin again. These show, for example, that the upward pull of the string on a (shown by the arrow-line), which represents the height and pressure of a gas-flame which a similar combination of harmonic sounds to that which exists in the plucked string would produce, if they formed together a vowel sound or air-wave, acting upon the stretched membrane behind a vowel-flame, is for a short time constant and strong, while the point b_1 moves from b to a;

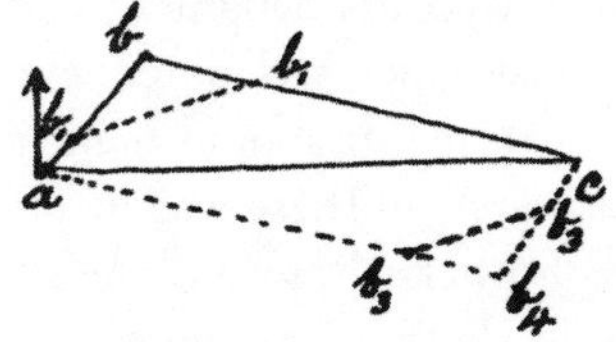

and it is afterwards constant, and acts downwards, while the same point moves from a to b_4.—The form of the gas-, or "vowel-flame" produced by the air-wave emitted from the plucked string of a monochord (or from the sharply-struck wire of a pianoforte), would therefore be the "curve of pressure" (fig 19 on the same page) which Prof^{r} Helmholtz has drawn in his book thus;—

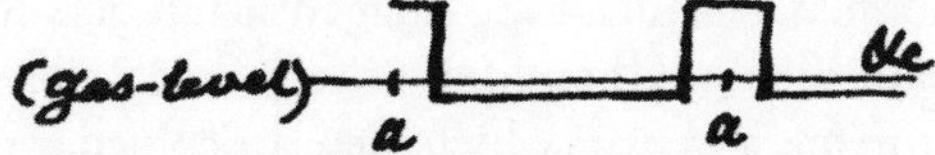

or a succession of square-topped flames, in which there is evidently no visible trace of the very distinct series of harmonic sounds w[hic]^{h} the practised ear, and the artifice of damping the string can prove to exist in the twang of the plucked wire of a monochord.—The "reading" of the vowel-flames in the "gas-vibroscope"[27] is, I suppose, very greatly complicated by these secondary forms of curves, and except in cases when some of the overtones are extremely prominent, I am afraid that it will be very difficult to interpret them, and to unravel their hidden meaning fully.*—The secondary curves however, explain, I think why the forms of the toothed flames are generally unsymmetrical thus

rather than

thus, as I at first expected to find them, from the combination of harmonic waves of short and of long periods!—The "twinkling" flame of this 'gas-vibroscope' is not, indeed, perhaps, such a good means of analyzing the vowel sounds as either the smoke-jets or the very sensitive "vowel-flame" which you used for the same purpose, and which I have occasionally tried here very successfully with bells and other very high-pitched notes, but could never make them sufficiently tall and slender to answer to *[1 word illeg]* words, or notes of my voice!—

I heard from Collingwood[28] that my Sisters[29] were fortunate enough to hear you describe the results of your expedition on board of the "Urgent", and to Oran,[30] at the Friday Ev[enin]g Lecture at the Royal Institution, early in last month,[31] with which they were very greatly delighted; and they enjoyed, and shared the sympathy of the crowd which welcomed you back after such a tempestuous voyage; and such a trying disappointment in the main object of the journey for all the observers![32]—There was, if possible more to do, and less time to do it, than in any former eclipse; and the amount of successful work which was accomplished is, after all a matter of real congratulations and surprise!—The accompanying stereoscopic view of the two photographs of the corona, by Mr Brothers at Syracuse, and the American astronomers at Xeres, in Spain, enables them to be compared together in a stereoscope, by opening first one eye and then the other; and the photographs (re-photographed, here, for the purpose) do certainly present striking points of similarity and coincidence of shape in some parts of the indented outline of the corona; although the prominences which I was obliged to retouch in the negatives because their red-colour (painted) on Mr Brother's C. D. V. photographs[33] which my Father sent to me, and from which these are copied, would unfortunately only print as black-mountains instead of as mountains of light, projecting from the moon!—They are therefore, unfortunately overdrawn in size, although still on the places where the "red-flames" are found in Mr Brother's small card photographs of the Eclipse in Sicily and Spain.

I am very glad to hear that Prof[esso]r Maxwell is elected Professor of Physics at Cambridge; where he will be a great acquisition to the University.—Sir W[illia]m Thomson[34] who told me of this good appointment being so well filled by an excellent master of Physics, and Mathematics, is himself very busily experimenting with Prof[esso]r Rankine on the stability of armour plated ships;[35] and some of their "compound-oscillations" of ships and seas especially when a case of "Vanishing Stability" occurs (!), represented by pendulums, and rolling half-cylinders with weights, are very beautiful as well as most interesting, and important results.

But I must close this long letter, with many apologies for taking up so much of your valuable time to peruse its unattractive pages;—And with best thanks again, and much regard, I remain,

Dear Prof[es]s[o]r Tyndall | Yours faithfully | A S Herschel

*Coal-gas (unburned) as well as hydrogen, both, mentioned in your review of the experiments (Nature Jan. 27th, '70) on "Haze and Dust"[36] gives, I find a very good "black cloud",—and an equally dark space in a clear glass shade, &c, as an excellent illustration.

* I have described the study of "these forms of flame-teeth", in my lecture (p. 422), as "surrounded with difficulties, which have not yet permitted me to use them successfully"!—

RI MS JT/1/H/84
RI MS JT/1/TYP/2/587–92

---

1. *"Diamagnetism and Magnecrystallic Action"*: Tyndall's work on diamagnetism; see letter 3370, n. 12.
2. *pressing occupations of this Session*: Herschel was professor of natural philosophy at Anderson's University, Glasgow until 1871. It changed its name to Anderson College in 1877 and is now part of the University of Glasgow.
3. *"Sound"*: Tyndall's 1867 work, *Sound*.
4. *"Heat as a Mode of Motion"*: Tyndall's 1863 work; see letter 3370, n. 7.
5. *the President*: James Young, the President of Anderson's University, Glasgow at the time.
6. *"Popular Course" lectures on Physics . . . Glasgow Artisans!*: Herschel gave lectures to the public and had as many as 250 students coming to these (*Fourth Report of the Royal Commission on Scientific Instruction and the Advancement of Science* (London: George Eyre and William Spottiswoode, 1874), p. 52).
7. *Annual Conversazione*: annual meetings for the Faculty of Physicians and Surgeons in Glasgow. Conversaziones were often informal meetings, comprised of a mixture of professionals and invited members of the public who would listen to lectures and discuss the issues of the day.
8. *the question of 'Dust and Disease'*: a reference to Tyndall's Friday Evening Discourse at the RI on 21 January 1870, 'On Dust and Disease' (*Roy. Inst. Proc.*, 6 (1870–72), pp. 1–14).
9. *the late eclipse expeditions*: probably a reference to the 22 December 1870 solar eclipse. Jules Janssen (1824–1907), an astronomer from France, escaped the siege of Paris in the Franco-Prussian War in a hot air balloon. Well-known for studying the chromosphere (part of the corona of the sun), he observed eclipses to gather his data. He went to Algeria,

but the eclipse was covered by clouds. American astronomer Charles Young (see n. 16) observed the solar eclipse in Spain. Tyndall traveled to Oran, Algeria from Portsmouth on the steam-ship *Urgent* to view this total solar eclipse, but the weather caused the expedition to abandon hope of witnessing the celestial event (*Ascent of John Tyndall*, pp. 270–2). See letter 3416. See J. Tyndall, 'Voyage to Algeria to Observe the Eclipse', in *Hours of Exercise in the Alps*, pp. 429–73 (this was also published later in *Fragments of Science for Unscientific People: A Series of Detached Essays, Lectures, and Reviews*, 5th edn (London: Longmans, Green, and Co., 1876), pp. 186–217). On eclipse expeditions in general, see A. S.-K. Pang, *Empire and the Sun: Victorian Solar Eclipse Expeditions* (Palo Alto, CA: Stanford University Press, 2002).

10. *whatever the curiously-shaped light of the Corona will, at last be found to be!*: The 7 August 1869 solar eclipse allowed Charles Young and William Harkness of the United States to observe a new emission line in the Sun's corona. During the solar eclipse of 12 December 1871, Janssen used spectroscopy to determine that the corona is, in fact, part of the Sun.
11. *My father*: John Herschel, who died about two months after this letter, on 11 May 1871.
12. *Mr Brothers, at Syracuse*: Alfred Brothers (1826–1912), a British photographer and FRAS. He took the photographs during the 22 December 1870 solar eclipse from Syracuse, Sicily, and they were regarded as the 'first really successful photographs of the corona' (*Science*, 8:194 (16 September 1898), p. 344).
13. *the American observers at Xeres*: Charles Young (see n. 9 and n. 16) and Richard Abbay (see n. 16), among others. Xeres is also Xerez, or Jerez de la Frontera, which is near Cadiz, Spain.
14. *Stereoscope*: 'an instrument for obtaining, from two pictures (usually photographs) of an object, taken from slightly different points of view (corresponding to the positions of the two eyes), a single image giving the impression of solidity or relief, as in ordinary vision of the object itself' (*OED*).
15. *measurement of the . . . Corona*: a reference to either the 7 August 1869 or 22 December 1870 solar eclipse.
16. *Professor Young and others*: Charles Augustus Young (1834–1906), an American astronomer who took photographs of the Sun's corona in 1869–70, and was well known for his work on the Sun and other general astronomical works. Rev. Richard Abbay, FRAS (1844–1927) was also observing this eclipse with Young. Abbay was a lecturer at Wadham College, Oxford, and a member of two eclipse expeditions.
17. *"Chemical Section" of the Glasgow "Philosophical Society"*: Glasgow's Royal Philosophical Society was founded in 1802 'to aid the study, diffusion, and advancement of the arts and sciences with their applications, and the better understanding of public affairs' (https://royalphil.org/, accessed 11 May 2020).
18. *proceedings of the Sicilian expedition*: see n. 12. Thorpe (see n. 19) gave this talk on 25 January 1871. T. E. Thorpe, 'The Eclipse Observations in Sicily,' *Proceedings of the Royal Philosophical Society of Glasgow*, 7 (1870–1), pp. 413–5.
19. *Dr Thorpe*: Thomas E. Thorpe (1845–1925), professor of Chemistry at Anderson's College,

Glasgow from 1870–4. He was known for accurately determining the atomic weights of metals. He participated in four eclipse expeditions from 1890–3. He was elected FRS in 1876, and was President of the BAAS in 1921.

20. *Professor Penny:* Frederick Penney (1816–69), a Scottish chemist and FRSE. He was Professor of Chemistry at Anderson's College, Glasgow from *c.* 1840.
21. *Professor Roscoe*: Henry Enfield Roscoe (1833–1915), a British chemist and university administrator. He revitalized Owens College in Manchester when he joined the faculty there in 1857 and helped establish the college as a leading British center for the study of chemistry. Roscoe's work in inorganic chemistry included the isolation of the element vanadium, but he was more widely known for his writing on education. He established Science Lectures for the People in Manchester, bringing Huxley, Tyndall and others to the city to lecture for the general public. He was Thorpe's teacher and mentor. He was elected to the Chemical Society in 1855, FRS in 1863 and knighted in 1884.
22. *a paged copy . . . Eclipse*: enclosure missing.
23. *the last number of its "Proceedings"*: see n. 18 and n. 22. Herschel's paper was published just before Thorpe's. See A. Herschel, 'Observations of the Recent Eclipse of the Sun in Spain', *Proceedings of the Royal Philosophical Society of Glasgow*, 7 (1870–1), pp. 405–13.
24. *"Analysis of the Vowel Sounds"*: referring to Herschel's paper, 'On Helmholtz's Analysis of the Vowel Sounds', *Proceedings of the Royal Philosophical Society of Glasgow*, 7 (1870–1), pp. 417–24.
25. *"Lectures on Sound"*: see n. 3.
26. *"Die Tonempfindungen, &c."*: first German edition of H. Helmholtz, *Die Lehre von den Tonempfindungen als Physiologische Grundlage für die Theorie der Musik* (Braunschweig: Friedrich Vieweg und Sohn, 1863). The first English edition was H. Helmholtz, *On the Sensations of Tone as a Physiological Basis for the Theory of Music*, trans. A. J. Ellis (London: Longmans, Green, and Co., 1875).
27. *"gas-vibroscope"*: an instrument for making vibrations of gas (or another substance) visible (*OED*).
28. *Collingwood*: the Herschel home near Kent.
29. *my Sisters*: Herschel was not clear to which of his eight surviving sisters he was referring.
30. *the results . . . Oran*: see n. 9.
31. *the Friday Ev[enin]g Lecture . . . last month*: Tyndall's Friday Evening Discourse at the RI on 20 January 1871, not in February, included observations from his trip to Oran, Algeria. See J. Tyndall, 'On the Colour of Water, and on the Scattering of Light in Water and in Air', *Roy. Inst. Proc.*, 6 (1870–72), pp. 189–99.
32. *such a trying disappointment in the main object of the journey for all the observers!*: The eclipse expedition to Oran, Algeria failed due to poor weather.
33. *C. D. V. photographs*: Carte-de-visites, or visiting cards, were a common way for people to leave notice that they had stopped by someone's home or office to visit with them. The photographs themselves were 'small photographic portraits, mounted on a card, 3 1/2 by 2 1/4 inches' (*OED*).

34. *Sir W[illia]m Thompson*: William Thomson was working in Glasgow at this time.
35. *stability of armour plated ships*: Thomson and Rankine were both part of a fifteen-member government committee that 'reported in 1871 that the "only method of bringing about a well-considered armour-plated ship was to have a central belt and raft ends with an underwater deck." Two naval members of the Committee, Admiral Ryder and Admiral Elliot, went much further in this direction, and advocated the entire abolition of side armour for the protection of buoyancy and stability, and to employ armour-plating only for the protection of guns and gunners. The water-line was to be protected by a cellular structure only' (W. E. Smith, *The Distribution of Armour in Ships of War* (London: Her Majesty's Stationary Office, 1885), p. 12).
36. *your review . . . on "Haze and Dust"*: J. Tyndall, 'On Haze and Dust', *Nature*, 1 (27 January 1870), pp. 339–42.

To James Clerk Maxwell [20 March 1871][1] 3388

From Professor Tyndall. | Monday.

MY DEAR MAXWELL—Why . . . did you run away so rapidly.[2] I wished to shake your hand before parting.—

Yours ever, | John Tyndall.

L. Campbell and W. Garnett, *The Life of James Clerk Maxwell: With a Selection from His Correspondence and Occasional Writings and a Sketch of His Contributions to Science* (London: MacMillan & Co., 1882), p. 381
Typed Transcript Only

1. *[20 March 1871]*: L. Campbell and W. Garnett (eds), *The Life of James Clerk Maxwell with a Selection from his Correspondence and Occasional Writings* (London: MacMillan & Co., 1882), p. 381, dated this between 15 and 20 March 1871. 20 March is the Monday.
2. *run away so rapidly*: This likely refers to a meeting at the RI, when Maxwell left quickly before Tyndall could greet him.

To Hector Tyndale 20 March [1871][1] 3389

20th March.

Dear Hector.

On Saturday night on my arrival here[2] I found waiting for you a letter[3] from Sir Frederick Pollock, addressed to my care. I forwarded it immediately by a porter.

I trust it reached you safely.

Debus informs me that you leave London today. I had hoped for a longer continuance of your visits. But doubtless all these things are ruled by Destiny; if so it must be destiny that causes me to wish you good speed & prosperity in your journey.

Yours faithfully | John Tyndall

To General Hector Tyndale

RI MS JT/1/T/1468
RI MS JT/1/TYP/4/1634

---

1. *[1871]*: This letter comes in the sequence after letter 3385 and before letter 3393 of this volume, during Tyndale's visit to London. See letter 3395 for clear dating to 1871.
2. *here*: LT noted this letter was written from the RI.
3. *a letter*: letter missing.

From Emilio Ruiz de Salazar[1] 21 March 1871 3390

Madrid | 21 de Mars de 1871

Mr John Tyndall

Monsieur et très honorable confrère: je n'ai pas reçu la réponse que j'attendais à ma lettre du 17 Janvier, et connaissant que cela dépendre de que vous ne l'aurais pas reçue je vous écris nouvellement. Je vous disais: <u>la bonne renommée</u> a ses charges et à présent j'en vous impose une en vous priant de me répondre à tôt qui vous *[*sera*]* possible en faveur que je sollicite.

Je suis depuis six années Professeur agrégé à l'université de Madrid de la faculté es sciences et pendant ce temps j'ai fait mes leçons de Géodésie (j'ai expliqué deux cours) de calcul d'Astronomie, étant chargé actuellement de la classe de mécanique naturelle qui n'a pas de Professeur numéraire par élection unanime de la Faculté.

Maintenant que vous connaissez mon rôle dans l'enseignement et mes séances depuis que j'ai reçu le doctorat en sciences j'espère que quand je fais appel aux sentiments de confraternité vous avez *[*très*]* *[*bien*]* pour me répondre et me satisfaire dans mes désirs.

La classe de Physique Mathématique de cette Université, la seule de ce genre qu'on explique en Spagne n'a pas de Professeur et notre gouvernement ignorant la difficulté de cette enseignement au lieu de nommer le Professeur d'entre les membres de l'Académie le fait sortir d'*[*oposition*]* c'est-à-dire va nommer à celui qui fait des expériences scientifiques de *[1 word illeg]* le plus estimables—

J'ai expliqué *[*rien*]* cette science ; mais *[*il*]* *[*en*]* n'a pas de personnes à consulter et la Physique Mathématique est [pourtantt] plus grand—Ma [position] actuelle m'oblige à faire mes exercices et dans cette [position] je vous demande vos conseils pour réunir donc les deux travaux.

- Une Programme raisonné de la Physique Mathématique

- Une mémoire sur la source de connaissance et les méthodes d'enseignement dans cette science—

Connaissant quelques de vos excellents travaux et surtout vottre très distingué compétence je vous en prie instamment que vous m'aidez dans mon entreprise. Je n'ignore que je n'ai pas aucun motif ni fondement pour vous demander une si grande faveur; mais si vous considérez ma position si vous pensez que si grâce vottre bonté je réussise je vous serai reconnaissant pour m'avoir fait [conquérir] une place dans le Profesorat, je crois que vous me satisferai tenant comte aussi de la dificultés que je trouve ici pour avoir des conseils et d'aide au rapport une science si peu connue en Spagne comme la Physique mathématique.

Aussi je désire qu'employant quelques heures à mes demandes; si tôt que vous auriez [fini] vous me répondrez aux suivantes:

10 Que est ce qu'on entend par un programme raisonné et un mémoire sur la source de connaissance et méthode d'enseignement dans la Physique Mathématique.

20 Mémoire détaillé de la développement (Programme 10)—[memoire]

30 Notice historique et bibliographique de la même science et degré de son développement dans le monde scientifique avec quel [caracter] on donne cet enseignement et avec quel [stances], noms de professeurs . . . établissement &&.

Dans cette renseignement je crois que vous me diriez qu'est ce qu'on entend par Physique mathématique, ses principes, sa base, sa marche, ses rapports avec d'autres sciences, connaissances mathématiques supérieurs qui *[1 word illeg]* et surtout que vous fixerez vottre atention au <u>Programme raisonnée</u> [où] je trouve plus de dificulté et que je crois on doit les faire [parctiquement], l'exposition et le raisonnement—

Je vous en prie de me répondre si tôt que vous pourriez car j'ai le temps trop limité déjà et étant sûr que vous le faisiez je vous remercie avec profonde reconnaissance en vous offrant tous mes services à Madrid où je serais bien heureux de vous rendre service dans quelque chose.

Agréez Monsieur l'assurance de ma profonde considération désirant que vous me considérez comme vottre très humble serviteur et très dévoué confrère

Emilio Ruiz de Salazar

Mon adresse, Professeur à l'Université de Madrid calle del Horno de la

Mata n° 12- 20–ou au Directeur du journal d'Instruction publique El Magisterio Espanol—*[*Olivo*]*—11—pre[mier]—

Madrid | 21 March 1871

Mr John Tyndall,

Sir and most honourable colleague: I did not receive the response I was awaiting following my letter of January 17th,[2] and knowing that it is contingent on you having not received it, I am writing to you again. I said to you: a good reputation has its burdens, and at present I am imposing one on you by asking you to respond as soon as is possible for the favour that I ask.

I have been a teaching Professor for six years at the University of Madrid in the faculty of science,[3] and during this time I gave my classes on Geodesy[4] (I gave two courses) and Astronomy calculations, being currently assigned to the natural mechanics class which does not have a full Professor appointed through unanimous election by the Faculty.

Now that you know my role in teaching and my lectures since I received my doctorate of science, I hope that when I appeal to your feelings of fellowship, you will *[*be well disposed*]* to answer me and to satisfy me in my wishes.

The Mathematical Physics course in this University, the only one of its kind that we teach in Spain, does not have a professor, and our government, ignoring the difficulty of teaching it, has decided to give the job to our opposition, that is, to whomever has done scientific demonstrations of the most estimable *[1 word illeg]* instead of naming the professor from amongst members of the Academy.

I have *[*not contributed much*]* to this science; but *[*there is no one*]* to consult about it and Mathematical Physics is nevertheless the greater for it—My current position requires me to do my work, and it is because of this *[*position*]* that I am asking for your advice, in order to combine both:

- A rational program of Mathematical Physics.

- An essay on the source of knowledge and the methods of teaching this science—

Knowing your work to be excellent and especially your competence to be renowned, I urgently ask you to help me in my undertaking. I do not deny that I have neither reason nor any grounds to ask you for such a great favour; but if you consider my position, if you think that I will succeed thanks to your goodwill, I would be grateful to you for helping me win a place in the Professoriate.[5] I believe that you will do this for me, though I also acknowledge how much difficulty I have had in getting advice and assistance in Mathematical Physics, a science so little known in Spain.

I also hope that if you are spending a few hours on my requests, as soon as you *[*finish*]* you will tell me about the following:

1st What do we mean by a rational program and an essay on the source of knowledge and the method of teaching in Mathematical Physics.

2nd Detailed essay on the development (Program 10)—*[essay]*

3rd A historical and bibliographic summary of the same science and the degree of its development in the scientific world, with what *[*character*]* we provide these teachings and with what *[*stances*]*, names of professors . . . establishment &&.

As part of this information, I believe you will tell me what is meant by Mathematical Physics, its principles, its base, its operation, its relationship with other sciences, superior mathematical knowledge that *[1 word illeg]*, and especially that you will fix your attention on the Rational Program which I found most difficult, and whose introduction and argument I think we must do them *[*with practicality*]*.

I ask you to answer me as soon as you can, since I have so little time already, and being certain that you will do this, I express my deepest gratitude by offering you all my services in Madrid, where I would be very happy to be of service to you in any way.

Please accept, Sir, the assurance of my deep devotion, hoping that you will consider me as your humble servant and dedicated colleague

Emilio Ruiz de Salazar

My address: Professor at the University of Madrid, that of the *Horno de la Mata* N° 12–20—or to the Director of the Journal of Public Instruction, *El Magisterio Espanol*[6]—*[*Olivo*]*—11—*[*number one*]*.

RI MS JT/1/D/121

1. *Emilio Ruiz de Salazar*: Emilio Ruiz de Salazar y Usátegui (1843–95), Professor of Mathematical Analysis at the University of Madrid. He was interested not only in mathematics and physics, but also in the connections that education had on social and political development, and he worked in Madrid to use education of the general public to improve society.
2. *my letter of January 17th*: see letter 3340, *Tyndall Correspondence*, vol. 11.
3. *I have been . . . faculty of science*: see n. 1.
4. *Geodesy*: the measuring or surveying of land (*OED*).
5. *win a place in the Professoriate*: Salazar was successful in this, whether or not Tyndall replied and complied with his request. He was substitute in the Chair of Mathematical Physics before he was appointed Chair of Mathematical Analysis in 1877 (see n. 1).
6. *El Magistero Español*: a magazine for teaching science to the public, founded by Salazar in Madrid in 1867.

To Madame Mohl[1] 22 March 1871 3391

22nd March 1871

Dear Madame Mohl.

The entanglements in which I find myself will, I fear, prevent me from doing what would be a real pleasure to me—that is to say, prevent me from calling upon you. But I wish to say a word or two to you regarding our friend & your kinsman Helmholtz[2]—

I wished very much to have the 2nd edition of his great work. Die Tonempfindungen[3] reproduced in England, and with that view urged the work on the attention of Longmans. They have been in communication with Vieweg,[4] and with English people competent to form an opinion regarding the probable success of the work in this country.

I send you the upshot of Longman's enquiry into the matter: they are not, I am sorry to say, favourable to the reproduction of the book as a whole.

But a very great admirer of Helmholtz[5] and a very competent man has offered to write a smaller book, based on the work of Helmholtz. It will be in fact a kind of condensation of Helmholtz—with all acknowledgement of course—this Longman is willing to publish; and if nothing occurs to impede the project, the new work would be ready in about a year.[6]

Helmholtz's "Populäre Vorträge"[7] I have strongly recommended Longman to publish. They will, I trust, if permission be granted, be for the most part translated by Dr. Debus, with whom I am in terms of close intimacy, and to whom I can lend a hand whenever it is needed.

Would you kindly, when the opportunity offers, convey this intelligence to your kinsman, and give him his wife, and daughter my kindest regards?

Yours most faithfully | John Tyndall

Archiv der Berlin-BAW 477

1. *Madame Mohl*: Mary Elizabeth Mohl (née Clarke, 1793–1883), an English writer, feminist and salon hostess in Paris. In 1847, she married the orientalist Julius Mohl.
2. *your kinsman Helmholtz*: Hermann Helmholtz (see also letters 3387 and 3545). Mohl was the aunt of Helmholtz's wife Anna Helmholtz, whose maiden name was Mohl.
3. *Die Tonempfindungen*: see letter 3387, n. 26.
4. *Vieweg*: Friedrich Vieweg and Son, in Braunschweig; the original publisher of Helmholtz's work. See letter 3432.
5. *admirer of Helmholtz*: Debus, see paragraphs below.
6. *new work . . . a year*: not clear that this was ever published.
7. *"Populäre Vorträge"*: H. Helmholtz, *Populäre wissenschaftliche Vorträge* (Braunschweig: Drück und Verlag von Friedrich Vieweg und Sohn, 1876).

To Thomas Archer Hirst [24][1] March 1871 3392

Friday | Mar *[date illeg]* 71

My dear Tom.

Unless drawn by some sweet influence which may pervade them here and there, I will not ask you to read any of these[2] but that addressed to Students.

I was ill & weak when it was written, and doubtless I could say something much better to an audience of young fellows. Still even in my strong mind this address[3] contains things worth uttering. But I would ask you to weigh it and decide whether it should be kept; and if kept whether it might not be thrown among the "minor papers."[4]

I shall be glad of course if you think it worth leaving where it is.

Yours aff[ectionate]ly | John

RI MS JT/1/T/974
RI MS JT/1/HTYP/578/1

1. *[24]*: This letter was written on a Friday, with '24' barely legible on the original letter. This letter was written just before letter 3397, so we placed this on the 24th.
2. *any of these*: parts of *Fragments of Science*. See n. 4.
3. *this address*: possibly J. Tyndall, 'Address to the Students of University College, London, on the Distribution of Prizes in the Faculty of Arts', in *Fragments of Science*, pp. 95–106.
4. *"minor papers"*: *Fragments of Science* concludes with a selection of 'Shorter Articles' (pp. 405–49).

To Hector Tyndale 25 March [1871][1] 3393

25th March.

My dear Hector.

If the weather be fine tomorrow I must have a day in the Country as my brain is out of order & needs renewal. Could we not manage to dine somewhere in the Country? Supposing you were to join me at the Star & Garter Richmond[2] at 6.30? I will try to get Debus to come also.

Yours ever aff[ectionate]ly | John.

RI MS JT/1/T/1469
RI MS JT/1/TYP/4/1691

1. *[1871]*: LT annotation, from the postmark. This letter also fits with Tyndale's visit to the England in March 1871; see letters 3376, 3378, 3380, 3385, 3389 and 3395.
2. *Star & Garter Richmond*: a hotel and restaurant in Richmond, a town in what was then the countryside around London.

To Edward Livingston Youmans 26 March 1871 3394

March 26, 1871.

MY DEAR YOUMANS: . . . [1] The desire for lecturing in America seems to be very strong. My relative, Hector Tyndale, who is now in this country, was the bearer of a very flattering proposal to me.[2] Suppose I ask *you* what would be expected of me were I to close with the terms suggested in your last letter?[3] I want to know the amount of slavery that will, under the contract, be inflicted on me.

I take it for granted that I should occupy no other position than that habitually accepted by such men as Emerson, Sumner,[4] Wendell Phillips,[5] Wendell Holmes.[6] I should not, of course, dream of becoming a traveling lecturer in England, and I should as little dream of doing so in America if the constitution of society were not such as to render the work of lecturing not unworthy of your own best men.

The best men in England, be it remembered, would engage in nothing of the kind.

E. A. Youmans, 'Tyndall and His American Visit', *Popular Science Monthly*, 44 (February 1894), pp. 502–14, on pp. 507–8
Typed Transcript Only

1. . . . : beginning of letter not published and missing.
2. *Hector Tyndale . . . very flattering proposal to me*: Tyndall did not accept Youman's invitation, but did accept Tyndale's, which the latter made on behalf of Joseph Henry of the Smithsonian Institution; see letters 3504, 3510, 3542, 3560, 3669, 3684 and 3702.
3. *your last letter?*: letter missing. However, see also letter 3175, *Tyndall Correspondence*, vol. 11.
4. *Sumner*: Charles Sumner.
5. *Wendell Phillips*: Wendell Phillips (1811–84), an American lawyer and a leader in the abolition of slavery in the US. He attended Harvard University, and its Law School, becoming a lawyer by 1834. Because of his protests, speeches and writings, he became known as abolition's golden trumpet. He ran for governor of Massachusetts in 1870, and lost.
6. *Wendell Holmes*: Oliver Wendell Holmes, Sr. (1809–94), an American physician and writer. He was educated at Harvard University and in Paris. He became the first professor of anatomy and physiology at Harvard in 1847, remaining there until 1882. He founded the *Atlantic Monthly* in 1857 and published a number of his writings there. He also wrote an essay about puerperal fever, affecting women who had just given birth.

To Thomas Archer Hirst 27 March 1871 3395

Monday M[ornin]g. | March 27/71

My dear Tom.

Hector who dined with me yesterday at the Star & Garter[1] wishes much to have you to dinner with him today at Wood's Hotel Furnival's Inn[2] at 6.30.

He & his friend M^r McLaughlin[3] came almost to your door last night with the view of asking you. but when there thought it too late—It was past 11.

Debus will be with him.

Go if you can. I would gladly do so but pledged myself long ago to D^r. Lushington.[4]

Yours aff[ectionate]^ly | John

RI MS JT/1/T/950
RI MS JT/1/HTYP/580/1

---

1. *Hector... Star & Garter*: Hector Tyndale; see letter 3393. See also letters 3376, 3378, 3380, 3385 and 3389 for more detail of Tyndale's trip.
2. *Wood's Hotel Furnival's Inn*: a hotel in Holborn, central London.
3. *Mr McLaughlin*: possibly John M. McLaughlin (n.d.), Tyndale's friend and author of *A Memoir of Hector Tyndale: Brigadier-General and Brevet Major-General, U. S. Volunteers* (Philadelphia: Collins, 1882).
4. *Dr. Lushington*: Stephen Lushington (1782–1873), an Oxford-educated English judge, earned his DCL in 1808. He was elected a Fellow of All Souls College in 1801 at the age of nineteen, but resigned the fellowship on his marriage twenty years later, as was required. He was called to the Inner Temple in 1806 and ultimately became a judge of the court of the arches in Canterbury in 1858. He was also an MP for the London borough of Tower Hamlets for over two decades. He died at the age of ninety-one after a short illness.

From John Ryan[1] 27 March 1871 3396

Leighlinbridge[2] | March 27th. /71.

Professor Tyndall

Sir,

It's a considerable time since I furnished you with a small account £0.17.5 obtained by order of your sister Miss Tyndall[3] the only reply I got was from Mrs Steuart saying you should not pay it how I forfeited her estimation I cannot tell. I cannot comprehend why a lady of her position who so punctually exacts her rights should interfere in my affairs more than I should interfere in hers. My bill is fairly due and if Miss Tyndall had her will and the means I should have been paid long since it would be useless for me to address her as I understand her letters are subject to espionage.[4] You Sir having disposed of the effects of your respected mother[5] as you thought proper it's to you I consider I have a right to apply for payment of my demand which I expect on consideration you will do by you forwarding me a money order for the amount will oblige

Yours obed[ien]t[ly]. | John Ryan.

RI MS JT/1/TYP/10/3404
LT Typescript Only

---

1. *John Ryan*: John Ryan (n.d.) was someone from Leighlinbridge.
2. *Leighlinbridge*: Tyndall's birthplace and hometown. Much of his family still lived there.
3. *Miss Tyndall*: Emily, or Emma, Tyndall.
4. *subject to espionage*: situation not further identified.
5. *your respected mother*: Sarah Tyndall.

To Thomas Archer Hirst 28 March [1871][1] 3397

March 28

My dear Tom

Please read from the bottom of page 440 to the bottom of page 442.[2]

At the end I have tried to give in a condensed form the notion of Goethe[3]—the original, in part, is underneath.

You will remember Goethe immediately afterwards speaks of the Allumfasser; the Allerhalter.[4]

I send you a couple of slightly different renderings—say which you would choose.

Read also, please, the Additional Remarks on Miracles[5]—It is new to you.

And especially look at the point where I speak of such & such things being demanded and taught in our seats of learning. God knows I have no grudge against Pusey[6] or any body else, but it may be thought that I am giving him a back handed blow here. I hardly think so—but I want your impression.

Aff[ectionate][ly]. | John.

You will run your eye over this in a few minutes—the printer is waiting for it—all the rest is pulled off.

RI MS JT/1/T/1000
RI MS JT/1/HTYP/580/2

1. *[1871]*: referring to *Fragments of Science*, which was published in April of this year.
2. *440 to . . . 442*: referring to the first English edition of *Fragments of Science*.
3. *at the end . . . Goethe*: *Fragments of Science*, p. 442. He quotes part of Goethe's poem 'Wer darf ihn nennen?' ('Who can call him?') from J. W. von Goethe, *Faust: Eine Tragödie* (Tubingen: J. G. Cotta'schen Buchhandlung, 1808), p. 228.
4. *the Allumfasser; the Allerhalter*: from Goethe, see n. 3. According to Tyndall's translation in *Fragments of Science*, p. 442, the All-enfolder; the All-upholder.
5. *Additional Remarks on Miracles*: in *Fragments of Science*, pp. 445–9.
6. *Pusey*: Edward Bouverie Pusey (1800–82), an Anglican church official and Regius Professor of Hebrew at Oxford University, from 1828. He was one of the leaders of the Oxford Movement in the Anglican church.

To Henry Scott[1] 29 March 1871[2] 3398

<u>The Acoustics of the Royal Albert Hall.</u>[3]
Royal Institution | 29th March, 1871[4]

My dear Colonel Scott,—

I think you are to be congratulated on the success, from an acoustic point of view, of the vast edifice[5] opened to-day.

I was placed in the lower tier of boxes at the end of the long diameter of the oval, opposite to the organ, and I heard there the singing and the music with admirable clearness.

I happened also to go up to the picture gallery in search of some friends, and there also the effect seemed exceedingly good.

When, some months ago, I heard you say that the Hall might be employed for the purposes of oratory, I confess I felt incredulous. But the very distinct reading of the Prince of Wales[6] to-day banished my incredulity. Had he faced the audience, and put on a little more steam, of which he had lots in reserve, he would have been better heard by an audience of 8,000 people than I could make myself heard some time ago by an audience numbering little more than 2,000 in the Philharmonic Hall at Liverpool.[7]

How the great Hall is to be employed I do not[8] know, but, as far as its acoustic properties[9] are concerned, I think you have demonstrated that it may be made entirely successful.

Faithfully yours, | (signed) John Tyndall.

Lieutenant. Colonel Scott, R.E.

RI MS JT/6/5/1[10]
The Orchestra, no. 393, 6 April 1871, p. 12

1. *Henry Scott*: Henry Young Darracott Scott (1822–83), an English Major-General in the Corps of Royal Engineers. He is best known for the construction of The Royal Albert Hall in London while working under the commission of the Great Exhibition of 1851 at South Kensington. The acoustics were a problem from the start, but later they installed a curtain under the true roof which fixed the issue. In 1875 he was elected FRS and throughout his career earned a number of medals for his engineering works. He and his wife, Ellen Selina Bowes, had fifteen children.
2. *29th March, 1871*: date noted on the published version, but not on the handwritten one.
3. *The Acoustics of the Royal Albert Hall*: letter forwarded by Scott to *The Orchestra* with a letter of his own; the two were published under the title 'The Echo in the Albert Hall'.
4. *Royal Institution | 29th March, 1871*: place and date inserted on the published version; see n. 2.
5. *vast edifice*: the Royal Albert Hall in London, which opened on 29 March 1871.
6. *Prince of Wales*: Prince Albert Edward of Wales (1841–1910) who became King Edward VII on the death of his mother, Queen Victoria in 1901. The title Prince of Wales is a title that is granted to the heir apparent to the British throne. Edward held the title of Prince of Wales for sixty years before becoming King.
7. *2,000 in the Philharmonic Hall in Liverpool*: The Philharmonic Hall was founded in 1840 in Liverpool. Tyndall was likely referring to the address he gave in Liverpool in 1870, for which see letter 3370, n. 5.
8. *do not*: There is a small difference between the handwritten and published versions: in the handwritten copy there is a contraction 'don't' and in the published version it is 'do not'.
9. *acoustic properties*: Tyndall was a member of the Organ Committee for the Albert Hall in 1868, and in 1869 was a member of the committee on the acoustics of the lecture theatre at the South Kensington Museum (*Ascent of John Tyndall*, p. 282).
10. *RI MS JT/6/5/1*: letter is not in Tyndall's handwriting.

From Juliet Pollock [March 1871][1] 3399

Rude Boreas[2]—Excellent friend,

Next Thursday Miss Watts the Welsh singer[3] I told you of dines with us. Will you come too at 7 o'clock.

Yours | Juliet Pollock

RI MS JT/1/TYP/6/2115
LT Typescript Only

1. *[March 1871]*: dated to March 1871 because of the reference to Megan Watts. Frederick Pollock recorded that she sang at their house in March 1871 (F. Pollock, *Personal Remembrances of Sir Frederick Pollock: Second Baronet*, 2 vols (London: Macmillan, 1887), vol. 2, p. 226).

2. *Boreas*: Pollock and Tyndall had nicknames for each other. She called him Boreas, he called her Eolia. In reference to Classical mythology, Eolia is the master of the wind and Boreas is the North wind. See also letters 3513 and 3514.
3. *Miss Watts . . . singer*: Megan Watts Hughes (1842–1907), a Welsh vocalist who became well known quite early in her life through giving concerts locally. A local committee raised money for her to be trained in singing, and she ended up at the Royal Academy of Music in London in 1864. By late 1871 she was married to a banker named Lloyd Hughes, she founded a home for boys and she continued singing. Between 1885 and 1891, she invented a musical device called voice figures, which measured the intensities of vocal sounds, and presented on it at the RS and the RI; she also published a scientific paper about the topic.

From Cecil Trevor 3 April 1871 3400

1. Professor Tyndall F.R.S. | Royal Institution | Albemarle Street
2. Sec[retar]y | T[rinity].H[ouse].[1]

Sir,

Referring to previous correspondence[2] rel[ating]. to the proposal to substitute Gas for oil as an illuminating power in Lighthouses,[3] I am &c to transmit for

1.—your information

2.—the information of the E[lder].B[rethren].[4] of the T[rinity].H[ouse].

copy of a report[5] by Messrs. D & T. Stevenson upon the recently printed Parliamentary paper[6] on the subject.

I am &c[7]

National Archives MT 10/131 H1739 2319

---

1. *T.H.*: Trinity House, the official authority for lighthouses in the United Kingdom and Gibraltar.
2. *previous correspondence*: see, for example, letter 3353, *Tyndall Correspondence*, vol. 11.
3. *the proposal . . . Lighthouses*: refers to a long-standing debate over the use of different fuels in Irish lighthouses (see *Ascent of John Tyndall*, pp. 282–5). See letters 3417 and 3555, for example.
4. *E.B.*: Elder Brethren is the governing body of the Trinity House.
5. *copy of a report*: enclosure missing. It is probable this report is D. and T. Stevenson, 'Report as to the Use of Gas for Revolving Lights', in 'Further Papers Relative to a Proposal to Substitute Gas for Oil as an Illuminating Power in Lighthouses', C 1511, Parliamentary Papers (1875), pp. 3–4.
6. *recently printed Parliamentary paper*: possibly a report by Tyndall from 7 February 1871. See J. Tyndall to Board of Trade, 7 February 1871, in 'Further Papers Relative to a Proposal to Substitute Gas for Oil as an Illuminating Power in Lighthouses', C 282, Parliamentary Papers (1871), pp. 30–3; letter 3353 in *Tyndall Correspondence*, vol. 11.
7. *I am &c*: noted in the margins 'Signed by Mr. Trevor | April 3rd 1871'.

To Elizabeth Dawson Steuart 5 April [1871][1] 3401

5th April

My dear Mrs Steuart,

Kindly pay this man[2] for me: I do not like any trace of this kind to be left hanging about Leighlin Bridge.

Yours ever | John Tyndall

I have been terribly busy; otherwise I should have written to you about other matters before now

Pray give the few shillings over to the poor.

[...][3]

RI MS JT/1/TYP/10/3404
LT Typescript Only

---

1. *[1871]*: LT annotation, but clear from letter 3396.
2. *this man*: John Ryan, see letter 3396.
3. *[...]*: Tyndall copied letter 3396 here.

From William Thomson 9 April 1871 3402

GLASGOW COLLEGE | April 9, 1871

Dear Tyndall

I hope to be able to persuade Helmholtz, Huxley, Maxwell, and Tait to come a cruise with me in the "Lalla Rookh" ($128x10^6$ grammes)[1] for a fortnight after the meeting in Edinburgh.[2] Will you join us? We would all be very glad to have the pleasure of your company.

Believe me | Yours very truly | William Thomson

RI MS JT/1/T/18
RI MS JT/1/TYP/5/1558

---

1. *the "Lalla Rookh" ($128x10^6$ grammes)*: Thomson's schooner yacht of 126 tons, which he liked to jokingly refer to in grams.
2. *the meeting in Edinburgh*: the BAAS meeting in Edinburgh, 2–9 August 1871 (*Brit. Assoc. Rep. 1871*), which Tyndall did not attend as he stayed in Pontresina. See letters 3494 and 3511, for example.

To the Editor of *The Times*[1] 11 April 1871 3403

Imagination in Science | Times, April, 13, 1871.[2]

Sir,—

The writer of the review of Mr Darwin's Descent of Man[3] honours me with the following reference:—"He (Mr. Darwin) does not hesitate, in accordance with Professor Tyndall's advice, to let it (imagination) take the place of science when the means and methods of science fail."

You will, I am persuaded, give me room to say that I should not recognize the notions I entertain regarding either science or imagination in the "advice" here ascribed to me.[4]

I am, Sir, your obedient servant, | (s[igne]d) John Tyndall. | Athenaeum Club,[5] April 11.

If you would care to publish the views of that highly philosophical thinker, the late Sir Benjamin Brodie,[6] as to the use of the imagination, here they are:—

"Lastly, physical science, more than anything besides, helps to teach us the actual value and right use of the imagination—of that wondrous faculty which, left to ramble uncontrolled, leads us astray into a wilderness of perplexities and errors, a land of mists and shadows; but which, properly controlled by experience and reflection, becomes the noblest attribute of man—the service of poetic genius, the instrument of discovery in science; without the aid of which Newton would never have invented fluxions, and Davy have decomposed the earths and alkalies, nor would Columbus have found another continent." Address to the Royal Society, Nov. 30, 1859.[7]

RI MS JT/6/5/1
Transcript Only[8]

---

1. *Editor of* The Times: John Thadeus Delane (1817–79), editor of the *Times* from 1841–77. He attended Oxford and earned his BA in June 1840. He was immediately hired by the *Times*. One of his biggest achievements during his tenure at the *Times* was increasing circulation of the paper almost three-fold.
2. *April 13, 1871*: Tyndall wrote his letter on 11 April, which is why this letter has this date, but it was published on 13 April (on p. 4).
3. *The writer of the review . . . Descent of Man*: 'Mr. Darwin On The Descent Of Man', *Times*, 7 April 1871, p. 3. The review was so bad that Tyndall was not the only one to write in response to it.
4. *"advice" here ascribed to me*: a reference to Tyndall's lecture 'On the Scientific Use of the Imagination', for which see letter 3370, n. 5.

5. *Athenaeum Club*: see letter 3373, n. 15.
6. *Sir Benjamin Brodie*: Benjamin Collins Brodie, first baronet (1783–1862), a British physiologist and surgeon. He was educated in London in the medical school and in 1803 entered St. George's Hospital as a surgical pupil. He was elected FRS in 1858 and was an active part of the RS's dining club as well as in the Assistant Society for the Improvement of Animal Chemistry (see N. G. Coley, 'The Animal Chemistry Club; Assistant Society to the Royal Society', *Notes and Records of the Royal Society of London*, 22 (September 1967), pp. 173–85). As a surgeon, he participated in a number of animal vivisections. He was interested in metaphysical questions, as well as philosophy.
7. *1859*: B. Brodie, 'Address to the Royal Society, 30 November 1859', in *Roy. Soc. Proc.*, 10 (1859–60), pp. 161–78, on p. 165.
8. *Transcript Only*: handwritten copy not in Tyndall's hand.

From Arthur Russell[1] 11 April 1871 3404

Robertson Terrace | Hastings | 11 April | 1871

Dear Mr. Tyndall

Accept my best thanks for the book[2] you have been good enough to send me which came just in time for my Easter holidays & from which I have already gathered much "zur Entwicklung meiner Weltanschauung"[3]—

I send you the passage from the Psalms I mentioned to you,[4] in which David[5] uses the same image as Goethe and I hope you will find an opportunity of appealing to this text, when anyone objects to the speech of the Erdgeist to Faust[6]—

Y[ou]rs sincerely | Arthur Russell

Psalm 102. v. 25, 26, 26.

They shall perish (heaven & earth) but thou shalt endure, they all shall wax old as doth a garment. And as a vesture shalt thou change them and they shall be changed; but thou art the same and thy years shall not fail.

RI MS JT/1/R/61

---

1. *Arthur Russell*: Arthur John Edward Russell (1825–92) became Lord Arthur Russell in 1872. He was a Liberal Party politician, and MP for Tavistock from 1857–85.
2. *the book*: *Fragments of Science*.
3. *zur Entwicklung meiner Weltanschauung*: to the development of my worldview (German).
4. *the passage from the Psalms I mentioned to you*: Psalms 102: 25–6.
5. *David*: David (fl. *c.* 1010–970 BCE), King of Israel and Judah, was the person credited with writing the book of Psalms in the Bible.
6. *the speech of the Erdgeist to Faust*: see letter 3397, n. 3. There are three additional Goethe quotations in *Fragments of Science* (pp. 70, 126 and 168), but none look to be the Earth's

spirit's speech to Faust (see J. W. von Goethe, *Faust: Eine Tragödie* (Tubingen: J. G. Cotta'schen Buchhandlung, 1808), p. 25). It is possible the reference is more subtle than a direct quote.

From Mary Egerton [13 April 1871][1] 3405

Mountfield Court, | Robertsbridge, | Hawkhurst.[2] | Thursday

My dear Mr. Tyndall,

Your letter[3] has given me pain—more pain than I dare say you intended, though you were angry & meant to punish me.—Why will you not believe that I mean what I say, neither more nor less! It was just because you would meet Hugh[4] as well as Charley[5] that I preferred your coming this week;—did that look like mistreating you? I was merely thinking aloud (or rather on paper.)

Hugh is a thoughtful boy for his age, & it is not every body with views like yours, that I should feel justified in encouraging him to apointe[6] with—not because I would keep him in the dark, but because I think those questions are better postponed till the intellect is ripe. But your mind is altogether of a different stump—you have none of that pregestious[7] spirit that loves to startle, and therefore, as I said, I trust you.

Then as to your book.[8] Have I not got every article in it, except perhaps one of the latter ones, whose title I do not quite recognize, unless it is the chapter out of "Mountaineering?";[9] & have we not all read them over & over, & gathered noble & beautiful thoughts from there? Ay, & a scripture interpretation from one of them, which has stuck to me ever since, as the true one! Shall I confess that the only reason I have not already sent for the work is, that from your kindness before, I could not help having a suspicion you might send it to me, & it is so much nicer having it from you, than from the shop!

Now, you must give it me, as a pledge of forgiveness, or rather a token of penitence, for it is much more I that have to forgive than you! To be considered as a narrow minded bigot by you, is what I do not feel to deserve!

If you preferred the solitude of the Surrey hills the first few days from London, (which I can quite understand), you might have now access to us on the Sat[urda]y till you returned to town; I had left that week quite open, though, thinking you would be sure to take a fortnight's holiday, I gave the preference to the following one.

I wonder if Mountfield went into your book? or was thrown away in a feh?[10]

I am so glad you put that in the Times today![11] I do get into such a rage with the people that will persist in misunderstanding you! I longed to write to that Darwin Reviewer myself I am afraid my letter w[oul]d not have been very craic![12]

Now Good-bye—I hope I have said nothing wrong today? I like to speak frankly to you, as to a friend, & to feel that you will understand me.

Y[ou]rs ever truly | MF Egerton

I have not thanked you for the promise of the Alpine book,[13] but I don't think I will have it if you will not give me the object.—& if you do, it w[oul]d be too greedy to want both!

RI MS JT/1/E/77

---

1. *[13 April 1871]*: Mention of Tyndall's letter (see 3403) to the *Times* and the reference to both *Fragments of Science* and *Hours of Exercise in the Alps* places this on Thursday, 13 April 1871.
2. *Mountfield Court, | Robertsbridge, | Hawkhurst*: the Egerton family home, located in Robertsbridge, a village in Sussex, in the south-east of England. Hawkhurst was the post town for Robertsbridge.
3. *Your letter*: letter missing.
4. *Hugh*: Hugh Edward Egerton (1855–1927) was the younger of the two boys. He became a historian who held the first chair of colonial history at Oxford. He was a fellow of All Souls, Oxford, from 1906.
5. *Charley*: Charles Augustus Egerton (1846–1912) was the oldest son of Edward and Mary Egerton, and likely named after his mother's father, Charles Pierrepont.
6. *apointe*: Egerton is using this in the sense of arranging for a meeting (*OED*).
7. *pregestious*: Egerton's original spelling; we assume she means prestigious.
8. *your book*: *Fragments of Science*.
9. *"Mountaineering?"*: J. Tyndall, *Mountaineering in 1861: A Vacation Tour* (London: Longmans, Green, and Co., 1862).
10. *feh*: exclamation of disapproval or disgust.
11. *you put that in the Times today!*: letter 3403.
12. *craic*: Irish for good or enjoyable; here Egerton must mean that her letter would not have been as enjoyable to read, or do the job that Tyndall's letter did.
13. *the Alpine book*: *Hours of Exercise in the Alps*.

From James Coxe[1] 15 April 1871 3406

General Board of Lunacy | Edinburgh April 15th 1871

My dear Dr Tyndall

You overwhelm me and mine with your kindness, and we are at a loss to say how much gratified we feel. Even the lady[2] shall read the book[3]—in whole or in part—as the owners of furnished apartments express themselves; and when you come to Edinburgh, in August,[4] as we do not doubt you intend doing, you shall put her to the test by examination. We hope you will not consider Kinellan[5] too inconveniently situated to serve as your domicile during the meeting. We shall try to have two or three pleasant companions for you, and if there be any one to whom you would like us to offer what we call in Scotland a "Howff" (from the German hof: gasthof) please name him, and we shall try and secure him. There will be a carriage at the disposal of the party, so that although we are two miles from the University[6] where the meetings will be held, it will not be more than twenty minutes in time. My wife and Cumming[7] greet you affectionately. The latter goes to his Highland banishment next month—So you will not get to see him unless you go to Mull;[8] but if you will be tempted to visit him he will kindle a huge—bonfire on the top of Ben More[9] that all the world may rejoice with him.

Yours most truly and with many—thanks | J. Coxe

RI MS JT/1/TYP/1/295
LT Typescript Only

---

1. *James Coxe*: James Coxe (1810–78), a Scottish physician and psychiatrist. He was FRSE from 1854; he was knighted in 1863 and in 1872 was elected President of the Psychological Association in Great Britain. As one of the two commissioners of the General Board of Lunacy for Scotland, he was an early advocate against restraint in asylums.
2. *the lady*: Mary Coxe (née Cumming 1806–75) was Coxe's wife.
3. *the book*: likely *Fragments of Science*.
4. *to Edinburgh, in August*: the BAAS meeting in Edinburgh in 1871 (*Brit. Assoc. Rep. 1871*). Tyndall did not attend.
5. *Kinellan*: Kinellan House, Coxe's residence in Murrayfield, near Edinburgh. See also letter 3497.
6. *the University*: Edinburgh University.
7. *Cumming*: probably William Fullerton Cumming (1804–92), a physician and author. He traveled heavily through Italy, Egypt, Greece and Turkey for his health. He was a member of the Royal Physical Society of Edinburgh and an associate member of the Egyptian Society of Cairo. See W. F. Cumming, *Notes of a Wanderer, in Search of Health, through Italy, Egypt, Greece, Turkey, up the Danube and Down the Rhine*, 2 vols (London: Saunders and Otley, 1839). See also letter 3497, n. 1.
8. *Mull*: an island off the west coast of Scotland.
9. *Ben More*: Beinn Mhòr, in Scottish Gaelic, is the highest peak on the Isle of Mull.

To William Adams[1] 18 April 1871 3407

18th April 1871

Dear Prof. Adams,

The Department of Science and Art[2] place in my hands the organization of a staff of examiners for the annual correcting of papers sent in by pupils in reply to questions drawn out by me.

Prof *[Goodeve]*,[3] Prof Guthrie,[4] Prof Carey Foster,[5] Dr. Debus and W. Barrett[6] will all I hope take part in the examination. Could you make it convenient to join them in lending me a helping hand?

Yours faithfully | John Tyndall

The pay is pretty good, on that point I can give you more information if you wish it.

Cornwall Records Office, Truro DD, AM/675

---

1. *William Adams*: William Grylls Adams (1836–1915), a British scientist and professor at King's College, London. He attended Cambridge University to study mathematics, and graduated twelfth wrangler in 1859. He became a professor of natural philosophy at King's College in 1865 and remained there until he retired in 1905. In London, he attended Tyndall's lectures at the RI, and later conducted experiments on lighthouses. He was elected FRS in 1872.
2. *The Department of Science and Art*: a body of the British government from 1853–99 which promoted education in science, technology and art in Britain. It was a direct result of the Crystal Palace Exhibition and Prince Albert's interest in the 'useful arts' of technology and science.
3. *Prof* [*Goodeve*]: possibly Thomas Minchin Goodeve (1820–1902), who held a number of professorial positions at King's College, London from 1846–92, including mathematics, machinery and natural philosophy. He was called to the Bar in 1862 and became a prominent patent lawyer.
4. *Prof Guthrie*: Frederick Guthrie (1833–86), a chemist and physicist. He was educated at UCL, and then in Heidelberg with Bunsen. By 1869, Guthrie had taken over Tyndall's lectureship in physics at the School of Mines. In 1871, he moved with the physics faculty to the Science Schools in South Kensington, where he received a suite of laboratories for his work. He was elected FRS in 1871, and helped to found the Physical Society in 1874. Upon Guthrie's death from throat cancer, Huxley was instrumental in securing a state pension for his widow and children.
5. *Prof Carey Foster*: George Carey Foster (1835–1919) was a chemist and physicist. He was educated at UCL, in the same laboratories as Guthrie (see n. 4). He also went to Glasgow

to work with William Thomson. In 1865 he was appointed professor of experimental physics and established a student lab for undergraduates, the first of its kind in Britain. He was elected FRS in 1869, and was Vice President of the RS in 1891–3 and 1901–3. He was active in the BAAS. While at UCL, he was an avid supporter of undergraduates and the equal rights of women. He retired in 1904.

6. *W. Barrett*: William Fletcher Barrett (1844–1925), a British physicist and psychical researcher. He took classes in chemistry and physics at the Royal College of Chemistry, London, and became assistant to Tyndall at the RI in 1863. He is best known for his work on 'sensitive flames' in acoustics and studies of the properties of metals, especially iron and iron alloys. He also did a lot of work in spiritualism, which brought him into conflict with Tyndall (see R. Noakes, 'The "Bridge Which Is between Physical and Psychical Research": William Fletcher Barrett, Sensitive Flames, and Spiritualism', *History of Science*, 42 (2004), pp. 419–64). Barrett was elected FRS in 1899, along with his memberships as FRSE and in the Royal Dublin Society. He was knighted in 1912.

To Jane Barnard 18 April [1871][1] 3408

18th April

My dear Miss Barnard.

The committee have allowed £25 to M. Petitjean.[2]

So, good Samaritan, you may congratulate yourself on at all events a measure of success. Yours ever | John Tyndall

Your aunt[3] thinks me a brute—but I am only a slave.

RI MS JT/1/T/119
RI MS JT/1/TYP/12/4200

1. *[1871]*: LT annotation; see also letter 3409, which is Barnard's response.
2. *M. Petitjean*: Tony Petitjean; see letters 3409, 3413, 3414 and 3418.
3. *Your aunt:* Sarah Faraday.

From Jane Barnard 19 April 1871 3409

Barnsbury Villa, | 320, Liverpool Road, | N.

My dear Dr Tyndall

The success that has attended your very kind exertions far exceed my hopes. I do not like to say anything in the matter to Mr. Petitjean as I hope & suppose the money will be handed to him by the Society.[1] I do not wish to appear more than can be helped in the affair.

I am afraid I shall not be at the R[oyal]. I[nstitution]. on Friday as we have a visitor staying with us, not learnedly inclined, so I cannot well leave home; It seems long since I was in the dear old place.[2] My Aunt[3] desires her kind regards, she knows how fully occupied you are;—she likes to hear of, or from you, always.

With many thanks to you for your kindness to me.

I am | yours very sincerely | Jane Barnard | 19 April 1871

RI MS JT/1/B/48

1. *the money . . . Society*: the RS. See letter 3414, n. 2.
2. *in the dear old place*: Jane and Sarah had lived in the RI when Faraday was alive.
3. *My Aunt*: Sarah Faraday.

From George Busk 19 April 1871 3410

32 Harley St | April 19 1871

My dear Tyndall

In answer to your request that I should give you my opinion respecting D^r W[illiam]. Budd's claims to admission into the Royal Society I can only say that I know no one in many respects more justly entitled to that honour.

Although his labours have of necessity been chiefly devoted to professional subjects or to sanitary subjects their results have been such as not only to have given him a very high position in his profession but at the same time to stamp him as deserving a high position in biological Science. His writings for the most part have had a wider scope than the mere improvement of medical practice. His observations for instance on Symmetrical Diseases,[1] published thirty years since perhaps one of the most original and valuable contributions to pathology ever made is replete with reflections of the highest interest and importance in a strictly physiological sense. The same may be said of his earlier paper on the Pathology of the Spinal Chord.[2] In fact in all Dr Budd's writings it is evident that his views & thoughts have extended far beyond the professional limits of his subject and have always embraced the more general biological considerations, which though really underlying nearly all medical enquiries are too often overlooked or disregarded.

For the reason therefore that he is as justly entitled to a high position as a man of Science as well as of undisputed eminence in his own profession I strongly feel that Dr Budd may most justly claim the honor of the Fellowship of the Royal Society[3]—

Believe me | Yours very truly | Geo[rge] Busk

RI MS JT/1/B/164
RI MS JT/1/TYP/1/180

---

1. *Symmetrical Diseases*: W. Budd, 'On diseases which affect corresponding parts of the body in a symmetrical manner', *Medico-Chirurgical Transactions*, 25 (1842), pp. 100–66, 306.
2. *the Pathology of the Spinal Chord*: W. Budd, 'Contributions to the pathology of the spinal cord', *Medico-Chirurgical Transactions*, 22 (1839), pp. 153–90.
3. *justly claim the honor . . . Royal Society*: He was indeed elected FRS on 8 June 1871.

From Henrietta Huxley 19 April 1871 3411

26 Abbey Place | St. John's Wood | April 19th. 1871

Dear Brother John

Many thanks for the tickets[1] which I shall try and make use of, if the children's[2] engagements of my time permit. But will you think me very troublesome for asking upon what days you discourse and the hour?

Hal[3] and I are so set up with our pleasant week at Liphook.[4] If you ever want a clean little inn—with civil hosts—and good cooking—let me recommend the "Anchor". And such walks—through fine country! I managed 5 miles on one day and 8 another—most of the latter was uphill, and on damp ground so that it was heavy walking. Hal and I both got into a bog. As I am writing in Abbey Place and Hal is at the School Board[5] I need hardly say we got out of it—the bog.

The children were sadly disappointed at not going to the Lubbocks[6]—through Lady Lubbock having the measles the more so as I was to be away from home—but I consoled them by sending them sight seeing in London.

Have you fully succeeded in your experiments of enabling people to endure the smoke incident to great fires?[7] Those are the results that make the ignorant believe there is some good in science.

"By their fruits you shall know them".[8]

Love to you (2nd best you know) | Yours very sincerely | Henrietta Huxley.

RI MS JT/1/TYP/9/2945
LT Typescript Only

---

1. *the tickets*: Tyndall probably gave her tickets to his lectures on Sound at the RI after Easter (see *Roy. Inst. Proc.*, 6 (1870–72), p. 188).
2. *the children's*: The Huxley children were: Jessie Oriana Huxley (1858–1927), Marian Huxley (1859–87), Leonard Huxley (1860–1933), Rachel Huxley (1862–1934), Henrietta (Nettie) Huxley (1863–1940), Henry Huxley (1865–1946) and Ethel Huxley (1866–1941). Their first child, Noel Huxley, died of scarlet fever in 1860 at the age of four.

3. *Hal*: her nickname for Thomas Huxley.
4. *Liphook*: a village in East Hampshire, England.
5. *the School Board*: Huxley had been elected to the newly created London School Board in 1870; he resigned early in 1872. See Desmond, *Huxley*, pp. 401, 410; see letter 3635.
6. *the Lubbocks*: John Lubbock and his family, including his wife, Ellen (1835–79).
7. *your experiments . . . to great fires?*: possibly referring to Tyndall's experiments with respirators. He discussed this with Darwin in letters 3367, 3368, 3369, 3373, 3374 and 3375. Tyndall also delivered a Friday Evening Discourse at the RI on 9 June 1871, 'On Dust and Smoke'; see *Roy. Inst. Proc.*, 6 (1870–72), pp. 365–76. For more on Tyndall's work on this subject, see letter 3620, *Fragments of Science*, pp. 316–7, *Ascent of John Tyndall*, p. 275 and J. Tyndall, 'On some Recent Experiments with a Fireman's Respirator', *Roy. Soc. Proc.*, 22 (1874), pp. 359–61.
8. *"By their . . . know them"*: Matthew 7:16.

To Thomas Henry Huxley [21 April 1871][1] 3412

Friday. | The Athenaeum.[2]

My dear Huxley,

I came down here purposely to express to you my approval of your last bit of work.[3] It was sorely needed, and it is well done.

I am not jealous after this of being only "Second best".[4] More especially as that depended purely on the accident of my not being on board the Rattlesnake[5] certain years ago.

Yours ever | and more power to your elbow, | J. T.

IC HP 8.86
RI MS JT/1/TYP/9/2940
Typed Transcript Only

---

1. *[21 April 1871]*: The first Friday after Henrietta Huxley's letter (letter 3411) is 21 April 1871.
2. *The Athenaeum*: see letter 3373, n. 15.
3. *your last bit of work*: unclear to which piece of work Tyndall is referring. Possibly a letter to the editor of the *Times*: T. H. Huxley, 'The Royal School of Mines', *Times*, 11 April 1871, p. 5. Huxley was responding to a letter concerning a plan to move the Royal School of Mines.
4. *"Second best."*: see letter 3411, which Henrietta Huxley signed 'Love to you (2nd best you know)'.
5. *on board the Rattlesnake*: Huxley met his future wife while in Sydney, during his 1847–9 voyage on HMS *Rattlesnake*.

From Jane Barnard 22 April 1871 3413

Barnsbury Villa | 320, Liverpool Road, | N. | 22 April 1871

My dear Dr Tyndall

Last night on my return home I found a note from Mr White[1] enclosing the cheque for Mr. Petitjean.[2] May I ask you yet to add to your kindness to me in this affair by telling me whether I should do anything more as to thanks & acknowledgments. I have written to Mr White, but merely to say I have received his note & enclosure. No doubt, Mr Petitjean will be anxious to shew his gratitude to you, & the Royal Soc[iety]. I should feel very sorry if through ignorance either he or I should appear remiss, & yet I feel that I have no right to thank anyone but you who have taken so much interest & *[trouble]* in the matter. I am sorry to tease you further but will add no more but that I am

Yours very sincerely & obliged, | Jane Barnard

P.S. I find on enquiry that Mr Petitjean is at the Convalescent Asylum Welton;[3] I will write to him early next week.

Dr Tyndall | &c &c &c

RI MS JT/1/B/49

1. *Mr White*: Walter White (1811–93), at first an attendant and librarian at the RS, then in 1861 was appointed assistant secretary. He wrote over 200 articles for *Chambers's Edinburgh Journal* and a number of other works about hiking and walking. He held his post at the RS until 1884, when he resigned in poor health.
2. *the cheque for Mr. Petitjean*: see letters 3408, 3409, 3414 and 3418.
3. *Convalescent Asylum Welton*: a place to recover from an illness or surgery; this one was in Welton, Lincolnshire.

To Jane Barnard 24 April [1871][1] 3414

24th April

My dear Miss Barnard.

There can be no impropriety in Mr. Petitjean thanking as cordially as he likes the Committee of the Scientific Relief fund. I think the best way would be to address his acknowledgement to the Secretary of the Scientific Relief Fund,[2] Royal Society, Burlington House.

That I think will be quite sufficient. With kind regards to you and your Aunt

ever faithfully Yours | John Tyndall

It gave me real pleasure to respond to your wishes in this matter | J. T.

RI MS JT/1/T/120
RI MS JT/1/TYP/12/4201

1. *[1871]*: letter refers to and follows letter 3413.
2. *Scientific Relief Fund*: founded by members of the RS in 1859 to help scientists and their families with financial difficulties. For more information about Tyndall's involvement in this fund, see *Ascent of John Tyndall*, p. 75. See also letters 3408, 3409, 3413 and 3418.

To the Editor of the *Daily Telegraph & Courier*[1] 23 April 1871 3415

THE PORTSMOUTH SCANDAL.

TO THE EDITOR OF "THE DAILY TELEGRAPH."

SIR—Mr. Serjeant Ballantine's[2] reference to me, as reported in your last number,[3] is without warrant or foundation.—Your obedient servant, JOHN TYNDALL. | April 23.

J. Tyndall, 'The Portsmouth Scandal', *Daily Telegraph & Courier*, 24 April 1871, p. 6
Typed Transcript only

1. *Editor of the* Daily Telegraph & Courier: probably Edward Levy (1833–1916). Levy was never the formal editor of this paper after his father bought it in 1855, but he was responsible for the direction of the paper from around 1870. He was the oldest of eight children whose father managed a printing business on Fleet Street. By 1871 the circulation of the paper sold almost 200,000 copies per day, easily out-selling the *Times*. He added the surname Lawson in 1875 by royal license. He became the first Baron Burnham in 1903.
2. *Serjeant Ballantine*: William Ballantine (1812–87), a lawyer, Serjeant-at-law, and honorary member of the Inner Temple. He argued a number of murder and other cases in London. He was acquainted with writers and playwrights, as he frequented the pubs in Covent Garden.
3. *your last number*: *Daily Telegraph & Courier*, 22 April 1871, p. 4. This article, also entitled THE PORTSMOUTH SCANDAL, had to do with drugs given to an Emily Vyse to obtain an abortion. Ballantine claimed that 'he had the report of Professor Tyndall, who said that the drugs alleged to have been administered to Miss Vyse were perfectly harmless, and not calculated to procure abortion'.

To Auguste de la Rive 25 April 1871 3416

Royal Institution | 25th, April, 1871.

My dear Friend

I am afraid that you will have expunged me from your list of friends I have been so tardy in replying to your last kind letter.[1] In fact I was about to reply to it on the moment, when some matters relating to my expedition to Oran[2] laid hold of me: and thus it has come to pass that I have not written to you till now.

I sent you a few days ago a collection of scattered essays:[3] I was obliged to put them together because they had already begun to publish them in America, and I wished naturally to do the work of selection and revision myself.

You are aware of course of our defeat in Algeria as regards the solar eclipse.[4] The day we reached Oran I burnt a hole in an instant through my hat by a pocketlens; the sun was as strong as on a July day in England: but the weather changed and remained bad throughout our stay.

Janssen was there with a young sailor: they had escaped from Paris[5] in a balloon. He is now in London.

No doubt you are aware of the changes impending with regard to the Royal Society. I was greatly grieved to have to oppose both Sabine[6] and Gassiot[7] in the appointment of a treasurer to fill the place of Miller.[8] But I never had the least doubt of the wisdom of the proceeding. By electing Spottiswoode we have secured a most excellent man, and have removed a great deal of discontent.

Sabine is exceedingly well, but cold to me. Gassiot, I grieve to say, is by no means well.[9] Had he been elected treasurer I do not know how he could have got through his duties.

I suppose you are aware that Airy succeeds Sabine as President.

I have not been able to investigate the magnetic question[10] you referred to. But this much I think is certain, that Airy though a man of the highest ability and candour, is sometimes rash in his conclusions. I know nothing that invalidates my belief in the periodicity of the effects referred to.

Dr Marcet[11] (the old man) told me a few days ago that you are coming to England. I shall be rejoiced to see you—and until then goodbye.

But I must ask you to present my kindest regards to Madame De la Rive,[12] to Mr De la Rive[13] and to Mr Lucien.[14]

Ever Yours | John Tyndall

RI MS JT/1/TYP/1/380
LT Typescript Only

---

1. *last kind letter*: see letter 3303, *Tyndall Correspondence*, vol. 11.
2. *expedition to Oran*: Tyndall had gone on the eclipse expedition to Oran, Algeria, for the 22 December 1870 solar eclipse. See letter 3384.
3. *I sent you . . . scattered essays*: likely referring to *Fragments of Science*.
4. *defeat . . . solar eclipse*: see letter 3384 for more details about this expedition.
5. *Janssen . . . Paris*: see letter 3387, n. 9.
6. *Sabine*: Edward Sabine was, at this point, the outgoing President of the RS. See also letter 3370, n. 14.
7. *Gassiot*: John Peter Gassiot (1797–1877), an English businessman and electrician. He was elected FRS in 1841 due to his work in electricity out of his home, where he had state-of-the-art apparatus available to other men of science. He was one of the founders of the Scientific Relief Fund to help scientists and their families who needed help (see letters 3408, 3413, 3414 and 3418).
8. *Miller*: William Allen Miller (1817–70), an English chemist and Treasurer of the RS from 1861 until his death. He was trained as a surgeon under his uncle, Bowyer Vaux, then in 1840 he went to Germany to study with Justus Liebig. He became a demonstrator of chemistry at King's College, London in 1841 and in 1842 earned his MD from the University of London. He was elected FRS in 1845 and was involved in the Chemical Society from its founding in 1841. He died in 1870, of apoplexy following a manic episode.
9. *Gassiot . . . no means well*: Gassiot must have been ill, but did not die until 1877, at his home.
10. *magnetic question*: De La Rive had written to Tyndall about an exchange with Airy concerning magnetic disturbances and sunspots in letter 3303, *Tyndall Correspondence*, vol. 11 (see n. 7 there).
11. *Marcet*: François Marcet (1803–83), a Swiss physicist and the son of Jane Marcet (1769–1858), who wrote *Conversations on Chemistry*. F. Marcet was known for an apparatus he invented for measuring the vapor pressure of water. De la Rive worked with F. Marcet on the specific heat of gases (See 'Auguste Arthur De La Rive', *Encyclopedia Britannica*, 11th edn, 11 vols (New York: Encylopaedia Britannica, 1911), vol. 7, p. 944).
12. *Madame De la Rive*: Louise Maurice (née Fatio, m. 1855).
13. *Mr De la Rive*: not further identified, but must be one of Auguste's family members.
14. *Lucien*: Lucien de la Rive (b. 1834), Auguste's son, also became a physicist and investigated the propagation of electrical waves ('Auguste Arthur De La Rive', *Encyclopedia Britannica*, vol. 7, p. 944).

To Cecil Trevor 28 April 1871 3417

28th April 1871.

Sir,

I have read the Report of the Mes[srs] Stevenson[1] which you did me the honour of sending to me on the 3rd of this month.[2]

From the earlier Reports of the Mes[srs] Stevenson it could not be inferred that they considered it in the least degree likely that gas would ever be advantageously employed for lighthouse purposes.

It is however an undoubted fact that notwithstanding the unfavourable opinion expressed by the Mes[srs] Stevenson, the applicability of gas to Lighthouse illumination has been reduced to demonstration in Ireland.[3]

What's now contemplated on the parts of shore to where skill and enterprise we owe this result, is the examination of the question whether gas is, or is not, applicable to Revolving Lights.[4]

Considering the importance of the subject, and its possibilities, I do not hesitate to recommend that arrangements be made at Rockabill[5] which shall permit me, when I go to Ireland,[6] to come to a definite conclusion as to the report in question.

Apart from the advantages already proved to exist, I think it highly desirable, when a solid prospect of advance presents itself, and where no extravagant outlay is demanded, to stimulate the inventiveness of lighthouse engineers generally, by the encouragement of experiments on the use of gas in Ireland.

I may say that there is no idea, qualification, or objection contained in the Report of the Mes[srs] Stevenson that had not already passed through my own mind. What they urge shall have due weight given to it. But the discussion of this new mode[7] of dealing with Lighthouse flashes,[8] which it was my object to excite when I mentioned it incidentally in a Report on a substantially different question,[9] has only confirmed my belief in the necessity of further experiment.

I am Sir | Your obedient Servant | John Tyndall

The Assistant Secretary | Harbour Dep[artmen]t | Board of Trade[10]

National Archives MT 10/131 H2197

---

1. *Messrs Stevenson*: David and Thomas Stevenson. See letter 3400, n. 5.
2. *Report . . . 3rd of this month*: see letter 3400.
3. *reduced . . . Ireland*: see letter 3400, n. 3.

4. *Revolving Lights*: Gas use in revolving lights in lighthouses was a debate that lasted for some time. In 1873, William Thomson published a piece, 'Lighthouses of the Future', in which he used Tyndall's arguments to show that gas could be used in a revolving light. See W. Thomson, 'Lighthouses of the Future', *Good Words*, 14 (December 1873), pp. 217–24, and *Ascent of John Tyndall*, pp. 284–5.
5. *Rockabill*: a lighthouse north-east off the coast of Skerries, Dublin County. It was operational by June 1860.
6. *I go to Ireland*: possibly a reference to his trip in June 1871; see letters 3475, 3476, 3480 and 3483.
7. *new mode*: see 'Further Papers Relative to a Proposal to Substitute Gas for Oil as an Illuminating Power in Lighthouses', C 1511, Parliamentary Papers (1875), pp. 3–4.
8. *Lighthouse flashes*: Lighthouses flash in intervals particular to each house, for the purposes of identification.
9. *Report . . . question*: unclear to what report Tyndall is referring.
10. *Board of Trade*: department of the British government concerned with trade and commerce.

From Jane Barnard [April 1871][1] 3418

Barnsbury Villa, | 320 Liverpool Road, | N.

My dear Dr Tyndall

I feel that it is not for me to thank you for the trouble you have taken, & the kindness you have shown to Mr. Petitjean;[2] but I must say what ever the result may be, that it has given me much pleasure.

I do not know whether it was from shame that I should have <*1 word excised*>, I hope not, but at any rate I am glad I was not at the R[oyal]. S[ociety]. on that occasion.[3]

Ever yours | very sincerely | Jane Barnard

Saturday Evening[4] | Dr Tyndall | &c &c &c

RI MS JT/1/B/78

1. *[April 1871]*: LT thought this would be before 19 April 1871 (see letters 3408 and 3409), but it is not clear when in April this would be.
2. *Mr. Petitjean*: see letters 3408, 3409, 3413 and 3414.
3. *on that occasion*: unclear to what she is referring.
4. *Saturday Evening*: possibly April 14.

From Mary Egerton [April 1871][1] 3419

Mountfield Court, | Robertsbridge, | Hawkhurst. | Sunday

Dear Mr Tyndall,

The book[2] has come & 1,000 0000 00000000 . . . thanks for it. I shall hold it as a pledge that henceforth, whatever I may say, you will never forget that first Canon of Criticism,—not to interpret words so as to be inconsistent with the context—That is, in the present case, with those feelings which you know well enough I have about you.

I recognize 11 out of the 13 longer articles as old friends; of the short ones I think I only knew the "Spirits"[3] one before, at least in their present form. Good you sent me those proof sheets you once talked of, I sh[oul]d have asked you, before putting in your most true & just observations on the so called miracle of fashion, to read the passage in Dean Stanley's[4] Sentences on the Jewish Church[5] about it (Vol: 1st p. 241 &c. and notes). I sh[oul]d think most really "instructed" Divines must now take his view, that the whole account is not meant for a historical record at all, but is a book of Song of Triumph, taken (as it says) from "the Book of father", which, from the only other reference we have to it, (in 2 Sam: 1ch 18v) seems to have been a collection of "Songs of the Gentiles". Kepler's remarks[6] on the subject are very striking for that age; the Dean quotes them at page 248. There—it is Sunday so you must put up with a little bit of Biblical Criticism!

I know you are an adept at the manufacture of "Puff Pastry" have you ever tried your hand at the simpler atchievement of making arrow-root? & produced only a gritty conglomeration? I have! and it is very wicked. I am irresistibly reminded of the process by the doctrine of the Evolution of life in the universe! We can't produce it because we can't stir hard enough & long enough! I am not going to waste time spoken in saying "I know you meant

<1 or more pages missing>

RI MS JT/1/E/89

---

1. *[April 1871]*: dated to April because of the discussion of *Fragments of Science*.
2. *The book*: *Fragments of Science*.
3. *"Spirits"*: 'Science and Spirits', in *Fragments of Science*, pp. 427–35.
4. *Dean Stanley*: Arthur Penrhyn Stanley (1815–81), Dean of Westminster and a leading liberal theologian at the time. He was elected FRS in 1863.
5. *Sentences on the Jewish Church*: A. P. Stanley, *Lectures on the History of the Jewish Church*, 3 vols (London: John Murray, 1863–70), vol. 1, *Abraham to Samuel* (1863).
6. *Kepler's remarks*: Johannes Kepler (1571–1630), a German astronomer, astrologer and

mathematician. He is well-known for his three laws of planetary motion that helped to confirm and spread the new Copernican astronomy. The remarks Egerton speaks of may have to do with the theories he put forth in his *Mysterium Cosmographicum* (1596) in which he attempted to use biblical passages to support heliocentrism. See P. Barker and B. R. Goldstein, 'Theological Foundations of Kepler's Astronomy', *Osiris*, 16 (2001), pp. 88–113.

From Auguste de la Rive 2 May 1871 3420

Cheltenham le 2 Mai 1871.

Très cher ami,

C'est à Cheltenham où je suis en visite auprès de ma fille, Madame Prevost, que je reçois votre bonne lettre du 25 avril. Je viens vous en remercier cordialement et vous dire que je serai demain soir à Londres où je ne puis rester que quelques jours.

J'irai vous chercher à la Royal Institution jeudi matin et je suis bien impatient de vous serrer la main. J'aurais un bien vif désir de voir vos belles expériences sur la décomposition des vapeurs par la lumière. Cela est-il possible? Je serais dans ce but à votre disposition le jour et le moment que vous voudrez. Vous me direz cela jeudi matin.

J'ai reçu au moment où je quittais Genève, votre beau volume renfermant la collection de vos articles & de vos discours de science populaire. Je vous suis bien reconnaissant de ce beau cadeau. Je n'ai, comme vous le comprenez, pas eu encore le temps d'y jeter les yeux; mais je l'ai prêté en partant à Casimir de Candolle à qui cela a fait grand plaisir. Il vous envoie ses compliments les plus affectueux. Merci mille fois de tous les détails intéressants que vous me donnez; il y a bien des points sur lesquels je suis impatient de converser avec vous.

Agréez, très cher ami, l'assurance de mes sentiments les plus affectueux.

Votre tout dévoué | Aug. de la Rive.

Tous vos amis de Genève me chargent de mille choses pour vous.

Cheltenham, 2 May 1871.

Very dear friend,

It is in Cheltenham where I am visiting my daughter, Madame Prevost,[1] that I received your kind letter of 25 April.[2] I am sincerely thanking you and letting you know that I will be in London tomorrow evening, where I will only be able to stay for a few days.

I will look for you at the Royal Institution on Thursday morning, and I am eagerly looking forward to shaking your hand. I have a very strong desire to see your beautiful experiments on the decomposition of vapours by light.[3] Is that possible? I would be at your disposal for this at any day or time that you would like. Tell me about it on Thursday morning.

When I left Geneva, I received your marvellous volume that includes the collection of your articles & your lectures for popular science.[4] I am very grateful to you for this beautiful gift. I have not had, as you may understand, the time to glance at it yet, but before leaving I lent it to Casimir de Candolle,[5] who greatly enjoyed it. He sends you his most affectionate compliments. Thank you so very much for all of the interesting details that you give me; there are many things about which I am looking forward to chatting with you.

Please accept, very dear friend, the assurance of my most affectionate sentiments.

Sincerely yours | Aug. de la Rive.

All of your friends from Geneva ask me for a thousand things for you.

RI MS JT/1/TYP/1/381
LT Typescript Only

1. *Madame Prevost*: Adélaide-Eugénie-Augusta Prevost (née de la Rive, 1838–1924).
2. *letter of 25 April*: see letter 3416.
3. *experiments . . . by light*: likely referring to Tyndall's work associated with 'On the Action of Rays of High Refrangibility upon Gaseous Matter' which he read to the RS on January 27, 1870. It was published in the *Phil. Trans.*, 160 (1870), pp. 333–65.
4. *marvelous volume . . . popular science*: *Fragments of Science*; see also letter 3416.
5. *Casimir de Candolle*: Anne Casimir Pyrame de Candolle (1836–1918), a Swiss botanist, who also had training in chemistry, physics and mathematics. He was well-known in the fields of plant systematics and plant physiology.

To unidentified 6 May [1871][1] 3421

6th. May.

My very kind and cordial Friend.

On the Friday a gentleman lectures here[2] whom it is my duty to support.

And on the Thursday I am bound to the dinner of a club[3] which I have shamefully neglected.

Had you said Monday the 22nd. or Tuesday the 23rd. I should have replied by a hearty 'Yes'!

Ever yours faithfully | John Tyndall

SGB/1/79. RGS-IBG

1. *[1871]*: 22 May fell on a Monday in 1871 (as well, of course, as a number of other years). We have assigned this letter to 1871 because Huxley delivered the Friday Evening Discourse at the RI on 19 May (see n. 2); we think it unlikely that Tyndall would have felt obliged to support any of the speakers on 19 May in the other years.
2. *a gentleman lectures here*: We believe that this was Huxley, who delivered a Friday Evening Discourse at the RI on 19 May 1871, 'On Bishop Berkeley and the Metaphysics of Sensation' (*Roy. Inst. Proc.*, 6 (1870–72), pp. 341–54).
3. *a club*: possibly the X Club, which met the first Thursday of each month.

From William Valentin 6 May 1871 3422

Enclosure 3, in No. 34.[1]
Royal College of Chemistry, London, | 6 May 1871.

Sir,

I HAVE the honour of submitting to you, for the information of the Elder Brethren of the Trinity House, the results I obtained in testing photometrically[2] the light produced by the first order lighthouse oil lamp,[3] as constructed for and in use in the lighthouses controlled by the Board at the Trinity House, and a new first order paraffin lamp, constructed by the engineer of the Trinity House, Mr. J. N. Douglass; also the photometrical results obtained with a new fourth order petroleum lamp, as compared with the light produced by an ordinary fourth order oil lamp.

I examined the lamps on two days, viz. on Saturday, 22nd and 29th April. I had placed at my disposal the very best photometrical apparatus, and every facility was afforded to me by the engineer of testing the lights thoroughly.

In accordance with the wish expressed by you, I directed my attention mainly to a comparison of the new lighthouse lamps with the lamps now in use, with a view of giving an opinion on their applicability and safety, and the comparative cost of the two lights, in order to enable the Elder Brethren to judge the more independently of the interesting series of experiments previously carried on by Mr. Douglass.[4] In addition, I also inquired into the chemical composition of the oils, viz. colza oil,[5] Young's paraffin,[6] and petroleum oil employed in the experiments; and, lastly I made a somewhat extended examination of a number of samples of petroleum oil, obtained by Mr. Douglass at my desire, from various distillers and petroleum merchants, with a view of ascertaining whether the new first order lamp would burn with equal facility mineral oils[7] of different composition so as to find out the oil best suited for the lamp as now constructed.

The colza oil employed in my experiments had a specific gravity of ·915 at 60 deg. F. The paraffin oil employed for burning in the new first order lighthouse lamp was Young's best oil, specific gravity ·815 at 60 deg. F., flashing point 132 deg. F., which distilled nearly entirely between 240 deg. And 572 deg. F., viz.:

30·2 per cent., distilling between	240 deg. and 380 deg. F.	
61·4 „ „ „	380 deg. and 572 deg. F.	
8·2 „ „	above 572 deg. F.	

The specific gravity of the three portions of oil was as follows:—

1st portion (240 deg. to 380 deg. F.) – – – – – –	·777
2nd portion (380 deg. to 572 deg. F.) – – – – – –	·824
3rd portion above 572 deg. F. – – – – – – – – –	·850

This clearly shows that the oil consists of a mixture of liquid hydrocarbons, differing comparatively very widely in density.

The oil which was burnt in the new fourth order light submitted to me for examination was a somewhat dark-coloured petroleum oil of specific gravity; ·806 at 60 F., flashing point 107 deg. F.

When submitted to distillation I found that it commenced to boil at 284° F., and that it distilled almost entirely between 284 and 572° F., viz.:—

42·6 per cent. distilling between	284 deg. and 380 deg. F.
51·2 „ „ „	380 deg. and 572 deg. F.
6·2 „ „	above 572 deg. F.

The last portion of the distillate solidified on cooling, owing to the presence of paraffin.

The specific gravity of the two first portions of the oil was as follows:—

1st portion (from 284 deg. to 380 deg. F.)– – – –	·770
2nd " (from 380 deg. " 572 deg. F.)– – – –	·821

The flashing point of the first portion of the distillate was below 80°F.

This shows, then, that the sample of petroleum, like most oils of the same class, is a mixture of a certain proportion of light oil with some heavier oil; and what is in favour of this special sample, of a small proportion also of solid paraffin, with the exception of the low flashing point, the oil is everything that could be desired.

I abstain from describing the mechanical construction of the lamps.

In taking the photometrical observations I left the lamps in charge of the attendants, who, however, placed themselves entirely at my orders, and obeyed all my directions. The results were checked by repeated trials, by testing the lamps against one and two sperm candles,[8] consuming a trifle less than 120 grains, viz., 118 and 117 grains per hour. The illuminating power is expressed in sperm candles, consuming 120 grains of sperm per hour; and in the summaries of the results which are subjoined I have restricted myself to the simplest figures, abstaining as much as possible from expressing one and the same thing by different terms and figures.

I saw the lamps filled and emptied, and I controlled the measuring of the oils. All the photometrical observations were taken by myself, each figure representing the average of at least 10 observations. On the first day's trial I allowed the lights to be kept burning with the largest flame compatible with the non-production of smoke on the chimney-glasses, in harmony with the trials made in Scotland. On the second, however, I had the lights reduced to a flame of about 3 to 3½ inches in height, so as to avoid the formation of a streaky flame as much as possible. Judging from what I saw on the first day's trial, I certainly think that oil, both colza and paraffin, passed unconsumed off through the high chimney shaft, as was shown by distinct clouds of smoke issuing from the top of the chimney. On the second day the burning of the lamps was so regulated that at no time smoke became ever perceptible.

I subjoin in a tabular form the detailed results of the different trials with the paraffin and colza oil lamps, commencing with the first order or four-wick lamps.

Comparative Illuminating Power produced by first Order or 4-wick Lamps, on | Saturday, 22nd April

————	First Order 4-wick Trinity Lighthouse Oil Lamp.	4-wick Paraffin New First Order Lamp.
Illuminating Power:		
1st experiment————	284·85 candles	292·93 candles.
2nd „————	278·53 „	261·26 „
3rd . „ ————	236·24 „	225·00 „
(after lowering lights so as to avoid smoking).		
Average—	266·54 „	259·63 „
Consumption of oil per hour———	11,565 grains	12,677 grains.
(The experiment extended over 4½ hours).	=·1805 gall.	=·2222 gall.
Cost of light per hour————Cost of one	7·58 *d.*	4 *d.*
sperm candle-light (unit) per hour–	·0084 *d.*	·0154 *d.*

Comparative Illuminating Power produced by first Order, or 4-wick Lamps, on | Saturday 29th April.

————	First Order 4-wick Trinity Lighthouse Oil Lamp.	New First Order 4-wick Paraffin Lamp.
Illuminating Power:	246·87 candles	248·51 candles.
1st experiment————	234·08 „	229·60 „
(the lamp was trimmed so as to avoid smoking).		
2nd experiment————		
Average—-	240·48 „	239·05 „
Consumption of oil per hour———	13,610 grains	12,961 grains.
(The experiment extended over five hours).	= ·2125 gall.	=·225 gall.
Cost of light per hour———	8·925 *d.*	4·05 *d.*
Cost of one sperm candle-light (unit) per hour–	·0371 *d.*	·0169 *d.*

It is evident from these experiments that in the first order, or 4-wick lamp, both colza and paraffin oil evolve *a light of practically identical intensity*, the consumption of paraffin oil being somewhat in excess over that of colza oil, viz. in the proportion of 100 : 114.

The *Average* Illuminating Power, indicated by the Five Experimental Trials, is:—

————	Oil Lamp.	Paraffin Lamp.
Illumination power————	256·11 candles	251·46 candles.
Average cost of light per hour———	8·25 *d.*	4·025 *d.*
Average cost of one unit of light (one sperm candle) per hour————	·03275 *d.*	·01615 *d.*

From these figures it follows that the comparative cost of colza and paraffin is as 1,000 : 488, and when corrected for illuminating power (in the proportion of 256·11: 251·46 candles) as 1,000 : 497; or, light for light, the paraffin costs only about half as much as the oil light.

I had also submitted to me a new fourth order lighthouse lamp, constructed for the burning of petroleum and I compared its illuminating power, &c. with that of the ordinary fourth order colza oil lamp. The following are the comparative results obtained in this trial.

————	Fourth Order Single-Wick Light-house Oil Lamp.	New Fourth Order Single-wick Petroleum Lamp.
Illuminating Power: Consumption of oil per hour—— (The lamp was kept burning during three hours and 25 minutes). Cost of light per hour——— Cost of one sperm candle-light (unit) per hour—————	11·63 candles 935 grains = ·0146 gall. ·6132 *d.* ·05272 *d.*	18·49 candles. 753 grains. =·01335 gall. ·2402 *d.* ·01299 *d.*

From these figures it follows that the smaller light, the fourth order lighthouse lamp, burns colza oil *less economically* (viz., in the proportion of ·1965 gallons, divided by 256·11 candles, to ·0146 gallons divided by 11·63 candles), consuming, in fact, ·000767 gallons of colza oil per unit of light (1 sperm candle) compared with ·00126 gallons, consumed in the small burner, for the same unit of light, during one hour's combustion, or, expressed in round numbers, in the proportion of 100 per large burner to 164 per small burner. This, I believe, is in accordance with the general experience of lighthouse engineers.

The comparative expense of colza and petroleum oil is as 1,000 : 392, and when corrected for illuminating power (in the proportion of 18·49 candles to 11·63 candles) as 1,000 : 245, or, light for light, the fourth order petroleum light costs only about one-fourth what the fourth order colza light costs.

Without venturing upon barren speculations as to a ratio of increase or decrease in the illuminating power, according as large or small quantities of paraffin or petroleum are consumed in lamps, lighthouse and other lamps, when so little is known with certainty, I cannot pass on without remarking that everything depends upon the proper construction of the lamp itself, upon the chimney, and upon the mode of supplying air to the light. These conditions require to be carefully studied, and when once known, it will probably be found that the illuminating power is proportional to the quantity of paraffin or petroleum oil consumed.

All that can be inferred with safety is, *that the petroleum lights are as yet far from perfect*, for, instead of yielding the light of 315 candles, the proportional increase which one would expect by glancing at the respective consumption of oil, the large first order paraffin lighthouse lamp which I tested yielded on an average only the light of 251·46 sperm candles, or, contrary to what holds good for colza oil lights, *small burners consume the petroleum more advantageously*, as yet, *than large lamps*. There is, in fact, room for considerable improvements in these large lamps as now constructed for the burning of mineral oils, improvements which can only be made by patient attention to mere mechanical details of construction.

This brings me to a most important point, viz., *the suitability of mineral oils, differing in specific gravity and in their chemical composition, for one and the same lamp*.

My experience with petroleum lights has taught me that a lamp may be constructed for burning *one* oil and may be found perfectly incapable of burning *other* oils, differing in composition, although perhaps of the *same* specific gravity. I had only a short time at my disposal on the second day of the trials, to put this to the practical test. Mr. Douglass supplied me, at my desire, with various samples of petroleum oil. After examining these in the laboratory, I fixed upon a sample which had almost the same specific gravity as the oil employed in these and the Scotch trials; upon the sample of petroleum marked No. 4A., of specific gravity ·816, flashing point 146 deg. F. When I submitted this oil to distillation I had found a considerable difference, compared with Young's paraffin oil, of specific gravity ·815. It commenced, in fact, to boil at 320 deg.; about three-fourths distilled below 572 deg. F., and had the specific gravity ·798. This portion of the oil flashed at 141 deg. F. The last fourth consisted of the heaviest hydrocarbons, among which chiefly paraffin, which were absent or nearly so in Young's paraffin oil, and from what I had seen in other experiments with petroleum lights, I knew that this oil would give a considerably higher illuminating power and a light of greater intensity, if it could but be burnt in the new first order lighthouse lamp. I found, however, that the oil, although giving a splendid light, at first charred the wick rapidly, the illuminating power receding almost visibly, and it was evident that the light would in a short time have extinguished itself altogether. This is likewise in harmony with what I observed before in the case of smaller lamps. No two petroleum oils can be burnt in one and the same lamp, unless their chemical constitution and not merely their specific gravities be alike.

With a view of instituting further experiments on this all-important point, which apparently has been overlooked by the different experimenters, I examined somewhat more fully the several samples of petroleum oil which Mr. Douglass had submitted to me. The following is a list of the samples of petroleum with their respective prices:

Number.	Price.	Number.	Price.
	s. d.		*s. d.*
Number 1——	1 6½ per gallon.	Number 4A.——	1 6 per gallon.
2——	1 8 "	5——	1 6 "
3——	1 4½ "	5A.——	2— "
4——	1 7 "	6——	1 5 and 2½ % discount.

I subjoin the results of the chemical examination in a tabular form:

TABLE showing Specific Gravity, Flashing Point, and Approximate Chemical Composition of Paraffin and Petroleum Oils.

NAME OF OIL.	Specific Gravity of Oil at 60° F.	Flashing Point. *Deg. F*	Temperature at which the Oil began to Boil. *Deg. F*	Per-Centage of Oil Distilling below 380° F. I. Distillate.	Per-Centage of Oil Distilling between 380° F and 572° F. II. Distillate.	Per-Centage of Oil Distilling above 572 F. III. Distillate	Specific Gravity of 1st Distillate	Flashing Point of 1st Distillate *Deg. F.*	Specific Gravity of 2nd Distillate	Specific Gravity of 3rd Distillate	Remarks
Young's Parrafin Oil·	·815	132	248	30·2	61·4	8·2	·777	. . .	·824	·850	
Petroleum Oils, No. 1·	·801	95	230	33·7	42·6	3·6	·767	below 80	·828	Semi-Solid	
" " 2·	·805	102	206	41·6	45·8	2·5	·766	below 80	·820	·855	
" " 3·	·795	95	284	46·2	44·0	9·8	·759	below 80	·813	Semi-Solid	
" " 4·	·806	107	284	42·6	51·2	6·2	·770	below 80	·821	Semi-Solid	
" " 4*a*	·816	146	320	..	76·8 {	14·0 | 9·2 }·	. .	. .	·793	Semi-Solid	
" " 5·	·796	90	150	36·8	37·2	6·0	·766	below 80	·823	Semi-Solid	The flashing point if the 2nd distillate was 141° F.
" " 5*a*	·783	111	248	80·4	18·8	·8	·771	101	·506	A clear fluid	
" " 6·	·794	95	240	43·2	48·0	8·8	·756	below 80	·811	Semi-Solid	Remarkably well purified; no blackening on distilling.

Several of these samples of petroleum must evidently be rejected at once, on the ground of low flashing points, viz., No. 1, 2, 3, 5, and 6. Two samples only being available in my opinion for use in lighthouse lamps, where perfect safety must always remain the principal condition, viz. No. 4 *a.* and 5 *a.* Now sample 4 *a.* rapidly chars the wick and cannot be burnt in the lamp, as it is now constructed, 5 *a.* promises well, and should be tried in the new lamp, but is expensive, its price being 2 *s.* per gallon.

I would submit, however, to the engineer of the Board, whether the first order lamp could not be altered in such a manner as to make it possible to burn in it a mineral oil containing a larger per-centage of the heavier hydrocarbons, such for instance as petroleum 4 *a*. It will no doubt be always a serious drawback to the adoption of lamps constructed for mineral oils, that only one oil can be properly burnt in them. For this and other reasons, already referred to, I am of the opinion *that in the absence of any superiority of the new first order lighthouse lamp consuming paraffin instead of colza oil, the new lamps should be used at first in a limited number of lighthouses only* till further experience and additional improvements may justify its general adoption on the ground of greater cheapness, and that great attention should be paid to secure a constant supply of an oil having the same chemical composition. I would strongly recommend, however, that mineral oils should at once be employed for the smaller lights, such as fourth-order lights, &c., where a marked superiority of illumination power, and a considerable saving in the consumption of oil, has already been established most satisfactorily.

I have, &c. | (signed) *William Valentin.*

Professor J. Tyndall, F.R.S., | &c. &c.

'Correspondence between Lighthouse Authorities and Board of Trade, relative to Proposals to substitute Mineral Oils for Colza Oil in Lighthouses', HC 318, Parliamentary Papers (1871), pp. 40–4
Typed Transcript Only

---

1. *Enclosure 3, in no. 34*: Valentin's letter and letter 3423 are presented in the Parliamentary Papers as enclosures in a letter from Trinity House to the Board of Trade. See Trinity House to the Board of Trade, 16 June 1871, 'Correspondence between Lighthouse Authorities and Board of Trade, relative to Proposals to substitute Mineral Oils for Colza Oil in Lighthouses', HC 318, Parliamentary Papers (1871), pp. 33–5.
2. *testing photometrically*: testing the intensity of the light given off.
3. *first order lighthouse oil lamp*: Lighthouse lenses came in various sizes, called orders, with the first order being the largest, up to twelve feet in height and six feet in diameter.
4. *experiments . . . Douglass*: explained in some detail in *Ascent of John Tyndall*, pp. 284. These were the product of the long-standing conflict about fuel for Irish lighthouses (see letter 3400, n. 3).
5. *colza oil*: a non-drying oil obtained from rapeseed.
6. *Young's paraffin*: an oil that was thin and light, which was useful as a lamp oil. Paraffin, a byproduct of petroleum refining, was invented by James Young.
7. *mineral oils*: oils derived from petroleum.
8. *sperm candles*: candles made from spermaceti, a wax taken from the spermaceti organ in the head of a sperm whale or bottlenose whale.

To Robin Allen 13 May 1871 3423

Enclosure 2, in No. 34.[1]

Sir,

I HAVE much pleasure in forwarding to you, for the information of the Elder Brethren, a very solid and exhaustive Report[2] from Mr. Valentin regarding the combustion of mineral oils in lighthouse lamps.

The principal points in the Report may be here indicated.

It is important to note that Mr. Valentin shows that the specific gravity of an oil is no certain guide as to its applicability to lighthouse illumination. Each of these mineral oils usually consists of a mixture of several different oils possessing different flashing points.

Comparative experiments were made with two Trinity four-wick lamps, the one burning colza oil, the other Young's paraffin oil. The illuminating powers of these two lamps were found practically equal, but, taking cost and consumption into account, the result was:—

I. *That, light for light, in the first order lamp, the cost of the paraffin light is about one- half that of the colza light.*

Two fourth-order lights were then subjected to examination. It was found that while the colza oil burnt less economically in the small lamp, the paraffin oil gained in economy. The result is:—

II. *That for a fourth order light the cost of the paraffin light is about one-fourth of that of the colza.*

In descending from the first to the fourth order, the illuminating power of the paraffin is considerably augmented; this is the same as saying that in passing from the fourth order to the first we do not gain light in proportion to the augmented quantity of oil. This absence of proportionality Mr. Valentin ascribes to the fact that the large burner has not yet reached its utmost attainable perfection. "There is," he says, "room for considerable improvements in the large lamps now constructed for the burning of mineral oils, improvements which can only be made by patient attention to mere mechanical details of construction."

I may say that it gave me pleasure to notice how small an alteration of the ordinary Trinity four-wick burner rendered it available for the combustion of the paraffin oil.

I have also to direct your attention to the limited range of power possessed by the lamps hitherto constructed as regards the combustion of different oils. An apparently slight change in the character of an oil suffices to introduce charring of the wick. Mr. Valentin's remarks on this head are contained on page 43; they render obvious the necessity of testing the oils in relation to the burners employed, before they are accepted by the Elder Brethren.

Mr. Valentin has also examined various samples of petroleum, and given an instructive tabular statement of his results (see page 43).

His concluding recommendation is:—

"That in the absence of any superiority of the new first order lighthouse lamp consuming paraffin instead of colza oil, the new lamps should be used, at first in a limited number of lighthouses only, till further experience and additional improvements may justify their general adoption on the ground of greater cheapness, and that great attention should be paid to secure a constant supply of oil having the same chemical composition. I would strongly recommend, however, that mineral oils should at once be employed in the smaller lights, where a marked superiority of illuminating power, and a considerable saving in the consumption of oil, have been already established most satisfactorily."

In this recommendation I concur, expressing the hope that the skill which has already turned the ordinary four-wick burner to such valuable account may be able greatly to perfect that burner, so as to bring the light which it emits more nearly into proportion with the quantity of oil which it consumes.

I have, &c. | (signed) *John Tyndall.*

Robin Allen, Esq., | &c. &c. &c. | Secretary to the Trinity House.

'Correspondence between Lighthouse Authorities and Board of Trade, relative to Proposals to substitute Mineral Oils for Colza Oil in Lighthouses', HC 318, Parliamentary Papers (1871), pp. 39–41

Typed Transcript Only

1. *Enclosure 2, in No. 34*: see letter 3422, n. 1.
2. *solid and exhaustive Report*: see letter 3422.

To Thomas Henry Huxley 17 May [1871][1] 3424

Royal Institution. | 17th. May.

My dear Huxley,

I had set my heart—and stomach—on having you to luncheon with me on Friday: but I must go to Herschel's funeral—need I say that a well-filled cellar—and a skillful housekeeper are at your disposal to prepare any and every mortal thing of which you may stand in need.

Yours ever | J. Tyndall.

IC HP 8.88

Typed Transcript Only

1. *[1871]*: John Herschel died on 11 May 1871; his national funeral at Westminster Abbey was Friday, 19 May.

To the Editor of the *Daily News*[1] [19 May 1871][2] 3425

To the editor of the Daily News.

Sir,

It was my sad privilege this day to stand in Westminster Abbey beside the grave of Sir John Herschel; and as I quitted it a thought which has occurred to me from time to time during the past week revived. Namely this: to ask your permission to say a few words regarding a sketch of Herschel's work and character which has recently appeared in your columns.[3] I pass over minor errors as to matters of fact, and of the graver moral questions raised in this remarkable article I shall confine myself to one. Your correspondent permits himself to write thus of Sir John Herschel. "This habit of flattery was his great fault. It may be doubted whether it is ever found apart from vanity or vicious craft. The latter of course bears no relation to Herschel; but it is impossible to suppose that a man who was always saying such smooth things, and abasing himself so unnecessarily, was not extremely sensitive to his neighbour's opinion of him. It was the great blemish of his character. It affected his honesty in his vocation (spontaneously adopted) of critic in the great reviews, and his manners as a gentleman."

I think it was in 1854, and in presence of a Friday evening audience at the Royal Institution, that Faraday introduced me to Sir John Herschel. From that hour to this, through the advancing years, his character has grown in beauty to me. As I knew him better respect ripened into reverence, and until I read the words of your correspondent, this feeling never encountered from the expressed opinions of others the slightest shock. During the past week I have tried to check and extend my data by reference to older men. I have conversed with many whose intimacy with Sir John Herschel extends far beyond the range of mine; and if their unanimous and indignant testimony be worth anything, I should hesitate to write the term which would most fitly describe the quoted words of your contributor. He may perhaps be able to make his position good; he may even have the courage to give his name; but as it now stands I must regard his article, notwithstanding its apparent warmth of appreciation, as embodying the most conspicuous personal wrong to which anonymous writing has of late years given birth.

I am &c. | John Tyndall.

RI MS JT/1/TYP/2/576
LT Typescript Only

1. *Editor of the* Daily News: Frank Harrison Hill (1830–1910). Hill was a journalist educated at London University, earning his BA with first class honors in 1851. He was supportive of Liberal politicians, and editor of the *Daily News* from 1869–86.
2. *[19 May 1871]*: Herschel's funeral, which Tyndall mentions in the first line of the letter, took place this day. See letter 3424 and Herschel's obituary in *Proceedings of the American Philosophical Society*, 12:86 (1871), pp. 217–23 (its author, Henry Field, mistakenly referred to 19 May as a Tuesday on p. 222).
3. *Herschel's work . . . your columns*: The original piece was printed in the *Daily News*, 13 May 1871, p. 5. It was also reprinted in *The Engineer*, 31 (19 May 1871) p. 345.

From Mary Egerton 20 [May 1871][1] 3426

45 Eaton Place, | S.W. | Sa[tur]d[a]y 20th

Dear Mr. Tyndall,

Your 2nd note[2] came just in time to prevent my sending one full of elaborate combinations for Monday (when we are going to the Geo[lo]g[ica]l Soc[iet]y meeting). If we do not go out of town, we might still perhaps effect a meeting on the 29th; or will Richmond Park[3] be a Bear Garden;[4] being Whit-Monday?[5]

If that lecture of Mr. Huxley's[6] is printed I sh[oul]d so much like to have it; it was most deeply interesting to me, but one wanted more. I felt bound to go & hear "Pantheism", being somewhat of a Pantheist myself.[7] Some parts I thought very fine & good—were you there? But argument felt cold after the scene of that morning![8] May I confess that I longed to read your thoughts as you looked into that grave![9] Did it not seem as if that Chapter of Corinthians[10] must be true? Now don't say as you do sometimes, that I "want to convert you". How could I think to influence a soul that I feel convinced has far more real sympathy with the Spiritual World than my own, if it would only allow it!

Y[ou]rs ever most truly | MF Egerton

RI MS JT/1/E/97

1. *20 [May 1871]*: dated Saturday 20 May. The letter refers to Monday, the 29th, as Whit-Monday, a holiday that was celebrated in England the day after Pentecost. According to historical calendars, this would fall in the year 1871.
2. *2nd note*: letter missing.

3. *Richmond Park*: a large park in southwest London, set aside as a deer park in the early seventeenth century. It remains the largest of the Royal Parks at around 2500 acres.
4. *Bear Garden*: 'a place or scene of strife and tumult' (*OED*). This term is used to refer to how busy or crowded Richmond Park would be on Whit-Monday.
5. *Whit-Monday*: see n. 1.
6. *lecture of Mr. Huxley's*: possibly referring to Huxley's Presidential Address to the BAAS in Liverpool (*Brit. Assoc. Rep. 1870*, pp. lxxiii–lxxxix). However, with Egerton, one cannot guess. Huxley's address was published separately as *Address to the British Association for the Advancement of Science, delivered by the President, Thomas H. Huxley, at Liverpool, September 14, 1870* (London: Taylor and Francis, 1870).
7. *"Pantheism," … myself*: Rev. James Harrison Rigg (1821–1909) delivered a lecture on Pantheism in London on 19 May 1871. See J. H. Rigg, *Pantheism. A Lecture, Delivered in Connection with The Christian Evidence Society, May 19, 1871* (London: Hodder & Stoughton, 1871). Rigg was a Methodist minister and educationist. In 1868 he was appointed the principal of the Westminster (Wesleyan Training) College and stayed there until 1903. He was a member of the first London school board from 1870–6, serving alongside Huxley. He fought for a broad educational curriculum for both boys and girls and for elementary education.
8. *scene of that morning!*: likely referring to the funeral of John Herschel, which was on Friday, 19 May. See letter 3424.
9. *that grave!*: John Herschel's grave at Westminster Abbey; see n. 8 and letters 3424 and 3425.
10. *Chapter of Corinthians*: This may refer to 1 Corinthians 15, which speaks about the resurrection of Christ, the resurrection of the dead and the resurrection of the body itself.

To John Lubbock 21 May 1871 3427

Royal Institution of Great Britain | 21st. May <u>1871.</u>

Dear Lubbock.

If I understand you aright in our half conversation on Friday the Royal Institution was mentioned by your friend. With reference to that Institution the following definite, permanent, and most important piece of service might be done.

A short time before his death the late Duke of Northumberland,[1] who had been so long president of the Institution, had requested to have estimates of the cost of a new laboratory placed in his hands, and had he lived, he, no doubt, would have erected such. At the present moment designs[2] are in my hands for new laboratories, which it is proposed to build out of the funds of the Institution. Now no laboratories in the world can point to greater work accomplished within them than those of the Royal Institution. The antecedents of the place are grand, and as far as I can see, the Royal Institution, meeting as it does a widespread want, and growing in strength annually, is likely to hold its own as regards scientific work for an indefinite period.

Here then is a rounded piece of work with which a benefactor's name might be honourably and permanently associated. Quite apart from my own connexion with the Royal Institution, I would say that I do not know a scientific purpose to which three, or four, or five thousand pounds could be more usefully applied than to the erection of these laboratories.[3] I think them likely to demonstrate to future times, not only the liberality but the wisdom of their founder.

I am not aware whether your friend[4] contemplates immediate action, but we must have our laboratories soon. It is purposed to build them next year.[5] Perhaps the obvious benefit which the act would confer upon a private institution, which as regards two great departments of modern science has enabled England to keep abreast of, if not above, the most favoured Continental nations, may induce your friend to render the bestowal of his benefit a thing of the present instead of a thing of the future.

Sir Henry Holland, to whom I read this note entirely concurs in its purport, D^r^. Bence Jones has views and aims regarding the future of the Institution which he will probably communicate to you.

Yours faithfully, J. Tyndall

BL Add MS 49643, ff. 152–3a

1. *Duke of Northumberland*: Algernon Percy, Fourth Duke of Northumberland (1792–1865), an English landowner, philanthropist, explorer and politician. He was one of the early aristocratic Englishmen to go to Egypt, where he collected over 2000 objects and began his decades long support of Edward William Lane. He partly paid for and traveled with John Herschel to Cape Town in 1834 to observe the Southern constellations. He was elected FRS (in 1818) and FRSE, and was a member of the Royal Geographical Society. He was President of the RI from 1842 to 1865.
2. *designs*: The finished laboratories were designed by architect Hamilton Edward Harwood (1824–72), and built by George Smith (n.d.).
3. *these laboratories*: successfully built in 1872. The RI had been slowly doing some renovations for years, but at this point they were able to build 'a two story laboratory . . . on the east side of the building on the site of the old basement chemical laboratory and the room housing the mineralogical collection on the ground floor'. The new labs were for chemistry in the basement and physics on the ground floor. These were the last major improvements to the building for twenty-five years (F. A. J. L. James and A. Peers, 'Constructing Space for Science at the Royal Institution of Great Britain', *Physics in Perspective*, 9 (2007), pp. 130–85, on pp. 162–3).

4. *your friend*: possibly Alfred Davis. Davis, upon his death in 1870, gave £2000 to the RI 'for the Promotion of Philosophical Researches'. It is possible they used this money for the project ('General Monthly Meeting, Monday, 4 July 1870', *Roy. Inst. Proc.*, 6 (1870–72), pp. 183–4).
5. *build them next year:* built in 1872. Bence Jones kept Tyndall apprised of the progress while he was in America on a lecture tour. See letters dated 2 July 1872, 16 July 1872 and 3 December 1872 in the forthcoming thirteenth volume of *The Correspondence of John Tyndall.*

To William Grove[1] 22 May [1871][2] 3428

Royal Institution of Great Britain | 22nd May

My dear Grove

Could I at all spare the time nothing would give me greater pleasure than to respond to your request.[3] But my engagements here up to the middle of June, and among the Irish Lighthouses afterward[4] which put the work you propose beyond my range.

I am truly gratified to learn that you like the "Fragments", and still more gratified to find that you have discovered in me the qualities which would prompt you to call me 'friend'.

Yours ever faithfully | John Tyndall

RI MS DPM/1/G

---

1. *William R. Grove*: William R. Grove (1811–96), a Welsh judge trained in Classics at Oxford, earned his BA in 1832. He was called to the bar in 1835 and in that same year joined the RI. In 1839 he reported on his new battery, which used nitric acid. In 1840, he was elected FRS. In 1844 he became a Vice-President of the RI, but resigned his post at the London Institution because of his legal commitments. In 1847 he won a Royal Medal from the RS for his work on the battery, and continued to be involved in scientific pursuits. In 1872 he testified in front of the Devonshire Commission in support of scientific instruction. In 1880 he was appointed to the Queen's Bench, a post he retired from in 1887. He died in 1896 at his home.
2. *[1871]*: dated to the first edition of *Fragments of Science*, published in 1871.
3. *your request*: letter missing.
4. *Irish Lighthouses afterward*: see letter 3400, n. 3.

From Charles Lyell 22 May 1871 3429

May 22d 1871

Dear Tyndall

I must thank you heartily for your manly protest in the Daily News[1] against that shameful & unfeeling attack upon the character of Herschel whom I had known well from before the time of his & my marriage[2]

It was quite as much the faint praise as the misrepresentation of his character which astonished me & I wish the authors name may be known as it has been attributed to persons of note who had nothing to do with it. Some day I should like to tell you more of what I know of Sir J. H.

ever faithfully y[ou]rs | Charles Lyell

RI MS JT/1/L/55
RI MS JT/1/TYP/3/856

1. *manly protest in the Daily News*: see letter 3425.
2. *his & my marriage*: Lyell married Mary Horner (1808–73) in 1832; Herschel married Margaret Brodie Stewart (1810–84) in 1829. They knew each other for over forty years.

To Charles Lyell 23 May 1871 3430

23rd May 1871

My dear Sir Charles

It was a very great gratification to me to receive your note[1] yesterday evening. The article[2] required an antidote. It was copied into the other papers—for instance into the 'Engineer'[3] which has a very large circulation.

Yours ever faithfully | John Tyndall

I had very distinct assurance on Sunday night that the author of the article is a lady. Erasmus Darwin[4] told me he knew it to be so

RI MS JT/1/T/1049
RI MS JT/1/TYP/3/840

1. *receive your note*: letter 3429.
2. *the article*: see letter 3425.
3. *the 'Engineer'*: see letter 3425, n. 3.
4. *Erasmus Darwin*: This may be either Charles Darwin's brother, Erasmus Alvey Darwin

(1804–81) or Charles Darwin's son, William Erasmus Darwin (1839–1914). Erasmus Alvey was the older brother of Charles Darwin. He went to Cambridge and Edinburgh to study medicine, but was considered too weak to take up the profession in the end. He lived a life of a confirmed bachelor while remaining close to his younger brother. William Erasmus was the first son of Charles Darwin and, famously, the subject of Charles's child psychology study that influenced his work *The Expressions of the Emotions in Man and Animals* (London: John Murray, 1872) and 'A Biographical Sketch of an Infant', *Mind*, 2:7 (July 1877), pp. 285–94. William was educated at Rugby and at Christ's College, Cambridge. He advocated for university education for all.

To Emily Peel 24 May 1871 3431

Royal Institution. | 24th May, 1871.

My dear Lady Emily,

From my perch in the Dress Circle[1] I saw Sir Robert[2] and you on Saturday night.

I therefore ask your acceptance of a book.[3]

But only your temporary acceptance; for I have directed the publisher to get one bound in a way more worthy of you.

Pray when you receive the better book hand over this one to that fine little fellow whose acquaintance I made at Drayton Manor,[4] and who loved to talk with me about the glaciers—young Master Stonor[5] I think it was.

I have just alluded to your goodness to me at page 316.[6]

Yours most faithfully, | John Tyndall.

RI MS JT/1/TYP/3/962
LT Typescript Only

---

1. *Dress Circle*: lowest and most expensive gallery seats above ground floor in a theatre (*OED*).
2. *Sir Robert*: Robert Peel.
3. *a book*: likely *Hours of Exercise in the Alps*.
4. *Drayton Manor*: home of Sir Robert and Lady Emily Peel.
5. *young Master Stonor*: possibly Francis Stonor, fourth Baron Camoys (1856–97). He was the son of Robert's sister.
6. *page 316*: Tyndall discussed Peel caring for him after an injury.

From Friedrich Vieweg and Son 24 May 1871 3432

Braunschweig d. 24n Mai 1871
Herrn Professor J. Tyndall | royal Institution | Albermaile Str. | London.

Hochgeehrter Herr!

Wir haben Ihnen noch unsern verbindlichsten Dank zu sagen für Uebersendung der Aushängebogen Ihres Buches „fragments of science".

Es hatte ein Hr Dr K. B Hofmann aus Wien sich erboten eine Uebersetzung des Buches für unsern Verlag anzufertigen. Derselbe hat aber unserer Aufforderung ein Stück der Uebersetzung als Probe an uns einzusenden bisher nicht entsprochen. Wir konnten uns nur dazu verstehen mit ihm in Unterhandlung zu treten, wenn wir uns durch Einsicht eines Stückes der Uebersetzung die Ueberzeugung verschaffen konnten, daß er die dazu erforderliche Sprachgewandtheit besaß, was namentlich bei oesterreichischen Gelehrten nicht unbedingt anzunehmen ist.

Mittlerweile haben wir mit Herrn Helmholtz verhandelt und jetzt von ihm die Zusage erhalten, daß er unter seiner Aufsicht und Mitwirkung die Uebersetzung Ihrer „fragments of science" veranlaßen will, ebenso wie früher „on sound und Faraday".

Wir zweifeln nicht, daß Sie mit diesem Abkommen ebenso zufrieden sein werden wie wir selbst. Die Uebertragungen der früheren Werke unter Herrn Helmholz' Mitwirkung haben eine so allgemeine Anerkennung in der litterarischen Welt gefunden und schon der Name des berühmten deutschen Gelehrten gereicht dem Buche zur Zierde und Empfehlung so daß seine Theilnahme an der Veröffentlichung dem Buche sehr förderlich sein wird.

Wir eilen Ihnen dies mitzutheilen und hoffen auf die Billigung dieses Abkommens von Ihrer Seite.

Mit ausgezeichneter Hochachtung | ergebenst | Friedr[ich] Vieweg & Sohn

Braunschweig, 24th May 1871
Professor J. Tyndall | royal Institution | Albermaile[1] St. | London.

Esteemed Sir!

We still have to give you our deepest thanks for sending us the corrected proofs of your book "Fragments of Science".

A Dr K. B Hofmann[2] from Vienna had offered to prepare a translation of the book for our publishing house. Up till now, however, he has not complied with our request to send us a part of the translation as a sample. We could only agree to enter into negotiations with him if we could convince ourselves, by seeing a part of the translation, that he possessed the linguistic elegance necessary for this, which cannot necessarily be assumed, especially with Austrians scholars.

In the meantime, we have been negotiating with Herr Helmholtz and have now received the commitment from him that he will arrange for the translation of your "Fragments of Science" to be done under his supervision and with his assistance, as earlier with "On Sound and Faraday".[3]

We do not doubt that you will be just as satisfied with this agreement as we ourselves are. The translations of previous works done with Herr Helmholtz's assistance have found such a general acceptance in the literary world, and the mere name of the famous German scholar will serve to adorn and commend the book, with the result that his involvement in the publication process will be very beneficial to the book.

We hasten to inform you of this and hope for the approval of this agreement from your side.

With excellent esteem | your humble servants | Friedr[ich] Vieweg & Son

RI MS JT/1/V/9

---

1. *Albermaile*: Albermarle Street, where the RI was at the time, and still is.
2. *Dr K. B Hofmann*: Karl Berthold Hofmann (1842–1922), an Austrian chemist and physician. He wrote, with R. Ultzmann (1842–89), the *Guide to the Examination of Urine: With Special Reference to the Diseases of the Urinary Apparatus* (the original in German is *Anleitung zur untersuchung der harnes, mit beonderer Berücksichtigung der Erkrankungen des Harnapparates* (Wien: W. Braumüller, 1871)). He was a professor and eventually Rector of the University of Graz.
3. *"On Sound and Faraday"*: Anna Helmholtz translated two of Tyndall's books, *Sound* and *Faraday as a Discoverer* (London: Longmans, Green, and Co., 1868) (*Ascent of John Tyndall*, p. 273). See also letters 3615, 3618, 3625, 3632, 3640 and 3649.

To [John Stenhouse][1] 26 May [1871][2] 3433

26th May.

My dear */Dr Stenhouse/*

You lectured here some years ago on your charcoal respirators.[3] Have you discovered or developed any new effects since that time? I shall have reason to refer to the subject in my next Friday evening discourse.[4]

Yours faithfully | John Tyndall

RI MS JT/1/T/1394

---

1. *John Stenhouse*: John Stenhouse (1809–80), a Scottish chemist and inventor. He attended Glasgow University, where he studied chemistry from 1824–8. He went to Germany to study with Justus von Liebig from 1837–9. In 1851, he went to St. Bartholomew's Hospital as a lecturer on Chemistry, a position he resigned in 1857. He was elected FRS in 1848, and

received a Royal Medal from the RS in 1871 for his work in chemistry. He suffered from rheumatism of the eyelids for the last four years of his life and had to live in a darkened room where he died in 1880. The addressee of the letter is illegible, but a note from the Librarian at the RI dated 24 October 1963 confirms Stenhouse as the recipient.

2. *26th May*: Tyndall refers to his Friday Evening Discourse at the RI, 'On Dust and Disease', from 1871.
3. *charcoal respirators*: possibly an RI lecture from 2 March 1855, 'On the economical applications of Charcoal to Sanitary purposes', *Roy. Inst. Proc.*, 2 (1854–58), pp. 53–5.
4. *Friday evening discourse*: 'On Dust and Smoke', delivered at the RI on 9 June 1871 (*Roy. Inst. Proc.*, 6 (1870–72), pp. 365–76), in which Tyndall 'came out firmly in favour of the germ theory of disease, regretting he had not done so publicly before' (*Ascent of John Tyndall*, 274–5).

To Friedrich Vieweg and Son 26 May [1871][1] 3434

26th May

Gentlemen

I fully share your pleasure in learning that my friend Prof. Helmholtz is willing to undertake the supervision of a translation of the Fragments.[2] Indeed it is more than an ordinary pleasure to find my book brought out under the sanction of a name second to none in the present scientific world.

I have published another little book[3] which has been rapidly diffused here in England, and a copy of which I have requested the Messrs Longman to send to you.

Faithfully yours | John Tyndall

TU Braunschweig/Universitätsbibliothek/UABS V1T : 63

1. *26 May [1871]*: dated from letter 3432.
2. *Helmholtz . . . Fragments*: see letter 3432.
3. *another little book*: *Hours of Exercise in the Alps*.

From Edward Sabine 26 May 1871 3435

13. Ashley Place | May 26. 71

Dear Tyndall:

I send you Dr. Copelands note,[1] which accompanied the little paper which you have sent[2] for Mr. Longman's perusal.[3] You can return it to me with the paper when Mr. Longman returns it.

I will only add that any thing you would say regarding the interest of the comparison of Nature in East Greenland & the Alps, (which is the subject of the paper) would be so much more influential & instructive than anything I could write, that I should greatly like to send a sentence or two from your pen, either in reference to, or in addition to anything from myself.

Always sincerely yours | Edward Sabine

RI MS JT/1/S/26
RI MS JT/1/TYP/4/1332

---

1. *D^r^. Copelands note*: note missing. Probably Ralph Copeland (1837–1905), an English astronomer who accompanied a German Arctic expedition to explore the east coast of Greenland in 1869–70. During the preparations for this journey, he earned his PhD. In January 1871 he was appointed assistant astronomer at the observatory at Birr Castle, Ireland. He was in charge of the large format telescopes, and elected FRAS in 1874. By 1896 he had been appointed Astronomer Royal for Scotland, director of the Royal Observatory and professor of astronomy at Edinburgh University.
2. *little paper . . . sent*: letter missing and paper unidentified.
3. *M^r^. Longman's perusal*: one of the two Longman brothers who ran the publishing house Longmans, either Thomas (1804–79) or William (1813–77). Tyndall had a long relationship with Longmans, having fourteen of his sixteen books published by this house.

From Mary Baring[1] 27 May [1871][2] 3436

Lordan | Romsey[3] | May 27th

Dear Professor Tyndall

Thank you so much for sending me such a beautiful book.[4] It was so kind of you thinking of me when you were so busy.

It was such a pity you could not come while M^r^. Carlyle was here. But I hope you will soon be able to.—I remain

yours affect[ionate]ly | Mary

Elton Hall, Ashburton Collection
RI MS JT/TYP/1/109

---

1. *Mary Baring*: Mary Baring (1860–1902), daughter of Louisa Caroline Baring (1827–1903), Lady Ashburton, art collector and philanthropist, and William Bingham Baring, the second Baron Ashburton (1799–1864). Tyndall met the Ashburtons in 1856 (*Ascent of John Tyndall*, pp. 108–9).
2. *[1871]*: Mary, or Maysie, was only eleven at the time. Further, her mention of Carlyle dates

this letter to 1871, when Carlyle had been visiting his good friends, the Ashburtons. He wrote a letter to his brother from Melchet Court on 13 May 1871 (*New Letters of Thomas Carlyle*, ed. A. Carlyle, 2 vols (London: J. Lane, 1904), pp. 278–9).

3. *Lordan | Romsey*: J. Lordan (n.d.) was a publisher based in Romsey, Hampshire, the location of Louisa Baring's estate Melchet Court. Construction of a new home on the estate was begun in 1862 by William Baring and completed in 1868 by his widow Louisa.
4. *book*: possibly *Fragments of Science* or *Hours of Exercise in the Alps*.

From Stephen Lushington[1] 28 May 1871 3437

Ockham Park | Woking Station, Surrey | May 28, 71.

Dear Tyndall

I have received your book[2] & your letter[3] & for both I sincerely thank you.

I am stationed here as I have chiefly been for 25 years[4]—a visit from you would be most acceptable any time will suit me Saturday & Sunday would most probably be most convenient to you.

Very truly yours | S. Lushington

RI MS JT/2/10/1211–2
RI MS JT/2/13c/1381

1. *Stephen Lushington*: see letter 3395 n. 4.
2. *your book*: probably *Fragments of Science*.
3. *your letter*: letter missing.
4. *stationed here . . . 25 years*: Ockham Park was his country home.

From Clementine Soret[1] 29 May 1871 3438

Genève | 29 Mai 1871

Cher Monsieur

Votre aimable envoi s'est croisé avec la douloureuse nouvelle, que mon mari vous adressait, il y a peu de jours; celle de la mort de son père, qui nous a été enlevé, après une agonie de quatre semaines et plus, et dont le départ, laisse un grand vide à notre foyer, qui était le sien depuis notre retour d'Allemagne. Ce triste événement et l'excessive fatigue et ébranlement (qui ont beaucoup augmenté mon état de souffrance constant et croissant depuis quelques années) m'ont empêché de vous écrire plus vite; mais soyez sûr qu'aucun état de souffrance, ni de tristesse, ne peut m'empêcher de sentir avec reconnaissance, un témoignage de souvenir, de votre part; et qu'au contraire, ils ont souvent apporté un rayon brillant, dans des périodes de jours pesants, et de nuits éternelles!

Quand pourrai-je vous le dire à vous même! Comment faire pour créer un glacier, dans nos environs, afin d'avoir quelque chance qu'il vous attire!...

Nous avons eu notre pauvre ami Regnault, tout l'hiver, comme vous le savez. Nous avons dû lui apprendre la mort de son Henri bien aimé; O quelle mission! C'est mon mari et Mr de la Rive qui ont été le lui dire, après que j'avais écrit à l'amie qui les a accompagné en exil, pour la préparer.—Il a été, et est, admirable de courage, de soumission, et de ressort. Mais vous devez avoir reçu une lettre de lui, parce qu'il m'en a envoyé une de vous, à lui traduire avant de vous répondre—En tout cas son adresse actuelle est Château de Lassigneux par Bellay, Ain—Il a passé ici il y a deux jours, allant chercher son fils aîné, dans la Forêt Noire; je pense qu'ensuite il sera obligé de consentir à aller à Versailles, où on l'appelle à corp et à cris... Pauvre Paris! Mme Quinet m'écrivait hier, de Versailles, que les siècles mis bout à bout—n' avaient pas renfermé jusqu'ici de pareils événements. Et cet incendie de Paris au pétrole, qu'en dire? Qu'en penser? Le coeur est bouleversé et navré en voyant le mal ainsi déchainé dans toute son horreur!...

Mon mari vous serre la main bien sûr de votre sympathie, et moi cher Monsieur je vous dis encore adieu et merci

Votre dévouée et recon<naissante> | Clémentine Soret

Geneva | 29 May 1871

Dear Sir,

Your pleasant letter[2] crossed with one sent a few days ago[3] by my husband[4] relating some painful news; that of the death of his father,[5] who was taken from us after more than four weeks in agony, and whose departure left a great void in our home, which had also been his since our return from Germany. This sorrowful event and the excessive fatigue and distress (which have greatly increased my state of constant suffering that has been increasing for years) prevented me from writing sooner. But be assured that no state of suffering, nor of sorrow, could keep me from appreciating a token of remembrance from you; and that on the contrary, these have often brought light to the heavy days and eternal nights!

When will I be able to tell you this in person; how to create a glacier nearby, for the chance that it might attract you!...

As you know, we had our poor friend Regnault[6] as a guest through the winter. We had to inform him of the death of his dear Henri;[7] Oh what a task! My husband and Mr. de la Rive had to give him the news after I wrote to the friend who followed them in exile[8] to prepare her.—He was, and still is, admirable for his courage, resignation, and resilience. But you must have received a letter from him,[9] because he sent me one of yours,[10] to translate for him, before he sends you a reply.—In any case, his current address is Chateau de Lassigneux par Bellay, Ain. He came here two days ago on his way to fetch his eldest son, in the Black Forest.[11] I believe he will then have to agree to go to Versailles where he is called upon with clamour. Poor Paris![12] M[me] Quinet[13] wrote yesterday from Versailles that until now no century, one after the other, had seen such events. And this arson of Paris using petroleum, what can be said? What can be thought? The heart is overwhelmed and saddened at the sight of evil thus unleashed in all its horror!...

My husband shakes your hand, sure of your sympathy, and I, dear Sir, bid my farewell and thank you,

Your devoted and grateful | Clémentine Soret

RI MS JT/1/S/83

---

1. *Clementine Soret*: Clementine Soret (née Odier, 1831–99), wife of Jacques-Louis Soret (see n. 4). The couple took in Regnault (see letter 3382, and n. 6 below) during the Franco-Prussian War.
2. *Your pleasant letter*: letter missing.
3. *one sent a few days ago*: letter missing.
4. *my husband*: Jacques-Louis Soret (1827–90), a Swiss chemist who discovered the element holmium in 1878 with a spectroscope. He was also known for identifying the structure of ozone as three oxygen atoms bound together.
5. *death of his father*: Nicolas Soret (1797–1871), not to be confused with the Swiss painter of the same name.
6. *Regnault*: chemist and physicist Henri Victor Regnault (see letter 3382, n. 1).
7. *dear Henri*: Regnault's son, the painter Henri Regnault, who was killed during the Franco-Prussian war of 1870–1. See letter 3382, n. 4.
8. *the friend who followed them into exile*: not identified.
9. *a letter from him*: letter 3382.
10. *one of yours*: letter missing.
11. *his eldest son, in the Black Forest*: his oldest son, Léon, who was convalescing in an asylum in a small town near Geneva, along with Regnault's other surviving children. The Black Forest is a region in south-west Germany.
12. *Poor Paris!*: reference to the destruction of Paris during the Franco-Prussian War.
13. *M[me] Quinet:* Hermiona Asachi (1821–1900), a Romanian writer, translator and intellectual. She married French politician and historian Edgar Quinet (1803–75) in 1852.

From Elizabeth Colling[1] 31 May 1871 3439

Hurworth, n[ea]r Darlington | May 31st 1871

To Professor Tyndall,

Sir,

As I fear that the letter I am about to commence will unavoidably prove a long one, I had best begin it by stating that I am a lady—not a young one—who have long been honoured with the kind and condescending correspondence of the late illustrious Sir John Herschel;—and I take the liberty therefore to express to you the high satisfaction I experienced in reading in the Daily News (only two days ago, for it is not a paper I am in the habit of seeing) your indignant protest against the odious and extraordinary attack on the veracity and moral rectitude of so excellent and eminent a man.[2]—a man so highly revered, so deeply and universally lamented. A friend had before sent me the paper which contained the notice of the funeral,[3] and I had read the passage you quote, with disgust and indignation, and readily guessed who wrote it—it being of a piece with the usual obituaries from the same trenchante pen—which rarely fails, even in the midst of apparent panegyric "to hint a fault, and hesitate dislike"[4]—if not to speak out more openly and rancorously, as in the present instance. However high the cup of flattery may be filled and frothed, for it is seldom that of hearty commendation, "some bitter o'er the brim its bubbling venom flings"[5]—Astonished and revolted at these strange imputations I expressed myself with a warmth to my private friends entirely congenial with that for which thousands of readers will thank you for having expressed yourself in the widely spread Journal in question. What! shall courtesy to all, and cordial and unselfish commendation where he saw that it was merited—the true test of a noble and generous soul—be stigmatized as "vanity and vicious craft" (whatever that term implies). He had little to gain truly, by vanity or craft in my case—and those who bring such accusations against others must be base and ignoble themselves. He gained indeed my fervent gratitude and profound veneration, if that was worth his gaining, not less for the goodness of his heart than for his high mental endowments. His departure was to me as the sudden extinction of a bright luminary which could never be rekindled, and the blank which he has left can never be supplied.—My acquaintance with him by letter (I never had a personal one) commenced, like yours, in 1854, in the October of which year, a most stupendous Meteor[6] passed over this place, which of course I expected to see noticed in all the papers; but such, to my astonishment, not being the case, I thought it a duty not to let so extraordinary a phenomenon pass unrecorded, and therefore, though unused to writing in newspapers, sent an account of it to a local one, a copy of which I forwarded to Sir J. Herschel, and another to the late Dr. Dick.[7] Both wrote to thank me, and Sir J. H. said he had placed the copy I had sent, "in the hands of an eminent Meteorologist"—and there, for the time, our communications ceased.—About a twelvemonth after, I published a book of poems entitled "Far & Near or translations and originals"[8]—the translations chiefly from the German, which, being, if not Herschel's mother-tongue at least that of his father, I thought it possible he might take an interest in them, and took the liberty to forward him a copy—but receiving no reply for a considerable time, I began to fear I had been too presumptuous. I was most agreeably surprised by receiving from him when I had ceased to expect it, a most frank and cordial apology for having (being much occupied at the time of receiving the book) laid it aside and forgotten it—and then proceeding to the kind commendation, an extract from which you will find in the printed Notices.[9]

Of course I returned a grateful response and an occasional correspondence has ever since continued; his last valued letter having been dated but a short time before his death, which (tho' he complained of having been "severely handled, this savage winter, by his old atmospheric enemy") was written in other respects with his usual cheerfulness upon general subjects, so that the melancholy intelligence came quite as an unexpected shock to me, and few things out of my own family have ever affected me so deeply.

In this state of mind, it may readily be conceived how jarring it was to my feelings to read such passages as you have quoted—or even, that ill-timed reprobation of the vulgar and irreverent women who appeared in gaudy colours instead of appropriate mourning on so sad and solemn an occasion—A reprobation which, however richly merited, had better have been reserved for a more fitting time and opportunity.—Pardon my enclosing the Notices[10] on my former and recent publications. I should be sorry to think that I was indebted for those kind commendations to "vanity and vicious craft"—but if so, the panegyrist is kept in countenance by many similar offenders, as most of my critics, you will observe, bestow at least equal encomiums!

And now I must beg of you to pardon all this egotism, detaining your attention so long; and, requesting your acceptance of the enclosed humble tribute to the memory of our revered mutual friend—I remain, Sir,

Y[ou]rs sincerely and respectfully | Eliz[abe]th Colling

P.S. "Eta Mawr"[11] is the name which I am known by in the literary world.—My address "Miss Colling | "Hurworth on Tees | "Darlington.

May I ask if you have ever heard of the above mentioned Meteor? If not I shall be happy to send you my account and sketch of it.

RI MS JT/1/C/40
RI MS JT/1/TYP/2/577–9

1. *Elizabeth Colling*: Elizabeth Colling (1798/9–1879), known by the pseudonym Eta Mawr, was an English hymn writer, poet and fiction author. Not much more is known about her life, but she lived in Hurworth-on-Tees, near Durham.
2. *Daily News . . . eminent a man*: see letter 3425.
3. *the paper . . . funeral*: see letters 3424 and 3425.
4. *"to hint a fault, and hesitate dislike"*: A. Pope, *An Epistle from Mr. Pope, to Dr. Arbuthnot* (London: printed by J. Wright for Lawton Gilliver, 1735), l. 204. The phrase seems to have passed into colloquial usage.
5. *"some bitter . . . venom flings"*: Lord Byron, *Childe Harold's Pilgrimage: A Romaunt* (London: John Murray, 1812), LXXXII, p. 49.
6. *a most stupendous Meteor*: While no newspaper account can be found, according to T. L. Phipson's *Meteors, Aerolites, and Falling Stars* (London: Lovell Reeve, 1867), p. 139, two men in Hurworth in October 1854 saw 'one of the most brilliant bolides witnessed in England'. A bolide is a meteor that explodes upon entry into the atmosphere.
7. *the late Dr. Dick*: possibly Thomas Dick (1774–1857), a British church minister and astronomer. He argued that science and Christianity were compatible and should be studied together.
8. *"Far & Near or translations and originals"*: published in London by Saunders & Otley in 1856 under the pseudonym Eta Mawr. Mawr means 'great' in Welsh.
9. *printed Notices*: see, for example, the notices in the back of E. Mawr, *The Story of Count Ulaski* (London: Provost & Co., 1870), after p. 286.
10. *enclosing the Notices*: missing.
11. *"Eta Mawr"*: see n. 8.

From Agatha Russell 31 May 1871 3440

Pembroke Lodge | Richmond Park. | May 31. 1871.

Dear Professor Tyndall

I must at once write to express my very best thanks for your most kind present which arrived here this morning. I am sure your book will make me long to go off to the Alps.[1] They must be so beautiful—

Mama[2] had written[3] to thank you beforehand, & now she thanks you still more as she knows what the book is.—Hoping that you will soon recover & that we shall see you here,

believe me, yours sincerely | Agatha Russell

RI MS JT/1/R/46

1. *most kind present . . . the Alps*: probably *Hours of Exercise in the Alps.*
2. *Mama*: Frances Russell.
3. *had written*: letter missing.

From Mary Egerton [May 1871][1] 3441

45, Eaton Place, | S.W. | Wed[nesda]y

"Scorn!"[2] Is it scorn to love every word of a book so that one cannot spare one? Is it scorn to be almost afraid to sit down to a book[3] for the engrossing exciting, interest it causes one? Not as the first reading only, but as to some part of it at the 20th? Perhaps Lord Romilly[4] had not learnt "Mountaineering" and the "Glaciers" by heart, and so had no drawback to his full enjoyment in the fear of not finding the whole; but as to being "grateful" for this & every-one of the kind things you have done by me, I will not yield to him or anybody else; & I believe you have it quite well, & only pretend to misunderstand me in order to tease me in revenge for my "Metaphysics".[5] Well, notwithstanding all I am my dear Mr. Tyndall

Y[ou]rs truly & gratefully | M F Egerton

Poor noble Bennen![6] After reading the sad story one seems to see the foreshadowing of his fate in his earnest thoughtful countenance! How awfully close that little incident of the watches[7] seems to bring the parallel between the two adventures!—& forces upon one the shuddering thought of what might have been!

RI MS JT/1/E/74

1. *[May 1871]*: dated to May 1871, when *Hours of Exercise in the Alps* was published.
2. *"Scorn!"*: The entire letter is likely referring to a previous discussion with Tyndall that we do not have.
3. *a book*: *Hours of Exercise in the Alps.*
4. *Lord Romilly*: John Romilly (1802–74), a British judge and politician. He was educated at Trinity College, Cambridge, taking his BA in 1823 and MA in 1826. He became a Liberal MP in 1832, but lost in 1835. He was well-known for taking good notes and dispatching with cases quickly in a back-logged court. He was promoted to the peerage in 1866. He died in 1874.
5. *my "Metaphysics"*: possibly a reference to the conversation referred to in n. 2.
6. *Poor noble Bennen!*: *Hours of Exercise in the Alps* includes a chapter on the death of Johann Bennen (see letter 3386, n. 2).
7. *little incident of the watches*: Tyndall had been in an avalanche himself in 1864. He and the members of his expedition were unhurt, but he could only find pieces of his watch chain (*Hours of Exercise in the Alps*, pp. 212–15).

From Philip Gosset[1] [May 1871][2] 3442

I rejoice to see your book[3] appear and sincerely hope you will do as you did in the "Gl[aciers]. of the Alps",[4] namely divide the excursions from the scientific part. This second part will probably explain several things on glaciers that I am at the present moment utterly unable to account for; for example: In August 1869 I passed the Oberaarjoch and Grünhornlücke with the Section "Aarau" of the Swiss Alpine Club.[5] Including guides and porters we were 22 in number. The Aletsch Gl[acier].[6] presented a most extraordinary appearance; its whole surface especially the middle part of the Glacier was covered with long series of stripes varying from 80 to 200 feet long, 6 inches apart, 2/3 of an inch deep and all converging at both ends into a nearly mathematical point; these series of stripes looked as if they had just been made (it had not snowed for several days, the Gl[acier] was not covered at the time with summer snow) I give you here a rough sketch; the stripes crossed the crevasses in several places and were every where parallel to the sides of Glacier.

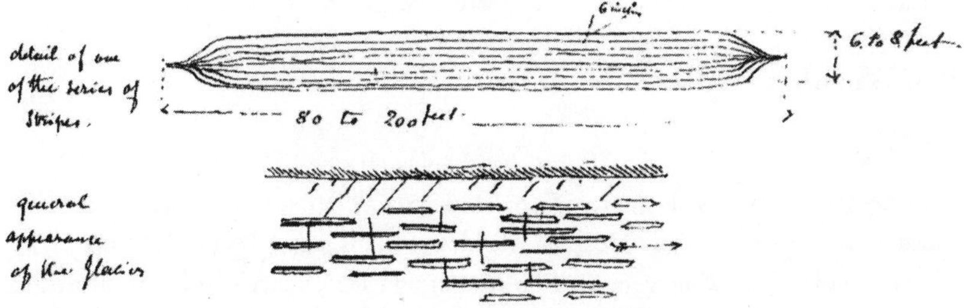

I was much puzzled by what had produced these stripes, so exquisitely drawn; are they perhaps in connection with the interior structure of the Gl[acier] and should I have found blue veins if I had looked better?

Believe me to be, dear Sir, | yours very sincerely | Ph[ilip] Gosset

P.S. As you will perhaps allude in your work to the extent of Gl[aciers]. in general, you may find page 180 Vol IV of the Alpine Journal a table that may be of use to you.[7]

RI MS JT/1/G/8

---

1. *Philip Gosset*: see letter 3386, n.1.
2. *[May 1871]*: letter dated to May 1871, the publication date of *Hours of Exercise in the Alps*.
3. *your book*: *Hours of Exercise in the Alps*.
4. *"Gl[aciers]. of the Alps"*: Tyndall's 1860 book *Glaciers of the Alps*.
5. *Oberaarjoch . . . Club*: mountain passes in the Bernese Alps that Gosset explored with the Aarau section of the Swiss Alpine Club.

6. *Aletsch Gl[acier]*: in the Bernese Alps.
7. *Alpine Journal*: The table to which Gosset refers describes the size of Swiss glaciers. See P. C. Gosset, 'The Inundations in Switzerland in 1868', *Alpine Journal*, 4 (February 1869), pp. 177–83, on p. 180.

To John Murray 1 June 1871 3443

[ans[were]d June 2][1] | 1st June 1871

My dear Sir,

Mr. Appleton,[2] the head of the firm Appleton & Co.[3] is in London. A friend of his[4] in America has written[5] to Mr. Huxley, Mr. Spencer, & myself suggesting that he is well worthy of an invitation to the Athenaeum. In duty to the club we would like to have the opinion of a judge so competent as yourself. Would you be good enough to inform me in confidence in advance whether you think the invitation would be an appropriate one?

Yours faithfully | John Tyndall

[1871 June 1 | Tyndall Prof. | Appleton & the Athenaeum][6]

NLS, John Murray Archive, Ms 41214

---

1. *answered June 2*: note in John Murray's hand; letter missing.
2. *Mr. Appleton*: likely William Henry Appleton (1814–99), but he had four brothers with whom he ran the company. He began clerking with his father, Daniel Appleton (1785–1849), and became a buyer for the Appleton firm in 1835, when he made his first trip to London. He became a partner in the firm in 1838, and the head of the firm in 1848. Appleton was the firm's London representative by 1853 and was a prominent figure in American and international scientific publishing.
3. *the firm Appleton & Co.*: Founded by Daniel Appleton in Massachusetts, the firm published its first book in 1821. The firm expanded to cover numerous topics, and by 1872 it was publishing *Popular Science Monthly*. In 1848 the firm was taken over by Appleton's five sons and became a prosperous trade publisher. In 1924 it became part of the Appleton Century Company (*ANB*).
4. *A friend of his*: friend not identified.
5. *has written*: letter missing.
6. *1871 . . . Athenaeum*: on the verso in Murray's hand. See letter 3373, n. 15.

From Rudolf Clausius 2 June 1871 3444

Bonn, 2/6/71.

Lieber Tyndall,

Du hast mich durch die Übersendung Deiner beiden neusten schönen Werke sehr erfreut, und ich danke Dir bestens dafür. Ich muss Dich wirklich bewundern, mit welcher Leichtigkeit Du bald streng wissenschaftliche Arbeiten machst, welche in irgend einem Zweige der Physik ganz neue Gebiete eröffnen, bald wieder in so schönen populären Darstellungen zur Verbreitung der Wissenschaft beiträgst. Ich habe Dir als Gegengabe nur eine sehr kleine Brochüre schicken können.

Ich habe vor einiger Zeit wahrscheinlich auf Veranlassung von Hirst, eine mir sehr ehrenvolle Einladung zur Versammlung der British Association in Edinburgh bekommen, und würde um so lieber gekommen sein, als Thomson diesmal Präsident ist; aber leider habe ich durch einen Stoss, den ich in der Schlacht von Gravelotte an das Knie bekommen habe, ein Leiden davon getragen, welches mir das Gehen erschwert, und noch immer nicht ganz gehoben ist. Ich darf daher, eine solche Reise noch nicht unternehmen. Sei so gut dieses auch Hirst mitzutheilen, und ihm meinen Dank für die freundliche Einladung auszusprechen. Vielleicht kommt Ihr beide einmal wieder nach Bonn, worüber wir uns sehr freuen würden. Oder gedenkt Ihr diesmal auf Eurer Reise das arme Paris zu besichtigen? Es ist doch wirklich so gekommen, wie Du vor längerer Zeit mir schriebst, dass die Franzosen, nachdem sie von Preussen nicht mehr bedroht waren, sich untereinander zerfleischten, und wer weiss, wie lange es noch dauern wird, bis sie wieder vollkommen zur Ruhe und zu gesicherten Zuständen gekommen sind. Ich bin sehr gespannt darauf zu erfahren, wie die Gelehrten, deren Schicksal uns vorzugsweise interessirt, diese schreckliche Zeit durchgemacht haben. Mit herzlichen Grüssen von meiner Frau und mir an Dich und Hirst,

Dein | Clausius.

Der kleine Johnny hat mich gebeten, auch einen Brief mit einlegen zu dürfen.

Bonn, 2/6/71.

Dear Tyndall,

You[1] gave me great joy by sending your two most recent, fine works,[2] for which I thank you kindly. I admire the ease with which you can produce rigorous scientific works, which open new fields in any branch of physics, and how you then contribute to the diffusion of science through fine popular presentations. I was only able to send you a very short brochure[3] in exchange.

I received a while ago, possibly through Hirst's mediation, a very honourable invitation to the meeting of the British Association in Edinburgh. I would have been delighted to attend, the more so now that Thomson presides, but unfortunately, I suffered an injury after an impact on the knee during the battle of Gravelotte,[4] which has not fully healed and makes it difficult for me to walk. Therefore, I cannot do such a journey yet. Please be so good to let Hirst know and to express to him my gratitude for the kind invitation. Maybe the two of you will come again to Bonn, which would make us very happy. Or are you planning to visit poor Paris[5] on the present journey? It has indeed turned out as you once wrote to me, and the French are tearing each other to pieces after no longer being threatened by the Prussians. Who knows how long it will be before they come to rest and attain stability. I am very anxious to hear how the scholars, whose fates interest us the most, have come through this horrible time. With kind regards from my wife[6] and me to you and Hirst,

Your | Clausius

Little Johnny asked me to also include a letter of his.[7]

RI MS JT/1/TYP/7/2322
LT Typescript Only

---

1. *You*: 'du' in the German, the informal address.
2. *two most recent, fine works*: *Fragments of Science* and *Hours of Exercise in the Alps*.
3. *very short brochure*: brochure missing.
4. *battle of Gravelotte*: a battle during the Franco-Prussian War, fought on 18 August 1870. Clausius organized an ambulance corps during the Franco-Prussian War, and his injury gave him a life-long disability. He won the Iron Cross for his bravery in battle.
5. *poor Paris*: referring to the state of Paris in June 1871. See also letter 3438.
6. *my wife*: Adelheid Clausius.
7. *Little Johnny asked me to also include a letter of his*: letter missing. Johnny is Clausius's son Rudolf John Clausius. Tyndall was his godfather; see letters 2151, 2157, 2195 and 2196, *Tyndall Correspondence*, vol. 8.

From Emily Peel 4 June 1871 3445

Whitehall | June 4. / 71

Dear Professor Tyndall,

I have made two journies from Geneva in the space of three weeks—On my return here the first time, three weeks ago, I found a small book[1] which you had kindly sent to me some time during the spring[2] and for which I was about to thank you when I was suddenly called back to Geneva, scarlet fever having broken out in my nursery there and the nurses having taken it.

My object was to bring the children away at once, in the hope that they might not be infected by the disease, and I have succeeded so far as they are all quite well and safely lodged here.

I merely mention these details to account for my silence which under other circumstances would have been unpardonable. Your charming volume[3] greeted me on my return home yesterday, and I cannot thank you sufficiently for bearing me so constantly in your mind. That I may have been the means of mitigating the weariness of your days of illness[4] will be a grateful thought for ever.

Believe me yours | sincerely Emily Peel

RI MS JT/1/P/54

1. *a small book*: likely *Hours of Exercise in the Alps*; see letters 3549 and 3553.
2. *sent . . . the spring*: letter missing.
3. *charming volume*: *Hours of Exercise in the Alps*; see letter 3431.
4. *days of illness*: see letter 3431, n. 6.

From Sarah Faraday 7 June [1871][1] 3446

Barnsbury Villa, | 320, Liverpool Road, N. | 7th. June.

Dear Dr. Tyndall

I have just cut open your book[2] and kind remembrance of us, and must write to thank you before I begin to read it, tho' it looks <u>very</u> interesting and the illustrations very striking, I expect to enjoy it much—but I am rather concerned that you should send it to us too for you must have so many calls upon you for such courtesies. You do keep yourself busy indeed writing so much in addition to all your other avocations. I begin to fear you will not have time to pay your promised visit to us but I shall not give you up yet for I know you have the will when you can find the time. This cold ungenial weather keeps me a prisoner and takes away my spirits but it is no use bemoaning, I leave Jane[3] to speak her own thanks as she has many more opportunities than I have

Believe me dear Dr. Tyndall | Your sincere friend | S. Faraday

RI MS JT/1/TYP/12/4183
LT Typescript Only

1. *[1871]*: *Hours of Exercise in the Alps* includes illustrations and was published in May of 1871, placing this letter in June of that year.

2. *your book*: *Hours of Exercise in the Alps*.
3. *Jane*: her niece, Jane Barnard.

From Alexander Herschel 7 June 1871 3447

Collingwood | Hawkhurst. | June 7th 71.

My Dear Professor Tyndall

I have received your very kind Testimonial Letter,[1] and I sent it today (the last at my disposal for enclosing any other Testimonials with my application)[2] to the Warden's Secretary;[3] the Dean of Durham;[4] and to Sir W[illia]m Armstrong,[5] Mr RS. Newall[6] and to Mr I. Lowthian Bell[7] (formerly the Mayor of Newcastle) who form the Committee of the new Science College.[8] At Sir W[illia]m Armstrong's recommendation (which I sought), about 15 copies of Testimonials were sent to the Warden's Secretary, today. The enclosed is a copy of the formal Application and of the accompanying Letters.[9] The Syllabuses, and a few Author's Papers which you have already seen, accompanied each printed Application. Since receiving, and printing your letter, yesterday, the much-desired letter from Professor Rankine[10] which you so properly recommended, reached me today, and 15 or 20 MS copies accompanied the other Testimonials! It will now be printed; and it shows very clearly the opinion you wished me to obtain from the Glasgow Scientific men. Prof[esso]r Grant's[11] Testimonial reached me by the same Post from Glasgow; and the same No of MS. copies of his letter were also sent.—I wrote to Sir W[illia]m Thomson, at the same time that I wrote to Prof[esso]r Rankine. But he must be travelling or otherwise unable to have answered, yet. His opinion would be a great help, if it were as favourable as Professor Rankine's, and I yet hope that it may not be wanting to my cause.—Some other Testimonials which have since arrived you will see in the accompanying "additional" printed pages.—I know not if I should go to Newcastle? It will be painful to wait events, when the last formal letters which I can send from here will be dispatched.—A few days may, however, perhaps, see me in London, on my way to Newcastle; and I will then call at the Royal Institution, to ask your further advice on my plans, and to thank you very warmly for the interest which you have already taken in them so heartily.—

The Warden's Sec[reta]ry writes to me today from Durham, that (before receiving my Testimonials) he "has received some other applications from gentlemen of very high attainments". He adds that the appointment rests almost entirely with the Dean of Durham. I cannot do better than to send him without delay the excellent Testimonials of my Glasgow Patrons!—

Believe me to remain | My dear Prof[esso]r Tyndall | yours very faithfully | AS. Herschel

RI MS JT/1/H/85
RI MS JT/1/TYP/2/593–3a

---

1. *Testimonial Letter*: letter missing.
2. *my application*: Herschel was applying to be the first professor of physics and experimental philosophy in the University of Durham College of Science, Newcastle upon Tyne, which he obtained later this year.
3. *the Warden's Secretary*: William Charles Lake (see n. 4) was the Warden, but his secretary is not identified.
4. *Dean of Durham*: William Charles Lake (1817–97), the Dean of Durham at this time.
5. *Sir W[illia]m Armstrong*: William Armstrong, First Baron Armstrong (1810–1900), a British industrialist and manufacturer. He studied law in London, but soon his interest in mechanics and science made him shift his work to studying the use and efficiency of water in machines. He started a company called W. G. Armstrong & Co., which specialized in hydraulic cranes. He was able to corner the market, thanks to his connections to family and friends. He also became an armaments manufacturer and sold guns to military units all over the world. In 1863 he became the President of the BAAS and was further involved in intellectual life in Britain. In 1871 he was on the employer side of a massive strike, and eventual worker lock out, in which over 6000 of his employees were asking for a nine-hour work day. As a rich and connected industrialist, he had influence at the university in Durham.
6. *Mr RS. Newall*: Robert Stirling Newall (1812–89), a British astronomer and engineer. In 1840, he filed a patent on wire rope, which was useful for manufacturing cable for use underwater. By 1851, his business was so successful that he had a monopoly on the market. He also had built a large telescope that was too large to be built in the factory. It was 32 feet long and weighed 9 tons. He took delivery of the telescope in 1871, but was unable to use it much because of clouds. Upon his death, it was given to the University of Cambridge, where his son was an astronomer. He was elected FRAS in 1864 and FRS in 1875.
7. *Mr I. Lowthian Bell*: Isaac Lowthian Bell (1816–1904), a British iron and steel manufacturer in Newcastle. His life and work were exemplary of the middle-class male industrialist. He was educated at Edinburgh, then the Sorbonne and other places on the Continent, coming back to Newcastle to study the operations of blast furnaces and rolling mills. He operated a few companies with his brothers, then on his own, and continued to improve the operations of furnaces in the industrial North. He was elected FRS in 1875 and received a number of awards from technical societies in recognition of his work.
8. *new Science College*: see n. 2. The College of Physical Science at the University of Durham was founded at Newcastle on Tyne in 1871. In addition, see J. Waite, 'The Science College at Newcastle', *Nature*, 3 (20 April 1871), pp. 485–6.
9. *the enclosed . . . Letters:* all missing. LT noted that, at the time of her typing, a 'copy of a strong testimonial to Alex H. from Rankine *[is]* enclosed'. Louisa wrote that the testimonial said, in part, that Herschel was one of the clearest and most skillful lecturers.
10. *Prof[esso]r Rankine*: William Rankine.

11. *Prof[esso]r Grant*: Robert Grant (1814–92), a Scottish astronomer. He largely taught himself Latin, Greek, mathematics and astronomy and in 1835 he observed Halley's Comet. He attended King's College, Aberdeen in 1839–40, but then moved to London to work as a bookkeeper. He continued his astronomical work, and by 1850 he was FRAS and in 1852 he published his *History of Physical Astronomy from the Earliest Ages to the Middle of the Nineteenth Century* (London: Baldwin). He became the chair of practical astronomy at Glasgow in 1860 and remained there. He was elected FRS in 1865 and was active in astronomical and scientific circles in Britain and the rest of Europe.

From George Henry Lewes[1] [before 9 June 1871][2] 3448

I wish to direct your attention to the experiments of Von Recklingshausen[3] should you happen not to know them. They are striking confirmations of what you say of dust and disease. Last spring, when I was at his laboratory in Würzburg, I examined with him blood that had been three weeks, a month, and five weeks, out of the body, preserved in little porcelain cups under glass shades. This blood was living and growing. Not only were the Amoeba-like movements of the white corpuscles present, but there were abundant evidences of the growth and development of the corpuscles. I also saw a frog's heart still pulsating which had been removed from the body (I forget how many days, but certainly more than a week). There were other examples of the same persistent vitality or absence of putrefaction. Von Recklingshausen did not attribute this to the absence of germs—germs were not mentioned by him; but when I asked him how he represented the thing to himself, he said the whole mystery of his operation consisted in keeping the blood *free from dirt*.[4] The instruments employed were raised to a red heat just before use, the thread was silver thread and was similarly treated, and the porcelain cups, though not kept free from air, were kept free from currents. He said he often had failures, and these he attributed to particles of dust having escaped his precautions.

J. Tyndall, 'On Dust and Smoke', *Roy. Inst. Proc.*, 6 (1870–72), pp. 366–7
Typed Transcript Only

1. *George Henry Lewes*: Lewes (1817–78), writer, philosopher and literary critic, as well as an amateur physiologist. He was well-known in London literary society and was friends with Charles Dickens, Herbert Spencer, and Charlotte Brontë. He lived with author Mary Ann Evans, known by her penname George Eliot, in a long-term committed relationship from 1854. Because of his work in physiology, including *Physiology of Common Life* (1859–60) and *Problems of Life and Mind* (1873–8), he was well known in scientific circles as well.

2. *[before 9 June 1871]*: This letter is from the *Roy. Inst. Proc.*, and comes just before the following letter (see letter 3449). Tyndall quotes this and letter 3449 in a lecture on 9 June, thus putting these letters before that date, but we are not sure exactly when.
3. *Von Recklingshausen*: Friedrich Daniel von Recklingshausen (1833–1910), a German pathologist, known for his study of neurofibromas.
4. *free from dirt*: This was key in all experiments in the spontaneous generation/germ debate, which is what Tyndall was working on at this time.

From Mr. Ellis[1] [before 9 June 1871][2] 3449

"I do not know," writes[3] Mr. Ellis, "whether you happened to see the letters, of which I enclose you a reprint,[4] when they appeared in 'The Times.' But I want to tell you this in reference to my method of vaccination as here described; because it has, as I think, a relation to the subject of the intake of organic particles from without into the body. Vaccination in the common way is done by scraping off the epidermis, and thrusting into the punctures made by the lancet the vaccine virus. By the method I use (and have used for more than twenty years) the epidermis is lifted by the effusion of serum from below, a result of the irritant cantharadine[5] applied to the skin. The little bleb thus formed is pricked, a drop of fluid let out, and then a fine vaccine point is put into this spot, and after a minute of delay it is withdrawn. The epidermis falls back on the skin, and quite excludes the air—and not the air only, but what the air contains.

Now mark the result—out of hundreds of cases of re-vaccination which I have performed, I have never had a single case of blood-poisoning or of abscess. By the ordinary way the occurrence of secondary abscess is by no means uncommon, and that of pyaemia[6] is occasionally observed. I attribute the comparative safety of my method entirely, first, to the exclusion of the air and what it contains; and, secondly, to the greater size of the apertures for the inlet of mischief made by the lancet.

J. Tyndall, 'On Dust and Smoke', *Roy. Inst. Proc.*, 6 (1870–72), p. 367
Typed Transcript Only

---

1. *Mr. Ellis*: possibly Robert Ellis (1819/20–82), a classical scholar, ordained priest and alpinist. He studied Hannibal's route through the Alps and got into a fierce debate over the matter with William Law, from 1854–6. The main reference to this particular Mr. Ellis in association with Tyndall is in *Hours of Exercise in the Alps*, p. 318. After Tyndall was injured on a mountain, he met Mr. Ellis about halfway down. He examined Tyndall and gave him treatment advice.

2. *[before 9 June 1871]*: dated according to the *Roy. Inst. Proc.* issue, from the 9 June meeting, which states that Tyndall received this note 'a day or two ago', putting the letter before that date. See letter 3448.
3. *writes*: The text of this letter is recorded in Tyndall's talk 'On Dust and Smoke' from 1871 (see the archival information above). The page in the *Roy. Inst. Proc.* begins with the following text, but we have removed it to this note, for possible context: '... and dirt, and the wisdom of avoiding them. The note is from Mr. Ellis, of Sloane Street, to whom I owe a debt of gratitude for advice given to me when sorely wounded in the Alps'. See n. 1 and *Hours of Exercise in the Alps*, p. 318, for the story of Tyndall's being wounded.
4. *a reprint*: a series of letters by Ellis on the subject of 'Re-Vaccination' appeared in the *Times* in 1871 (January, February, March and June).
5. *cantharadine*: a potent blistering agent.
6. *pyaemia*: blood poisoning.

From Emil du Bois-Reymond — 11 June 1871 — 3450

Berlin, 17 Victoria Str. | June 11th, '71

My dear Tyndall,

You really overwhelm me with presents. In the trouble and excitement of last year I did not find leisure to thank you for your Researches on Diamagnetism,[1] and I still felt a sting of remorse about it every time I thought of you or when the book caught my eye on the shelves of my library; and lo! now you pour a stream of lava on my head (to improve on the scriptural simile)[2] in the shape of two handsome volumes more![3]

Your "hours of exercise" have been hours of delightful recollection for me, and have stirred up within me the kakoethes scandendi.[4] Since my tour in Switzerland and Italy in 1869, when I sprained my ankle on the Mont Pers,[5] I have left Berlin only for two short trips, and I now long for trying my foot on ice and crag once more. Although I would hardly be able to follow you in one of your greater excursions, I would be extremely happy to meet you in the field of action. The misfortune with me is that my vacation begins too late and that the true climbing season is regularly over just when I got into the proper condition. One thing interested me highly in your accounts of accidents with avalanches. I was not aware of the fact of the men buried in the avalanche being emprisoned in the snow by regelation at the instant of arrest, just as if they were caught in a mould of plaster.[6] It is one of the most curious instances of regelation on record, and I was the more struck by it because there is a sort of analogy between this fact and the beautiful quicksand-experiment which I communicated to our friend Bence Jones some time ago, and which no doubt he reported in your presence. The analogy, however, is not a perfect one; for in the case of the avalanche the snow particles do not press against each other because they are all carried down by gravity at an equal rate, while in the case of the quicksand the grains of sand do not press one upon the other because the tendency to fall is counter-acted by the pressure of the water being stronger on the under surface of the grains than on the upper one, in consequence of the upwards motion of the water.

I read in your "Fragments of Science" with a professional interest, being something of a popular lecturer myself in these parts of the world and much I find to learn and to admire in them.

How much I should like to have a talk with you about the events of last year. Poor France![7] I have little hope for her, and my prophecies so far have but too well been fulfilled.

Helmholtz is now fairly at work in his new laboratory in the University,[8] and of course his presence here will inaugurate a new period of scientific life in Berlin. Radau[9] and König[10] are still here. I do not know whether König will return to Paris or settle here for good. I do not think, however, that he would do the latter unless on dire compulsion. He is intensively frenchified. Old Poggendorff, Dove[11] and Riess[12] are alive and well, Quinke[13] very downcast on account of Helmholtz having got Magnus[14] place. Good-bye my dear Tyndall, believe me

yours faithfully | E du Bois Reymond

RI MS JT/1/D/150
RI MS JT/1/TYP/7/2439–40

---

1. *Researches on Diamagnetism*: J. Tyndall, *Researches on Diamagnetism and Magne-crystallic Action* (London: Longmans, Green, and Co., 1870).
2. *lava . . . simile*: from Proverbs 25: 21–22: '[21]If your enemies are hungry, give them food to eat. If they are thirsty, give them water to drink. [22]You will heap burning coals of shame on their heads, and the Lord will reward you'.
3. *two handsome volumes more*: *Fragments of Science* and *Hours of Exercise in the Alps*, both published in 1871.
4. *kakoethes scandendi*: also cacoethes scandendi, or the urge to climb mountains (Latin).
5. *Mont Pers*: known today as Munt Pers; a mountain in the Bernina Range of the Alps, overlooking the Morteratsch Glacier.
6. *emprisoned . . . mould of plaster*: frozen in the snow and ice instantly.
7. *events . . . Poor France!*: referring to the Franco-Prussian War. See also, for example, letters 3370, 3387 n. 9, 3382, 3416, 3438, 3444 and 3451 for more discussion about the war and its impact on the practice of science.
8. *Helmholtz . . . University*: in his new position at Humboldt University in 1871 as professor of physics.
9. *Radau*: Jean Charles Rodolphe Radau (1835–1911), a Prussian astronomer and mathematician. He worked in Paris on the *Revue des deux Mondes* from 1866 with Moigno (see letter 3370) and was the co-founder of the *Bulletin Astronomique*. In 1871, he received a

PhD in honor of his work on mathematics, especially the three-body problem which has to do with solving problems for the movements of three physical bodies in relation to each other. He worked on lunar theory, and in 1892 won the Prix Damoiseau from the Académie des Sciences in France and was elected a Fellow in 1897 for this work. He was also elected FRAS in 1905.

10. *König*: Karl Rudolph Koenig (1832–1901), also known as Rudolf Koenig, a German physicist who studied acoustic phenomena. He studied under Jean Baptiste Vuillaume from 1851–8 and became a master violin maker. He then opened his own business as an acoustical instrument maker for scientists. He collaborated with Victor Regnault (see letter 3382, for example) in 1866 and in 1868 he earned an honorary doctorate from the University of Königsberg. In 1882 he traveled to America and gave public lectures in Toronto.
11. *Dove*: Heinrich Wilhelm Dove (1803–79), a Prussian physicist and meteorologist. He found that cyclones and hurricanes rotate in different directions in the Northern and Southern hemispheres. He was also well-known for his work on electricity and magnetics. He was elected FRS in 1850, and won the RS's Copley Medal in 1853. He was the rector at the Friedrich-Wilhelms-Universität in Berlin in 1871. The Dove crater on the moon is named after him.
12. *Riess*: Peter Theophil Riess (1804–83), a German physicist known for his work in electricity and magnetics. While he never took an academic position, he published a number of books, and prolifically in the *Annalen der Physik und Chemie*.
13. *Quinke*: Georg Hermann Quincke (1834–1924), a German physicist known for his work in chemistry and experimental work on reflection of light. He developed an interference tube that would show the interference patterns of sound waves.
14. *Magnus*: Heinrich Gustav Magnus (1802–70), a German physicist, Professor of Physics, and Director of the Physical Cabinet at the University of Berlin. He was well-known for his teaching as well as his experimental physics. He published a number of books and articles. In the 1860s, Tyndall had a dispute with Magnus about the absorption of radiant heat, though their relationship remained amicable. See *Ascent of John Tyndall*, pp. 166–8, 172–3, 179.

From Gustav Wiedemann 11 June 1871 3451

Leipzig im Fridericianum | *[d]* 11ten Juni 1871

Lieber Freund!

Nehmen Sie meinen herzlichsten Dank für die so freundliche Uebersendung Ihrer letzten schönen Werke, deren Lectüre mich und meine Frau in gleichem Maaße erfreut. Sie wissen, wie hoch ich Ihre mit so viel Erfolg gekrönten Bemühungen schätze, neben der streng wissenschaftlichen Thätigkeit auch zu der Einbürgerung unserer Wissenschaft in weiteren gebildeten Kreisen beizutragen. Es ist dieses Streben auch bei uns um so mehr zu fördern, als bisher in vielen Kreisen die Classicität in Folge der vorhin nicht ganz unberechtigten, überwiegend philologischen Bildung unserer Jugend die Alleinherrschaft ausübte. Nehmen Sie also nochmals für Ihre neuen Leistungen *[1 word illeg]* den besten Dank und besten Glückwunsch.

Der Krieg mit seinen schweren, aber auch wieder erhebenden Eindrücken hat die wissenschaftliche Arbeit bei uns sehr zurückgedrängt.—

Wie Jeder bei uns Hand anlegte, wo es irgend möglich war, habe auch ich längere Zeit ein großes Lazareth dirigirt—. Dadurch ist dann auch der Druck meines, im Manuscript fertigen Galvanismus verzögert worden. Ich hoffe indeß, die Vollendung des Ganzen wird jetzt schnell [gefördert] werden. Inzwischen hat mich sehr unverhofft ein Ruf von Carlsruhe nach Leipzig entführt, in einen wissenschaftlichen Kreis & eine Thätigkeit [wo ich hin zu] [gehen] wünschte. Ich habe die Professur für physikalische Chemie angenommen, wo ich, alten Jugendwünschen entsprechend, in beiden Gebieten arbeiten, wirken kann. Hoffentlich sehen wir Sie auch hier einmal, nachdem in Carlsruhe nur meine Frau in einem unruhigen Moment der Abreise die Freude hatte, Sie zu begrüßen. Sie finden in Hotel Wiedemann, N° 3, gegenüber der Bürgerschule, stets ein bescheidenes Zimmer für sich bereit.

Vieweg schreibt nun so eben, daß er sehr gern eine Uebersetzung Ihrer [letzten] hours of exercise publiciren möchte und ersucht mich, mich derselben anzunehmen. Haben Sie etwas dagegen einzuwenden? Es würde mich sehr erfreuen, für die Deutsche Veröffentlichung Ihres <hervorragenden> Buches Sorge tragen zu können. Vielleicht würde meine Frau die Sache in die Hand nehmen, womöglich mit Frau Helmholtz gemeinschaftlich. Ich erwarte indeß erst Ihre Antwort, ehe ich mit den beiden Damen conferire.— Ihr zweites Werk wird etwas schwieriger zu publiciren sein, vielleicht sollte man wenigstens einzelne Ihrer Vorträge erst auswählen. Bitte, sagen Sie mir darüber Ihre Meinung.

In aufrichtiger Ergebenheit & mit bestem Gruß von meiner Frau | freundschaftlichst | Ihr | G. Wiedemann.

Leipzig, in the Fridericianum | 11th June 1871

Dear friend!

Accept my warmest thanks for so kindly sending your latest fine works,[1] the reading of which delights me and my wife in equal measure. You know how much I appreciate your such successful efforts, along with your strictly scientific work, to also contribute to the introduction of our science in wider educated circles. This striving has to be supported all the more over here too, as, in many circles up till now, Classicism has exercised absolute power as a consequence of the [previously] not entirely unjustified predominantly philological education of our youth. Accept, therefore, my best thanks once again and best congratulations for your new achievements.

The war,[2] with its difficult but also edifying effects, has set scientific work over here back very much.—

Just as everyone over here lent a hand wherever it was possible, I also directed a large military hospital for a considerable time—. Because of this then, the printing of my Galvanism,[3] which was already finished in manuscript, has also been delayed. I hope, however, that the completion of the whole thing will now be *[moved along]* quickly. In the meantime, the offer of a chair has carried me off very unexpectedly from Karlsruhe to Leipzig, into a scientific circle & a job *[where I]* wished *[to go]*. I have accepted the professorship in physical chemistry[4] where, fulfilling old dreams from my youth of working in both fields, I shall be able to have some effect. Hopefully we shall see you here too some time, after my wife alone had the pleasure of greeting you in Karlsruhe, and even then while she was busy about to leave. You will always find a modest room ready for you in Hotel Wiedemann, No 3, opposite the Bürgerschule.[5]

Vieweg[6] has just written now that he would very much like to publish a translation of your *[latest]* Hours of Exercise and is requesting me to take care of it. Do you have any objection to this? I would be very delighted to be able to take care of the German publication of your *[excellent]* book. Perhaps my wife[7] would take the matter in hand, possibly together with Frau Helmholtz.[8] I shall wait, however, for your answer first before I confer with the two ladies.—Your second work[9] will be somewhat more difficult to publish; perhaps one should at least select some of your lectures *[first]*. Please, tell me what you think about this.

In sincere devotion & and *[with]* best regards from my wife | in greatest friendship | Your | G. Wiedemann.

RI MS JT/1/W/49

1. *latest fine works*: likely *Hours of Exercise in the Alps* and *Fragments of Science*.
2. *The war*: Franco-Prussian War. See letter 3450, n. 7.
3. *Galvanism*: Wiedemann was well-known for his work *Lehre von Galvanismus und Elektromagnetismus*, which was first published in 1861.
4. *professorship in physical chemistry*: Wiedemann held this position at Leipzig until 1887, when he moved to physics.
5. *Bürgerschule*: roughly the equivalent of the British secondary modern school. This was a municipal secondary school that prepared its pupils for practical occupations in commerce or in the trades, rather than for university study.
6. *Vieweg*: Friedrich Vieweg; see letter 3432.
7. *my wife*: Clara Wiedemann; see *Ascent of John Tyndall*, p. 207 and p. 272.
8. *Frau Helmholtz*: see letters 3432 and 3434.
9. *second work*: *Fragments of Science*.

From Julia Herschel[1] 12 June [1871][2] 3452

Upton Court | Slough. | June 12th

My dear Mr Tyndall,

I hope you will forgive my troubling you with an unnecessary & perhaps foolish letter, but I cannot resist the wish I have to endeavour in however poor a fashion, to express my share of our heartfelt thanks to you for your true & honest words of interest in our brother's career, & the most kind help you have also given him.[3] Just now[4] we feel such kindness so deeply, & he himself is no less truly sensible of it—I need not add how anxiously we hope & trust he may succeed in gaining this Professorship, but if it is not to be, at all events we can be & shall always be grateful for the many proofs of real approval & friendship he has received on all sides, and—help him to bear the disappointment by not being too much disappointed ourselves.

I am sure I ought to apologise for thus troubling you—the fact of my having only this m[ornin]g left home, its glory & sunshine for ever fled! & being about to leave England for some months in a few days, can be my only excuse.

Believe me | yours very sincerely | Julia Herschel

Upton Court. Slough. | June 12th

RI MS JT/1/H/119
RI MS JT/1/TYP/2/594

1. *Julia Herschel*: Julia Herschel (1842–1933), the ninth of twelve children of John Herschel and Margaret Brodie (see letter 3466, n. 1), therefore the sister of Alexander. She married in 1878 but had no children of her own.
2. *[1871]*: LT dated this to 1876, but here Julia is referring to Alexander Herschel's application to the University of Durham (see letter 3447), placing this letter in 1871.
3. *our brother's . . . given him*: Alexander Herschel; see letter 3447.
4. *just now*: John Herschel died on 11 May of this year; see letters 3424 and 3425.

To Mary Aitkin[1] 13 June 1871 3453

13th June 1871

My dear Mrs. Aitkin

Accept this book[2] from me, which has been intended for you for some time.

With love to your uncle[3] | Believe me | Yours most faithfully | John Tyndall

Wellcome Library MS 7777/14

1. *Mary Aitkin*: Mary Carlyle (née Aitkin, 1848–95), Thomas Carlyle's niece. During the last decade of her life she lived with and cared for him, running his household and assisting him in his writing. In 1879 she married her cousin Alexander Carlyle.
2. *this book*: either *Fragments of Science* or *Hours of Exercise in the Alps*.
3. *your uncle*: Thomas Carlyle.

To Richard Owen 13 June 1871 3454

Royal Institution of Great Britain | 13th June 1871

My dear Owen.

Will you pardon me for addressing you in this free frank way. It is prompted by the pleasure I experienced last night in noticing you and Huxley conversing together.[1] At the outset I should have relieved you both of what seemed to be a difficulty had not a still small voice whispered to me that this unforeseen arrangement might be the means of improving your relations to each other;—perhaps of dissolving the estrangement which has so long subsisted between you.[2]

I earnestly trust that you will allow the spirit of peace and friendliness which operated last night to have further scope. Huxley I know is capable of raising himself above all these personal mists, and by meeting you amicably in future I cherish the confident hope that you are capable of meeting him in the same spirit. Do so pray.—

There will be rejoicings among those who have the interests of science at heart over the consummation.

Believe me dear Owen | Ever truly yours | John Tyndall
Professor Richard Owen

RI MS JT 2/10/472–3a

1. *pleasure ... conversing together*: at a dinner at the Merchant Taylors' Hall, in London. The Hall itself is the seat of the Worshipful Company of Merchant Taylors, one of the Great Twelve (see M. Davies and A. Saunders, *History of the Merchant Taylors' Company* (Leeds: Maney Publishing, 2004)). The Hall can still be used for private parties and dinners. This dinner was associated with the Company itself and Tyndall arrived late to see that Owen and Huxley were seated next to each other. He saw the men talking, Huxley with a 'good humoured expression'. Huxley related to Tyndall that he and Owen had talked about being friendly with each other again. See Journal, pp. 1377–8; letters 3458, 3461 and 3463.

2. *improving . . . you*: referring to their troublesome relationship which had really begun long before it became public in 1860. Owen argued that human brains had parts that ape brains did not, and so humans were a completely unique animal. Huxley argued that humans, apes and other primates were very closely related and humans were not unique. The This became known as The Great Hippocampus Debate (or Question). See also letter 3458, n. 1.

From Acton Ayrton 14 June 1871 3455

14 June 1871.
[H.M. Office of Works | Whitehall Place][1]

My dear Sir

Having imputed the Clock Tower[2] I find greater facilities than I expected for fixing and attending to the light[3] but as this is so much of an experiment I think the best course would be to ask the person[4] who you stated had invented the new machine[5] whether he would have it put up with the lamp &c on the understanding that if it answered he should sell it for a sum to be fixed provided the House sanctioned the expense

He would have the advantage of a splendid advertizement which might recompence him for taking part in the experiment & if he agrees & you think the new machine is sufficiently advanced I should be glad if he could put himself in communication with M[r] Galton[6] in this office the Director of Works who will make the requisite arrangements.

I am | Yours faithfully | Acton S Ayrton

Prof. Tyndall | &c &c &c

RI MS JT 2/10/1194–6

1. *[H.M. Office of Works | Whitehall Place]*: LT annotation.
2. *Clock Tower*: referring to the clock tower at Parliament. The tower is commonly called Big Ben, even though Big Ben is technically the nickname of the Great Bell in the striking clock.
3. *the light*: later named the Ayrton Light. It was eventually installed in 1885 to show Queen Victoria at Buckingham Palace if either house of Parliament was sitting after dark.
4. *the person*: William Ladd.
5. *the new machine:* a type of light; see also letter 3460.
6. *Mr Galton*: Douglas Galton. Ayrton appointed Galton early in 1870 to the post of Director of Public Works. His job consisted of overseeing new construction, especially with an eye to cost.

From Archibald Hamilton 14 June 1871 3456

Trillick | Omagh[1] | June 14/71

Dear Sir,

Having lately read your Lectures on Sound,[2] with much gratification, & having myself spent much time & trouble upon the subject, (partly with the hope of communicating some of its pleasures to the Deaf) I take the liberty of asking one or two questions upon the subject.

In the first place I know that that part of Sir I. Newton's Principia which treats of Sound[3] is admitted to be very unsatisfactory & defective.[4] But surely every written thought of so great a thinker deserves the most attentive study—Where can a student find the best commentary on the line of thought which was in Newton's mind, & which is perhaps not very happily expressed in the received Edition of the Principia?

Secondly, I want to know, where can I find a good popular account of those sounds which may be termed essentially un-musical or inarticulate? I mean such sounds as are caused by slackening or tightening a string or by increasing or diminishing its length while it is sounding.

To my ear, increasing the tension gives a peculiar plaintive sound, something like the wail of the gnat, while the contrary gives rather a blunt cheerful sound.

Now I believe myself that these sounds are truly and independently perceived by the ear, in virtue of an implanted power of detecting (within certain limits) minute variations in the pressure of the air.

But I can quite conceive an opponent arguing thus: "No, for the string does not continuously slacken or tighten, but with fits or intervals of repose, & thus the effect on the surrounding air is that of a very rapid succession of musical notes, a cadence or run &c &c". This is of course a priori possible when one does not know the law of elasticity in the string. It is a point for physicists to decide:—But if it should turn out that the string does slacken or tighten continuously, (though perhaps not just as the pressure) & if the air likewise transmits the motion of the string to the ear, then it seems to me that it is certain from hence that the ear must have an independent power of registering pulsations of the air whose formula is entirely incommensurable with that of the ordinary musical note. (I ought to say, its formula is incompatible & its structure may be called incommensurable, as much as the circle with the triangle).

If we express the ordinary musical note by T = const[ant]. where T means the tension of the unit string which would produce this sound, then the other will be expanded by T = variable or if we say for the musical note D = 0, meaning by D the first Difference of x of successive maxima or minima or crest of waves &—then for the other we shall have D ≠ 0, however it may be varied.

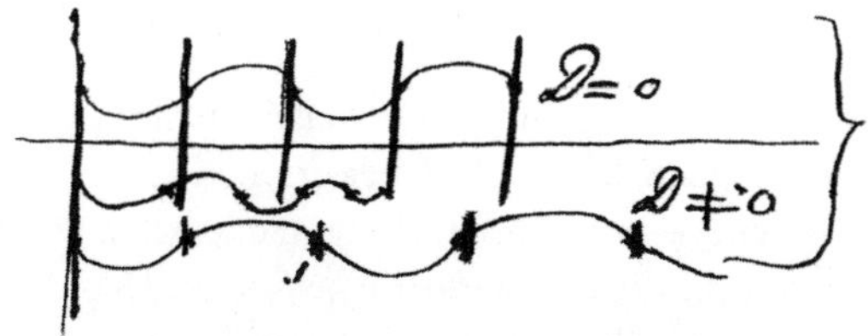

I believe moreover that these inarticulate sounds are of the utmost importance in the right understanding of Nature, being far more common & abundant than musical sounds, which are only an infinitely particular case of them. But I should not be surprized to learn that they are more easily destroyed or at least more rapidly lost than musical sounds, & therefore less permanent, as they are certainly less pleasing. I can readily understand likewise that the presence of that Harp of many strings[5] which Anatomists speak of, in the Ear, must add very greatly both to the powers of that member in taking cognizance of musical notes & combinations—& also to the pleasure which may be produced by a skilful & the contrary by an unskilful combination—also that musical & articulate sounds may be promoted & encouraged by it in the development of the voice &c. &c.

But I believe that quite independent of this there must be a power in the ear which I have ventured boldly to develop in the accompanying sketch of a system of Hearing.

I remain | yours faithfully in the Lord | A. H. Hamilton Clk[6]

RI MS JT/1/H/11

1. *Trillick Omagh*: small village in what is now Northern Ireland.
2. *Lectures on Sound*: Tyndall's *Sound*.
3. *Principia . . . Sound*: Newton postulated that the rules he applied to the movement of waves in fluids applied to both light and sound. In Book II, he postulated the speed of sound, and argued that the speed of sound could change depending on how much water was in the air (*Principia*, Book II, p. 181–2).
4. *defective*: Newton's theory did not take into account the effect of temperature in the sound wave itself. Laplace figured this out by 1816.
5. *"Harp of many strings"*: possibly a reference to the poem 'A Harp of Many Strings' by American poet George Henry Calvert. See G. H. Calvert, *Anyta and Other Poems* (Boston: E. P. Dutton, 1866), pp. 51–5.
6. *A. H. Hamilton Clk*: Clk is likely short for Clerk; possibly the position he held. This letter is almost identical to a later letter Hamilton sent to Tyndall. See also letters 3499 and 3600.

From Georg Krebs[1] 14 June 1871 3457

Geehrtester Herr Professor!

Nehmen Sie nicht ungütig, daß ich so frei bin einige Zeilen an Sie zu richten & entschuldigen Sie besonders, daß ich mich dabei der deutschen und nicht der englischen Sprache bediene;—es ist zu schwierig, namentlich wenn es sich um wissenschaftliche Gegenstände, Kunstausdrücke *[etc etc]* handelt, in einer fremden Sprache zu schreiben, selbst wenn man das in dieser Sprache Geschriebene oder Gedruckte ohne Schwierigkeit versteht.

Ihr Buch über die Wärme, welches in Deutschland mit so großem Eifer gelesen wird, berührt auch an einzelnen Stellen die in der neusten Zeit viel besprochenen Siedeverzüge (retards of ebullition), sowie die Explosionen der Dampfkessel nach den neueren Ansichten.

Allerdings liegt es wohl nicht in dem *[Plane]* Ihres Buches diesen Gegenstand ausführlicher zu besprechen, sondern nur das Wesentliche hierüber anzudeuten und es wäre deshalb Anmaßung von meiner Seite, wollte ich den Wunsch aussprechen, Sie möchten auch meinen Arbeiten über diese Gegenstände, welche sich theilweise an die von Dufour (Pogg. Ann. Bd 124) anlehnen, in Ihrem Werke einen Platz *[verstatten]*. Andrerseits freilich wäre dieser Wunsch bei dem großen Gewicht, welches man in unserem Lande auf Ihre Arbeiten legt, wohl verzeihlich.

Ich spreche nicht von den in meiner ersten Abhandlung (Pogg. Ann. Bd CXXXIII, pag 673) über Siedeverzüge enthaltenen 4 Versuchen, obwohl dieselben gerade das große Publicum, wie ich mich bei öffentlichen Vorlesungen überzeugt habe, in nicht geringes Erstaunen versetzen; man verwundert sich, daß Wasser, welches (in dem Zustand des Siedeverzugs) eben noch ganz regungslos war, sogleich darauf (durch Erschütterung, Erhitzung, Einführung von Gas, oder plötzliche Druckverminderung) in explosives Sieden geräth. Erwähnenswerth aber (neben den Versuchen von *[Douny]*) dürfte es sein, daß es mir gelungen, durch Vermischen von Wasser & starkem Weingeist & nachheriges Eindampfen bis auf eine kleine Quantität, das Wasser so luftfrei zu machen, daß es fast auf 200°C erhitzt werden kann, ohne zu sieden; und daß es dann entweder in explosives Sieden geräth, oder <u>nur an der Oberfläche mit großer Geschwindigkeit verdunstet ohne eine einzige Dampfblase zu entwickeln.</u> Was ferner die Dampfkesselexplosionen anlangt, so erwähnen Sie, daß Dampfkessel gewöhnlich dann explodiren, nachdem sie einige Zeit in Ruhe gestanden. Ich habe nun hierüber zahlreiche Versuche angestellt & es ist mir wiederholt vorgekommen, daß ein Dampfkessel (von Glas), nachdem er zur Ruhe gestellt worden, nach einiger Zeit ohne weiteres Zuthun plötzlich von selbst explodirt ist, weil das Wasser allmählig in den Zustand eines sehr starken Siedeverzugs gekommen war (Pogg. Ann. Bd CXXXVIII, pag 439).

Sollten Sie vielleicht die Güte haben wollen in einer neuen Auflage Ihres Werkes dieser meiner Arbeiten auch nur mit einem Worte zu gedenken, so würde ich Ihnen sehr dankbar sein. Zugleich erlaube ich mir Ihnen eine kleine Abhandlung über das Sieden des Wassers, sowie meinen dritten Aufsatz über Siedeverzüge aus Pogg. Ann. blos als Ergänzung zu übersenden, da Ihnen ja diese Annalen selbst zur Hand sind.

Schließlich möchte ich mir noch die Frage erlauben, ob alle Ihre Arbeiten über singende Flammen in dem Philosophical Magazine erschienen sind. Ich will selbst einige Untersuchungen hierüber anstellen & einige *[1 word illeg]*.

Hochachtungsvoll | Dr Georg Krebs.

Wiesbaden 14/VI 71. | (Preußen)

Most honoured Professor!

Please do not take it unkindly that I am taking the liberty of sending you some lines & excuse in particular the fact that in doing so I use the German and not the English language—it is too difficult, especially when dealing with scientific subjects, technical terms *[etc etc]*, to write in a foreign language, even if one understands what is written or printed in that language without difficulty.

Your book on heat,[2] which is read in Germany with such great enthusiasm, in some places also touches on the retards of ebullition[3] that have been much discussed in recent times, as well as the explosions of steam boilers according to the more recent views.

However, it is probably not within the *[plan]* of your book to discuss this subject in greater detail, but to hint at only the essential element about it and it would therefore be presumptuousness from my part if I wanted to express the wish that you might also *[allow]* a place in your work for my papers on these subjects, which partially follow those by Dufour[4] (Pogg. Ann.[5] Vol 124).[6] On the other hand, however, this wish would probably be excusable, given the great store that one sets on your papers in our country.

I am not talking about the 4 experiments contained in my first article on the retards of ebullition (Pogg. Ann. Vol CXXXIII, p. 673),[7] although these especially astonish the wider public greatly, as I have convinced myself when giving public lectures; people are amazed that water which (when in a state of retarded ebullition) was quite still just now, immediately afterwards undergoes explosive ebullition (through shaking, heating, addition of gas, or a sudden drop in pressure). However, it might be worth mentioning (along with the experiments by *[Douny]*)[8] that I succeeded, by mixing water & strong ethyl alcohol & subsequently evaporating it down to a small quantity, in making the water so free of air that it can be heated to nearly 200°C without boiling; and that it then either undergoes explosive ebullition or evaporates very quickly on the surface only and without forming a single bubble of steam. As far as the explosions in steam boilers are also concerned, you mention that steam boilers usually explode after having rested for some time. Now, I have done numerous experiments on this & it has repeatedly happened to me that a steam boiler (made of glass), after being set to rest, suddenly exploded by itself after some time, without further assistance, because the water had gradually come to a state of very high retarded ebullition (Pogg. Ann. Vol CXXXVIII, p 439).[9]

Should you possibly be so kind as to remember this paper of mine in a new edition of your work, even if only with one word, then I would be very grateful to you. At the same time, I permit myself to send you a short article on the boiling of water, as well as my third essay on retarded ebullitions from Pogg. Ann, just as a supplement, as, of course, you have these Annals already at hand.[10]

Finally, I would like to permit myself to ask if all of your papers on singing flames[11] have appeared in the Philosophical Magazine.[12] I want to do some experiments on them & *[1 word illeg]* some myself.

Yours faithfully | Dr Georg Krebs.

Wiesbaden[13] 14/VI 71. | (Prussia)

RI MS JT/1/K/33

---

1. *Dr Georg Krebs*: possibly Georg Krebs (n.d.), author of *Leitfaden der Experimental Physik für Gymnasien und zur Selbstbelehrung* (*Guide to Experimental Physics, for Gymnasia and Self-Instruction*) (Wiesbaden: J. F. Bergmann, 1881) and *Lehrbuch der Physik- und Mechanik für Real- und hohere Bürgerschulen, Gewerbschulen und Seminarien* (*Physics and Mechanics for Real- and Higher 'Burger' and Industrial Schools*) (Wiesbaden: J. F. Bergmann, 1882).
2. *Your book on heat*: *Heat Considered as a Mode of Motion* (London: Longmans, Green, and Co., 1863). Possibly referring to later editions: second (1865), third (1868), or fourth (1870); or German translations.
3. *retards of ebullition*: slowing down the boiling process.
4. *Dufour*: Louis Dufour (n.d.), a Swiss physicist who worked with De La Rive, Soret and others, and discovered the Dufour effect.
5. *Pogg. Ann.*: *Poggendorf's Annalen,* referring to the *Annalen der Physik und Chemie* when Johann Christian Poggendorf was the editor from 1824–76. It is useful to note that when new editors would take over editorship of the journal, they would restart the volumes from number one, alongside the continuous numbering. We have used the continuous numbers here, in order to standardize citations.
6. *my papers… Vol. 124)*: L. Dufour, 'Ueber das Sieden des Wassers und über eine wahrscheinliche Ursache des Explodirens der Dampfkessel', *Annalen der Physik und Chemie*, 200:2 (1865), pp. 295–328.
7. *(Pogg. Ann. Vol CXXXIII, p. 673)*: G. Krebs, 'Versuche über Siedverzüge', *Annalen der Physik und Chemie*, 209:4 (1868), pp. 673–7.

8. *[Douny]*: possibly the Mr. Douny referred to by Faraday in his Friday Evening Discourse at the RI, 'On certain Conditions of Freezing Water', on 7 June 1850, covered in *Athenaeum*, 15 June 1850, pp. 640–1.
9. *(Pogg. Ann. Vol CXXXVIII, p 439)*: P. Groth, 'Ueber Krystallform und Circularpolarisation und über den Zusammenhang beider beim Quarz und überjodsauren Natrium', *Annalen der Physik und Chemie*, 213:3 (1869), pp. 433–42.
10. *a short . . . Annals*: Copies are missing and we are unable to identify which articles Krebs refers to here.
11. *your papers on singing flames*: possibly J. Tyndall, 'On the Action of Sonorous Vibrations on Gaseous and Liquid Jets', *Phil. Mag.*, 33 (1867), pp. 375–91. This was the only one of Tyndall's papers on this topic in this magazine.
12. *Philosophical Magazine*: founded in 1798 and one of the oldest continuously published English-language scientific journals.
13. *Wiesbaden*: a city in what is now western Germany.

From Richard Owen 14 June 1871 3458

Sheen Lodge | Richmond Park.

14 June, 1871.

My dear Tyndall,

Professor Huxley disgraced discussion on a scientific difference by imputing falsehood on a matter in which he differed from me.[1] Until he retracts the imputation as publicly as he made it I must continue to believe that, in making it, he was merely imputing himself.

Believe me, | very sincerely yours, | Richard Owen.

RI MS JT 2/10/472–3b

---

1. *scientific difference . . . from me.*: referring to the Great Hippocampus Debate (or Question). On 19 March 1861, Owen delivered a lecture at the RI arguing that the gorilla's brain did not have a human cerebrum or hippocampus (published as 'The Gorilla and the Negro', *Athenaeum*, 23 March 1861, pp. 395–6). In 1862, Huxley argued that Owen was lying about the structure of ape brains (*Evidence as to Man's Place in Nature* (London: Williams & Norgate, 1863)). At the Cambridge meeting of the BAAS in 1862, Huxley had an ape brain dissected to demonstrate that it had a hippocampus. Huxley believed this was evidence of Owen being a 'mendacious humbug' and felt he had to 'get a lie recognized as such' (Desmond, *Huxley*, pp. 292–311). See also letter 3454.

From Edward Sabine 14 June [1871][1] 3459

Wednesday night | June 14

Dear Tyndall,

I enclose a note from Mr. Whymper.[2] I fear it is too late for any help from the Gov[ernmen]t. Grant this year. But if you think his object deserves such encouragement, it might still be aided, and in sufficient time, by a grant from the Donation Fund.[3] That Fund you know is at the absolute disposal of the Council;[4] and there is no one who would be listened to on such a subject so effectively as yourself.

I think you have still the Greenland Lecture?[5]

Sincerely yours | Edward Sabine

BL Add MS 53715, f. 3

---

1. *June 14 [1871]*: dated to 1871 because the letter relates to letters 3462, 3464, 3470, 3471, 3481 and 3487. 14 June 1871 was a Wednesday.
2. *note from Mr. Whymper*: note missing.
3. *Donation Fund*: a fund established in 1828 by an initial £2,000 donation from William Hyde Wollaston (1766–1828). Wollaston was a chemist, physicist and physiologist and was known for his work with platinum and with discovering palladium and rhodium between 1800 and 1809. He was elected FRS in 1793, was its secretary from 1804–16 and was president briefly in 1820. He established the Donation Fund at the RS to award money for 'the promotion of experimental research', but this should not be confused with the Wollaston Fund of the Royal Geological Society. See *The Dictionary of National Biography*, vol. 62 (London: Smith, Elder & Co., 1900), p. 315. See also M. B. Hall, *All Scientists Now: The Royal Society in the Nineteenth Century* (New York: Cambridge University Press, 2002), pp. 39–40, 146–7.
4. *the Council*: the Council of the RS.
5. *Greenland Lecture*: This is possibly referring to William Bradford's lecture on Greenland at the RI on 16 June 1871: W. Bradford, 'On the Esquimaux and Ice of Greenland', *Roy. Inst. Proc.*, 6 (1870–72), pp. 377–8. Whymper had been to Greenland in 1867 and may have given a lecture on it to the RS, but this is unclear.

To Acton Ayrton 15 June 1871 3460

Royal Institution | 15th. June 1871.

My dear Sir,

The machine to which your note[1] refers is, as I explained to you at our interview, a new one. It has not yet been sufficiently tested to warrant its unqualified recommendation.

I should therefore propose that the maker be bound by agreement to furnish a light equal say to that furnished by 60 or even 50 ordinary Voltaic elements.

That he should guarantee the proper action of the machine for a certain period—say during a single Session of parliament—undertaking to take it back if it should break down during this interval.

The maker of the instrument is Mr William Ladd of Beak Street, Regent Street. I have no doubt the consideration as to advertizing urged in your note would have due weight with him, but the statement of these considerations lies I think outside of your province on the present occasion.

The most powerful machine of the character we are now considering is one constructed by Mr Holmes[2] for the Trinity House. Mr Holmes was also the first who turned mechanical power to account in the production of a light fit for such purposes as you are now contemplating

Mr Holmes is therefore another resource in case of need.

faithfully yours | John Tyndall

The first Commissioner of Works.

RI MS JT/2/13c/1380
LT Typescript Only

1. *your note*: letter 3455.
2. *Mr Holmes*: possibly Frederick Hale Holmes (b. 1812). Holmes was a professor of Chemistry at the Royal Panopticon of Science and Art. He worked on electric lighting, and his 'magneto-electric apparatus' was installed in Trinity House's South Foreland Lighthouse in 1858 ('Lighthouse Illumination by Magneto-Electricity', *The Dublin Builder*, 15 August 1864, p. 14).

To Richard Owen 15 June 1871 3461

Royal Institution of Great Britain | 15th June. 1871

My dear Owen,

I had hoped against hope:—for it has saddened me beyond measure to witness the conflict and consequent estrangement between two such men as you and Huxley.

Excuse me if the desire to see you friends should have rendered me too free in my interference.[1]

I need hardly say that my writing to you was my own proud act. To this hour Huxley knows nothing about it.

Believe me | Yours faithfully | John Tyndall

Prof. Owen | &c. &c.[2]

RI MS JT/2/10/476–7a

1. *Excuse . . . my interference*: letters 3454, 3458 and 3463.
2. *Prof. Owen | &c. &c.*: Tyndall wrote a note to himself: 'Knowing that Huxley was disposed to meet Owen more than half way and hoping to avoid the probable consequences of any advances on his part. I wrote to Huxley these 3 lines'. Here he copied letter 3463.

To Edward Whymper 15 June 1871 3462

The Royal Society, | Burlington House, London, W. | 15th. June 1871.

My dear Sir,

Sir Edward Sabine has forwarded to me (with a note from himself) your letter[1] respecting a proposed series of observations on the Veined Structure of Glaciers.[2] I had the pleasure of reading your letter, and the note which accompanied it, this day to the Council of the Royal Society. I expressed at the time my strong desire to promote a series of observations suggested by so eminent a glacialist as Principal Forbes[3] and to be carried out by yourself. The Council authorised me to communicate with you on this subject, and I can only ask you to rely upon my willingness to forward your wishes,[4] if you can give me any data, on which I can found a conscientious support.

I shall be glad either to see you or to hear from you sometimes between the present date and the 20th of this month. I go immediately afterwards to Ireland, and it will be necessary to communicate, in writing, the result of our conference to the Council on this day fortnight.

I am, my dear Sir, | faithfully yours | John Tyndall.

Journal JT/5/15b/217
LT Typescript Only

1. *forwarded to me . . . your letter*: original Whymper letter missing. See letter 3459.
2. *observations on the Veined Structure of Glaciers*: Whymper included a long discussion on this matter in his *Scrambles amongst the Alps*, pp. 356–60.
3. *glacialist as Principal Forbes*: James David Forbes (1809–68), a Scottish physicist and glaciologist. He was Principal of the United College of St. Andrews from 1860 until his death. He had first postulated the importance of the veined structure of glaciers, or the fact that there are vertical lines of ice in glaciers, demonstrating the process by which they form. Tyndall and Forbes had a long-lasting dispute on this subject which began in 1857 (*Ascent of John Tyndall*, pp. 114–21).
4. *your wishes*: see letters 3464, 3470, 3471, 3481 and 3487.

To Thomas Henry Huxley 16 June 1871 3463

16th. June 1871.

Dear Hal.

As regards Owen[1] I think I would let things take, for the present, their old course.

Yours ever | John Tyndall.

RI MS JT 2/10/476–7a
IC HP 8.89
Typed Transcript Only

1. *As regards Owen*: see letters 3454, 3458 and 3461.

From Edward Whymper 17 June 1871 3464

19 Canterbury Place. | Lambeth Road. S.E. | June 17. 1871.

My dear Sir,

I beg to acknowledge the receipt of your letter of the 15th June,[1] written on the part of the Council of the Royal Society.

It will perhaps be best for me to communicate with you by letter on the matter to which it refers[2]—I think that I shall be able to explain myself with sufficient precision in writing, and that your time will be spared by the adoption of that course. I will take care that you receive a communication upon the subject before the 25th ins[tan]t,[3]

I am, my dear Sir, | yours very obediently, | Edward Whymper.

Professor John Tyndall | F.R.S. etc. etc.

RI MS JT/1/W/36

1. *your letter of the 15th June*: letter 3462.
2. *the matter to which it refers*: the veined structure of glaciers (see letters 3462 and 3459).
3. *before the 25th ins[tan]t*: instant, meaning of the current month; see letter 3470.

To D. S. Thomas[1] 19 June 1871 3465

19th June 1871

Dear Sir,

My projected expedition to America,[2] which is still a mere project, will certainly not be carried out this year. There is therefore I suppose no need to hasten the matter which you have done me the honour of mentioning to me.[3]

I may say that in photographs which they have recently taken of me Mess[rs]. Elliott[4] & Fry[5] have been very successful. If either of those would answer your purpose I should be very glad to place it in your hands.

If however you wish for something special for the "Graphic"[6] (from which I often derive pleasure) I shall endeavour to meet your wishes.

Faithfully yours | John Tyndall | D. S. Thomas Esq[ui][re] | &c &c &c

Staatsbibliothek zu Berlin (Stabi) Slg Darmstaedter 1855 Tyndall Be. 1–58, 54

---

1. *D. S. Thomas*: possibly related to the Thomas family who owned and began the *Graphic* (see n. 6).
2. *projected expedition to America*: Tyndall went to America from October 1872 to February 1873 for a multi-city lecture tour.
3. *mentioning to me*: letter missing.
4. *Elliott*: Joseph John Elliott (1835–1903), a photographer, along with Clarence Fry (see n. 5). Their firm, Elliott & Fry, was established in 1863 at 55 Baker Street. They were responsible for thousands of portraits between 1863 and 1962, when the firm was acquired by Bassano & Vandyk. The National Portrait Gallery owns all the surviving negatives.
5. *Fry*: Clarence Edmund Fry (1840–97), a photographer and Elliott's business partner (see n. 4).
6. *the "Graphic"*: a weekly illustrated British newspaper, which ran from December 1869 to July 1932. It competed with the *Illustrated London News*.

From Margaret Brodie Herschel[1] 20 June 1871 3466

Collingwood 20 June | 1871.

Dear Professor Tyndall,

I must now return to you the two interesting letters enclosed,[2] having ventured to take copies of them to send to Alexander[3]—to whom the Dean of Durham[4] wrote in nearly the same terms, so I ventured to reply to him explaining the cause of Alex[ande][rs] absence in Norway—

I am happy to say also that Sir William Thomson, who had only just returned from Lisbon, wrote to the Dean without losing a Post "expressing a very high opinion of your fitness for the Office" as he says in a private letter to Alexander, who will be much gratified at this explanation of a silence which he could not understand—

Whether success follow all these kind efforts of his friends or not, I am quite sure that we all feel most grateful for them, & that we have the utmost confidence in the honour & conscientiousness of those who have to decide—

I have seen such occasions myself, often enough to know the extreme difficulty of balancing equal claims, & the painfully anxious care to come to a just & right conclusion—and in any case, this trip to Norway will have done dear Alexander good service in strengthening his health after his winter's labours & his great sorrow,[5] for his next winter's work, wherever that may be.

Hoping that you too may enjoy a Summer's Holiday,

believe me, | my dear Professor Tyndall | yours sincerely | M. B. Herschel.

RI MS JT/1/H/120
RI MS JT/1/TYP/2/595

---

1. *Margaret Brodie Herschel*: Margaret Brodie Herschel (1810–84), widow of John Herschel, mother of Alexander Herschel. She was the main artist of the botanical drawings on her and John's trip to the Cape of Good Hope from 1834–8.
2. *two interesting letters enclosed*: letters missing. These probably referred to A. Herschel's application to be the first professor of physics and experimental philosophy in the University of Durham College of Science, Newcastle upon Tyne (see letter 3447, n. 2).
3. *Alexander*: Alexander Stewart Herschel, her son.
4. *Dean of Durham*: see letter 3447, n. 4.
5. *great sorrow*: sorrow over his father's death on 11 May 1871; see letters 3424 and 3425.

From Marcel Croullebois 21 June 1871 3467

1871 | A Monsieur le Professeur Tyndall,

Monsieur,

Je vous remercie de tout mon cœur de votre bienveillant accueil et de l'assistance que vous avez bien voulu m'accorder auprès des savants anglais.

J'ai vu M. Wheatstone qui m'a montré ses ingénieux appareils d'optiques, plaçant chaque expérience sous les yeux et dans la main du lecteur, pour me servir de votre juste expression empruntée à votre préface du Son.

Je désirerais vivement voir vos appareils, devenus classiques en France, pour étudier le diamagnétisme des corps. Je passerai la semaine prochaine tout entière à Londres avant de me rendre à Edimbourg et à Dublin. Mais auparavant, je veux aller voir immédiatement l'Université de Cambridge et celle d'Oxford.

Je vous serais reconnaissant de vouloir bien me donner des lettres d'Introduction et de Recommandation auprès des professeurs de ces Universités ou des personnes influentes dans l'enseignement et dans la Science.

J'ai transmis au Ministre de France l'accueil et le concours, que vous m'avez prêtés, et surtout le souvenir vivace et chaleureux que vous avez conservé en sa faveur.

J'aurai l'honneur de me présenter demain dans l'après midi à l'Institution Royale, non pour vous faire définitivement mes adieux, mais pour vous adresser mes remerciements.

Daignez agréer l'expression sincère de mon respect et de ma reconnaissance.

Marcel Croullebois | London, ce 21 juin 1871

1871 | To Professor Tyndall,

Sir,

I thank you from the bottom of my heart for your warm welcome and the assistance that you were willing to provide me with the English scientists.

I met Mr Wheatstone,[1] who showed me his ingenious optical instruments, placing every experiment under the eyes and in the hands of the reader, to use an expression borrowed from the preface of your Sound.[2]

I would dearly love to see your instruments, now classics in France, to study the diamagnetism of bodies.[3] I will spend the whole next week in London before going to Edinburgh and Dublin. But before this, I want to see the Universities of Cambridge and Oxford.

I would be grateful to you if you could give me letters of introduction and recommendation[4] for the professors of these universities or influential persons in teaching and Science.

I have notified the French Minister[5] of the welcome and support you gave me, and above all of the vivid and warm memory that you keep of him.

I will have the honour of going to the Royal Institution tomorrow afternoon, not to say goodbye for good but to express my thanks to you.

Please accept the sincere expression of my respect and gratefulness.

Marcel Croullebois | London, 21 June 1871

RI MS JT/1/C/59

---

1. *Mr Wheatstone*: Charles Wheatstone (1802–75), an English inventor, best known for his Wheatstone bridge, used to measure electrical resistance. He also developed methods in telegraphy that were useful in underwater environments, invented the stereoscope and was

known for his other work in optics and sound. He was elected FRS in 1836 and knighted in 1868.

2. *Sound*: Tyndall's *Sound*, p. vii.
3. *instruments . . . diamagnetism of bodies*: instruments Tyndall used for experiments described in his *Researches on Diamagnetism and Magne-crystallic Action* (London: Longmans, Green, and Co., 1870).
4. *letters of introduction and recommendation*: see letter 3469, for example.
5. *the French Minister*: Jules Simon (1814–96), a French philosopher and politician; he was a republican and anti-radical. Simon held the post of minister of education from 1871 until 1873. See also letter 3557.

To [Norman Lockyer][1] 22 June 1871[2] 3468

Sir

The results of the examination[3] of 1871 had not reached me when your letter[4] arrived. But I am now in a position to make some response to the questions which the Duke of Devonshire[5] and the Commission of which his Grace is Chairman have done me the honour of submitting to me.[6]

My relation to the South Kensington examinations is this: I set all the examination papers, organise staff of gentlemen as assistant examiners; divide the staff into two portions who devote themselves to the two subdivisions of experimental physics. A principal assistant examiner takes charge of the Honours paper & of the checking & division of a certain fraction of the other papers in each subdivision. With him I communicate in all needful cases. The principal examiners also when called upon to do so report through me to the Department[7] on the results of the examination, referring to any features which may require future attention.[8]

1. In reply then to the first question[9] submitted to me the two principal assistant examiners[10] are unanimous in the conclusion that through the agency of such examinations as that just completed a great amount of sound scientific knowledge is diffused throughout the country. This I may add was also my own conclusion at the time when I read through the papers myself.

2. The second question submitted to me is this: "What in the evidence afforded as to the practical nature of the teaching?" It is difficult to answer it, for the term practical is eminently vague. If by it is meant teaching by aid of experiments with suitable apparatus then the teaching is defective in many cases. The teachers throughout the country have never as yet had any systematic training in practical experimentation; and although here and there a man may may become an experimenter through the force of natural bent, it is not to be expected that teachers in general are of this class. This therefore is an

object to be arrived at. The founding of an institution or of institutions in which teachers should be taught the use of apparatus sufficiently cheap, simple & effectual to be introduced with profit into class instruction.[11]

I would here however remark that although no scheme of instruction in physics is even approximately complete without illustrative experiments, an able teacher in the absence of apparatus can do a great amount of good. It is possible by the judicious use of the blackboard and chalk to convey clear conceptions of various parts of physics; so clear indeed that should the pupil afterwards witness the experiment he shall witness that of which had previously a sound, though not perhaps complete conception. Indeed, even where apparatus exists, the performance of experiments ought to go hand in hand with their diagrammatic exposition.[12]

3. The third question has reference to the amount of 'cram'. This word I find is differently interpreted but I take the significance of the word to be knowledge which bears the same relation to the mind as indigested food does to the body. This interpretation excludes the idea that all knowledge of experimental science which does not embrace the experiments themselves is necessarily cram. For as I said before by proper descriptions and diagrams clear conceptions may be conveyed, and the mind of the pupil may be exercised in a highly salutary manner. On the other hand "cram" I hold to perfectly compatible with the performance of experiments. For experiments may certainly be conducted in such a manner as to address the senses without laying any real hold of the intellect.

In depth, this standard the amount of 'cram' revealed by the late examination is far more than it ought to be. The detection of cram must ever depend upon the sagacity of the examiner.[13] Many teachers evidently set before themselves as the prime object of their efforts the augmentation of their funds. And they manifestly entertain the notion that this object is more likely to be gained by augmenting the number of the Candidates sent up than by any at more excellent instruction.[14] Many candidates have appeared at the recent examination where their incapacity must have been known to their teachers, provided of course that the teachers themselves are not incapable. Some method ought to be devised which should not only deprive such teachers of the advantages expected to arise from mere numbers, but also of subjecting them to a definite penalty for the perfectly fruitless trouble and expense which they entail upon the Department.

The detection of Cram must depend upon the sagacity of the examiner. And it is for this reason that I have I have sought to obtain as far as possible the cooperation of gentlemen, of experienced in teaching and distinguished in science.

4. With regard to the fitness of candidates to become teachers I would

say that knowledge and teaching power are two different things. The latter implies the former, but the former by no means implies the latter. Teaching power is to some extent a question of character which is only very imperfectly revealed by written examination. The chances undoubtedly are that the youth who expresses himself in writing with clearness and straightforwardness, will prove effective as a teacher.[15]

5. I think the bad spelling may, for the most part, be traced to the more youthful candidates. Still cases now and then arise in which knowledge, sound in quality and considerable in amount, is found associated with bad spelling. This defect, however, is a great nuisance and ought to be stringently dealt with.

While commending much of the work accomplished by the teachers and recognising in this examination many creditable results, I think it ought to be the aim of the Department to employ the less satisfactory returns with a view to lessening the incompetence or neglect which these returns imply.

I have the honour &c | John Tyndall.

RI MS JT/4/7b/270–3
John Tyndall to Norman Lockyer, 22 June 1871, 'Royal Commission on Scientific Instruction and the Advancement of Science. Vol. I. First, supplementary, and second reports, with minutes of evidence and appendices', C 536, Parliamentary Papers (1872), pp. 52–3

1. *[Norman Lockyer]*: Lockyer was the secretary of the Royal Commission on Scientific Instruction and the Advancement of Science, which sat from 1871–4. He signed the letter from the Royal Commission on Scientific Instruction and the Advancement of Science that Tyndall was answering. See J. N. Lockyer to Professional Examiners for Science, 15 June 1871, 'Royal Commission on Scientific Instruction and the Advancement of Science. Vol. I. First, supplementary, and second reports, with minutes of evidence and appendices', C 536, Parliamentary Papers (1872), p. 48. Tyndall's copy of this letter is missing.
2. *22 June 1871*: We have based this letter on Tyndall's handwritten draft, consulting, as needed, the published version of the letter. The published version establishes the date. See Tyndall to Norman Lockyer, 22 June 1871, 'Royal Commission on Scientific Instruction and the Advancement of Science. Vol. I. First, supplementary, and second reports, with minutes of evidence and appendices', C 536, Parliamentary Papers (1872), pp. 52–3.
3. *the examination*: Tyndall was an examiner for the School of Mines at South Kensington.
4. *your letter*: letter missing.
5. *Duke of Devonshire*: William Cavendish (1808–91), the seventh Duke of Devonshire; Cavendish chaired the Royal Commission on Scientific Instruction. He was educated at Eton and at Trinity College, where he was second wrangler and where he earned his BA and MA in 1829. Cavendish served as the MP for Cambridge until 1831, but was ejected

because he supported the Reform Bill. A supporter of education, he was Chancellor of the University of London from 1836–56, then Chancellor of Cambridge from 1862 until his death.

6. *the questions . . . to me*: Lockyer, as secretary of the Commission, addressed a list of questions to the Professional Examiners for Science on 15 June 1871; see n. 1.
7. *the Department*: the Science and Art Department.
8. *A principal assistant . . . future attention*: The published version of the letter is substantially different: 'Two principal assistant examiners are appointed, who take charge of the honours papers, and who also check and revise the other papers. With these two gentleman I have been recently in communication regarding the questions which you have submitted to me' (Tyndall to Lockyer, 'Royal Commission on Scientific Instruction and the Advancement of Science. Vol. I. First, supplementary, and second reports, with minutes of evidence and appendices', C 536, p. 52).
9. *the first question submitted to me*: 'What opinion have you formed as to the amount of good which is done by the South Kensington System generally, and the value of the examination in the different grades?' ('Royal Commission on Scientific Instruction and the Advancement of Science. Vol. I. First, supplementary, and second reports, with minutes of evidence and appendices', C 536, p. 48).
10. *two principal assistant examiners*: not identified, but see letter 3407 for a discussion of Tyndall's organization of a staff of examiners.
11. *The founding . . . instruction*: The letter published in the Parliamentary Papers reads, 'It is, therefore, an object to be aimed at, the founding of institutions in which teachers should be instructed in the use of apparatus sufficiently cheap, simple, and effectual, to be introduced with profit into class instruction' (Tyndall to Lockyer, 'Royal Commission on Scientific Instruction and the Advancement of Science. Vol. I. First, supplementary, and second reports, with minutes of evidence and appendices', C 536, p. 53). Tyndall may have in mind the Science Schools just being opened in South Kensington under Huxley's instruction.
12. *I would here . . . exposition*: Tyndall wrote and struck out, 'I have myself taught a class of boys the elements of mechanics by mean of diagrams'.
13. *The detection . . . examiner*: This sentence does not appear in the published version.
14. *Many teachers . . . instruction*: The published version clarifies this sentence: 'Many teachers evidently set before themselves, as the main object of their efforts, the bettering of their own pecuniary estate; and a good number of them appear to entertain the notion that this end is more likely to be gained by increasing the number of candidates than by aiming at more excellent instruction' (Tyndall to Lockyer, 'Royal Commission on Scientific Instruction and the Advancement of Science. Vol. I. First, supplementary, and second reports, with minutes of evidence and appendices', C 536, p. 53).
15. *prove effective as a teacher*: Tyndall wrote and struck out: 'If he be able to express his mind with clearness in writing to his examiner, the chances are that he will be able to express himself verbally with clearness to his pupils'. The published version has 'the chances are that the youth who expresses himself in writing with clearness and straightforwardness,

will also prove efficient as a teacher' (Tyndall to Lockyer, 'Royal Commission on Scientific Instruction and the Advancement of Science. Vol. I. First, supplementary, and second reports, with minutes of evidence and appendices', C 536, p. 53).

To George Gabriel Stokes 22 June [1871][1] 3469

Royal Institution of Great Britain | 22nd June.

My dear Stokes

I have received a very flattering introduction[2] of M. Croullebois from the Astronomer Royal.[3] M. Croullebois is in the country on an educational mission, and I thought you would not take it amiss if, at his request,[4] I give him this note[5] to you.

Yours faithfully | John Tyndall

Prof Stokes | &c &c &c

CUL SC, Letter T585

1. *[1871]*: see letter 3467 for Croullebois's letter and the explanation of his coming to England.
2. *flattering introduction*: letter missing.
3. *Astronomer Royal*: George Biddell Airy.
4. *at his request*: see letter 3467.
5. *this note*: refers to the present letter, which is introducing Croullebois to Stokes.

From Edward Whymper 22 June 1871 3470

19 Canterbury Place | Lambeth Road S.E. | June 22/71.

My dear Sir,

In the spring of 1866, the late Principal J. D. Forbes[1] urged me to endeavour to find out more about the 'veined structure' of glaciers,[2] which he then, and, I believe, until his death considered, was very much in want of elucidation. After thinking the subject over, it seemed to me that its difficulties were so considerable that it would be useless to attempt to grapple with them except in a thorough manner, and that it would be necessary to scrutinize and to follow out the gradual transition of snow into glacier-ice, from beginning to end, in at least one glacier. Superficial examination was almost worthless, for it was well known that the veined structure, or structures, existed in glacier ice above the snow-line; and hence it appeared that the only effectual procedure would be to sink a number of pits or trenches through the superincumbent snow,[3] commencing at the very birth-place of the glacier, and watching its growth and structural development as it descended to the lower regions. This opinion I still entertain.

I consequently left England at the end of July, with the intention of sinking several pits in the Stock glacier,[4] which descends towards the north-east from the Col de Valpelline.[5] Wretched weather and miserable workmen retarded the work, and only one pit was sunk in the time at my disposal. This was little more than 22 feet in depth, and although it threw scarcely any light upon the veined structure, it yielded information of a novel and interesting character about kindred subjects. Most unwillingly I left the excavation just at a time when it promised to give more valuable information than it had done previously, and since then I have never been able to rescue the work,—for a reason which has been already mentioned to Sir Edward Sabine.[6]

Principal Forbes approved the method by which I proposed to investigate the subject, but I am unaware that he considered it preferable to any other, and it is only due to him to say that it was not suggested by him. Moreover, I must alone be held responsible for the opinion that the continuance of these experiments is likely to furnish valuable results. I mention the name of my lamented friend only to explain the origin of what might otherwise seem a rather unaccountable proceeding upon my part. Although I had long entertained the opinion that the structures which are met with in glacier ice required further elucidation, I certainly should not have meddled with questions which you and he had made especially your own had it not been for his intervention.

Your structures seem to have been described under the head of "the veined structure". One, which you have described and figured upon p. 381 of your Glaciers of the Alps,[7] and (it seems to me) referred to again upon p. 394. Another, which you have described with considerable precision upon p. 376 of the same work. The print I will call no. 1, and is crudely represented in the annexed figure, the other I will call no. 2, and is represented in fig. 2. I cannot doubt that these structures have different origins, and that it will be necessary, sooner or later, to find a new term for one or the other. In the mean time let me call them, for the sake of brevity, nos. 1 & 2.

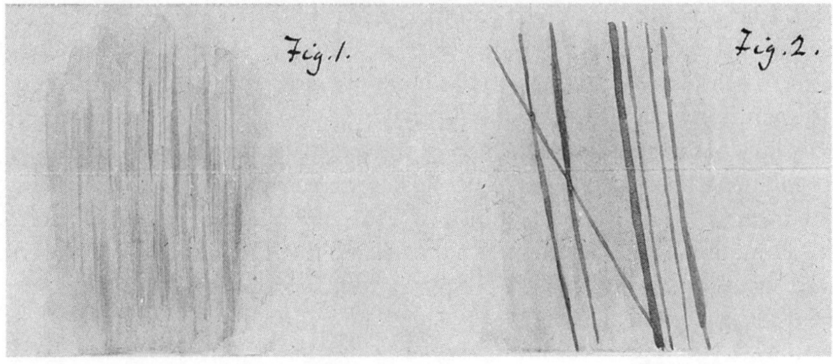

depths, but I do not care to quote my readings, as they were, without a doubt, falsified by the wind. I am not sure, moreover, that it is possible under any circumstances to obtain correct readings of snow temperature in the way that they were taken. The recorded temperatures, anyhow, must have been influenced by the surrounding air. If they were correct they proved that the lower strata were warmer than the upper ones.

We must now quit the region of facts, and descend to that of surmises and conjectures. The differences in the quality and in the tone of the snow of the first three below the surface were sufficiently marked to suggest that we saw in them snow belonging to three different years. The unanimous opinion of the four men was, that the uppermost 11 inches belonged to 1865-6, the next 10 inches 1864-5, and the next 16 inches to 1863-4. In this matter they were not, perhaps, altogether incompetent judges. I am doubtful, however, whether their opinion was correct, and incline to the idea that the uppermost 11 inches had fallen during the summer of 1866, and that the succeeding 10 inches *may* have been all that remained of the preceding winter's snow. Whatever surprise may be felt at so small a depth being considered as representing a year's fall, must be modified when it is remembered that the position at which the pit was sunk could scarcely have been more exposed. We had evidence that a mere fraction only of the snow that

[break in original facsimile]

fell remained *in situ—the* wind tore it away in sheets and streams. It will be remembered, too, that no inconsiderable amount passes off by evaporation. If other pits had been sunk to the north and to the south of the pass, we should probably have found in them a greater depth of snow between each of the horizontal layers of pure ice. This is mere conjecture, and it may be taken for what it is worth. It is more important to note—1. (a) That the fine layers or strata of pure ice were *numerous* towards the surface ; *(b) disappeared* as we descended ; *(c)* and that the lower strata were, upon the whole, much thicker than those towards the surface. 2. That the thickness of these strata of pure ice amounted to nearly one-tenth of the mass that we were able to penetrate. 3. That, below the depth of 15 feet, vertical glacification began to show itself. Upon each of these subjects I will now venture to offer a few remarks.

1 *(a.) The fine horizontal layers or strata of pure ice were numerous towards the surface.* All of these layers had been formed by weathering *at* the surface. It is usual, even during the winter, for considerable periods of fine weather to succeed heavy snow-falls ; and in these periods the surface of the snow is alternately melted and refrozen, and, at length, is glazed with a crust or film of pure ice. This, when covered up by another snow-fall, and exposed as in the section, appears as a bluish horizontal line drawn through the whiter mass. The snow between any two of these layers (near the surface) did not therefore represent a year's snow, but it was the remnant, and only the remnant, of a considerable fall, between whose deposition, and that of the next stratum above, a considerable interval of time had probably elapsed.

(b.) The fine strata disappeared as we descended. I imagine that this was a

result of pressure from the superincumbent mass, but I leave to others to show the exact manner in which these finer strata were got rid of. Is it possible to liquefy by steady pressure a plate of ice (say, one-tenth of an inch in thickness) placed in the interior of a mass of snow, without liquefaction of the snow?

(c.) The lower strata of pure ice were, upon the whole, thicker than those *towards the surface.* This, doubtless, was a result of vertical pressure. The strata grew under pressure. But why should some grow and others disappear ? I presume that the *finest* ones disappear, and that the stouter ones grow. Can it be shown experimentally that it is possible to liquefy by steady pressure a fine plate of ice placed in the interior of a mass of snow, and at the same time, under the same conditions, to thicken another and stouter plate of ice ?

2. *These horizontal strata of pure ice amounted in the aggregate to nearly one-tenth of the thickness of the mass that we penetrated.* It was perfectly well known prior to 1866 that the upper snows (which give birth to glaciers) were pervaded with strata of pure ice, and a host of observers had written before that date upon stratification of snow and of glacier. It may be questioned, however, whether any had an idea of the very important amount of glacification that is effected by superficial weathering, and subsequent thickening of the strata through vertical pressure. A search through the works of the principal writers on glaciers has failed to show me that any person imagined that one-tenth of the mass, or anything like that amount, was composed of <s>trata of pure ice.

[break in original facsimile]

There are two points in regard to these horizontal strata of pure ice that are worthy of consideration :—(a) Does not their existence, and especially the existence of the fine layers towards the surface, conclusively disprove the idea that the production of glacier-ice is greatly promoted by infiltration of water from the surface ? *(b)* Can these numerous strata of pure ice (some of which are of such considerable thickness, and extending over large areas) be *obliterated* in the subsequent progress of the glacier ? If so, how are they obliterated ? Or is it not reasonable to suppose that these thick strata of solid ice must continue to exist, must continue to thicken under pressure, and must supply many of those plates of pure ice which are seen in the imperfect ice of the glacier, and which have been referred to at different times and by various persons as the 'veined structure ?'

3. *Below the depth of* 15 *feet the appearances which I have ventured to term vertical glacification were first noticed.* Were they accidental ? or will they be found at or about the same depth in all other places ? Into what would those appearances have developed at a greater depth ? What produced them ? These questions may perhaps be answered one day by future investigators.

You see that a variety of questions are raised here. Let me confine myself, however, to the bearing of the facts that were observed upon the veined structure. There *[are numerous]*[8] strata of pure ice (or ice of a density of about

0.9178) can they be subsequently obliterated? Is it possible for these strata (which at the summit of the Col de Valpelline were parallel to the surface of the snow) to remain parallel to the surface for any considerable distance. In the case of a glacier that does not at any part of its course descend torrentially (each as the upper parts of the glacier d'Argentière, for example), is it possible for them to remain parallel to the surface? Must not these strata of ice crop out in consequence of the motion of the glacier and the ablation of its surface? And if they crop out, in which way could these strata of ice be distinguished from the blue veins of the veined structure (no. 2)?

[break in original facsimile]

by yourself 14

Again, It has ~~also~~ been urged^that "the blue veins of glaciers are not always, nor even generally, such as we should *expect* to result from stratification. The latter would furnish us with distinct planes extending parallel to each other for considerable distances through the glacier ; but this, though sometimes the case, is by no means the general character of the structure." With this observation I agree. It amounts, however, only to saying, that it is impossible to consider that all of the blue veins have their origin in the stratified beds of snow and ice from which glaciers are born. Any person who has been close to an "ice-fall" on one of the principal Alpine glaciers, and observed the great *séracs* lurching forward, with the primitive beds remaining parallel, or nearly so, to the surface of the glacier, must feel that it is extremely improbable that the masses will be so re-compacted lower down as to "furnish us with distinct planes extending parallel to each other for *considerable distances.*" It will be felt that some of the *séracs* will be so smashed up that the original structure will be got rid of ; that others, which descend more gently, will remain intact, but will settle down with their beds more or less inclined to the horizon ; and that it will be a very extraordinary chance if the *dip* of the strata of any two of the masses coincides within many degrees.

Upon these grounds I believe that many of the veins of the veined structure of glaciers are nothing more than the upturned layers of blue ice which are formed upon and between the beds of snow that are deposited in the higher regions.* I am far from thinking that the occurrence of the whole of the veins of blue ice which are found in glaciers should be accounted for in this way. I do not believe that the combinations of different varieties of ice that are found in glaciers, which have been referred to by various authors as the *veined structure,* can be accounted for in two or even in three ways. Avoiding disputed points, I will observe that there are at least two other modes by which many veins of blue ice are undoubtedly produced in glaciers.

First, by water freezing into crevasses. I have seen hundreds of crevasses in Greenland nearly full of water ; never *quite* full : the water seldom came within two or three feet of the surface of the glacier. I have seen the entire surface of the water in such crevasses frozen and freezing. I have seen the water sometimes frozen solid at one end and remaining liquid at the other end ; and in the walls of icebergs I have seen sections of crevasses that have been

* Sometimes, probably thickened by pressure.

nearly filled with water, in which the water has been frozen solid.* These veins in icebergs are frequently one to three fee thick, and can be seen at several miles' distance. If veins of blue ice are not formed in the Alpine glaciers in the same manner, it is only because there are outlets from the crevasses by which the water escapes. It is rare to see a crevasse even partly filled with water in the Alps.†

Secondly, by the closing together of crevasses. The unequal motion of the parts of a glacier causes crevasses continually to open and to close up ; and the walls of these crevasses, whether 12,000 feet or more above the level of the sea, or whether only 5000, all become weathered and more or less coated with pure ice. Even narrow crevasses in the high regions, well bridged with snow, are not exempt. The warm air of midsummer penetrates the chasms, and, assisted by the percolation of snow water, glazes the walls from top to bottom. The superficial coatings of ice which are thus formed upon the sides of crevasses vary greatly in thickness according to circumstances—in a single crevasse they may range from a thickness of less than an inch to more than a foot.‡ The crevasses close up ; the surfaces of their icy walls are brough into contact ; they regele, and the coalesced films will then appear as veins of pure ice in the generally whitish mass of the glacier. When one considers the myriads of crevasses when there are in any glacier, and the incessant openings and closing up that goes forward, it is easy to see that a large portion of the veins if pure ice which constitute the veined structure‸ of glaciers must be considered as the *scars of healed crevasses*. (No 2)

I have not clai any originality for explanation, but I unaware that it been offered by any else.

* These veins in icebergs are frequently seen intersecting each other. Dr. Rink has shown this in an illustration in his *Grönland Geographisk og Statistisk*, vol. i. 1852.

† Charpentier long ago advanced the option that the motion of glaciers was promoted by freezing of water in crevasses. His notion is commonly regarded as explodded, but there may by something in it after all.

‡ The same thing is to be noticed in regard to the blue veins of the veined structure. The veins frequently thin out and are lost, or swell into lenticular masses. This is best seen when the veins are regarded in vertical sections of the glacier.

You will of course understand that I neither assert that the primitive beds remain parallel to the surface, nor that they do not do so. It is a subject that seems to me to require investigation, and since I have observed (or believe that I have) the planes of the bedding remaining parallel to the surface in some icebergs in Disco Bay (N. Greenland)[9] which had come from a glacier at least 20 miles long, I advance any opinion upon the subject with the very greatest hesitation.

Neither do I for a moment suppose that the strata of ice of the primitive bed, or veins formed by water freezing with crevasses, or the veins which must be formed by the closing together of crevasses, whether separately or collectively, account for the occurrence of the whole of the veins of the veined structure no 2. The very puzzling example of veined structure which is represented in the annexed figure is sufficient to prevent any such alternative.

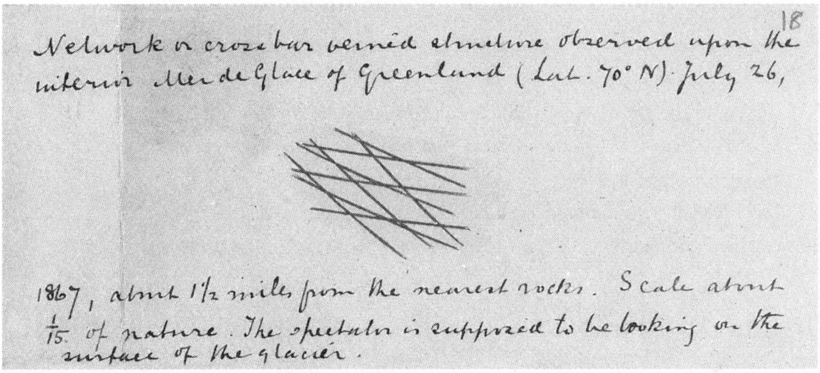

18

Network or crossbar veined structure observed upon the interior Mer de Glace of Greenland (Lat. 70° N). July 26,

1867, about 1½ miles from the nearest rocks. Scale about 1/15 of nature. The spectator is supposed to be looking on the surface of the glacier.

In regard to the structure no. 1. Is it possible that the appearance figured at the lower portion of the Col de Valpelline section represented, in embryo, the structure you have figured upon p. 394 (Glaciers of the Alps). If they did not, I ask again, with what would they have developed at a lower depth?

I have now, Sir, mentioned to you a few of the doubts and perplexities which may possibly be removed by a repetition of the experiment which was initiated upon the Col de Valpelline in 1866. The time at my command does not permit me to enter into further details, nor is it, perhaps, necessary that I should do so—I desire to expose considerable sections of the interior of a glacier, at different parts of its course, and to record whatever may be observed. This can only be done by a considerable expenditure of time and money. The time I can give, but the money I cannot afford to give. If you are of opinion that I am not an unfit person to conduct the work, you will, I am sure, after the expression contained in your letters of the 15th & 18th[10] make such representations to the Council of the Royal Society, as will lead to the operations being carried out in a thorough manner. You are, of all persons, the most competent to offer an opinion upon the amount that will be required, and I need not add that any suggestions which you may be pleased to make will be very highly esteemed by

Your obedient servant, | Edward Whymper

Dr. J. Tyndall F. R. S.

BL, John Tyndall Papers, Add MS 53715: 1.4v–19

---

1. *J. D. Forbes*: see letter 3462, n. 3.
2. *'veined structure' of glaciers*: see letters 3462 and 3464.
3. *superincumbent snow*: the snow that is on the top of a glacier, exerting pressure on it.
4. *Stock glacier*: the Stockji Glacier in Switzerland, near the Matterhorn.
5. *Col de Valpelline*: a pass in the Alps from which you can see the Matterhorn and the Stockji glacier.
6. *already mentioned to Sir Edward Sabine*: see letter 3459. The reason is likely that he does not have enough money.
7. *Glaciers of the Alps*: Tyndall's 1860 book *Glaciers of the Alps*.
8. *[are numerous]*: Whymper repeats this language in his subsequent book, *Scrambles amongst the Alps*, p. 442.
9. *Disco Bay (N. Greenland)*: Disko Bay, in western Greenland.
10. *your letters of the 15th & 18th*: see letter 3462; letter of the 18th is missing.

To Edward Sabine 23 June 1871 3471

Royal Institution | 23rd . June. 1871.

My dear General

Mr. Whymper has sent me an account of his proposed experiments,[1] and in view of those immediately sought, I have ventured to recommend to the council that a Grant of £50 (the sum mentioned in Mr. Whymper's note to yourself)[2] from the Donation Fund be placed at his disposal.

It would have given me great pleasure to communicate this to Mr. Whymper himself. But circumstances which have just come under my notice prevent me[3] from doing so, and cause me to ask you to be good enough to do it for me.

It might be worth while to mention to Mr. Whymper that some of the appearances he describes are as distinct from the true "veined structure" as the pointing and bedding of rocks from their true cleavage. It is highly desirable to keep the question of true structure apart from the spurious veining which arises from other causes.[4]

Ever faithfully yours | John Tyndall

RI MS JT/1/T/1270
RI MS JT/5/15b/219

1. *an account . . . experiments*: see letter 3470.
2. *Mr. Whymper's note to yourself*: referenced in letter 3459.
3. *circumstances . . . prevent me*: Whymper was extremely critical of Tyndall's mountaineering in his most recent book, *Scrambles amongst the Alps*. See letter 3525, n. 7 and n. 8.
4. *It might be . . . other causes*: originally included as a post-script by Tyndall who then drew an arrow up to include it as part of the letter.

To George Gabriel Stokes 23 June [1871][1] 3472

Royal Institution of Great Britain | 23rd. June.

My dear Stokes

M^r^. Whymper, in a communication of considerable length,[2] has set forth his reasons for supposing that the observations he contemplates would present interesting results.

The observations refer to the veined structure of glacier ice; and although M^r^. Whymper includes in his communication under the term "structure" appearances which have no more to do with it than the pointing of rock has to do with their true <u>cleavage</u>, still I am of opinion that a sum of £50 would not be ill expended on promoting the object he has in view.

Even though but little may come of the observations as regards the structure itself, the incidental observations which are sure to be made and recorded, are I thought of sufficient interest to warrant the outlay.

I would therefore recommend that a sum of £50 from the Donation Fund be placed at the disposal of M^r^. Whymper.

faithfully yours | John Tyndall

Prof. Stokes DC.L.[3] | Sec[retary]. <u>R.S.</u> &c[4]

RS Miscellaneous Correspondence 9.220

1. *[1871]*: letter dated to 1871 because it discusses Whymper's correspondence and proposed experiments. See letters 3459, 3462, 3464 and 3470.
2. *a communication of considerable length*: see letter 3470.
3. *Prof. Stokes DC.L.*: Stokes was awarded an honorary Doctor of Civil Law from Oxford in 1857.
4. *Sec[retary]. R.S. &c*: Stokes was one of the Secretaries of the RS from 1854–85, after which he became President.

To James Clerk Maxwell 27 June 1871 3473

[page cut off, first visible line illeg][1]

Or o'er the foliage lifting summits dreary
Which the bold eagle chooses for his eyrie.
Mute presences! who quell the gazer's spirit
A measure of the strength your crags inherit
Fit nurses of the thoughts which fill like granite
Upon the caked traditions of our planet.[2]
In front of you he stood when day was failing
Watching the listed clouds through azure sailing
Tracking the swallow in its course so cheery
Clearing the stilly air with wing unweary.
With eye and heart sublime as fell the glory
From the red west upon their ridges hoary.

CUL MS.Add.7655.7655/V/1 /3(vii)

1. [page cut off, first visible line illeg]: envelope dated 27 June 1871 from Cambridge (presumably forwarded then to Maxwell in Scotland; Tyndall was observing at Howth Bailey that day but had been in London a couple of days previously). This poem was a version of one written in May of 1849 called 'Brave Hills of Thuring'. A different version of the poem appeared earlier in the *Preston Chronicle* under a pseudonym; see Wat Ripton, 'A Whitsuntide Ramble', *Preston Chronicle*, 16 June 1849, p. 3. The poem is also at RI MS JT/2/13b/434–5. See *Ascent of John Tyndall*, p. 322, n. 27. For a full discussion of the different versions of the poem, see J. Tyndall, 'Brave Hills of Thuring', in R. Jackson, N. Jackson and D. Brown (eds), *The Poetry of John Tyndall* (London: UCL Press, 2020), pp. 177–8.
2. *Fit nurses . . . planet*: These two lines were interleaved from the bottom of the original letter, with an arrow up to their appropriate spot.

To Elizabeth Dawson Steuart 29 June [1871][1] 3474

Royal Institution. | 29th June.

My dear Mrs Steuart,

I shall make an effort to reach you to-morrow:[2] but I am so unacquainted with the routes that I do not know whether I shall be successful—If successful I may ask you to give me a bed. Is not that impertinence?

Yours ever faithfully | John Tyndall

If at the same time I could see the two youngest sons[3] of Caleb Tyndall[4] it might be useful to them—But perhaps you and I can arrange in conversation every thing needful regarding them.

RI MS JT/1/TYP/10/3394
LT Typescript Only

1. *[1871]*: letter dated to 1871 because of Tyndall's travel plans.
2. *effort to reach you to-morrow*: Tyndall was in Ireland for some work on lighthouses and was trying to meet Steuart. See also letter 3401.
3. *two youngest sons*: see letter 3379, n. 10.
4. *Caleb Tyndall*: see letter 3379.

To Cecil Trevor 29 June 1871 3475

Kingstown,[1] Ireland, | 29th June 1871.

SIR,

I BEG to inform you that on the evening of Monday the 26th I conducted a series of observations on the new form of light[2] proposed by Mr. Wigham[3] and Captain Hawes.[4]

In a shed at Howth Baily[5] and at some distance below the fixed light an annular lens was placed; the lens rested on wheels which moved along a circular rail, and by this motion the action of an ordinary revolving apparatus was imitated.

The lamp was placed at the centre of the circle formed by the rail.

The station of observation was on Kingstown Pier, about five miles distant from the lens of Howth Baily.

When the observations began the ordinary 28-jet burner was already shining from the fixed apparatus above. On igniting a similar 28-jet burner below, the vast superiority of the light through the annular lens over the fixed light was immediately demonstrated. This of course might be anticipated from the construction of the lens, which squeezes the beam together in *all* directions instead of in a vertical direction only, as in the fixed apparatus.

A burner of 108 jets in the fixed apparatus produced, at the distance mentioned, an effect sensibly equal to that of the 28-jet burner in the annular lens.

Under ordinary circumstances, as the lens passes in front of the observer, the beam is carried along with it, the edge of the beam finally passing the observer, and leaving him in darkness until the arrival of the next succeeded beam.

In the method proposed by Mr. Wigham and Captain Hawes the beam, in passing the observer, is divided into a series of distinct flashes. The use of gas renders this easy of accomplishment. It is only necessary to open and close a cock at the proper intervals. When the cock is closed the flame is sensibly extinguished, a small by-way only supporting a flame so minute as entirely to escape observation from a little distance; when the cock is open the full flame flashes forth.

Several series of observations were made in which various angular velocities were imparted to the annular lens. The motion was in the first series so regulated as to corresponded to a series of revolving panels succeeding each other at intervals of a minute. During each passage of the beam 10 to 12 flashes impinged upon the eye of the observer.

It will be remembered that the motion of the lens and the turning on and off of the cock, being regulated by the hand, strict mechanical uniformity was not to be expected.

Other series were conducted in which the velocity of rotation corresponded to that of a revolving apparatus with panels succeeding each other at intervals of three quarters of a minute, half a minute, and a quarter of a minute respectively.

The conclusion arrived at from these experiments is that the proposed method promises to furnish a light of remarkable power and individuality. The total quantity of light that reaches the observer's eye by the method of flashing is probably not much more than half of that which reaches it during the passage of the unbroken beam; but the division of the beam into a succession of powerful shocks to the retina is, perhaps, more calculated to arouse the attention of the mariner than the passage of the steady beam itself.

In other words, by reducing, in accordance with the ingenious proposal of Mr. Wigham and Captain Hawes, the quantity of gas by nearly one half we rather improve than prejudice the character of the light as a signal.

The flashes observed were of sensibly uniform intensity, and they vanished with great rapidity towards the limit of the beam.

I assured myself by actual trial that there was no difficulty whatever in taking the bearings of the flashing light.

It may be worth remarking that the effect of the flashing light is heightened by the fact that during the interval of darkness between every two flashes the eye, which soon becomes partially paralysed by a continuous light, recovers its sensibility, and is thereby rendered more capable of appreciating the succeeding shock.

On the evening of the 27th fresh observations were conducted from the same point. On this evening arrangements were made for a steady advance in the number of jets both above and below. In the fixed light, for example, a 28-jet burner was in action when a 28-jet burner was in the focus of the annular lens; a 48-jet burner was in the fixed apparatus when a 48-jet burner was in the focus of the annular lens; and so of the other burners up to the largest, which embraces 108 jets. As might be expected the advantage derived from the augmentation of the burners was greatest in the fixed apparatus.

The observations of the 27th confirmed those of the 26th as to the marked efficiency and individuality of the flashing light.

To give this light a fair trial, and to give practical seamen an opportunity of testing its value, it is necessary that it should be exhibited in an actual revolving apparatus. The lighthouse of Rockabill[6] is at hand, which offers every facility for the extension and completion of the observations.

The Messrs. Edmundson,[7] I am informed, are willing to erect at Rockabill the works necessary for the observations, and to remove them in case of disapproval, for a sum of 100*l.* In view of the results likely to flow from it I do not hesitate to recommend the Board of Trade that this proposal be accepted.

And I would also recommend that the sanction of the Board be given sufficiently early to permit of the erection of the works before the end of August, when I propose to return to Ireland with the view of presenting to the Board a concluding report on this important subject.

It is to be remarked, finally, that, even if the system of flashing should not be adopted, the gas-flame is more than competent to fulfil all the purposes of the oil-flame.

Additional Remarks.—I think it not unlikely that in foggy weather some advantage will be gained from the intermittent atmospheric glow corresponding to the intermittent ignition of the light. The rise and fall of this glow were very marked on the evening of the 27th when a rather thick haze swathed Howth Baily. | July 5th 1871.

I have, &c. | (Signed) JOHN TYNDALL.

Cecil Trevor, Esq., | Harbour Department, | Board of Trade.

'Further Papers Relative to a Proposal to Substitute Gas for Oil as An Illuminating Power in Lighthouses', C 1151, Parliamentary Papers (1875), pp. 9–10
Typed Transcript Only

---

1. *Kingstown*: coastal town near Dublin now known as Dún Laoghaire.
2. *new form of light*: part of Tyndall's work for the Board of Trade and Trinity House in deciding what kind of fuel to use for lighthouses in Ireland. See also letters 3400, 3417 and 3521, and *Ascent of John Tyndall*, pp. 284–5.
3. *Mr. Wigham*: John Richardson Wigham (1829–1906), a Scottish lighthouse engineer who invented gas lighting for lighthouses. He was apprenticed to his brother-in-law Joshua Edmundson (d. 1848) who ran a hardware and manufacturing business. Upon Edmundson's death, Wigham took over and began focusing on gas lighting for private homes. The system in the Howth Baily lighthouse was invented and installed by Wigham. Wigham was a member of the Royal Dublin Society, the Royal Irish Academy, an associate member of the Institution of Civil Engineers and a fellow of the Institution of Mechanical Engineers.
4. *Capt. Hawes*: E. H. Hawes (fl. 1870s), Captain, R.N., the Inspector of Irish Lights. His name appears on a number of reports for Irish lighthouses during this period. He was

previously Assistant Inspector, and succeeded Captain E. F. Roberts, R.N., upon his death (J. Tyndall, "The Story of our Lighthouses," *Nineteenth Century*, 24 (July 1888), pp. 61–80, on p. 73).

5. *Howth Baily*: a lighthouse at Howth Head, Ireland.
6. *the lighthouse of Rockabill*: a lighthouse built in 1860 on one of two rocky islands in the Irish Sea, to the north-west of Dublin.
7. *Messrs Edmundson*: the business originally run by Joshua Edmundson (see n. 3), which continued to be known as Joshua Edmundson & Co.

To Cecil Trevor 29 June 1871 3476

Royal Institution of Great Britain | 29th June 1871.

My dear Sir,

I hope you will overlook the elisions & corrections of this Report.[1] It is written under some difficulty, and as time is precious I wish to see it as soon as possible in your hands.

Very faithfully yours | John Tyndall

Cecil Trevor Esq[ui]re | &c &c &c

National Archives MT 10/131 H3055

1. *this Report*: see letter 3475.

From Henry Bence Jones [late June/early July 1871][1] 3477

Dear friend

The Emperor of Brazil[2] sent a messenger to say that he wished to come to the Institute to you some day. Sir H Holland told the messenger we were shut up. He will be here on the 3d.[3] Let me know if you are inclined to entertain an Emperor for a private show & I will try to get it fixed.

If not don't answer this

Y[ou]rs | H B J

RI MS JT/1/J/103
RI MS JT/1/TYP/2/722

1. *[late June/early July 1871]*: dated to late June or early July 1871 because of the dates of the Emperor of Brazil's journey to Europe. He and his wife traveled to and around Europe

from 25 May 1871 to 31 March 1872. He was in England from early June to 12 August 1871. See n. 3, below, for further information on dating.

2. *Emperor of Brazil*: Pedro d'Alcantara; see n. 1 for the dating of his visit. He was elected FRS during this trip and also made an honorary member of the RI (*Roy. Inst. Proc.*, 6 (1870–72), pp. 384, 422; *Roy. Soc. Proc.*, 20 (1871), pp. 48, 139).
3. *3d*: must be 3 July. Tyndall was in Switzerland by August 3 and Bence Jones knew this, so would not have asked him to meet then. See also letters 3478 and 3501.

From Henry Bence Jones [early July 1871][1] 3478

Dear friend

The Emperor[2] wishes quite privately alone to see some of your sights on heat sound & light He will come any evening about 8 1/2 to 9 When you are disposed will you write to the Brazilian Minister

Chevalier J C De Almeida Areas[3] 25 Mansfield Street to tell him you are ready

The Emperor was very much obliged to you for your civility

Ever y[ou]rs | H Bence Jones

I telegraphed to you to day.

Saturday

I mentioned to Cottrell that you w[oul]d some evening shew your best heat, light & sound experiment I did not say who to.

Just got your letter from Gorey[4]

RI MS JT/1/J/104
RI MS JT/1/TYP/2/721

---

1. *[early July 1871]*: see letter 3477, n. 1, for notes on dating to early July. This is possibly 1 July, a Saturday.
2. *The Emperor*: Pedro d'Alcantara; see letters 3477 and 3501.
3. *Chevalier J C De Almeida Areas*: José Carlos de Almeida Arêas, Viscount of Ourem (1825–92) was the Plenipotentiary Minister (Ambassador) of Brazil in London from 1868–73. He had been a lawyer, trained at the São Paulo Law Academy and was the only Viscount of Ourem.
4. *Gorey*: Gorey, Ireland. Letter is missing, but it was likely from when Tyndall was there for the lighthouse work in June of 1871 (see letters 3474 and 3475, for example), placing this letter in early July.

From Alexander Buchan[1] 1 July 1871 3479

Scottish Meteorological Society | General Post Office Buildings | Edinburgh 1 July 1871

Dear Sir

10,000 thanks for the copy of Hours of Exercise in the Alps you kindly sent me which has been an abounding source of pleasure to me for some weeks past.

To me in particular, the short note on clouds[2] is valuable and many paragraphs scattered through the Books. The sentence "Winds blow and waters roll in search of a stable equilibrium" exactly expresses the principle in faith of which I have worked for years. It is surprising that this truism in physics is set aside for this extraordinary principle:—Winds originate atmospheric unstable equilibrium,—not from the differences of temperature and vapour they bring about,—but from their mere mechanical action. This view is held so persistently by many great names in General Science, that occasionally while in indigestive states I despair of a general change of opinion till the inevitable operation of physical law effects it.

Again thanking you very heartily for the pleasure you have given me in reawakening recollections of a botanical tour in the higher Alps in August 1858, and in suggesting many thoughts which did not occur to me

I am | Yours sincerely | Alexander Buchan

RI MS JT/1/B/138
RI MS JT/1/TYP/1/165

1. *Alexander Buchan*: Alexander Buchan (1829–1907), a Scottish meteorologist, well known for establishing the map used in weather forecasting today. His first scientific interest was botany, which took him on the 1858 trip to the Alps he mentioned in this letter. He was the Secretary of the Scottish Meteorological Society from 1860 until his death. He was elected FRSE in 1869 and FRS in 1898.
2. *the short note on clouds*: in *Hours of Exercise in the Alps*, pp. 405–12.

To Cecil Trevor 3 July 1871 3480

Royal Institution of Great Britain | 3rd July 1871

My dear Sir

The Irishmen[1] evidently thought that the application required renewal because they could not guarantee the erection of the apparatus in the few weeks prior to my quitting England.

But I purpose going over in August[2]—the end thereof—They will then have every thing in readiness for me.

Do you need that Report I wrote to you?[3] If so, and if you would kindly let me have it for a short time, I would make a few slight alterations on it.

Yours faithfully | John Tyndall

C. Cecil Trevor Esq

National Archives MT 10/131 H3055 2290–1

---

1. *Irishmen . . . apparatus*: see letter 3475, toward the end of which Tyndall discusses getting a structure built in time for his August trip.
2. *in August*: Tyndall was able to travel there in September (*Ascent of John Tyndall*, p. 285).
3. *Report I wrote to you?*: see letters 3475 and 3476.

From Edward Sabine 4 July [1871][1] 3481

July 4.

Dear Tyndall:

I enclose a note[2] from Mr. Whimper.[3] Who did you entrust with the application to be made to the Wollaston Fund?[4] They are funds enough to the good.

Sincerely yours | Edward Sabine

Prof. Tyndall.

RI MS JT/1/S/39
RI MS JT/1/TYP/4/1333

---

1. *[1871]*: LT annotation, and dated to 1871 to place it within the Whymper conversation. See n. 3.
2. *enclose a note*: letter missing.
3. *Mr Whimper*: Edward Whymper. See letters 3462, 3464, 3470 and 3471 for Whymper's proposed work and funding questions.
4. *Wollaston Fund*: see letter 3459, n. 3.

To Richard Strachey[1] July [1871][2] 3482

5th July

My dear Strachey

Allow me to bid you a hearty welcome to your own shores.[3] I trust we have now laid hold of you for ever.

always yours | John Tyndall

BL Add MS 60631, f. 112

---

1. *Richard Strachey*: Richard Strachey (1817–1908), a scientist and administrator in India. He was in the Engineer Corps in India, responsible for building British Indian infrastructure such as the Ganges Canal in 1843. He spent time working in England throughout his career, and by this point, had been released from his duties in India, having been promoted to Major General. He was the father of a number of members of the Bloomsbury group, such as Lytton, Oliver and Philippa Strachey. Their lives are the subject of B. Caine's *Bombay to Bloomsbury: A biography of the Strachey Family* (Oxford: Oxford University Press, 2005).
2. *[1871]*: It is difficult to date this letter, but it is possible that this was written by Tyndall in 1871, due to Strachey's return to England sometime in the late spring of 1871. Strachey went again to India in 1877 for two years, and then came back to England permanently in 1879.
3. *own shores*: England.

To Cecil Trevor 5 July [1871][1] 3483

Royal Institution of Great Britain | 5th July.

My dear Sir,

Although it is not very beautiful "copy" I think the printer, should his services be invoked, will find no difficulty in deciphering the Reprint[2]

faithfully yours | John Tyndall | C. Cecil Trevor Esq. | &c &c &c

National Archives MT 10/131 H3953

---

1. *[1871]*: dated to 1871 due to the report (letter 3475).
2. *Reprint*: copy not included. This is in reference to the report Tyndall sent to Trevor; see letters 3475, 3476 and 3480.

To Edward Livingston Youmans 5 July 1871 3484

ROYAL INSTITUTION, *July 5, 1871.*

MY DEAR YOUMANS: I am equally desirous with my friend Mr. Spencer to do all that is possible to extend and establish the arrangement[1] you are making with English authors, and which will practically amount to international copyright.

Let me also add the expression of my personal recognition of the excellent service which you have already rendered to English authors in America.

Most faithfully yours, | JOHN TYNDALL.

J. Fiske, *Edward Livingston Youmans: Interpreter of Science for the People. A Sketch of His Life, with Selections from His Published Writings and Extracts from His Correspondence with Spencer, Huxley, Tyndall and Others* (New York: D. Appleton, 1894), pp. 274–5
Typed Transcript Only

---

1. *the arrangement*: Youmans published the 'International Scientific Series' in 1871, which consisted of publishing a number of important works of science at the time in a number of modern languages. This included works by Tyndall and Spencer. At the time, the absence of an international copyright agreement meant that many books were published in pirated editions outside their country of origin, but the arrangements for the series were such as to ensure English authors received payment for their work. See L. Howsam, 'An Experiment with Science for the Nineteenth-Century Book Trade: The International Scientific Series', *British Journal for the History of Science*, 33 (2000), pp. 187–207; B. Lightman, 'The International Scientific Series and the Communication of Darwinism', *Journal of Cambridge Studies*, 5 (December 2010), pp. 27–38; and B. Lightman, *Victorian Popularizers of Science: Designing Nature for New Audiences* (Chicago: University of Chicago Press, 2009), pp. 378–9.

From George Gabriel Stokes 5 July 1871 3485

Cambridge 5th July 1871

My dear Tyndall,

I read your letter of the 25th[1] as intended to back an application from Mr Whymper—, not as an application itself, and even now I can hardly think I misunderstood you. I therefore put the letter into my pocket, intending to bring it on along with Mr Whymper's application. None such however has come in, and that leaves me to doubt whether your letter might not possibly have been intended for a direct application, though there is so much of apparent allusion to a document[2] supposed to be in possession of the reader that I can hardly think it. There will probably be another meeting of Council[3] after the 15th so it would be well to tell Mr Whymper that if he means to apply he had best do so, or, if your letter was all that was meant to come before the Council, you had best write me a line to say so.

Whether there will be another meeting of Council depends probably on the time of confirmation by H[er].M[ajesty].[4] of the new statutes relating to 3 of the Public Schools[5]

Yours very tr<uly> | G. G. Stokes

RI MS JT/1/S/253
RI MS JT/1/TYP/4/1415

1. *letter of the 25th*: letter 3472. It is possible Stokes wrote 25th, but meant 23rd.
2. *allusion to a document*: see letter 3472.
3. *Council*: Stokes was a member of the Council of the RS, the governing body which makes statutes and other decisions for RS.
4. *H[er].M[ajesty].*: Queen Victoria.
5. *Public Schools*: Likely referring to the impact of the 1868 Public Schools Act. See *Ascent of John Tyndall*, pp. 291–2 for Tyndall's appointment to the governing body of Harrow School.

To Emil du Bois-Reymond 6 July 1871 3486

6th July 1871.

My dear DuBois

It would give me immense pleasure to meet you in the Alps,[1] and were it not for the meeting of the British Association[2] which will probably call me back I would try and make our motions coincide this very year. I am going on Monday or Tuesday next to Pontresina[3] but early in August the association meets, and as excellent William Thomson with whom I quarrelled years ago because we did not know each other, presides, I feel a strong desire to support him. I shall probably finish my vacation by a scamper in the Scotch highlands.

Last year the occurrences of the war[4] so fascinated me that I never quitted the Bel Alp. I had three telegrams daily,[5] and newspapers from England, Germany and Switzerland. No public document ever affected me so profoundly as the manifesto of the King[6] issued just before the actual commencement of the war. There was in it such a ring of conscious right, strength and resolution. I read it on a hill top, and so much was I moved by it that had I been young I should have then and there offered myself to Germany and fired my shot in the cause that I held to be just and holy. As it was I returned to the Bel Alp and wrote out a cheque for a large portion of my total worldly possessions purposing to send it direct to the King himself. But reflection caused me to see what an infinitesimal drop in the bucket my contribution would be and so I suspended it.

But the calamities of France have of course mollified the indignation raised by her atrocity in commencing the war. What the issue is to be no man can tell. If she possesses the stuff that is purified by trial she may come out of the ordeal nobler than before. But the only thing now certain is that she possesses a great deal of stuff incapable of purification, and only fit for combustion. Still the country may be sound, and one knows personally so many noble Frenchmen, that one would fain[7] hope France possesses a sufficient number of them to secure her salvation. It would, in my estimation be a wholesome sign of the times if they were to choose the Count of Paris[8] or the Duc d'Aumale[9] as ruler, for it would be an indication that their eyes were open to the noble quality of those truly excellent men.

I am glad you like the 'Fragments' and the 'Hours of Exercise'. Helmholtz, to whom remember me right heartily, Vieweg tells me[10] will superintend a translation of the former, while Wiedemann has similarly undertaken the latter. Bence Jones is well. The political world is here in chaos over the Military[11] and the ballot bills:[12] did I not think the core of England sound, and that a capacity of righting what is wrong is still latent amongst us, I should despair of the country. If statesmanship be a work of <u>thought</u> how can it be accomplished amid this universal parliamentary gabble! I am sick of it.

But I will not afflict you further.

Ever yours dear DuBois | Your sincere friend | John <u>Tyndall</u>

RI MS JT/1/T/403
RI MS JT/1/T/1526
RI MS JT/1/TYP/7/2440

---

1. *meet you in the Alps*: Tyndall is writing in response to letter 3450.
2. *meeting of the British Association*: Tyndall did not attend the 1871 meeting of the BAAS that took place in Edinburgh from 2 to 9 August. He stayed in Switzerland. See also letter 3498, for example.
3. *Pontresina*: Switzerland; for more information about this trip, see letters 3494, 3495 and 3497.
4. *the war*: the Franco-Prussian War.
5. *three telegrams daily*: letters missing.
6. *manifesto of the King*: not further identified.
7. *fain*: willingly (*OED*).
8. *Count of Paris*: Prince Philippe of Orléans, Count of Paris (1838–94), the heir to the French throne before the 1848 revolution which saw the founding of the Second Republic of France and the leadership of Louis-Napoleon Bonaparte (1808–73). Philippe served in the American Civil War with Union General George McClellan (1826–85). He was refused permission to serve in the Franco-Prussian War, and afterwards, despite the popularity of the house of Orléans, he refused to become the king.
9. *Duc d'Aumale*: Henri Eugène Philippe Louis d'Orléans, Duke of Aumale (1822–97), served as captain of infantry in the French Army at the age of seventeen. He was refused permission to serve in the Franco-Prussian War, but by 1872 was returned to the Army as General of Division. In 1883, he was denied his Army position, having been a member of a reigning French family. He was exiled from, but later was allowed to return to, France.
10. *Vieweg tells me*: see letters 3432 and 3434.
11. *Military*: There were a number of army reforms during this period, and it is possible Tyndall is referring to the Cardwell Reforms, which took place between 1868 and 1874. These sought to abolish flogging as a punishment during peacetime (but not during active

service), withdrawing troops from colonies that could govern themselves and getting rid of money for new recruits. They also sought to professionalise the Army, to be more like the Prussian Army, and abolish the gentleman-soldier.

12. *ballot bills*: The Ballot Act of 1872 provided for having secret ballots in local and parliamentary elections in the UK, which was one of the requests of the Chartist movement from the 1830s to the 1850s. It was enacted on 18 July 1872 (Parliamentary Archives HL/PO/PU/1/1872/35&36V1n103).

From Edward Sabine — 7 July 1871 — 3487

Dear Tyndall

The money is now payable to M^{r}. Whymper,[1] as you will see by the enclosed.[2] You have acted so kindly to M^{r}. Whymper in the matter[3] that I think you will like to acquaint him accordingly.

Sincerely yours | Edward Sabine | July 7.

<Hereafter LT Typescript Only>

Her Majesty's Printing Office | London E.C. | 7 July 1871

Dear Sir Edward,

Dr Sharpey quite agrees with me in concurring with your suggestion about Mr Whymper. I have therefore sent him the Cheque this morning.

very sincerely yours | W. Spottiswoode.

RI MS JT/1/S/46
RI MS JT/1/TYP/4/1334

1. *payable to M^{r}. Whymper*: see letters 3470 and 3481.
2. *the enclosed*: LT noted that the enclosed was on the same piece of paper.
3. *so kindly to Mr. Whymper in the matter*: see letters 3462, 3464 and 3470.

From Walter White[1] — 7 July 1871 — 3488

Royal Society | July 7/71

Dear Sir

I forward the inclosed from Prof. Stokes.[2]

When he wrote it he was not aware that by authority of our President & Treasurer a sum of £50 has been advanced to Mr Whymper.[3]

your obed[ien]^{t} serv[an]^{t} | Walter White

Prof. Tyndall | &c &c

RI MS JT/1/W/33

1. *Walter White*: see letter 3413, n. 1.
2. *inclosed . . . Stokes*: possibly letter 3485.
3. *to Mr Whymper*: see letter 3487, for example.

To Emil du Bois-Reymond 8 July 1871 3489

8th. July 1871.

My dear DuBois.

Let me introduce to you Dr. Youmans,[1] an excellent American Gentlemen who has done good Service in promoting the Cause of Science in America; He knows you of course through your renown,—and if you heard his exclamations of admiration in looking at your portrait (which is suspended in my room) you would I am persuaded receive him with all kindness.

Ever my dear DuBois | yours faithfully | John Tyndall

RI MS JT/1/T/404
RI MS JT/1/TYP/7/2441

1. *Dr. Youmans*: see letter 3484.

From Agatha Russell 9 July 1871 3490

Pembroke Lodge. | Richmond Park. | July 9. /71.

Dear Professor Tyndall

We thought you had made a mistake when you wrote 'climber' in Mama's book;[1] but we shall not put in 'author' instead; as Mama[2] will send you the book & beg you to be so kind as to do it yourself. We are very sorry you have not been able to come here before going abroad. It must be delightful to be starting off to the snow mountains; & it is very kind of you to say you will drink my health on the top of one of them. Mama desires me to say that my brother Rollo's[3] essay, of which he spoke to you has only just come back, so that he could not send it; but he thinks perhaps you will kindly look over it a future time.

Believe me, very sincerely yours | Agatha Russell.

RI MS JT/1/R/47

1. *Mama's book*: *Hours of Exercise in the Alps*.
2. *Mama*: Frances Russell.
3. *my brother Rollo*: Francis Albert Rollo Russell (1849–1914), an English meteorologist. He was educated at Oxford and became well-known for leading the cause of ameliorating the foul London coal smog. He was elected to the Royal Meteorological Society in 1868, at the age of nineteen.

To Elizabeth Dawson Steuart 11 July 1871 3491

Royal Institution | 11th. July 1871.

My dear Mrs Steuart,

I have just received your note[1]—had it been a post later it would have missed me.

The fact that some heavy demands have been made upon my resources of late strengthens your notion that this ought to be a <u>loan</u>[2]

Yours ever | John Tyndall.

RI MS JT/1/TYP/10/3405
LT Typescript Only

1. *your note*: letter missing.
2. *this . . . loan*: see letter 3474. It is likely that Tyndall is talking about his relative Caleb Tyndall.

From Giuseppe Bianconi[1] 19 July 1871 3492

Io professo molta riconoscenza alla S.V. per la lettera sua 7.corr.e per la benevole accoglienza accordata alla mia mem[a] intorno alla flessibilità dal ghiaccio. Io sono felice che questa occasione mi abbia procurato l'onore di una relazione con Lei, i cui lavori altamente stimabili ho da lungo tempo letti con tanto piacere e profitto; ed ai quali io, e mio figlio abbiamo avuto spesso incontro di riferirci nelle nostre pubblicazioni. Mi è poi gratissimo di presentare a Lei le nostre più vive congratulazioni per servigi eminenti che Ella reca alla scienza colla felicità delle sue esperienze, e colla invidiabile lucidità del suo stile. Con profondo rispetto mi protesto.

Bologna 19. Lug. 1871

Suo dev[o]. obbl[o]. sero[re] | G. Giuseppe Bianconi

PS. Ella riceverà questa lettera dopo il suo ritorno, ma sin da oggi io lo seguo nel suo viaggio alle Alpi co'miei voti, per la prosperità dei suoi studi, e di Lei medesima.

G. G. B.

I am deeply grateful to Y[our]. L[ordship]. for your latest letter, dated the 7th of this month,[2] and for the favorable reception that you accorded my memoir on the flexibility of ice.[3] I am pleased that this opportunity has brought me the honor of having a relationship with You, whose highly-esteemed works I have read with great pleasure and profit for quite some time, and to which I and my son[4] have had past occasion to cite in our publications. I am further pleased to extend to You our warmest congratulations for the prestigious contributions that You bring to science with your experience and with the enviable lucidity of your style.

With deep respect I declare myself

Bologna 19. Jul[y]. 1871 | Your loyal obliged servant | Giuseppe G. Bianconi

PS. You will receive this letter after your return, but from now I will follow You on your journey to the Alps with my wishes for the prosperity of your studies and equally of Yourself.

G. G. B.

RI MS JT/1/B/97

---

1. *Giuseppe Bianconi*: Giovanni Giuseppe Bianconi (1809–78), an Italian natural historian, zoologist and geologist. He was a professor of natural history and director of the museum of natural history at the University of Bologna from 1842–64. He was also the author of anti-Darwinian works.
2. *your letter . . . month*: letter missing.
3. *memoir on the flexibility of ice*: G. Bianconi, *Esperienze intorno alla Flessibilità del Ghiaccio* (Bologna, 1871).
4. *my son*: Giovanni Antonio Bianconi (1841–75), an Italian naturalist, and the son of Giuseppe Bianconi. He was educated in mathematics, but contributed to the fields of entomology and geology.

From Lucy Wynne[1] 19 July [1871][2] 3493

Westwood Lodge | Sydenham | July 19

My very dear Mr. Tyndall

I am so very much obliged to you f[o]r the book you have sent me. It was so kind of you thinking of me. I think you will be glad to hear of my happiness. I am going to be married to Mr. O'Brien of the 60th. Rifles, & we know each other f[o]r so many years that our marriage promises to be a very happy one. If you were in town I sh[oul]d bring Mr. O'Brien to see you as I know you w[oul]d like him and he you.

My Father[3] is very happy ab[ou]t it too, & we are to be married in ab[ou]t 3 weeks & then go to Switzerland. Is there any chance of meeting you there?

Yours dear Mr. | Tyndall very aff[ectionate]ly. | Lucy H. Wynne

RI MS JT/1/W/111
RI MS JT/1/TYP/5/1866

1. *Lucy Wynne*: Lucy Harriette Wynne (d. 1932), the daughter of Major-General George Wynne. As a Lieutenant, George Wynne was in charge of a part of the Irish Survey of the Royal Engineers. Lucy Wynne married Major Aubrey Stephen O'Brien (1837–98) of the 60th Rifles in 1871 (see n. 2) and they had two children. In 1899 she legally changed her name to Lucy Harriette de Vere by Royal License. Tyndall knew her because her father had been in charge of the Ordnance survey in Ireland (see *Ascent of John Tyndall*, pp. 5–7, 281).
2. *[1871]*: dated to 1871 because Lucy Wynne was married 10 August 1871 and became O'Brien.
3. *My Father*: see n. 1.

To Henry Bence Jones [23 July 1871][1] 3494

Krone Gasthaus Pontresina | Les grisous | Switzerland. | Sunday morning

My dear Friend

I have been here exactly a week. I had a fair passage to Paris, and a good look around the city during the evening I spent there. Its brilliancy was greatly dimmed, and moral depression seemed to accompany the twilight-gloom of the dimly lighted streets. Of course I visited the places where the work of destruction was most pronounced. It was a very sad sight. Next morning I started to the station counting on the accuracy of Bradshaw which showed the trains passing as usual between Paris and Bale. I had a dull day of about 14 hours and in the end was obliged to put up for the night at Belfort. The presence of the Prussian troops makes a singular impression: it was most odd to see them sitting about the Platform at Amiens, standing guard at the various stations of the eastern railway, and occupying Belfort in force. There are about 5000 of them in Belfort; and they march, and counter march to fife and drum as if they were the natural lords[2] of the soil. Assuredly it must be bitter to France. I like the German fighting, it was heroic to the last degree, but I have my doubts as to the ultimate wisdom of those last demands. The time may come when military genius may be on the side of France, and if so the memory of those last crushing penalties may turn out to be a dynamic power on the side of France. And how glorious it would have been for humanity if Germany could have felt herself able to be magnanimous. But the German men know their own business best. They doubtless have weighed all these things, and they have shown such wisdom in other matters that they may lay claim to the presumption of wisdom in these final arrangements also.

I managed to get from Belfort to Coire in a single day, and the day following I crossed the Albula pass and was landed for the night at Samedan.[3] On the Sunday morning I drove up here, which is less than 4 miles distant from Samedan. I met some of my old guides,[4] and learnt that an expedition up the highest peak of this region the Pitz Bernina[5]-- was contemplated. Mr Puller,[6] a member of the Athenaeum club,[7] wished to make the expedition. I wished to try my London legs and here was an opportunity. I wished to revolutionise my life—here was a chance. So offering to take my share of the expenses of the expedition matters were arranged between Mr Puller and myself. On the Monday we went along the Morteratsch glacier to a hut on the mountain side at a place called Boval. Here we turned into a bed of hay where I should have slept were it not for the snoring of one of the guides. He blew his trumpet without intermission from 8o'c[lock]. to *[midnight]*,[8] and all this time I had to listen to him. At 12o'C[lock]. the guides rose and prepared some tea. At 20 minutes past one, we were harnessed and already upon the glacier. The heavens were crowded with stars, and nowhere was a cloud to be seen. After passing first a smooth portion of the ice we got upon a broken *[glacier]* and had some difficulty in threading our way round its crags and chasms, hidden and revealed, long snow-slopes surrounded the glaciers; and this year the mountains are particularly heavily laden with snow. In fact a few such years by giving increased nutriment to the glaciers would soon restore them to the magnitude from which they have been shrinking for many years. After hours of snow work, rendered slow by the depth and other circumstances, we got upon a rocky spur of the mountain. It stretched down like a wall; and fell in almost sheer precipices right and left. It was deeply hacked, and our progress along it for a time was slow that the guides at one place called a halt and held a consultation as to whether the expedition must not be given up. I confess I was in mortal anxiety about my companion, and should have been quite satisfied to return. We went, however, on. To the wall ascended some very steep rock work, and to this a ridge of excessive sharpness against one side of which the heavy snow had been blown, while at the other the mountain fell in almost a sheer precipice for some thousands of feet. On we went sinking to the knees in the snow, and having the edge of the ridge at about the height of our hips to the right. The summit remained long distant, but we gradually neared it and finally found ourselves upon a point as high as the summit to all appearance, but separated from it by a thin arête of rock and snow. Along this we crept, and a little before noon the highest point of the Pitz Bernina was won.

It was warm and pleasant at the top, we remained there about an hour with the Alps around us, and commenced the decent. I was anxious all day long, not for myself, for I felt quite fit to cope with the work; but for my companion. At times he seemed very anxious; but he managed to be careful, and we had no mischance throughout the day. Down the snow slopes, on to the glacier, where we were for a time imprisoned by séracs[9] and crevasses we reached the hut of Boval at 6 1/2 P.M. having seen 17 hours on the expedition. We had afterwards to descend for nearly two hours by the Morteratsch glacier. So that it was really a day of 19 hours. Now this was not bad for an old fellow who had no sleep and a hay bed the night previous. For was I much tired. I was sunburnt; but next morning I could hardly detect a trace of stiffness in my legs.

On the Wednesday I read and rested, and on the Thursday I strolled about in a small way, you know the mark that was on my hand. It became larger and uglier during my journey from England. A quantity of fluid had exuded from it, this had dried and an ugly scabby surface was the result. On Thursday I particularly marked its boundary, so as to assure myself whether it was extending or riot. Its edges then seemed angry. On passing various fountains from which water spouted I placed the spot underneath the spout and allowed the water to strike it forcibly. This I did several times. At night I put the golden ointment on it, securing the linen by straps of court plaster. On the Friday morning I removed the covering, when lo! I found three fourths of the scab lifted up, and a beautiful new skin underneath. The whole scab is now gone, and as far as I can see the spot is perfectly healed. There is no exudation, no itching, no pimples, but a fresh healthy new skin covering the place.

What is the cause of this sudden cure? like all your medical questions it is a complicated one. Your science[10]—if it can be called a science—needs the best heads in the world. But if the heads be good the training for the most part has been defective; and it will continue so until its cultivators have been brought practically to separate the elements which may contribute to a result, and to distinguish the essential from the nonessential, in other words until they have been bound alike in chemistry and experimental philosophy.

I still think that I shall be drawn back to Edinburgh.[11] Various incentives draw me hitherward. But I shall certainly not be there at the beginning of the meeting. If I get there by the 5th so as to see the winding up it is as much as I shall be able to manage. These however are as yet thoughts without consolidation. I shall certainly remain here eight days longer.

Give my kind regards to Lady Millicent[12] and the young ladies.

Yours ever | John Tyndall

RI MS JT/1/TYP/3/808
LT Typescript Only

---

1. *[23 July 1871]*: see letter 3495 for the same storyline and for dating to Sunday, 23 July 1871.
2. *lords*: LT crossed out 'sons' and inserted this notation.
3. *Samedan*: a town near Pontresina in the Swiss Alps.
4. *old guides*: for example, Peter Jenni.

5. *Pitz Bernina*: Piz Bernina. See letter 3495, n. 15.
6. *Mr Puller*: see letter 3495, n. 16.
7. *Athenaeum club*: see letter 3372, n. 15.
8. *8o'c to*: In letter 3495, Tyndall said he laid in the hay for four hours.
9. *séracs*: blocks or towers of glacier ice, formed by the intersection of crevasses; the *OED* cites Tyndall as having first used this term.
10. *Your science*: Bence Jones was a physician.
11. *Edinburgh*: for the BAAS meeting. Tyndall did not attend. See letter 3498.
12. *Lady Millicent*: Millicent Acheson.

To Thomas Archer Hirst 23 July 1871 3495

Krone Gasthaus | Pontresina | les Grisons Switzerland | 23rd July 1871.

My dear Tom,

I have been just a week at Pontresina. Last Sunday[1] I came here, and today I spend the Sabbath rest so far as to write instead of scampering. I had a fair passage to Paris; and naturally during my short stay there looked after the destructive effects of the late struggle.[2] A great change had manifestly come over Paris; moral depression on the part of the people appeared to go hand in hand with the physical twilight of the dimly illuminated streets. The Tuileries[3] and this environment were all in gloom. The great block of houses between the Rue Castiglione & the Rue Royale were gutted & tottering to their fall. The stump of the Vendome column[4] stood in the place. The Place de la Concorde[5] with its statues & fountains seemed forsaken. One fountain was torn by shot, and one of the statues also reduced to rubbish. The hotel de Ville[6] was a mass of lovely ruins. I hardly met a human being near it. I strolled in to the Champs Elysees; the singers were again at their work, and an old actor or two were grimacing on the slope: but the people appeared to be thinking of something else than the singing and the acting. The brilliant Paris appeared in fact to be in sackcloth and ashes;[7] let us hope that the Gods will give her grace to profit by the discipline.

It made an exceedingly odd impression upon me to see Prussian soldiers occupying the platform at Amiens, and all the principal stations along the eastern railway. At Belfort there are about 5000 of them and they march to fife and drum as if they were the first on free soil. I looked at them with profound interest, as examples of what heroic earnestness and endurance can accomplish. But their presence in France must be terribly bitter. Could they have felt strong enough to deal magnanimously with France at the end of the war, humanity assuredly would have been the gainer. But they are men, exposed to the chances of men, and if they have not acted a chivalrous part, doubtless in their own estimation they have acted a prudent one.

It is hard however to compare the substantial advantages which they have secured to themselves with the force which the memory of these very things may confer upon their antagonists should military genius again appear in France.

Trusting to Bradshaw[8] I went to the station hoping as usual to get to Bale[9] in a single day. I had to spend the night at Belfort;[10] and was lucky not to have to spend it in the streets.

On the Friday I got as far as Coire;[11] and on Saturday crossed the Albula to Samedan.[12] On the Sunday morning I took an Einspanner[13] from Samedan to Pontresina.[14]

There I found an expedition organized for the Pitz Bernina.[15] M^r^. Puller,[16] a member of the Athenaeum Club,[17] wished to ascend the mountain & had already bespoke guides. Jenni was one of them & Jenni urged me to accompany them. Accordingly I arranged to take due part of the expense and to join in the attempt. I wished in fact to suddenly revolutionise my life, and here was a chance of doing so.

On the Monday I had a scamper up the Roseg Valley, and on the afternoon of the same day went to Boval,[18] about 4 hours up the Morteratsch glacier: There I lay for four hours in hay wishing for sleep but really listening to the blowing of Jenni's trumpet.[19] at 1.20 by starlight we started; worked through broken glaciers, along heavy snow slopes, over hacked rocky spines, along sharp arêtes with wild precipices to the right; and snow slopes almost equally forbidding to the left and at a little before 12 got to the top of the mountain.

I thought of you dur[in]g the day; for in all dangerous places I was in mortal anxiety, not for myself but for my companion. But though apparently often alarmed he made his footing sure, and no mishap occurred during the expedition.

Toward the end of the day we stood a good chance of being imprisoned on a broken ice fall amid séracs[20] and chasms hidden & revealed. We reached Boval at 6½, and adding to this nearly two hours for the descent of the Morteratsch you have a day of 19 hours—not bad for untrained London legs.

Still next morning I was hardly able to detect a trace of stiffness in my limbs.

You know the spot upon my hand: it waxed uglier up to Thursday last, but it is now completely gone & fresh new skin covers the place.

Give my love to Lilly[21]—I may still go to Edinburgh; though certainly not for the beg[innin]g of the meeting.[22] I will stay here another week.

ever affectiona[te]ly John

I have had <u>one</u> enclosure[23] from you since I came away | <u>J. T.</u>

BL Add MS 63092, ff. 74–5
RI MS JT/1/HTYP/582

---

1. *last Sunday*: 16 July 1871.
2. *the late struggle*: the Franco-Prussian War. See also, for example, letters 3370, 3382 and 3438.
3. *Tuileries*: Palais des Tuileries was a sixteenth-century palace that stood along the right bank of the Seine until it was burned in May of 1871 by members of the Paris Commune after the Franco-Prussian War. When Tyndall saw it, it was a shell, with no roof. It stood that way until it was demolished in 1882.
4. *Vendome column*: a column commemorating the French defeat of the Austrians at Austerlitz. Completed in 1810, and made of melted-down Austrian cannon, it sits in the Place Vendôme.
5. *Place de la Concorde*: a large public square in Paris, between the Tuilieries and the Champs-Élysées.
6. *hotel de Ville*: the headquarters of Paris government since the fourteenth century. During the Franco Prussian War in 1870–1, it was the headquarters of the Paris Commune, and parts of it were destroyed.
7. *sackcloth and ashes*: (Biblical) clothed in sackcloth and having ashes sprinkled on the head as a sign of lamentation or abject penitence (*OED*). For example, in the Old Testament, see Esther 4:3; and in the New Testament, see Matthew 11:21.
8. *Bradshaw*: shorthand for the series of railway guides produced by George Bradshaw (1801–53); there were guides for continental as well as for British trains.
9. *Bale*: Basel, Switzerland, near the French border. Tyndall was hoping to make it all the way to Switzerland, but had to stop along the way. For the next few sentences, he traces his path to Pontresina for Hirst. One could follow him on a current map.
10. *Belfort*: a town in France, along the Swiss border.
11. *Coire*: Chur, a town in Switzerland in the Rhine Valley.
12. *Albula to Samedan*: Albula pass to Samedan, a town near Pontresina in the Swiss Alps.
13. *Einspanner*: a one-horse carriage.
14. *Samedan to Pontresina*: Pontresina is about six kilometers (about four miles) from Samedan.
15. *Pitz Bernina*: Piz Bernina, the highest mountain in the Eastern Alps.
16. *Mr. Puller*: Christopher Puller (n.d.), a fellow climber and member of the Athenaeum Club (*Ascent of John Tyndall*, p. 276).
17. *Athenaeum Club*: see letter 3372, n. 15.
18. *Boval*: an Alpine peak near the Morterasch glacier. There was also a hut there which opened in 1865 and Tyndall is likely referring to the hut.
19. *Jenni's trumpet*: apparently, Jenni snored.
20. *séracs*: see letter 3494, n. 9.
21. *Lilly*: Emily 'Lilly' Anna Hirst (1853–83) was Hirst's niece, the daughter of his brother

William Aked Hirst (b. 1829) and Rachel Anne Hirst (née Burkitt). Lilly traveled with and was devoted to her uncle throughout her life and appears throughout his journals. On 10 November 1877 she married Richard Methley. She died after the birth of her son Cyril in October 1883.

22. *Edinburgh . . . meeting*: see letter 3486, n. 2 and letter 3498. Tyndall did not attend.
23. *one enclosure*: missing.

To George Ginty[1] 30 July [1871][2] 3496

Dear George Ginty,

A letter written to me by your excellent mother[3] has been sent after me to Switzerland, and as I have forgotten her address in Birkenhead, I write to you, taking it for granted that any letter addressed to you at Mr Laird's Works[4] will be sure to reach your hands.

Will you express my regret to your mother that I was obliged to quit London without seeing her. I should have been glad to see her on the Saturday previous to my coming away,[5] but she no doubt found that I was busy and did not come to me. Indeed on that day I had to prepare for a visit of the Emperor of Brazil;[6] but I could have afforded ample time for a little conversation with her.

I know very well what the upshot of that conversation would have been. It would have been all directed to the securing of the welfare and happiness of her children.

I do not know a mother to whom her children owe a deeper debt of love and gratitude than your mother's children owe to her. Her love for you has always been unbounded and as you know more and more of the ways of this world you will learn more and more how precious the love of a mother is. And it will add to your happiness to think when she is gone, that you did nothing to render her life unhappy, as it assuredly will add to your regrets should the memory occur to you that you have not behaved toward her as her tender care of you, and her great anxiety regarding your welfare ought to prompt.

I hope and trust that you will prove a good and loving son to her. She naturally wishes to see you all following in the footsteps of your father,[7] who made himself respected wherever he went, as much through the uprightness of his character, as the energy of his mind.

I do not think I have anything more to say to you but be good enough to give my kindest regards to your mother.

Very faithfully yours | John Tyndall. | Switzerland 30th July

RI MS JT/1/TYP/11/3725–6
LT Typescript Only

1. *George Ginty*: George Ginty (b. 1851), one of Margaret and William's three sons; they also had two daughters.
2. *30 July [1871]*: Tyndall refers to both going to Pontresina and meeting the Emperor of Brazil in this letter, dating it to 1871. See also notes 5 and 6, below.
3. *letter . . . your excellent mother*: letter missing. His mother was Margaret Ginty (see letter 3501).
4. *Mr. Laird's Works*: John Laird (1805–74), a Liverpool shipbuilder. He took over the business from his father, William. He was educated at the Royal Institution, Liverpool. The business began to specialize in prefabricated riverboats, used by explorers in the late nineteenth century to explore Africa and the Middle East. He retired from his shipbuilding business in 1861 and it was then run by his three sons, William, John and Henry, and the company was called Laird Brothers until 1903.
5. *the Saturday previous to my coming away*: Tyndall left London for Pontresina on the 12th, making Saturday the 8th.
6. *Emperor of Brazil*: see letters 3477 and 3478.
7. *your father*: William Gilbert Ginty (*c.* 1820–66), Tyndall's friend and colleague from the Ordnance Survey of Ireland. They worked together there from 1839–42. Margaret was his wife. They had lived in Rio de Janeiro from 1855 until William's death. For more, see the entry for Margaret Ginty in the Biographical Register.

To Mary Egerton 31 July 1871 3497

My dear Lady Mary

You have doubtless ere this gone to visit your Scotch friends;[1] your arrangement I think was to go to them first and afterwards to Edinburgh. I will therefore enclose this to Hirst[2] and thus I hope it will reach you. I have been here a fortnight: a portion of the weather has been good, and when good weather favours this region it is very glorious. I passed through France on coming here, and though I knew that German soldiers still occupied part of France that knowledge did not lessen the strangeness of the impression of seeing them at the station at Amiens, along the various stations of the Chemins de fer de l'Est, at Belfort in force,[3] where they marched through the streets to the sound of fife and drum as if they had been lords of the soil. Paris I found in a kind of twilight gloom. It was a sad contrast to the brilliancy of former times, and still if there be any stamina in the people, the present sackcloth and ashes[4] may presage a purer and manlier state of things than that which previously existed. I arrived here on a Sunday, and on Monday learnt from an old guide of mine that an attempt on the Pitz-Bernina, the highest of the peaks in this region was contemplated.[5] I was permitted to form one of the party, which consisted of a gentleman whom I sometimes see at the Athenaeum Club,[6] two guides and myself. We had some coverlets sent to a hut about four hours from this place, and some provisions also. In that hut we spent the Monday night. We had hay, and I should have slept were it not for the nasal trombone of one of the guides. But as it was I rested. It was rather a doubtful thing for me to do, for such an excursion demands some previous training. We started a little after one o'clock with the heavens crowded with stars, and the hacked outlines of the mountains relieved by the sky behind them. First over glacier, then amid broken séracs,[7] with chasms between them; then after long unravelling emerging upon the open snowfields. The work was somewhat heavy for the snow was deep. After hours of this work we hacked our way round a boss of ice with a deep chasm below us and came upon a rocky spur of the mountain. It was like a ruined wall, being perfectly precipitous on both sides. Along this we worked for a time; then up some precipitous rock acclivities, and finally came upon the long steep sharp ridge which led to the summit of the mountain. A knife-edge of snow had been formed, to the left of which the substance gave us footing while to the right the rock fell sheer in merciless precipices. Steadily creeping upward the summit by degrees became nearer, and a little before midday we were on the top. We had some difficulty in disentangling ourselves from the séracs on returning, and might have been imprisoned there for the depth of snow and other causes rendered us far later than we ought to have been; as it was it turned out a day of 19 hours pretty hard work. I was glad to find that next day the stiffness consequent on the exertion was barely perceptible.

And since then I have had some modest excursions—more modest as regards height and labour, but also more modest as regards appliances, for no guides were involved. One day, accompanied by a friend I went up the Pitz Croutsch,[8] a very fine snowy mass of the second order. It is quite a different sensation cutting your own steps, and working warily along the arêtes without a guide from the mere following of an experienced nature. It is like making an original experiment instead of repeating the experiment of another. To me it is surprising to see the small amount of originality developed by English climbers, Girdlestone[9] is the only man who has shown anything of the kind. I will not say it is of the most prudent kind, but it is infinitely more creditable to him than hanging on to the skirts of a guide to whom you delegate the skill of scenting out the proper way, and the labour of leading, your own inventive powers being for the time perfectly torpid. And this torpidity—this absence of independence to which they have been accustomed, is preached up by some of our best known climbers.

However I did not intend, when I began, to write you a philosophy of climbing. As I write the hail patters with fury against my bedroom window. It is very cold. Hirst urges me[10] strongly to avoid the botheration of Edinburgh, and I think I will remain away among these mountains for another fortnight and then return to England. It would be a great pleasure to me, as indeed it has been a strong temptation to meet you and "May"[11] at the Association meeting, and to have a little scamper with you in the Highlands afterwards. But I was absolutely compelled to run away from London when I did. The air and habits of the place become deadly to me towards the end of June, and life there is to me no life at all—but a shamming of life. I therefore impatiently leave it, and once here it will be better to finish my holiday, which I intend to make a short one—in the Alps and then return to my work in London.

We had the shock of an earthquake here a few days ago.

Give my love to "May"—of course in the sense as she sent hers to me[12]—for I do not intend her to snub me. She will doubtless meet some fine fellow soon—perhaps she has met him already, who can venture to offer her the genuine article.

Yours ever | John Tyndall

Pontresina 31st. July 1871 | If I wait to read this through I shall miss the post. | John Tyndall

RI MS JT/1/TYP/1/391–4
LT Typescript Only

---

1. *your Scotch friends*: possibly at Kinellan, home of James and Mary Coxe (see letter 3406, n. 5). Later, William Fullarton Cumming (letter 3406, n. 7) occupied the house.
2. *enclose this to Hirst*: If there was a letter to Hirst on the same day, it is missing.
3. *Amiens . . . in force*: see also letters 3494 and 3495 for a similar retelling of this story.
4. *sackcloth and ashes*: see letter 3495, n. 7.
5. *Pitz-Bernina . . . contemplated*: see also letters 3494 and 3495 for a similar retelling of this story.
6. *Athenaeum Club*: see letter 3372, n. 15.
7. *séracs*: see letter 3494, n. 9.
8. *Pitz Croutsch*: Piz Corvatsch, one of the peaks in the Bernina range, where Tyndall was at this time. LT noted that this spelling is from the original.
9. *Girdlestone*: Arthur Girdlestone (1842–1908), an English clergyman and climber. His book *The High Alps without Guides: Being a Narrative of Adventures in Switzerland* (London: Longmans, Green, and Co., 1870) was controversial, possibly because he did not use guides for his climbing and instead prayed with his porters before the climb began (*Ascent of John Tyndall*, p. 236).
10. *Hirst urges me*: letter(s) missing.
11. *"May"*: Egerton's daughter, Mary Alice "May" Egerton.
12. *sent hers to me*: letter(s) missing.

To William Thomson 2 August 1871 3498

Pontresina les Grisons | Switzerland | 2nd August 1871.

My dear Thomson.

You begin your labours today in Edinburgh[1] and I cannot suppress the desire to wish you from my mountain lodging a thorough and entire success. I would give a great deal to be in Edinburgh tonight—and indeed I fully intended to be there, but London at the beginning of July is as deadly to me as the miasma of the Campagna,[2] and this year as usual I was forced to run away from it.

We had a small shock of an earthquake here a few days ago.[3] I was sitting at midday in a little temporary auberge[4] which has been erected near the foot of the Morteratsch glacier, eating my bread and sardines, and drinking a glass of English beer, when a violent tremor, accompanied by a loud noise seemed to pass under me. The little house shook, the woman of the house rushed out exclaiming 'ein Erdbeben!'[5]—the man followed her repeating the exclamation. The thought had not occurred to me until they expressed it. My notion was that a huge mass of earth had detached itself from some adjacent part of the mountain and had thundered down producing both noise and trembling. But nothing of this kind occurred. Moreover the houses of parvenus[6] 4 miles distant were also shaken, and indeed the tremor was felt and the noise heard at treble this distance from the Morteratsch auberge, so that no doubt can remain that the thing was a thud of those forces which will probably occupy some of the attention of the Edinburgh meeting.

Again most heartily I wish I was near you. Were I as strong as Tait, and equal like him to beer and tobacco ad infinitum, I should never have come to Switzerland at all but gone to Edinburgh instead. The post starts from this place at half past 9. It is now 9 and therefore time to close this note.

I was horribly burnt yesterday. The sun after days of hail and rain shone out with extraordinary power: the snow was new & in its most dazzling condition—I had already peeled off once from the ascent of the Pitz Bernina,[7] and I thought myself induratio[8]—I shall however have to undergo a second scaling. I wore the blue spectacles[9] yesterday—it was a day of deep blue in the firmament: but when the spectacles were removed the sky for a considerable time seemed as dirty as if the smoke of Edinburgh was diffused through the air. In fact the eyes had been paralysed to the blue, and an undue account of the complementary colour mingled with consciousness when the spectacles were taken off.

Believe me dear Thomson | yours ever faithfully | John Tyndall

CUL Papers of Lord Kelvin, Add MS 7342, T627

1. *Edinburgh*: the BAAS meeting. Thomson was presiding as President of the Association, but Tyndall did not attend. See letter 3497, for example.
2. *miasma of the Campagna*: The Campagna region of Italy was well known for having a 'bad air' that made people ill, and travelers had complained about it for centuries. By the turn of the twentieth century, it had been discovered that the cause was a particularly deadly form of malaria, borne by mosquitos in the area (see L. Beaven, '"Grave of Graves": The Responses of Grand Tourists to the Roman Campagna', in G. Bonifas and M. Monacelli (eds), *Southern Horrors: Northern Visions of the Mediterranean World* (Newcastle: Cambridge Scholars, 2013), pp. 79–94).
3. *earthquake . . . days ago*: see letter 3497, where he mentioned this to Mary Egerton.
4. *auberge*: an inn (*OED*).
5. *ein Erdbeben*: an earthquake.
6. *parvenus*: 'a person from a humble background who has rapidly gained wealth or an influential social position'; usually used in a derogatory manner (*OED*).
7. *Pitz Bernina*: Piz Bernina, a peak in the Alps. Tyndall detailed this climb in a number of letters, such as 3494, 3495 and 3497.
8. *induratio*: Latin for 'hardened'.
9. *blue spectacles*: using colored glass for eye protection, like sunglasses.

From Archibald Hamilton 2 August 1871 3499

Trillick | Omagh | Aug 2/71

Dear Sir[1]

Having lately read your Lectures on sound, I take the liberty of writing to you for some information on the subject, as it is one in which I take a deep interest.

I wrote a letter & drew up a brief outline of a Theory of the Senses which I venture to propose, especially with regard to that of Hearing—but after I had made this up for the Post, it was somehow put aside in the course of moving furniture &c. &c. &c. I have not found it since, but can easily re-work it.

The principal questions which I want to get answered are these:

1. What is the best commentary on that part of Sir I. Newton's Principia which treats of sound? & what do you suppose to have been Newton's true idea on the subject?

2. Where can I find the best description of those sounds which may be called essentially <u>unmusical</u> or <u>inarticulate</u> if there be any such? such as are caused by the continuous tightening or slackening of a string while it is sounding.

A great deal seems to me to depend on ascertaining whether the Ear in such a case does really hear the Sound whose register on my proposed barometer would be.

Where the base line with arrows represents the lapse of Time, & the Elevation above it the tension of the string. The only way in which this can be denied (as it appears to me) is by supposing that owing to some result of the properties of elastic bodies, the continuous increase resolves itself into a succession of small increments rapidly succeeding

each little step being parallel to the baseline; that is of equal Tension—

Your Lectures contain a great number of very beautiful & interesting experiments: but I should greatly wish to see a direct normative register of the varying pressure produced by Sound: a thing which I suggested (if I recollect 5 years ago to Prof. Stokes—& proposed to apply to registering a great variety of natural sounds, as the murmuring of streams &--

If you can spare time to reply to these questions[2] I shall feel greatly indebted

I remain | Yours faithfully in the Lord | A. H. Hamilton Clk[3]

Prof. Tyndall

RI MS JT/1/H/12

1. *Dear Sir*: This letter is almost identical in content to his first letter to Tyndall in this volume, letter 3456. See that letter for notes.
2. *these questions*: see also letter 3600.
3. *Clk*: see letter 3456, n. 6.

To Thomas Archer Hirst 4 August 1871 3500

MR. MATTHEWS[1] and Mr. Froude had supported long rectangles of ordinary ice at the two ends, weighted them in the centre, and thus caused them to bend. The ice employed, if I recollect right, was of a temperature some degrees below the freezing point, and in my little Alpine book[2] recently published I expressed a hope that similar experiments might be made with glacier ice. I have been trying my hand at such experiments. The ice first employed was from the end of the Morteratsch Glacier, and when cut appeared clear and continuous. A little exposure, however, showed it to be disintegrated, being composed of those curious jointed polyhedra into which glacier ice generally resolves itself when yielding to warmth. Still, when properly supported and weighted, a long stout rectangle of such ice showed, after twelve hours, signs of bending.

I afterwards resorted to the ice of the sand cones, which, as you know, is unusually firm. From it rectangles were taken from three to four feet long, about six inches wide, and four inches deep. Supported and weighted for a considerable time, no satisfactory evidence of bending appeared; the bars broke before any decided bending took place. Smaller bars were then employed. Two of these were placed across the mouth of an open square box, their ends being supported by the sides of the box. They formed a cross, and a clear interval of at least an eighth of an inch existed between them where they crossed. The upper one was carefully weighted with a block of ice; after two hours it had sunk down, and was found frozen to the under one. They were then separated, and one of them was allowed to remain supported at the ends and weighted by ice at the middle. In a few hours it had bent into a curve, the versed sine of which from a chord uniting the two ends was, at least, two inches. In fact, *when the rectangles are thin*, and the weight carefully laid on, flexure commences very soon, and may be cautious manipulation be rendered very considerable. I think Mr. Froude told me that in his experiment the molecules were "in torture," and that they in great part recovered their positions when the weight was removed. In the foregoing experiments the flexure was *permanent*.

I tried to bend the rectangle just referred to back again by reversing its position and weighting it with the same block of ice. But whether owing to my want of delicacy in putting on the weight, or through the intrinsic brittleness of the substance itself, it snapped sharply asunder.

I left in your hands when quitting London an exceedingly interesting paper by Prof. Bianconi,[3] in which are figured the results of various experiments on the bending of, I think, lake ice. The foregoing experiments on glacier ice confirm his results.

August 4 | JOHN TYNDALL.

J. Tyndall, 'On the Bending of Glacier Ice', *Nature*, 4 (5 October 1871), pp. 447–8
Typed Transcript Only

---

1. *MR. MATTHEWS*: William Mathews (1828–1901), a British mountaineer, botanist, land surveyor and a founder of the Alpine Club in 1857. He was President of the Alpine Club from 1869 to 1870. He made a number of first ascents in the Alps. He published the

article 'Mechanical properties of ice, and their relation to glacier motion', in *Nature*, 1 (24 March 1870), pp. 534–5.

2. *my little Alpine book*: See *Hours of Exercise in the Alps*, p. 359. Also, see *Ascent of John Tyndall*, pp. 325–6 for more about this experiment.
3. *Prof. Bianconi*: see letter 3492, n. 1. See the whole letter for a discussion of the paper.

From Margaret Ginty 9 August 1871 3501

4 Herningford Terrace | Birkenhead | Aug. 9th. 71.

Dear Dr. Tyndall,

Georgie[1] has shown me a letter which he received from you.[2] How can I thank you, I really do not know—but I am deeply grateful, for your letter came at a time to do him a great deal of good and have more effect on him than any thing I could say or do—Georgie is most affectionate and I know loves me very dearly, and when he makes mistakes I must excuse him sometimes, because he has been all his life living among rough boys. He left me when he was a very little fellow, only four years old, he is very fond of me, I know, and thinks there is no one like me in the world, but like many boys, he has a very poor opinion of a woman's judgment. I have to deal with him with the greatest caution. He likes his work and has plenty of energy. Mr Bevis[3] I think likes him. When I was in London I went to Claridges[4] to see some friends of mine. It happened that nearly all the ladies and gentlemen comprising the Emperor's suite[5] were intimate friends of mine particularly the Baron de Bono Retiro.[6] He told me that when they came to Liverpool that he would speak to Mr Laird[7] about Georgie, he did even more for he got the Emperor to speak, who asked Mr Laird to push him on as much as he could—The result of this kindness, is in Georgie being sent down from the drawing office to the work shop where his hours of work will be longer and the work much harder. This he likes. It shows they are going to take some interest in him, for generally they allow the Apprentices to do as much or little as they like. They are nearly all of them the sons of rich men, my boy is different, and it is important that he should gain the good opinion of our Rio friends for to Brazil he must ultimately go—To my great surprise I received an invitation to the Reception given by the Emperor and Empress[8] the day they were here—They both spoke to me, Oh! so kindly as if I was an old friend some years ago in Rio, when I was a different person, they were not half so gracious. I was glad when it was over and I was quite pleased that I did not burst into tears and disgrace myself when I heard them speak so kindly—At the same time I would not be a person of consequence for the world or have to undergo such an ordeal again. Of course I have to thank the good Baron de Bono Retiro for all this kind condescension. I began to read a book of yours called "Hours of exercise in the Alps". I could not read any further than that first frightfully narrow escape you had[9]—Indeed I don't admire you a bit for running such fearful risks. Why should you venture into such dangerous places? I am sure it is very wicked to do so. Why not leave this work to be done by some one whose life is not so valuable as yours? I would not even read your book, I will be very glad when I hear you are safely away from those cruel mountains. Pray excuse this long letter and thanking you hundreds of times for your kind letter to Georgie which has done a great deal of good

I remain | dear Dr. Tyndall | Yours very truly | M. Ginty

RI MS JT/1/TYP/11/3664–4a
LT Typescript Only

1. *Georgie*: see letter 3496, n. 1.
2. *letter . . . from you*: letter 3496.
3. *Mr Bevis*: not further identified.
4. *Claridges*: a luxury hotel in London's Mayfair neighborhood.
5. *Emperor's suite*: Pedro d'Alcantara; see letters 3477 and 3478.
6. *Baron de Bono Retiro*: possibly Luis Pedreira do Couto Ferraz, Visconde do Bom Retiro (1818–86). He was deputy general, president of the province of Rio de Janeiro, councilor of state and senator of the Empire of Brazil from 1867 to 1886.
7. *Mr Laird*: see letter 3496, n. 4. It is not clear if Margaret means John or one of his sons, William, John or Henry, in this letter.
8. *Empress*: Teresa Cristina (1822–89), Princess of the Two Sicilies and Empress of Brazil. She and Pedro were married by proxy in 1843, and then in person in 1844. She was a dutiful Empress and was known as the Mother of the Brazilians.
9. *narrow escape you had:* She may be referring to any number of stories in *Hours of Exercise in the Alps*.

From Louis Pasteur 10 August 1871 3502

Paris, 45 rue d'Ulm | 10 août 1871

Monsieur et très éminente confrère,

M. Dumas, sachant que j'ai l'intention de prendre en Angleterre un brevet d'invention pour un procédé nouveau de fabrication de la bière, m'a engagé à m'adresser à vous et à <recourir> à vos bons offices, parce que vous étiez très versé <dans> la connaissance de ces sortes d'affaires. Je prends donc la liberté de venir vous demander <si> vous consentiriez à être mon mandataire <dans> cette circonstance et si vous auriez la <*1 word missing*> de vous occuper, soit de la visite du <*1 word missing*>, soit de la surveillance d'une exploitation pour tel ou tel mode que nous réglerons au préalable d'après vos indications.

Si vous acceptez cette proposition, <je> vous prie de me faire connaître quelles seraient les formalités que j'aurais <à> remplir et vos conditions personnelles, c'est-à-dire votre part dans les bénéfices.

Il se pourrait que votre position vous empêchât de vous charger <de> cette affaire. Dans ce cas, je vous <serai> fort obligé de me donner le nom <d'une> personne à laquelle je pourrais <*1 word missing*> en tout sécurité.

Je suis déjà muni du brevet pour <la> France. Dès que j'aurai reçu votre <*1 word missing*> que je désire aussi prompte qu'il est possible, je vous enverrai, s'il y a lieu, une procuration notariée et le texte du brevet pour l'Angleterre. Vous reconnaîtrez que les avantages ont immenses.

Veuillez agréer, monsieur et très éminent confère, l'assurance de mes sentiments de haute considération et de bonne confraternité.

L. Pasteur | Membre de l'Institut de France et de la Société Royale de Londres

Paris, 45 Ulm Street | 10 August 1871

Sir, my very eminent colleague,

Mr. Dumas,[1] knowing that I have the intention of submitting a patent in England for a new method of brewing beer,[2] asked me to contact you and to <seek> your good offices, because you are well versed <in> the knowledge of such things. Therefore, I take the liberty of asking you to be my authorized representative <in> this circumstance, and to take care of either the visit of the <*1 word missing*> or the monitoring of operations under various conditions that we will settle beforehand according to your specifications.

If you accept this proposal, please let me know the formalities that I will have to handle, and your personal conditions, in other words, your share of the profits.

It may be that your position will prevent you from getting involved in this business. In that case, I will be most obliged if you could give me the name <of a> person that I can safely <*1 word missing*>.

I already have the patent in France. As soon as I receive your <*1 word missing*>, which I would like as soon as possible, I will send you, if necessary, a notarized proxy and the text of the patent for England. You will see that the benefits are immense.

Please accept, sir, my very eminent colleague, the assurance of my deepest regards and good fellowship.

L. Pasteur | Member of the French Institute and the Royal Society of London

RI MS JT/1/P/3

1. *Dumas*: Jean-Baptiste-André Dumas (1800–84), a French chemist and one of Pasteur's colleagues and mentors. He was also the father-in-law of Charles François Hervé Magnon (see letter 3624, n. 1). Dumas was elected permanent secretary of the Academy of Sciences in 1868.
2. *brewing beer*: Pasteur had become well known in France for his new process of removing micro-organisms from liquids, what we know as pasteurisation. In the spontaneous generation debate against Felix Pouchet, Pasteur demonstrated that by boiling a hay and water mixture, he could prevent the growth of micro-organisms. The patent he mentions here is a patent for pasteurizing beer. In 1873, Pasteur received an American patent (US135245A) for a brewing process that seemed to make beer-brewing more reliable and efficient, but it is unclear if his patent application was filed in Britain.

To Henry Bence Jones 13 August [1871][1] 3503

Pontresina. Sunday, 13th, Aug.

My dear Friend

You and Hirst combined to keep me away from Edinburgh,[2] and you succeeded. I am not sorry though I certainly wished much to turn upon the occasion a perfectly friendly face toward Thomson, for whom I have a great regard. Had I gone however, I should not be in the enjoyment of my present health and I do not know that I could have done anything to augment the interest of the meeting. That interest, if one may judge from the Times reports, has not been as great as usual.[3] I see the Times here in a fragmentary way: for the English are so numerous that the paper has to be cut up and its scraps fought for. The reports, at first meagre, appear to have dwindled into nothing: or it may be that the particular pages which contain the report have fallen of late into other hands than mine. I suppose the opening address was fairly reported.[4] It amused me and interested me. Nobody but a man of Thomson's brilliancy of intellect would have alighted upon that idea of peopling the earth by organic seed derived from meteorites; but I think that few men having conceived the notion would have ventured to enunciate it thus publicly. There is something so supremely unreal in it. Why should not the seeds be developed in the earth as well as in the meteorites, and if the fiat of a creator be necessary why should it not apply to the earth as well as to meteoric matter? Perhaps the report I have seen is an incomplete, or an incorrect one; but if it be correct I cannot help wishing that Thomson had continued to muse over it in private instead of embodying it in his address. What an enormous influence early education exercises on some men! I dare say it was such an influence that caused Thomson to take refuge in this hypothesis rather than accept the notion that life came direct from what has been indirectly termed inorganic matter. I dare say amid his thousand other avocations he has had but little time to dwell in the presence of this problem; if he had I hardly think he would have tried to dispose of it by a mere brilliant conceivability.

That letter in the Times looks very like little Lionel:[5] I wonder that those religious men take such pleasure in fluttering round the margin of misstatements, to use a mild word. The letter notwithstanding this tampering with the truth contains one or two good and amusing points. The bias of the Times in this matter is very clear, what a pity it is that there are no means of bringing the blush to the cheek of a newspaper, after the error which it espoused and the truth which it opposed have been demonstrated! Perhaps however such opposition fulfils a useful though unknown purpose, and that the best philosophy is to accept it as one of the ingredients of our intellectual life. Is it not a fact that the mechanical irritation of substances in themselves without nutrition, may set the bowels going, and thus end a fit of constipation? A tightness of the brain may be similarly removed by the irritation of the Times.

This I think is a very beautiful place so much so that I thought it wiser to remain here than to take a scamper through the Tyrol, which I at first intended. The English chaplain here is Mr Wade,[6] who has a parish near, or in Soho in London. The Dean of Durham[7] is also here. A few days ago we sallied forth for a real day's excursion over the glaciers and snow-fields. Such guidance as there was fell upon myself; and not having had much acquaintance with the glaciers this year I led them into scenes which would have been dangerous had they not been met with caution. It was an anxious job to get the party over one snow bridge in particular; for the crevasse which it spanned was profound, and the snow was tender to the last degree. But with careful roping and engineering the matter was accomplished. I like Lake's deportment very much: he never flinched for a moment, and was manifestly perfectly willing to commit himself to any venture which the prudence of his leader recommended! I like coolness, and courage, and trust, under such circumstances, but nervousness doubt and hesitation are exasperating. All the party however behaved well. We lost a pretty little dog, or more correctly the dog lost himself through falling into a crevasse. Science however took good care of the church.

That day we had a good variety of work; over glaciers; upon rocks; again over higher glaciers; and again up high rocks: then down along far stretching down slopes, and over green and stony alps to the Bernina Road; striking it near a Wirtshaus[8] where we had a delicious bottle of champagne, good coffee and superb cream. A carriage met us here and drove us back to Pontresina.

I think I will allow myself another week here and then return to my work in London.

Yours ever | John Tyndall

RI MS JT/1/TYP/3/805
LT Typescript Only

1. *[1871]*: dated to 1871 from letters 3494 and 3495.
2. *You and Hirst . . . Edinburgh*: see letters 3494 and 3495 for some of this conversation.
3. *Times reports . . . as usual*: for example: 'The British Association for the Advancement of Science', 4 August 1871, *Times*, p. 8; 'The British Association', 5 August 1871, *Times*, p. 12; and 'The British Association', *Times*, 12 August 1871, p. 4.
4. *the address*: William Thomson's address in which he investigated the question of whether evolution happened in biology, and 'nailed his colours to the mast of intelligent design' (*Ascent of John Tyndall*, p. 277). See also letter 3529 and 'Address of Sir William Thomson', *Brit. Assoc. Rep. 1871*, pp. lxxxiv–cv.
5. *little Lionel*: The letter is likely the Letter to the Editor of the *Times* entitled 'Vitality of this World', printed on 8 August 1871, p. 4. It was signed 'Your obedient, but unfortunate, servant, VITALITY'. 'Little Lionel' may be the microscopist Lionel S. Beale (1828–1906) who published *The Mystery of Life: an Essay in Reply to Dr. Gull's Attack on the Theory of Vitality in his Harveian Oration for 1870* (London: J. & A. Churchill, 1871).
6. *Mr. Wade*: Nugent Wade (1809–73), the rector of St. Anne's in Soho from 1846, and made the canon of Bristol in 1872. He received his BA from Trinity College, Dublin in 1829, and received his DD from Oxford University in 1843. He helped to turn St. Anne's into a parish of new Anglo-Catholics in London at the time.
7. *Dean of Durham*: William Charles Lake; see letter 3447, n. 4.
8. *Wirtshaus*: an inn or tavern.

From Joseph Henry 14 August 1871 3504

(<u>Copy</u>)[1] | Smithsonian Institution[2] | Washington Aug. 14. 1871.

My dear Prof. Tyndall:

I have just received a letter from my friend M^r. Lesley, Secretary of the Am[erican] Phil[osophical]. Society of Philadelphia, informing me that he has learned that you could be induced for the gratification of your numerous American readers to visit this country[3] provided it should not appear that you are seeking to make money by the operation.

If this information be correct, I beg leave to tender you in behalf of the Smithsonian Institution a cordial invitation to visit The United States, a missionary as it were in the cause of abstract science, to vindicate its claims to popular appreciation and to government support.

You need have no care in regard to money matters as all your expenses will be liberally provided for without the least suspicion of mercenary motives on your part. You should not forget however, that money is power; that it can create most ingenious apparatus, and assist in solving problems, the solution of which, without its aid, would be impossible.

I start this afternoon on a visit on Light-House duty to the Pacific Coast to be absent about six weeks. I should be happy on my return to learn that you have accepted the invitation, and that I may have the pleasure of welcoming you to my house as an honored guest, while you remain in Washington.

We shall require to know how long you can be absent from London, and what parts of this country you would prefer to visit. Of course you will bring with you all the apparatus necessary for the illustration of your lectures, and a skilled assistant.

Any information on the subject, you may require will be furnished in my absence by M^r. W. J. Rhees[4] the Chief Clerk of the Institution.

I am my dear Professor, | very truly your friend, | Joseph Henry | Sec[retar]'y S.I.

Prof. John Tyndall, | Royal Institution, | London

RI MS JT/1/H/75

---

1. *(Copy)*: The current letter was likely enclosed in letter 3507, and is the copy to which Tyndale refers in that letter.
2. *Smithsonian Institution*: The Smithsonian Institution was founded in 1846 with a bequest from British mineralogist James Lewis Smithson (1764–1829) to create 'an establishment for the increase and diffusion of knowledge among men'.
3. *you could . . . visit this country*: The present letter begins a conversation about this trip; see letters 3507, 3510, 3519, 3542, 3543, 3558, 3589, 3594, 3639, 3669, 3670, 3684, 3685, 3702 and 3703. For continued correspondence regarding Tyndall's trip to America as well as letters from during the trip, see the forthcoming thirteenth volume of *The Correspondence of John Tyndall.*
4. *W. J. Rhees*: William J. Rhees (1830–1907) was Henry's personal secretary from 1853, then Chief Clerk of the Smithsonian Institution by 1855. He left the position in 1870 to pursue a stationery business in Springfield (state unknown). He returned to the Smithsonian after less than a year. In 1891 he became the first keeper of the Archives until his death in 1907. He wrote a definitive biography of James Smithson, from whose bequest the Smithsonian came to exist.

From Mr. Holman[1] 17 August 1871 3505

Extract from Mr Holman's letter of the 17th August 1871[2]

"In making up the Inventory of Ginty's[3] effects, the Judge decided that Mrs Ginty, as "usufructuaria"[4] was liable to a tax of 10 per cent of the benefit derived from the property. The rent of the house was assessed at £140, so that on this she has to pay £14 per annum. She has also to pay 10 per cent on the interest of the money found at Ginty's death—and which amounts to about £1.12.0—so that in all she has to pay £15.12.0 per annum. The payment of this annual Tax can be remitted, by paying into the Treasury a lump sum of £164, otherwise she will have to pay £15.12.0 as long as she lives." "I made several attempts to get Mrs Ginty relieved of this Tax, but all to no purpose—I then tried to get her off part of it, but with the same result."

The tax has had to be paid from 1868 onwards—so Mr Holman had to deduct 3 years taxes from the remittance made to Mrs Ginty last May.

The 85 Gas shares which cost somewhere about £22 or £23 each are now worth £33 each.

The Brazilian Gold bonds, which cost about £4800 are now worth £5.400

The estate is now composed of

85 gas shares worth	£2.805
48 Gold bonds ,, ,	5.400
	£8.205

Yielding about £494 per annum less £15.12.0 for Taxes in Brazil.

The Cassino[5] share has been sold for £45, which has been sent home in a Bill due 20th Instant, and which is in possession of Mr Worthington.[6]

RI MS JT/1/TYP/11/3666
LT Typescript Only

---

1. *Mr. Holman*: not identified, but presumably he later married Kathleen Ginty, one of Margaret Ginty's daughters. The other daughter, Fanny, wrote to Tyndall on 5 December 1882 that 'I presume that it will be with you to appoint another Trustee, and I therefore venture to suggest that, as my brother-in-law Mr Holman does not care to be involved in it, my husband should be named to succeed Mr James' (RI MS JT/1/TYP/11/3682). Fanny's husband was John Owen Unwin. 'Mr James' was Bartlett James (n.d.), manager, or "Gerente," of the Imperial Gas Works Company in Rio de Janeiro in Brazil, established by Baron de Maua, where he worked with William Ginty, who was the engineer, or 'engenheiro' (*Almanak Administrativo, Mercantile Industrial da Corte e Provincia do Rio de Janeiro para o anno de 1865* (Rio de Janeiro: Eduardo & Henrique Laemmert, 1865), pp. 414–5). According to Fanny's 1882 letter, when Ginty died in 1866, James had taken on the role of executor to his will.
2. *Extract... 1871*: This letter was sent originally from Holman to Margaret Ginty, who then sent part of it to Tyndall, which is included between the quotation marks here. The rest we assume are Tyndall's notes. Ginty's letter to Tyndall is missing.
3. *Ginty's*: William Gilbert Ginty (see letter 3496). Tyndall assisted in the management of the Gintys' estate upon William's death in 1866.
4. *"usufructuaria"*: usufructuary, or beneficiary (*OED*).
5. *Cassino*: a town in central Italy where the Gintys may have owned property.

6. *Mr Worthington*: Worthington (n.d.), a solicitor, of Sale, Worthington and Shipman (or sometimes Sale, Worthington, Shipman and Seddon) in Manchester, England.

From Hugh Malet[1] 23 August 1871 3506

Joyce Grove Nettlebed Henley on Thames[2] 23 August 71

Sir, I trust the subject of this letter will be a sufficient excuse for troubling you. I have just read "Hours of Exercise in the Alps" with very much pleasure. You use thoughts and examples which have been used by myself. I have come to conclusions, but you have not; you are kept away from the legitimate results of natural facts, because you believe in Philosophers, who "with good reason, wander through molten worlds" (p 291). No one has followed out this hypothesis to the end. There is no legitimate proof of the stability of any one evidence invoked to support it. They talk of an increasing heat in the earth, till at 25 miles depth the hard rocks would be in a molten state, and the whole interior a perpetual furnace. This evidence is untrustworthy—no two mines or wells, on equal levels, of equal depths, give the same temperatures, we have hot springs at great heights, and great depths, we have warm mines at 3000ft, but the Mont Cenis tunnel[3] at 5400ft *[1 word illeg]* only *[1 word illeg]*, and we have the freezing point in the ocean five miles nearer the supposed central fire than the level land. Geologists have quite forgotten that man has never delved below deposits, that there is gas in these deposits, and that there is a constant pressure upon them. It is this pressure on certain materials which produces local heat, while the materials sometimes produce heat of themselves. All the calculations on the density of the earth, in reference to a supposed internal heat, are useless as evidence of it, because there are no data to go upon. All the astronomical experiments from an internal fire are founded on imaginary sidereal arrangements, the analysis therefore cannot be depended on. I defy any one to prove that there is any thing on the earth's surface conclusive of an inherent internal fire in the earth.

You allow that the alps are stratified, the very ledges to which you cling prove that formation, and at P230 you say that "they were in whole or in part once beneath the sea." You see the truth, but you reject it, because you hang on the theory of mountains rising out of the earth by an unknown force; you allow "that the protuberance of the Alps could hardly have been pushed out without dislocation and fracture."[4] You see these phenomena every where before your eyes, but following the lead of others, you forget that subsidence would produce these effects more surely than a "push up"; there is no law in nature for a forcible ejection of masses, and you allude to the law of gravitation which has in reality produced all the phenomena you observe—local upheavals have a law of their own—The present condition of the Alps tells you that some places were originally softer than others, you understand landslips, and you know that subterranean erosion may produce unsupported masses; I think your own unbiased judgement will lead you to the conviction, that all the displacements and dislocations you remark were caused by subsidence, due to a known law, and not to ejection, for which there is no law unconnected with subsidence. There are higher mountains in the world than the Alps with stratified tops, your imagination will direct you to the "discernment of the connection" (p 67) the formation could not have existed without water, water therefore covered the whole. There has been no necessity for any other forces than those which now exist, and with which we are familiar, for the formation of mountains, as for the sea reaching its greatest depths. The annexed diagrams will help you to understand, what I humbly conceive to have been the position of the two elements water and earth—

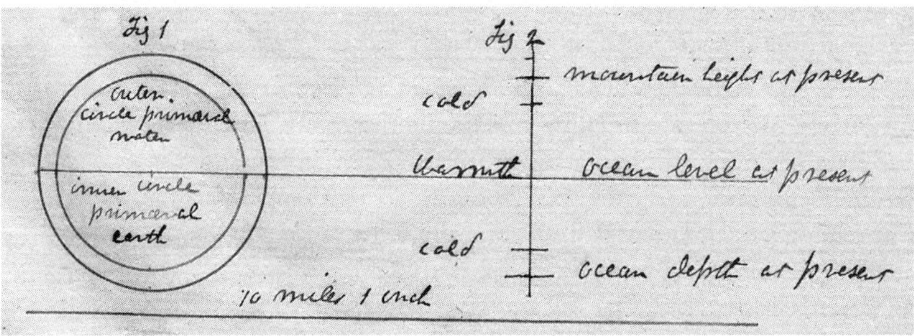

a current worked, and winds blew, the earthly matter was worked up, exactly as we see it worked up now; as water raised up matter on which it rested, it retired from that which it had raised, till it has, in places, attained a depth of 5 miles, and has left its original deposits 5 miles higher, than its present surface level. We have no proof of any entire cataclysm in nature, every thing seems to have been done gradually, and the only convulsions are local. You have viewed from lofty peaks the outstretched earth below, you allow that the hills have contributed to its formation, destruction was perpetually working around you, and this has been going on for uncounted ages, the soft material has long since gone, and water has worked its way through the hard rocks of the Via Mala,[5] as it has done through the deep stratifications of the Colorado.[6]

While the elements are perpetually destroying, the ocean is still building up, we cannot calculate its labours, or tell to what height our mountains once extended, we get into a distance beyond legend, but your imagination will connect the points. You have visited no heights where water has not flowed, and no man can say how much higher the Alps were (my one mile scale in the diagram is imaginary) but to whatever height these or any other mountains extended, we may assume, from present evidence, that warmth follows the sea level, it is cold below, and cold above, but yet deep in the deposits of our temperate zone earth we find remnants of tropical life. Will you follow out the idea, and show that your beautiful glaciers were not always there, and that life may have sprung up from beyond the measure of our present heights. The Alpine pinnacles are only the remnants of the vast plateau, from which the materials fill to form the foundations of a great part of Europe. The highest point of the Himalayas and the Andes are only what are left. Who will venture to measure what they once were, in reference to present levels, and who will venture to fix the elevation on which the first forms of life budded into existence. Many have tried the path, but all have failed to reach the end. You have probably reached it as nearly as any one, but you did not see it. You say, "The air indeed is filled with floating matter"—You know that the dust in the sun beam contains life, water and earth contain warmth, you have been, perhaps you still are, a follower of *[2 words illeg]* (note p. 191[7]) and it is from explanations of heat, moisture, and germs that life comes. Man never has, and never will trace the first, but of one thing we may be sure, there is a wisdom far surpassing that of man in all the laws of nature. Nothing comes to an unprepared abode and nothing comes without an instinct prepared for its position. No man can testify to these points better than yourself, you are gifted with physical power, and mental ability. You have used both to examine nature, yet you have rejected your own unconfessed convictions; will you think again, and confess the truth? I have done so to the extent of my poor ability, but a more practised pen is wanted to bring back those who are in error. I have no other object in writing than truth, and I feel that you are on the same track, though your confidence in false guides leads you astray.

I must now close a letter already too long for reading and am

Yours Faithfully | H. P. Malet

To Professor J. Tyndall &c &c &c.

P.S. I have said that I have had thoughts and used examples as you have done. I refer you to "The Interior of the Earth"[8] Messrs Hodder[9] & Stoughton.[10] 27 Paternoster Row. At Mudies[11] & 15 Old Broad Street.

I am the discoverer of a hill in India called Matèran.[12] Ma is majestic the compound Matè an elevation and ran means a waste or jungle has Matterhorn any hidden meaning?

BL, John Tyndall Papers, Add MS 53715: 1.20–22v

---

1. *H. P. Malet*: Hugh Poyntz Malet (1808–1904), a British civil servant in India in the middle of the nineteenth century. He wrote several books on India, including *Lost Links in the Indian Mutiny* (1867) and *The Interior of the Earth* (1870).
2. *Joyce . . . Thames*: Joyce Grove is a mansion in Nettlebed, Oxfordshire.

3. *Mont Cenis tunnel*: actually the Fréjus Rail Tunnel, which goes through the Alps. The tunnel took over the traffic that had previously been present on the Mont Cenis pass, through the Alps. See also letter 3533.
4. *"that the protuberance . . . fracture"*: see *Hours of Exercise in the Alps*, p. 230.
5. *Via Mala*: an ancient path along the Hinterrhein River in Switzerland that goes through a narrow and dangerous gorge.
6. *Colorado*: referring to the Colorado River and the path it takes through the Grand Canyon in the United States.
7. *note p 191*: this note in *Hours of Exercise in the Alps* reads: 'Eight years ago I was evidently a sun-worshipper; nor have I yet lost the conviction of his ability to do all I have ascribed to him'. Here, Tyndall ascribes to the sun the powers of making the mountains, glaciers, waters and gravity.
8. *"The Interior of the Earth" Messrs Hodder & Stoughton*: H. V. Malet, *The Interior of the Earth* (London: Hodder & Stoughton, 1870).
9. *Hodder*: Matthew Henry Hodder (1830?–1911), an English publisher. He was apprenticed at the age of fourteen to the publishers to the Congregational Union. He partnered with Thomas Stoughton (see n. 10) in 1868 to buy out his employers and start his own publishing firm. They published mostly theology, sermons and devotional works, and their secular titles conformed to evangelical standards. He was active in local missionary work and with the YMCA in England.
10. *Stoughton*: Thomas Wilberforce Stoughton (1840–1917), an English publisher who partnered with Matthew Hodder (see n. 9). Stoughton ran the financial side of the publishing firm, but tended to agree with Hodder on the types of books they would publish.
11. *Mudies*: a subscription circulating library in London from 1842–1937.
12. *Matèran*: Matheran, a hill station in India near Mumbai. Malet identified this place in 1850 as useful for the British, and it became a hill station for British residents to escape the heat of the city.

From Hector Tyndale 26 August 1871 3507

Philadelphia August 26th 1871

My dear John.

I arrived at home on the 17th July being and finding all well.[1] The family being glad to get your photographs and your friends to whom I delivered your books very thankful to you for your remembrances.

I should have written sooner but have been waiting until I could say something definite about the subject of our last conversation, namely your coming out here to lecture.[2] From about the 1st July until near October nearly all "the World" of society here leave their homes and go to watering places &c more or less distant. Immediately after my return I saw my, and your, friend

Prof. J. P. Lesley[3] and had a long talk with him about you and your coming here. Prof. Lesley himself left here for awhile (his family is absent) and has been away once or twice since. He wrote to several, fifteen or twenty, of the most prominent of the scientific men & others (mostly the former) stating that you were thinking about coming out to lecture here but that you were as yet undecided, as you dreaded to come with any thought of "exhibiting" to make money, and that an invitation from men of science & others here would probably decide you to come. The answer from the most of these have come to him and are now in my possession and as soon as Lesley writes the invitation to you and appends the names of the invitees I will send those answers, together with the invitation, to you.[4] Among the gentlemen he wrote to was Prof. Joseph Henry, Sec[retar]'y of the Smithsonian Institution, who stands as the foremost man of science in America, as the Smithsonian Institution itself stands far, very far ahead of all other scientific institutions of the country. In fact is the centre of American Science. Prof. Henry sent a reply saying that his <u>institution</u> would at once invite you to come over & lecture under their auspices. Subsequently, in Lesley's absence, he sent a copy of a letter of invitation of the 14th inst[ant]. which he had addressed to you. Only last evening the copy of that letter of invitation came into my hands and to day I enclose it herein.[5] I do this as it is possible the original may have miscarried. It is likely, however, you have already replied to Prof. Henry. Had I known of or rather seen this letter sooner I would have written to you before to explain matters somewhat.

Now excuse me if I proffer some advice which agrees moreover with the judgment of Prof. Lesley, who is the only person I have spoken to on the subject and so coincide in opinion. I should advise you to come over on the invitation of Professor Henry, it being official, and propose your lectures, your assistants, apparatus &c upon that invitation. <u>The Smithsonian will defray all your expenses; it being entirely responsible</u> but I <u>presume</u> nothing more is intended than that. This will relieve you from loss of money. I would make no other arrangements with any one, leaving your further course here open for your future judgment & decision. The more I have thought of it the more reason I have found to respect your dislike to any appearance (even) of mercenary motives. So said, the Smithsonian invitation (the highest by far in the land) takes all this away—it is <u>bona fide</u> and is given with my strong desire and hope that you will accept it. <u>After you arrive here there will be plenty of time & business men to arrange for other courses if you then see fit. One of your assistants, or if you bring but one then some other person should come with you as Private Secretary, known by that name and, of course, acting for you in all business arrangements</u>. This will save you from great worry and offensiveness—Of course you will decide on all points yourself but <u>your</u>

private Secretary will be the medium of all your business. I suppose you might use two assistants and one of these could well act in the capacity named, but I should advise by all means that you have such an official with you.

In regard to Prof. Henry's suggestions as to the value of money &c, which from him is only intended kindness to you. I think he means that if when here you choose to lecture otherwhere and for pay, he would advise you to do it, leaving it, as I think, entirely with yourself as to other lectures. That Institution, being the central one, has great powers to influence other places and institutions throughout the whole country.

Since my return Mr. Pugh,[6] who is the "professional" business man for Lectures of whom I spoke to you has called on me several times. He wishes very much to make an arrangement with you but I have said only that you were undecided and that if you came at all it would, probably, be under the advice of your Scientific friends here.* Please let me know if there is anything I can do for you in this or any other matter—at any rate let Lesley as well as Prof. Henry know your decision as soon as made[7]—and tell me frankly what I can do.

The replies to Lesley, of which I spoke in the first lines, are of course to him and are of different construction from letters sent yourself as they mostly merely require Lesley to add their names to your invitees in accordance with his suggestions. They are from men of great note here in every case and I am gratified by the almost perfect unanimity of desire to have you come out here.

Since my return I have had a letter from Dr. John Tyndall, our Gorey Cousin—his Great Grandfather William,[8] was brother of your & my Great Grandfathers i.e. of Matthew[9] and John,[10] who went from Wexford Co:[11] to Carlow of all which I have begun (!) a history! I have not yet answered his letter which is to my shame as well as regret for I esteem him and his family very much. When you write remember me to all of them & more particularly to your sister.[12] I will write him soon I hope.

Please remember me to Profs. Hirst & Debus who I hope are well.

*In relation to Mr Pugh I would say that he repeated what I quoted, as from him, to you in regard to figures of money &c to be made by a lecturing tour. But the more I think of it the better I am pleased that if you come you will do so, probably, with the invitation of your friends and without any thought of making money. Of all this however you must think and decide upon after your coming.

With me Mrs Tyndale hopes to have you at our house when you come. She sends her respects and remembrances.

There is no clue yet discovered to the murd<er> of my brother Sharon.[13] Good bye and God bless you | Affectionately your cousin | Hector Tyndale

To | Professor John Tyndall | London

I expect to write again in four days. | H.

Lesley's address is

Prof: J. P. Lesley &c | American Philosophical So[ciet]y | Fifth below Chestnut St | Philadelphia | U.S.

RI MS JT/1/T/63

---

1. *I arrived . . . all well*: Tyndale had visited London earlier in the year; see letters 3376, 3378, 3380, 3385, 3389 and 3393.
2. *your coming out here to lecture*: for additional letters on this topic, see 3504, 3510, 3519, 3542, 3543, 3558, 3589, 3594, 3639, 3669, 3670, 3684, 3685, 3702 and 3703.
3. *J. P. Lesley*: see letter 3504.
4. *those answers . . . to you*: see letters 3510 and 3519.
5. *enclose it herein*: see letter 3504.
6. *Mr. Pugh*: probably Thomas Burnett Pugh (1829–84), a publisher in Philadelphia, on Chestnut Street. In 1862, he signed the 'Petition for the Colored People of Philadelphia to Ride in the Cars', and he corresponded with Emerson and Frederick Douglass about public speaking engagements in the city. His firm published books about religion, science and other topics for the general public.
7. *your decision . . . made*: see letters 3542 and 3543.
8. *William*: William Tyndall (?) (n.d.) was the great-grand uncle of both John and Hector, but otherwise not further identified.
9. *Matthew*: Matthew Tyndall (?) (n.d.) was the great-grandfather of John or Hector, but otherwise not further identified.
10. *John*: John Tyndall (?) (n.d.) was the great-grandfather of John or Hector, but otherwise not further identified.
11. *Co:*: Tyndale often used a colon to indicate an abbreviation.
12. *your sister*: Emily, or Emma, Tyndall.
13. *murd<er> of my brother Sharon*: Sharon Tyndale (1816–71), the brother of Hector and the Secretary of State of Illinois from 1865–9. He was robbed and murdered outside his home in Springfield, Illinois, in the early morning hours of 29 April 1871. His murder was never solved (N. Bateman and P. Selby, *Historical Encyclopedia of Illinois and History of Tazewell County*, 2 vols (Chicago: Munsell, 1905), p. 532).

To the Editor of the *Times* 31 August 1871 3508

THE ALPS AGAIN. | TO THE EDITOR OF THE TIMES.[1]

Sir,—Three young students of the Polytechnic School at Zürich had been staying at Pontresina, and on Monday, the 21st of August, one of them, despite the opposition of his companions, quitted the village with the avowed intention of climbing the Pitz Ischierva[2] alone.[3] The mountain is the neighbor of the Pitz Morteratsch, and, like it, rises between the Morteratsch and Rosegg glaciers. On Tuesday, the 22nd, the young man had not returned. Three guides went out to seek him, and on Tuesday night it was known that he had not been found, but that the traces of a single traveller had been observed upon some old snow which filled one of the crevasses of the Rosegg glacier. An interview was immediately sought with Peter Jenni, the most experience guide in Pontresina, and with his aid a party of seekers was organized. On the Wednesday morning 12 were assembled, according to appointment, at the Misauna châlet[4], near the foot of the Ischierva, and were scattered thence in parties, some over the mountains to scan its precipices, and some over the Rosegg glacier to examine its crevasses. The glacier party found the traces in the old snow. Being surrounded by hard ice they were isolated, and up to the present time their real origin is unknown. Clouds lowered upon the peaks and rain fell upon the glacier. The seekers worked across it, estimating the chances of accident, and examining fissures as came in their way. They ultimately found a temporary shelter under the boulders of the lateral moraine, and subsequently down to a little hut beside the glacier, where they huddled together till the rain ceased. While halting here an active and willing young guide, named Gross,[5] who, in company with one of the comrades of the missing youth, had gone to the top of the Ischierva that day, joined the party. He had found upon the summit, trodden into the snow, a parchment envelope containing a leaf of a note-book on which was written in pencil that F. Bodmer,[6] the person sought, had reached the top on Monday, at 2 p.m. It is usual thus to note an arrival at an Alpine summit; but, as if warned by a sense of danger, the writer went on to describe the route by which he proposed to return. It was certain, therefore, that the young man had reached the top, and that if disaster had occurred, it must have been during his descent.

The day was too far spent to push the inquiry further, so the party descended to the Misauna châlet, where they were joined by others, who had been unsuccessful in their search. Here, unfortunately, an ill-advised remark by one of the leading guides introduced anger, bad language, and general disunion among the men. It was, however, to some extent stilled, and on reaching Pontresina M. Saratz, ex-president of the Engadin[7], was requested to use his influence to re-establish concert. This he did efficiently, and on the following day accompanied the party himself. Further traces of a single traveller were found on the fine rubbish of a moraine leading down from the Ischierva. These traces were said to resemble those observed on the snow near the summit, and they were followed down to the Rosegg glacier. They seemed, therefore, to point to the crevasses of that glacier as the tomb of the missing man. A subsequent day's search added nothing to what had been previously known.

But the proof of identity between the tracks below and those above was by no means satisfactory; neither was there sufficient surety that the upper regions of the Ischierva had been properly explored. The chasms of the Rosegg, moreover, merited a more searching examination. It is not easy to see into a deep crevasse from its edge, as the necessary leaning over imperils the observer. It was therefore proposed to throw ladders across the fissures, and from the ladders to look directly into the depths. Two parties, each furnished with a ladder, and all under the general control of Jenni, undertook this duty, while the writer, with the young man Gross, already referred to, and an excellent guide named Hans Grass[8], undertook the examination of the higher portions of the mountain.

Both parties had been held back for a day by rumours,—first that the young man had been seen at Sarnaden,[9] and afterwards that he had turned up at a village near Poschiavo[10], at the other side of the Beruma Pass. But the rumours upon examination melted into nothing. Early on Sunday morning, therefore, the two parties just referred to started, the one up the Ischierva the other up the Rosegg glacier. Quitting the Misauna châlet, the former party went straight up towards the overhanging Misauna glacier. Grass, after some time, bore to the right, while the others worked upwards along the right side. These at a certain place crossed the glacier, the whole party thus coming under the leadership of Grass. They ascended for a time among excessively riven ice and snow, and were finally fairly stopped by the pitfalls. An attempt was made to turn the chasms by resorting to the adjacent rocks, but those were for the most part of clear-cut granite, and refused a passage. The party had accordingly to re-descend for some distance to cross the glacier, to mount along the other side until the particular crevasses which stopped them were placed below them, there to recross the glacis and continue the ascent. Here the ridge of the Ischierva was to their right, and had a human body fallen from the ridge it could hardly have escaped their observation. Along the seldom-trodden though beautiful snow-fields of the Misauna glacier they continued to ascend, constantly scanning the adjacent slopes, and at length bent round to the right to get upon the ridge. Here the covered crevasses became so frequent and so perilous that the rope, which had been previously unemployed, was resorted to. Along the snow ridge the party proceeded to the top.

The faint traces of the lost man were still in the snow near the summit, and it seemed that in descending he had not pursued quite the same route as that by which he had ascended. Nothing, nevertheless, was observed at the summit. In descending, the searching party followed such gullies of the mountain as might be expected to tempt a traveller into danger, and they closely scanned the adjacent precipices from bottom to top. Once or twice Grass detached himself from the party to look into a more than ordinary repellent hole. But it was all to no purpose, they slowly hopped down the mountain and met at the bottom of the Rosegg glacier the party which had been engaged among the fissures, and which had been equally unsuccessful. Five days had already been spent in searching, all hope had vanished among the leading guides; on Sunday evening, accordingly, the investigation came to an end.

The visitors at the three hotels of Pontresina, the Krone, the Rosegg, and the Weisse Kreuz, behaved very liberally. The English Chaplain, Mr. Wade,[11] undertook the collection at the Krone, Mr. Crocker[12] undertook that at the Rosegg, while the collection at the Kranz[13] was made by Dr. Schwerin, of Berlin.[14] A sum of 461f.[15] was gathered together, which formed an ample renumeration for the guides. The payment was entirely the result of private effort, for the Commune of Pontresina took no action in the matter. M. Saratz, however, incurred a share of the expenses, he also gave his personal services for a whole day, and in various other ways forwarded the investigation.

The name of the lost youth was Frederick Bodmer; he was 18 years and two months old, and was the only child of his father, who is a teacher in a secondary school at Neumünster, near Zürich. His father describes him as having but one defect—"a breakneck rashness."

J. T. | Pontresina, Aug. 31.

Times, 4 September 1871, p. 6

---

1. *EDITOR OF THE TIMES*: John Thadeus Delane; see letter 3403, n. 1.
2. *Pitz Ischierva*: Pitz Tschierva, a mountain in the Bernina Alps, south of Pontresina, between the Roseg and Morteratsch valleys.
3. *Three young students . . . alone.*: Tyndall told or alluded to this story several times. See letters 3511, 3512, 3513, 3515, 3516 and 3536.
4. *Misauna châlet*: a popular châlet for early Alpine expeditions.
5. *Gross*: not further identified.
6. *F. Bodmer*: Frederick Bodmer.
7. *M. Saratz, ex-president of the Engadin*: Saratz (n.d.) was a member of the wealthy Saratz family, who founded the town of Pontresina, and also owned a popular guest house, now hotel.
8. *Hans Grass*: (n.d.) an Alpine guide who aided on this mission. Not further identified.
9. *Sarnaden*: possibly a typo for Samedan, a town near Pontresina in the Swiss Alps.
10. *Poschiavo*: a small town near the Piz.
11. *Mr. Wade*: see letter 3503, n.6.
12. *Mr. Crocker*: not further identified.
13. *Kranz:* more than likely a typo for Kreuz, the third Pontresina hotel.
14. *Dr. Schwerin, of Berlin*: not further identified.
15. *461f.*: francs, the conversion rate of which is unclear.

From Mary Egerton August [1871][1] 3509

Inverness | Aug[u]st.

Dear Mr. Tyndall,

When I wrote from Edinburgh,[2] (which letter by the bye I sh[oul]d think you never got through, at least I could not read it myself when it was done, & I w[oul]d not have sent it, only I had no better paper, & was anxious to send some acknowledgement of your kindness in writing[3] from far away,) but to return from this long parenthesis, we had then nearly given up our Highland trip, so I cannot resist sending you a P.S. to tell you we have accomplished it after all! We[4] have had lovely weather both in Ross shire & Skye,[5] which was peculiar good fortune. We are not important enough to make you conceited, or I would not tell you how often you were wished for! Something it was, "I wish he were here, what a scramble we would have," or another—"I wish he were here to explain these ice markings & glacier work." When all was bright & lovely we wished for you to share our enjoyment—and when "the 'way' was cold & dark & dreary"[6] we wanted you to raise our spirits & refresh our energies. I am very glad to have seen Skye; Loch Coruisk[7] surrounded by those weird dark purple Cuchullin Mountains,[8] is very fine, we saw it from a plateau like a small Görnergrat[9] where the ice ribs were visible even to the most uninitiated eyes. The whole country must be most interesting to a geologist I longed for one to give a theory of its strange formations. The whole island looks as if it had just risen from the sea & slanted—tilting up to let the water run off. Much of the interior is dreary to the last degree. Give me desolation on a grand scale. There is something incredibly exalting in a scene like the Grimsel[10] for instance; but these miles upon miles of rolling hills, & wide swampy valleys, are to me depressing in the extreme—and when you do come upon a hovel or two the inhabitants look more like half animated lumps of peat than anything else. The Quiraing[11] proved a less formidable climb than I expected, & is a wonderful place. I sh[oul]d have liked to see the other Basaltic rocks & caves, but they are so far off, and a parcel of women cannot go & take the chance of a night's lodging in a hayloft or some such place & a very dirty one too! But Loch Maree[12]—that is beautiful! Fine bold crags down to the water, fringed with patches of foliage, birch & mountain ash, & fern; fairy islands with old Scotch firs among their rocks, reflected on the deep blue lake, a real fine "Ben"[13] looking down right into it, & the purple mountains beyond. You would have enjoyed it, or rather you will enjoy it when you go. (I think you said you had not seen it?) And if you land on Isle Maree, where the curious old Burying ground is,[14] don't stay there, but go on to the largest island,[15] whence the view is I think, twice as fine, & which has only the natural picturesque growth of firs upon it, instead of a thicket of plantation. That place I sh[ou]^d like to see again—but for Skye, though I am exceedingly glad to have seen it—I feel inclined to say like the Frenchman asked to go out hunting—"J'ai été",[16] Tell Mr Hirst we enquired for him but could not hear of him;[17] I wonder if they went there after all? The sail from Portree[18] was lovely, & all that Ross shire railway is through splendid scenery—fine *[1 word illeg]* crags; which are so much more attractive to the eye that the shaly pyramids which seem the alternative to flat table shaped hills in Skye, except in the Cuchullin.

You are so associated with mountains that it seemed a necessity to tell you of what we have seen, but I only meant to write a line. We hope next week to see Glenroy, & Glencoe, en route to Oban,[19] & then pay two visits, for which thanks[20] to sea breezes & bridges, our personal appearance is not just now in the fittest state! I suppose you are returned home by this time. We shall be at Mountfield I hope early in Sept[emb]er. (unless I am lodged in the Queen's bench[21] first! This multiplying everything by 4, is awfully expensive work!) I hear people are stifled in England[22] here we have fresh October weather. Sometimes colder than pleasant! The 22 miles drive over the moors to the Quiraing was bitter, in a high dog-cart[23] with a hideous East (North)? winds in our face. I must put off the rest of our adventures till we meet when I shall hope to hear some more about your proceedings (though under protest!) I wonder if you met any of our various friends in the Engadine.[24]

Y[our]rs ever most truly | M. F. Egerton

RI MS JT/1/E/70
RI MS JT/1/E/99i

1. *[1871]*: This letter is a combination of two fragments. See letter 3497, and n. 20 below for more context.
2. *from Edinburgh*: letter missing.
3. *kindness in writing*: may refer to letter 3497.
4. *We*: Egerton's companions have not been identified, but possibly include her daughter Mary ('May') Egerton.
5. *Ross shire & Skye*: Ross-shire is a county in the Scottish Highlands that borders Inverness-shire. Skye is an island off the coast of northwest Scotland.
6. *"the 'way' was cold & dark & dreary"*: paraphrased from 'The Rainy Day', a poem by Henry Wadsworth Longfellow, the first line of which reads 'The day is cold, and dark, and dreary'; see H. W. Longfellow, *Ballads and Other Poems*, 2nd edn (Cambridge: John Owen, 1842), pp. 111–12.
7. *Loch Coruisk*: an inland, fresh water loch on the Isle of Skye.
8. *Cuchullin Mountains*: Cuillin, a range of mountains on the Isle of Skye. The main ridge is known as Black Cuillin and has a slightly purple glint in the sunshine.
9. *Görnergrat*: a rocky ridge overlooking the Gorner Glacier in the Swiss Alps.

10. *Grimsel*: Grimsel Pass is a mountain pass in the Swiss Alps.
11. *Quiraing*: a landslip in the mountains on the Isle of Skye in northwest Scotland.
12. *Loch Maree*: the fourth largest freshwater loch in Scotland. It contains over sixty islands and is now part of a national nature reserve.
13. *Ben*: a mountain-peak; the word is Gaelic in origin (*OED*).
14. *Isle Maree . . . Burying ground is*: Isle Maree is an island near the north shore of Loch Maree. It has the remains of a chapel, graveyard and the possible remains of the eighth century hermitage of Saint Maol Rubha.
15. *the largest island*: Eilean Sùbhainn, which contains a loch of its own that has an island in it, a situation that happens nowhere else in Britain.
16. *"J'ai été"*: I have been [there].
17. *Tell Mr Hirst . . . him*: see letter 3497. Tyndall seems to have sent a letter for Hirst to Egerton.
18. *Portree*: the largest town on the Isle of Skye.
19. *Glenroy, & Glencoe, en route to Oban*: towns in the northwest of Scotland.
20. *thanks*: From this point to the end, the letter is a fragment that had been separated from the original letter in the archives, but combined here.
21. *the Queen's bench*: a superior court in England.
22. *in England*: see letter 3497; the Egertons were in Scotland.
23. *dog-cart*: a horse-drawn vehicle for sporting, with no roof or protection from the elements.
24. *Engadine*: the high Alpine valley in which Tyndall had spent most of his summer in 1871.

From Joseph Henry, et al.[1] 1 September 1871 3510

September 1st. 1871.

To Professor John Tyndall | of London

Dear Sir

Being informed that you contemplate a visit to the United States,[2] where your works have made you widely known, and believing, that not only those specially engaged in your own department of Science, but a multitude of persons intelligently interested in the reception and diffusion of useful knowledge, will hail with pleasure any opportunity to know personally one whom they already so highly esteem,

We join in inviting you to afford this opportunity, by delivering a course of lectures in the Chief Cities of the Union, at your own convenience, and on whatever subjects you deem best.

We think that, in so doing, you will aid our efforts in furthering what, as we are well aware, you yourself have much at heart, the growth of a healthy taste for scientific knowledge.

Trusting that nothing untoward may prevent the accomplishment of our wishes, and assuring you of a hospitable reception, we remain | Dear Sir, | Most truly Yours. &c.

✓Joseph Henry of Washington
✓Charles Sumner of Washington
✓Montgomery C. Meigs[3] of Washington
✓N.H. Morison[4] of Baltimore
✓George B. Wood[5] of Philadelphia
✓Fairman Rogers[6] of Philadelphia
John F. Frazer[7] of Philadelphia
✓James C. Booth[8] of Philadelphia
✓Joseph Leidy[9] of Philadelphia
Peter Lesley of Philadelphia
✓Alfred M. Mayer of south Bethlehem, Pa.
✓William C. Cattell[10] of Easton, Pa.
✓F. A. P. Barnard[11] of New York City.
✓James Hall[12] of Albany, N.Y.
✓Frederick D White[13] of Ithaca, N.Y.
✓James D. Dana[14] of New Haven, Connecticut.
✓Theodore Lyman[15] of Boston.
✓Josiah D Whitney[16] of Cambridge Mass, & San Francisco, Cal.
✓Ralph Waldo Emerson, of Concord. Mass.
✓Louis Agassiz, of Cambridge, Mass.

RI MS JT/1/H/76

---

1. *Joseph Henry, et al.*: This letter was enclosed in letter 3519.
2. *a visit to the United States*: see, for example, letters 3504 and 3507. Tyndall travelled to and through the United States for a lecture tour from October 1872–February 1873. Plans for this trip are also discussed in letters 3519, 3542, 3543, 3558, 3589, 3594, 3639, 3669, 3670, 3684, 3685, 3702 and 3703. For continued correspondence regarding Tyndall's trip to America as well as letters from during the trip, see the forthcoming thirteenth volume of *The Correspondence of John Tyndall.*
3. *Montgomery C. Meigs*: Montgomery Cunningham Meigs (1816–92), an American architect and engineer and army officer. He was educated at West Point, graduating in 1836, and became a member of the Corps of Engineers. In this role he worked on the dome and wings of the United States Capitol Building in Washington, D. C. He became Quartermaster General of the United States Army in 1861, serving throughout and after the American Civil War until 1882.
4. *N.H. Morison*: Nathaniel Holmes Morison (1815–90), the first provost of the Peabody Institute, a music conservatory and college preparatory school in Baltimore, Maryland.

He graduated from Harvard University in 1839. He became the Provost-Librarian of the Peabody in 1867 until his death.

5. *George B. Wood*: George Bacon Wood (1797–1879), an American physician and writer. Educated at the University of Pennsylvania, where he earned his BA in 1815 and MD in 1818, he eventually became the professor of materia medica at the medical school at Pennsylvania in 1835. At this same time, he was a physician at the Pennsylvania Hospital. He held institutional positions throughout his life, including the president of the College of Physicians of Philadelphia, the American Philosophical Society and the American Medical Association.
6. *Fairman Rogers*: Fairman Rogers (1833–1900), an American civil engineer and philanthropist. He attended the University of Pennsylvania, graduating in 1853, and returned to teach civil engineering from 1855 to 1871. He fought in the American Civil War at Antietam and Gettysburg, and was also an officer in the Army Corps of Engineers. He was a founding member of the National Academy of Science in 1863. He was influential in building a number of Philadelphia buildings, such as the building for the Pennsylvania Academy of Fine Arts on Broad Street.
7. *John F. Frazer*: John Fries Frazer (1812–72), a professor of natural philosophy and chemistry at the University of Pennsylvania and later the Vice Provost of the institution. He graduated from the University of Pennsylvania in 1830 and earned his law degree in 1838, studying a number of subjects including law, but never practicing. He studied in the laboratory of James Booth (see n. 8), and assisted in the geological survey of Pennsylvania in 1836. He was elected to the American Philosophical Society in 1842, becoming its president in 1855. In 1863, he became a founding member of the National Academy of Sciences.
8. *James C. Booth*: James Curtis Booth (1810–88), an American chemist and the refiner at the US mint in Philadelphia from 1849 to 1887. He graduated from the University of Pennsylvania in 1829 and opened a chemistry teaching laboratory in Pennsylvania in 1836. With Frazer (see n. 7), he worked on the geological survey of Pennsylvania from 1836 until 1838. In 1849 he was appointed to his position at the mint while maintaining his new position as a professor in the chemistry department at Pennsylvania. He became a member of the American Philosophical Society in 1839; he was the president of the American Chemical Society from 1883 to 1885.
9. *Joseph Leidy*: Joseph Leidy (1823–91), an American paleontologist, parasitologist and human anatomist. He studied at the University of Pennsylvania, earning his MD in 1844 with a dissertation on comparative anatomy of the vertebrate eye. In 1848 he was elected to the Academy of Natural Sciences in Philadelphia, and, as an early American supporter of Darwin's theory of evolution, pushed for his election to the body (Darwin was indeed elected, in March 1860). Leidy preferred to study rather than practice medicine and anatomy, and he made a number of strides in these fields. He wrote more than 500 original pieces. He was a professor of anatomy at the University of Pennsylvania for thirty-eight years.
10. *William C. Cattell*: William Cassady Cattell (1827–98), an American Presbyterian minister, and a president of Lafayette College, in Pennsylvania, from 1863 until 1883. He was

a professor there of Latin and Greek from 1855 to 1860, when he resigned his post to become the first pastor of the Pine Street Presbyterian Church in Harrisburg, Pennsylvania. In 1863 he came back to Lafayette where he remained until his poor health forced him to resign. He was a member of a number of theological societies in the United States, and president of the Presbyterian Historical Society.

11. *F. A. P. Barnard*: Frederick Augustus Porter Barnard (1809–89), the tenth president of Columbia College, now University, in New York from 1864 to 1888. He was educated at Yale University, matriculating in 1824, being the youngest of that class at the age of fifteen. He excelled in mathematics and science. In 1832, he began his career at the New York Institute for the Deaf and Dumb, as he had begun to lose his hearing early in life. He was a professor at the University of Alabama from 1837 to 1854, and at the University of Mississippi from 1854 until his resignation in 1860, just before the outbreak of the American Civil War in April 1860. Even though he enslaved a number of people, he sympathized with the North and moved to Washington, D. C. He worked in a variety of fields, including chemistry, physics and mathematics.
12. *James Hall*: James Hall, Jr. (1811–98), an American paleontologist and geologist. He was educated at Rensselaer Polytechnic Institute (RPI), graduating with his BA in 1832 and his MA in 1833. He remained at RPI to teach chemistry and geology. He became a geologist in the large geological survey of New York, and in 1842 became the first state paleontologist. In 1893 he became the New York State geologist. He was a founding member of the National Academy of Sciences and a president of the Geological Society of America. He was an avid fossil collector and his laboratory building from 1857 still stands in Albany, New York. He is an author of over 300 scientific works.
13. *Frederick D White*: Andrew Dickson White (1832–1918), a founder and the first president of Cornell University in Ithaca, New York from 1868 until 1885. Frederick was one of Andrew's sons, but it is unclear why the name Frederick is listed here; it may be a nickname. Andrew was a historian, politician and diplomat. He was educated at Yale University and traveled widely throughout Europe. He is known for expanding college curriculum, and used the Morrill Land Grant Act of 1862 to help establish Cornell. Here, he started the first department of Electrical Engineering in the US.
14. *James D. Dana:* James Dwight Dana (1813–95), an American geologist. He was educated at Yale, and after graduating received an assistantship in the chemical laboratory at Yale. He went on the Wilkes Expedition of 1838–42, which circumnavigated the Earth, focusing on Polynesia and Antarctica. He wrote three of the reports about the expedition, *Zoophytes*, *Crustacea* and *Geology*. In 1856 he became the Silliman Professor of Natural History at Yale and remained there until his retirement in 1890. His work as a major American geologist and the data in his reports from the Expedition tended to support Darwin's theory of evolution.
15. *Theodore Lyman*: Theodore Lyman (1833–97), an American naturalist and military officer during the American Civil War. He was also a US Representative for Massachusetts. He was educated at Harvard University from where he graduated in 1855. He then went to study with Louis Agassiz at the Lawrence Scientific School, where he graduated in 1858.

After service in the Civil War, he became the State Fish Commissioner of Massachusetts. His life became devoted to preserving fish and their breeding grounds.

16. *Josiah D Whitney*: Josiah Dwight Whitney (1819–96), an American geologist. He was educated at Yale, graduating in 1839, and then worked on the state geological survey of New Hampshire. In 1842, he moved to Europe and spent the next five years studying with naturalists there. When he returned to the United States in 1847, he went to work in Michigan on the Lake Superior survey. In 1860, he became the state geologist of California (until 1874), and from 1863 to 1867 he oversaw a geological survey of California, which included geologist Clarence King. In 1865 he moved to Harvard to take up the Sturgis-Hooper professorship, where he remained until his death.

To Henry Bence Jones 4 September [1871][1] 3511

Royal Institution. | 4th, Sep.

My dear friend

I have just got back, swept with a sigh past you yesterday, but with a hope that I should return to see you as soon as I had got my travel soiled body cleansed and repaired. I stayed on at Pontresina for longer than I intended: but a week was consumed in the search of a young polytechnic student who came to grief, through his own rashness, on one of the neighbouring mountains. We did not find him either alive or dead.[2] It pulled my heartstrings to quit the mountains during such glorious weather, and my regret was not lightened by experiencing the intolerable heat of the lowlands. But this rain has sweetened and refreshed everything. I had some notion of staying a day or two in Paris: but when I set my thoughts on home I find it difficult to deflect them. I want to get some work done, and it is hard to do it. Ireland calls me over to look at a new lighthouse: The Trinity House calls me to look after paraffin as a means of illumination;[3] and here waiting for me is a solid prism of letters, official documents, and books, 3 feet long, 1 foot high and 1 foot wide: most of them demanding some notice. How pleasant it must have been for a philosopher to live in the good old days when printing was a difficulty and postage was dear! Still it is better to have too much than too little work. What I complain of is not the quantity but the variety—the incessant deflecting of thought from subject to subject no time being left to it to make any thing permanently great.

I hope when I see you to see you better than you were when I quitted you. Have you no book on hand? Or can you not fill your time by augmenting the interest of what you have already undertaken? I had become proud of my sea powers; but yesterday though I was not ill was not a time of rejoicing with me.

Kind regards to Lady Millicent and the young Ladies[4]

Yours ever | John Tyndall

My hand is not yet well. I find exposure to the mountain sun bad for it. The spot had all but vanished; but through the exposure during those days of search it was rendered angry. I also got a chill, or more probably I was the victim of an epidemic cold which ran through the whole of the visitors. Still I am very strong and very brown. But my cold causes me to cough much at nights. A Turkish Bath will I hope set it all right.

RI MS JT/1/TYP/3/798
LT Typescript Only

---

1. *[1871]*: dated to 1871 due to Tyndall's discussion of the search for Frederick Bodmer and his hand (see letter 3512).
2. *A week . . . or dead*: Frederick Bodmer. See letter 3508 for more discussion of this.
3. *Ireland . . . illumination*: See letter 3521, for example. Hirst recorded in his journal that Tyndall returned from Ireland on the 14th, 'with his left hand swelling and gathering' (Brock and MacLeod, *Hirst Journals*, p. 1912).
4. *Lady . . . ladies*: Millicent Acheson, his wife, and their children.

To Thomas Archer Hirst 4 September 1871 3512

[Royal Institution][1] | Sep[tembe]r 4th | 1871

My dear Tom

Sweet thanks to you for your pretty little note[2] which I found waiting for me here[3] in company with three cubic feet of other printed and written matter. I quitted Pontresina on Thursday;[4] got to Chur; got to Basel; got to Paris—travelling by night in preference to suffering the scorching heat of the day. Got to London last night. And today I have been chastening into order my chaos of gifts and correspondence. It made my head reel to look at it, and I sighed for the peace of my mountain home. I remained much longer than I intended at Pontresina; a week of the time however was spent in a work of mercy—viz in an endeavour to find the body of a lost boy—a student in the Polytechnic school in Zürich.[5] We failed in our search. A brief account of the matter with my initials appended appears in today's Times for which you may give two pence if you feel disposed to read the stuff.[6] I confess that was today heartily weary of my own notoriety. For during my absence Longman sends me the papers that contain notices of my books. Newspapers, Magazines and Scientific Journals. America has also furnished her instalment. And I opened and opened until I was weary of the task. They are wonderfully kind on the whole. In fact the only bit of nastiness was that of M^r^. Leifchild in the Athenaeum.[7] I found with your note one from Lady Mary[8] in which she does not mention her meeting with you. It was written before you met.[9] She has a fine sympathetic nature which comes out best on paper. For she is so tremulous that one's presence sometimes disturbs her Spontaneity. They are true and good people and the more I have seen of them the better I have liked them. I caught an epidemic cold in Pontresina, and during that search I exposed myself; the consequence is hard coughing at night but I am strong and brown. My spot on the hand[10] after dwindling to vanishing showed up again on exposure to the burning sun during those days of search. It is by no means so well as it was 3 weeks ago. But I ignore it as much as possible—that is I avoid rubbing it, though it sometimes solicits friction—and so I hope it will run its course & disappear. I am hoping to have a Turkish Bath today. By assiduous work I have almost got through my books and letters; and by the time they have all been got through my luncheon will have melted down and I shall be ready for the Turkish Bath.

I saw Lubbock's name in association with that of M^r^. & M^rs^. Grant Duff[11] in the list for a hotel at Neufchatel. What a complete dissolution of his associations!

I must now cut myself up warily for the next two or three days and divide myself fairly among my friends:—but friendship must not prevent me from getting some work done.

I hope you will come back[12] as strong as I am, and that when you do come back you will find yourself more free from tangle than I am. Kiss Lilly[13] for me—

ever affectionately yours | John

RI MS JT/1/T/687
RI MS JT/1/HTYP/583

---

1. *Royal Institution*: LT annotation.
2. *your pretty little note*: letter missing.
3. *here*: the RI, when he got back from the Alps.
4. *Thursday*: 31 August 1871.
5. *student . . . Zürich*: Frederick Bodmer.
6. *the stuff*: 'The Alps Again', *Times*, 4 September 1871, p. 6. See letter 3508.
7. *M^r^. Leifchild in the Athenaeum*: John R. Leifchild (1815–n.d.) reviewed *Hours of Exercise in the Alps* in the *Athenaeum*, 3 June 1871, pp. 679–80. He was the author of a number of books about coal and coal-mining.
8. *one from Lady Mary*: Mary Egerton; possibly letter 3509.
9. *before you met*: referring to the Edinburgh meeting of the BAAS (see, for example, letter 3498).
10. *spot on the hand*: see letter 3495.
11. *M^r^. & M^rs^. Grant Duff*: Mountstuart Elphinstone Grant-Duff (1829–1906), a liberal

politician and colonial government official. He was educated at Oxford University and was called to the bar by the Inner Temple in 1854. He was elected to the position of Lord Rector of Aberdeen University in 1866. He was a member of several gentlemen's intellectual clubs: the Athenaeum Club (see letter 3372, n. 15), the Cosmopolitan Club, the Literary Society, and Grillion's Club, and he founded the Breakfast Club in 1866. Anna Julia Grant-Duff (d. 1915) was the only daughter of Edward Webster. She and Grant-Duff were married in April 1859. They had four sons and four daughters.

12. *come back*: Hirst stayed in Scotland visiting with friends, traveling around, hunting and walking a lot. He returned to London on the 11 September (Brock and MacLeod, *Hirst Journals*, p. 1912).
13. *Lilly*: Hirst's niece; see letter 3495, n. 21.

To Juliet Pollock 4 September 1871 3513

[Royal Institution][1] | 4th. Sept. 1871

My dear Friend.

Last night I returned from Switzerland—when there I thought twenty times of writing to you, but it would have been all like the crackling of thorns under a pot. Crackling gossip and I hope not of a good kind. I stayed all the time at Pontresina: a village planted amid glorious scenery. But it is overflooded with people—All the hotels filled, for the quiet Alpine village of a few years ago is now a place of hotels, and in them the quantity of eating that goes on occupies space and noise within doors to the exclusion of most of the other comforts of life. I stayed far longer than I intended. A week must be put down to a work of mercy—for a young fellow was lost and I spent a week with the guides seeking him, first in the hopes that he was alive, and secondly when that hope vanished, seeking for his body. We did not find it.[2] It pulled my heart strings to part from the mountains in such glorious weather. In July we had the inclemency of winter, but towards the end of August deep unclouded heaven, from day to day. Nevertheless it is good for me to be here; for pleasure is not to be enjoyed with the whispers of duties unfulfilled coming across it. And my duties now are heavy: I am claimed in various places & directions officially, and there is a pile of books documents & friendly letters which have come during my absence and which measure a yard in length one foot in width, and one foot in height. Think of that mass to sift and notice & reply to! I sometimes wish to get into quiet quarters and rid myself of the whole affair. But it is better to have too much than too little work. Nothing festers me so much as insufficient occupation. So I will tend to my labours, and now & then halt in the midst of them to bless my friends, & among them Eolia.

Yours ever | Boreas[3]

RI MS JT/1/T/1154
RI MS JT/1/TYP/6/2132

1. *Royal Institution*: LT annotation.
2. *A week . . . find it*: see letter 3512 for an explanation for this.
3. *Eolia . . . Boreas*: see letter 3399, n. 2.

From Juliet Pollock and Frederick Pollock — 6 September [1871][1] — 3514

Mrs. Mill's | Clovelly[2] | N. Devon | is my address | Sept[embe]r 6th.

My dear Boreas[3]

Will you still like your friend when you meet her transformed into a chamois?[4]—because that is an infallible event of which you must make the best whether you like it or not: living here one must either become a climbing animal or an inanity. I prefer to become the chamois—and rapidly my feet are turning into little hoofs which consider a steep incline their natural position. The village is a long steep narrow staircase with cottages on either side by way of banisters, you can talk across, shake hands across, smile across, or frown across, according to your disposition. The people are kindly, friendly & honest, and for the most part, smile and chat a great deal. At top of the staircase, the rich woods of Clovelly court, at bottom the sea, the quay, the rough pier, where the fishers congregate, and wooded rocks stretching to the sea. a curious beautiful place; all full of fishermen & their families; children playing; dogs barking, & kittens frolicking. The house we occupy belongs to John Mill fisher & boat builder[5]—his wife & daughter in law[6] manage for us, with our own Polly:[7] they are devotedly attentive: & their cooking is perfect: everything is simple unpretending and good. It is like an old fashioned farm house inside, with deep window seats and small paned windows; rather dark, but very snug and there is a pervading sense of friendliness & absence of gentility—a presence of freedom, frankness & cordiality all over the place, which is to me really delightful. The dog is not afraid, the child is not afraid, the kitten, who is an orphan, is not afraid, and lives now in the palm of my hand: it did live before I came mostly in an old smoking cap left here by some reading & smoking young man. This kitten appeared suddenly one day in the midst of the fisherman's family at dinner, requested to be fed; was fed; purred, & was at once adopted as a man & a brother by the house man, the house wife, the house child, the house dog, & the kitten of more advanced years already in possession, with whom it goes to roost in the smoking cap every night, over my head, occasionally leaving it for a grand scamper.

We had rain for our first arrival & my husband had a severe bilious attack. But yesterday the sun shone, the husband became convalescent, and at night, Fred[8] arrived, and is here till tomorrow at noon, when he leaves us to go volunteering. It has begun to rain again to day, but we hope it may clear up presently & then we shall set forth for the beautiful grounds of Clovelly court where in the midst of quiet woods you overlook a deep blue sea.

Why should you not come and be a goat too? every body can always get a room at short notice; and every body is a friend to every body. You; always & every where a friend very dear to us!

So good bye till we see your legs pass our low windows. You will find brandy & water or brandy without water at discretion. Your true | Eolia[9]

Yours[10] received yesterday thanks for it.

I read the Indian Evidence Bill[11] & accompanying papers in the train. It is a grand piece of work and an example for us in England. The chapter of commentary without a title (which I suppose must be a fragment of some projected work) explains the general theory of legal evidence & the principles of the Bill as working out the theory. I think it might be interesting to you.

I shall be in town all Friday & may perhaps see you again.

Yours ever | F. P.[12]

RI MS JT/1/P/195

---

1. *[1871]*: dated to 1871, in response to letter 3513.
2. *Clovelly*: a harbor town in Devon.
3. *Boreas*: see letter 3399, n. 2.
4. *turned into a chamois?*: Clovelly is a town with steep streets and walking around may feel like climbing. A chamois is an antelope that occupies the highest peaks of the Alps and Pyrenees, so is very good at climbing.
5. *John Mill fisher & boat builder*: not further identified.
6. *his wife & daughter in law*: not further identified.
7. *our own Polly*: this could be a family member with the nickname Polly, or their maid; not otherwise identified.
8. *Fred*: Frederick Pollock (1845–1937), lawyer, eldest son of Juliet and Frederick Pollock's three sons.
9. *Eolia*: see letter 3399, n. 2.
10. *Yours*: letter 3513.
11. *Indian Evidence Bill*: the Indian Evidence Act, passed in India by the Imperial Legislative Council in 1872. The Act made rules of criminal evidence uniform for all residents of India, regardless of caste, religion or socio-economic status. Fred Pollock, a lawyer, had a copy of this from Tyndall. See letter 3516, n. 4.
12. *I read . . . F. P.*: This note was appended to Juliet's original letter. It is not separate, as it is a post script, so it has been left here. 'F. P.' is Frederick Pollock, the eldest son.

From Sarah Faraday 7 September 1871 3515

Laurel Cottage Holmwood Common | Sep. 7th. 1871.

Dear Dr. Tyndall

My Niece[1] reminds me of my deficiencies in neglecting to answer letters and I have intended to thank you for yours[2] for some time but could not till your return[3]—it is always a pleasure to hear of you and I am very glad now to hear that you have returned safe home. We are enjoying this sweet country in some degree but I shall be thankful to be at home again, I am so dependent upon creature comforts that a hard seat or a steep stair takes away a great deal from my pleasure, but I must not trouble you with the troubles of old age. You see I am not much in the cue for writing. We brought your book[4] with us and when I am able to bear the excitement caused by your graphic descriptions we read some portion. And a letter in the Times[5] which is very sad.

Farewell dear Dr. Tyndall and believe me always | Your sincere friend | S. Faraday | Dr. Tyndall

LT Typescript Only
RI MS JT/1/TYP/12/4186

1. *My Niece*: Jane Barnard.
2. *thank you for yours*: letter missing.
3. *your return*: Tyndall returned from Switzerland on 3 September.
4. *your book*: *Hours of Exercise in the Alps*.
5. *a letter in the Times*: see letter 3508.

From William Frederick Pollock 7 September 1871 3516

Clovelly | North Devon | 7th Sep. 1871

My dear Tyndall

That good fellow Fred[1] has come here to see his parents for a couple of days—before going to serve his country in the autumn campaign.[2] He cannot be certain of seeing you in passing thro' London, & I therefore wish to ask whether I may keep for a little time Fitzjames Stephen's[3] India documents, which Fred had from you.[4] From the look I have had at them, they have greatly interested me, & the Evidence Act[5] seems to be a masterpiece in its way. But I have had no time to read Stephen's remarks; or do more than look at certain parts of the Act where I expected to find novelties, or familiar difficulties dealt with. & if you do not immediately want the papers it will be a great favour to give me time to study them. & I will return them to you by Post.

I made a bad start here—with a bilious attack—from which I am now recovered. & we have had bad weather, but Miladi[6] is well, and stronger than she has been for some time. & I also mean to set about enjoying myself.

Don't you think the Light on Lundy Island[7] must want looking at. That would fairly bring you here.

I am sorry for you that your last week at Ponte Resina, had to be spent in such a melancholy way.[8]

Yours ever | W. F. Pollock

RI MS JT/1/TYP/6/2136
RI MS JT/1/P/239

---

1. *Fred*: Frederick Pollock, their eldest of three sons. See letter 3514 for much of this context.
2. *autumn campaign*: Frederick Pollock was admitted to the Bar in 1871, which is what he could be referring to.
3. *Fitzjames Stephen's*: James Fitzjames Stephen (1829–94), an English judge and writer. He was educated at Trinity College, Cambridge, receiving an undistinguished BA in 1851. While there, he was elected to the Apostles, which is a debate and discussion group. He was called to the Bar in 1854 and was also an active journalist, writing over 300 pieces for *The Saturday Review*. He spent time in India from 1869 to 1872 and was able to pass three legislative acts, including the Evidence Act (1872), the Contract Act (1872) and another reformed code of criminal procedure. When he returned to England, he tried similar reform, but failed. All three of his sons became lawyers.
4. *documents . . . you*: see letter 3514, n. 11.
5. *Evidence Act*: In 1872, Stephen was able to get this act passed; see letter 3514, n. 11.
6. *Miladi*: referring to Juliet Pollock.
7. *Lundy Island*: largest island in the Bristol Channel, off the coast of North Devon, where the Pollocks were staying.
8. *Ponte Resina . . . way*: referring to Tyndall's last week in Pontresina and his search for the lost climber. See also 3508, 3511, 3512 and 3513 for more details.

To Jane Barnard 8 September 1871 3517

8th Sept[embe]r 1871

My dear Miss Barnard.

It was kind and right of you to send me your aunt's letter.[1] I am sure it will do both her and you good to be away from London: and assuredly you have around you a lovely country—none that I know lovelier at this season of the year. Next Monday I run over to Ireland to inspect a Lighthouse,[2] but I hope to return in the course of next week. How pleasantly we might live did the sense of duty never intervene. I am sure without it to compel me I should not be now in London

Yours ever faithfully | John Tyndall

RI MS JT/1/T/113
RI MS JT/1/TYP/12/4201

1. *your aunt's letter*: Sarah Faraday; probably referring to letter 3515.
2. *inspect a Lighthouse*: see, for example, letters 3521, 3522 and 3540.

To Henry Bence Jones 9 September [1871][1] 3518

Royal Institution. | 9th, Sep.

My dear Friend

Your note[2] was far too short and your reason for it no reason at all. I would run down to you to day were it not that I must be off to Ireland to inspect a lighthouse early on Monday morning. For Sunday the 17th, I have been snapped up by Miss Moore,[3] but on the following week I shall run down and see you.

The house front is waiting for decision as to the colour of the wash: I think it ought to be very sober. First because sober colours are most dignified and secondly because they bear the dirt better than brighter ones.[4]

Sir Henry[5] has just been here. He had a fine time in Ireland. The generation succeeding that which he visited in 1810, and to whom he seems to be a historic character, received him with acclamation and entertained him with the utmost hospitality. Only think of his revisiting the Island after 61 years! He had a terrible passage home, but seems to have cared nothing about the tossing.

I shall return from Ireland very soon. I have got into my work, or at least am nibbling about the edge trying to get into it. The brain gets stiff through long retirement from action, I suppose as we grow older it requires a longer period of drill.

The Athenaeum[6] is dismal. I went there on my return from the Continent to get a bit of dinner, but found all in chaos. The committee might have made a little arrangement for us, instead of sending us adrift in this way.

I should like to know the grounds of your objection to sugar, I abstained from it once for 6 weeks and found it a mere affliction without any result. What should I take in the morning if I were to give up tea? Brandy and

Potash! Now as far as my self observation goes, and it has been very close, tea in the morning never did me any harm. I am quite willing to *[forego it]* at all other times: in fact for the last four months I have not had a cup of tea except at breakfast. It would take some evidence to make me believe that the beverage at breakfast is injurious to me.

I never saw London so empty, the streets look quite deserted.

Sir Henry's next trip will be to Lisbon[7] whether he goes on, I believe, the 27th. He hopes to be accompanied by Lovel.[8]

And now I think I must pull up. Keep yourself strong and joyful until I see you. My presence will make the strength and joy permanent!!

Kind regards to Lady Millicent and all the Ladies.

Yours ever | John Tyndall

RI MS JT/1/TYP/3/799
LT Typescript Only

---

1. *[1871]*: dated to 1871 because of his trip to Ireland for the lighthouse, which he mentioned in letters 3511 and 3517.
2. *Your note*: letter missing.
3. *Miss Moore*: either Harriet Jane Moore (1801–84) or Julia Moore (1803–1904), daughters of Harriet Moore (1779–1866) and James Carrick Moore (1762–1860); James was a surgeon and promoter of vaccination. Julia was a friend of Tyndall's as was the rest of her family. Harriet, a member of the RI since 1852, was a British watercolor artist who painted scenes of Faraday's apartment, study, and laboratory at the RI during the 1850s (F. A. J. L. James, 'Harriet Jane Moore, Michael Faraday, and Moore's Mid-Nineteenth-Century Watercolours of the Interior of the Royal Institution', in James Hamilton (ed.), *Fields of Influence: Conjunctions of Artists and Scientists, 1815–1860* (University of Birmingham Press, 2001), pp. 111–28). The Moore family often entertained a number of well-known scientists at their home in Wimbledon, such as Charles Babbage and Tyndall (*Ascent of John Tyndall*, pp. 109, 144–5).
4. *The house front . . . sober ones*: Tyndall was likely referring to the construction happening at the RI; see letter 3427. See also letters 3537 and 3541.
5. *Sir Henry*: Henry Holland.
6. *Athenaeum*: see letter 3372, n. 15.
7. *to Lisbon*: we have not found any reference to Holland traveling to Portugal in 1871. He did visit Iceland in August; Mary Egerton refers to this visit in letter 3547. See also A. Wawn, '"The Courtly Old Carle": Sir Henry Holland and Nineteenth-Century Iceland', *Saga-Book*, 21 (1982–5), pp. 54–79.
8. *Lovel*: LT could not read this name and this person is not further identified.

From Hector Tyndale 9 September 1871 3519

Philadelphia Sept 9th 1871

My dear John.

I wrote you last on the 26th Ult[imo]:[1] & enclosed copy of an invitation to you from the Smithsonian Institution by Prof: Henry,[2] Sec[retar]'y of that Institution. I had hoped to have sent the general invitation, which is here-inclosed[3] & to which I alluded in that letter, before this time but there have been delays in receiving replies from all those written to and in getting the matter into shape. Besides Prof. Lesley who has done all the work & arranged it, has been out of town, but yesterday he wrote that invitation & appended the names.

By this mail, in another envelope, I have sent all the notes of the gentlemen[4] who signed the invitation, thinking you would like to keep them. These notes, being only replies addressed to Lesley, are not phrased as if written to yourself of course. There was delay, too, in waiting for Agassiz' second reply (you will find two notes from him) as he is very busy preparing to go on the Government Scientific expedition to the Pacific,[5] to start in a short time. Prof: Agassiz evidently thought at first that some lecture agent or speculator had indirectly suggested to Lesley this invitation as a means of getting you here and then using it as an advertisement! But he was undeceived at once.

The names on the invitation are of the strongest and among the strongest men of the United States, as you will see, and they are given in the most cordial and kind way to you. This, even if you had not the particular official invitation of Prof: Henry, will make your visit here easy & pleasant, placing you just where you desired—beyond any suspicion of mercenary motives on one hand or of over-desire or self-thrusting on the other. The whole matter has been managed in the most gentlemanly way by Lesley—who could do no other way—and you may rest assured that you are in no manner compromised. You will not, probably make much money (but this you did not desire) but you have the warmest and a real invitation from men of Science and gentlemen which, as a gentleman, you can accept, knowing that there will be at least no pecuniary loss to yourself while your visit here will be made very agreeable by the hospitality of the invitees and many others who will be glad to see you. If you will pardon it, I would again suggest, as in my last,[6] that you come out on the invitations given you & firstly of the Smithsonian; that you make arrangements through that Institution (all other arrangements after your arrival here) and at same time write to me or to Lesley as to your conclusions and wishes—also I again suggest that one of your Assistants (if you have two) act as your Private Secretary—this will save you a world of trouble & annoyance. There are three names on the invitation from whom there are no notes as, living here, they were asked personally and gave, like all the others, cordial & gratifying replies. I enclose with the notes of reply, another list[7] of the inviters with their titles, professions &c, which, in some of the cases, it is possible, you may not know and it might be desirable you should know. Hundreds or thousands of other names can be had but these are enough I think. I have spoken to many of our best people of your probable coming and all were much pleased—hoping to meet and hear you lecture. In addition to the eighteen men of science, I had Emerson's & Sumner's added (the first through Lesley as all the others were) as an additional gratification to you. I hope I did right.

Lesley tells me that Agassiz is a very great admirer & friend of Emerson's. I am further convinced that it would not do for you to come out here to lecture on a "business tour" as the lecture agents say.

Mrs Tyndale sends her respects. With remembrances to Prof Hirst & Debus and to your Sister Emma and our Gorey relations (to whom I must write soon). I am affectionately

Your Cousin | Hector Tyndale

If I can be of any service write to me at once without hesitation

Prof. John Tyndall

RI MS JT/1/T/64

---

1. *26th Ult[imo]:*: 26th ultimo, or of the previous month (August). Tyndale often used a colon in abbreviations.
2. *invitation . . . Henry*: letter 3504. These letters are part of a conversation about Tyndall's plans to lecture in the United States; see also letters 3510, 3542, 3543, 3558, 3589, 3594, 3639, 3669, 3670, 3684, 3685, 3702 and 3703.
3. *hereinclosed*: letter 3520.
4. *notes of the gentlemen*: letters missing.
5. *Government Scientific expedition to the Pacific*: the Hassler Expedition to South America and the Galapagos, December 1871–August 1872 (see C. Irmscher, *Louis Agassiz: Creator of American Science* (New York: Houghton Mifflin Harcourt, 2013), chapter 8).
6. *as in my last*: letter 3507.
7. *another list*: letter 3520.

From Hector Tyndale[1] [9 September 1871][2] 3520

Prof Joseph Henry | Washington { Physicist. Sec[retar]y Smithsonian Institution for the diffusion of useful knowledge. | Member National Academy U.S. &c

Hon. Charles Sumner | Washington { U.S. Senator from the State of Massachusetts. &c

Gen[e]r[a]l Mon[tgomer]y C. Meigs[3] | Washington { Quarter Master Gen[e]r[a]l U. S. Army | Constructor U.S. Capitol at Washington &c

N. H. Morison L. L. D.[4] | Baltimore { Provost of the Peabody Institute Baltimore &c

George B. Wood M D.[5] | Philadelphia { Distinguished Medical Author | President American Philosophical Society Philadelphia (the oldest scientific Ins. In the U.S. founded by Franklin.) &c Principal Editor U.S. Pharmacopeia

Prof: Fairman Rogers[6] | Philadelphia { Member National Academy | U.S. Prof. Civil Engineering | Pennsylvania University (or University of P[ennsylvani]a) &c

Prof: John F. Frazer[7] | Philadelphia { Member National Academy U.S. | Prof. Chemistry University of Penn[sylvani]a &c

Prof: Joseph Leidy[8] | Philadelphia { Hon[orary]. Memb[e]r National Academy | Prof. Comparative Anatomy University of Penn[sylvani]a. Very distinguished paleontologist

Prof: James C. Booth[9] | Philadelphia { Member American Philosophical So[ciety]. Philad[elphi]a. | Chemist U.S. Mint. &c

Prof: Peter Lesley | Philadelphia { Member National Academy U.S. | Secretary American Philosophical Society &c (Geologist. Author)

Prof. Alfred M. Mayer | South Bethlehem Penn[sylvani]a { Prof. Physics Lehigh University Penn[sylvani]a &c

William C. Cattell[10] | Easton Penn[sylvani]a { President Lafayette College Penn[sylvani]a &c

F. A. P. Barnard[11] | New York City { Member National Academy U.S. | President Columbia College, New York City, &c

Prof: James Hall[12] | Albany N. Y. { Member National Academy U.S. | Geologist of the State of New York &c

Frederick D. White[13] | Ithica N.Y. { Member Philosophical Society Philad[elphia] | President Cornell University, Ithica New York. &c

Prof: James D. Dana[14] | New Haven—Conn[ecticu]t { Member National Academy U.S. | Professor Geology Yale College. &c

Theodore Lyman[15] of | Boston—Gentleman { Member American Academy of Arts & Sciences

Prof: Josiah D. Whitney[16] | Cambridge Mass: & San Francisco Cal[ifornia]: { Member National Academy U.S. | Prof: Geology School of Mines Harvard College Cambridge Mass: Geologist of the State of California. &c

Ralph Waldo Emerson | Concord Mass: You know him as well as I

Prof. Louis Agassiz | Cambridge Mass You know him much better than I

RI MS JT/1/H/76

---

1. *Hector Tyndale*: note at the top of list: 'This list was probably sent by Hector Tyndale. In Hector Tyndale's handwriting'.
2. *[9 September 1871]*: enclosed with letter 3519 with all the notes from the Americans; see letter 3519, n. 4.
3. *Gen[e]r[a]l Mon[tgomer]y C. Meigs*: see letter 3510, n. 3.
4. *N. H. Morison L. L. D.*: see letter 3510, n. 4.
5. *George B. Wood M D.*: see letter 3510, n. 5.
6. *Prof: Fairman Rogers*: see letter 3510, n. 6. Tyndale often used a colon in abbreviations.
7. *Prof: John F. Frazer*: see letter 3510, n. 7.
8. *Prof: Joseph Leidy*: see letter 3510, n. 9.
9. *Prof: James C. Booth*: see letter 3510, n. 8.
10. *William C. Cattell*: see letter 3510, n. 10.
11. *F. A. P. Barnard*: see letter 3510, n. 11.
12. *Prof: James Hall*: see letter 3510, n. 12.
13. *Frederick D. White*: see letter 3510, n. 13. Note that 'Ithaca" has been misspelled as 'Ithica'.
14. *Prof: James D. Dana*: see letter 3510, n. 14.
15. *Theodore Lyman*: see 3510, n. 15.
16. *Prof: Josiah D. Whitney*: see 3510, n. 16.

To Cecil Trevor 18 September 1871 3521

Royal Institution, | 18th Sept[embe]r 1871.

Sir,

Subsequently to my reports[1] on the application of gas to fixed lights, as illustrated at Howth Baily,[2] the Commissioners of Irish Lights applied to the Board of Trade for the means of testing the applicability of gas revolving lights. The proposed experiments involved the purchase & suitable mounting of an annular lens, and in view of the results already obtained in Ireland I did not hesitate to recommend to the Board of Trade[3] that the modest sum needed for this important object should be placed at the disposal of the Commissioners.

The lens was accordingly purchased and mounted, the preliminary experiments were contrived & executed, and immediately before my summer vacation this year I went to Ireland to witness and to report upon the results[4]. These were so far satisfactory as to warrant the further recommendation that the experiments should be rendered practical & demonstrative by having them executed in a Lighthouse, instead of in a temporary shed, where motion by the hand had been made to supply the place of machinery. Rockabill, being less than two hours steam-distance from Kingstown Harbour, was the light-house chosen for the experiments.

The recommendation being acceded to by the Board of Trade, during the summer months the necessary gas-works were erected on the isolated rock on which the lighthouse stands, such trifling additions being made to the revolving apparatus as would place the action of the light altogether under the control of mechanism.

Having been informed a short time ago of the completion of the work I again visited Ireland,[5] and I have now the honour to render to the Board of Trade an account of observations made there on the nights of the 11th, 12th, & 13th of September respectively.

Rockabill has been hitherto lighted by the four-wick Trinity lamp, burning oil of Colza. The revolving apparatus, as stated in a former report,[6] performs a complete rotation in 96 seconds; & as during this time eight lenticular panels pass before the eye of an observer, each panel sending forth its beam, it follows that the time-interval between the arrival of the axes of two successive beams is 12 seconds.

It will moreover be remembered that the gas-burners hitherto employed are of five different powers. The first burner has a diameter of 3¾ inches and embraces 28 distinct jets; the second a diameter 5½ inches and 48 jets; the third a diameter of 7½ inches and 68 jets; the fourth a diameter of 9¼ inches and 88 jets; the fifth a diameter of 10½ inches and 108 jets; the diameter of the four-wick Trinity Burner is 3 11/16 inches being practically the same as that of the 28-jet gas & burner.

On the evening of the 11th we steamed from Kingstown in the "Princess Alexandra," passed Howth Baily, and keeping the island of Lambay[7] between us and the shore attained a point equidistant from Howth Baily and Rockabill, and 9½ miles from both. The stars were visible, but the ocean was hugged by a thickish haze which rendered the lights at both stations much dimmer than they would have appeared had the weather been clear.

Up to 10.55 p.m. the customary oil-lamp had been permitted to burn at Rockabill: at this hour, and in accordance with a pre-arranged programme, the oil-lamp was removed and the 28-jet burner put in its place. A distinct augmentation of the light announced the introduction of the gas.

At 11.10 p.m. 20 jets were added to the burner, (a few seconds only being necessary for this change) thus raising the light to 48 jets. A still further augmentation of the light was observed.

A point of considerable importance may be noted here. As before stated the beams succeed each other at Rockabill in intervals of 12 seconds. With the oil-lamp, and with the 28 jet burner, this rapid rotation gives to each beam almost the character of a flash which quickly passes from the eye of the mariner. But the augmentation of the diameter of the burner, by widening the beam, materially affected its duration. The dark sectors were in fact diminished and the luminous sectors increased in angular width, so that besides the augmentation of its light the beam continued to act for a longer period upon the observer's eye.

At 11.20 p.m. the 68-jet burner was ignited: the duration of the beam was still further increased; the light was perceptibly so, though not to a very striking extent.

At 11.25 the 88-jet burner was ignited and at 11.30 the 108-jet burner. The duration of the beam was still further augmented in these cases, being sensibly proportional to the width of the burner; but the brilliancy of the beam was not apparently increased.

Remarks.

Supposing no alteration whatever to be made in the mechanism, or in the time of rotation, hitherto existing at Rockabill, we may draw, in passing, two practical conclusions from the foregoing observations.

1. That by substituting for the four-wick oil-lamp a gas-burner of practically the same size we obtain a better light.

2. That by the substitution of a gas-burner of higher power the light is not only still further augmented, but the duration of each particular beam is increased. Every mariner would, I doubt not, regard this increase of duration as an important improvement of the light.

By a former investigation it was shown that as regards Ireland the 28-jet burner is cheaper, while the 48-jet burner is dearer than the four-wick oil-lamp. Hence by the introduction of the former at Rockabill we should gain both in economy and efficiency; by the substitution of the latter we should gain considerably in efficiency but at an increased expense.

The above remarks are introduced at this place with the view of connecting them more closely with the data on which they are based. But the observations on the night of the 11th did not end here. At 11.35 the ordinary oil-lamp was again lighted and permitted to burn till 12.20. At 12.20 the 28-jet burner was ignited, when the superiority already recorded was once more observed.

The time of rotation of the revolving apparatus was now altered. Permitting the weight which moved the mechanism to remain unchanged, by a simple modification of the wheel-work the interval between two successive beams was augmented from 12 seconds to a minute, the duration of each beam being augmented in the same proportion. But instead of being permitted to pass steadily over the observer each beam, by a simple mechanical arrangement which caused the alternate extinction and re-ignition of the gas, was converted into a series of flashes, following each other at intervals of one second. I thought the succession too rapid—in fact tantalising to the eye.

At 12.35 the 48-jet burner was introduced, its beams being cut up in the same manner. As might be expected from the augmented breadth of the burner, & consequent widening of the beams, the number of flashes yielded by each beam was increased, but, as in the former case, the succession was, in my opinion, too rapid to be effectual.

At 12.45. the 68 jet-burner was ignited; and here the breadth of the beam was such as to admit of an interval of 2 seconds instead of one between every two successive flashes. The change was a marked improvement, these new flashes being distinct and powerful.

The same remark applies to the 88-jet and the 108-jet burners; the number of flashes yielded by the beam augmenting as the diameter of the burner from which it issued increased.

All these observations were made from a fixed position. At 1.10 a.m. the 28-jet burner being ignited a series of observations similar to those referred to and explained in a former report[8] was commenced. The method of observation it may be remembered involves the extinction of the gas as each alternate panel passes in front of the observer's eye. The interval between two flashes is thereby doubled, the quantity of gas consumed being reduced to one half. The action was such as might be inferred from a priori considerations. The flashes succeeded each other in accurate order with an interval of 24 seconds. We steamed over a considerable arc round the light-house, and from every position occupied by the vessel the same steady performance of the light was observed.

But though the interval of 24 seconds is a short one the passage of the beam was in this case so rapid that the saving of the gas would be accompanied by a too great deterioration of the light. In the light that has hitherto existed at Rockabill the quickness of succession has atoned for the shortness of duration, and I should not therefore, for the sake of saving half the gas, be disposed to recommend the introduction of this method of alternate darkening at Rockabill.

But if, while the period of rotation remained unaltered the duration of the flash could be materially increased, then, doubtless, the method of extinguishing the gas as each alternate panel passes the observer could be made to yield an excellent light. With the 68-jet burner, for example, a much more effective light than that now exhibited at Rockabill might, I doubt not, be obtained.[(x)]

On this and the subsequent evening I had the great pleasure of being accompanied by Sir Leopold M^c^Clintock,[9] and on both occasions took advantage of his experience in verifying my conclusions in regard to the power and the fitness of the lights

---

A brief recapitulation will be useful here. On the night of the 11th the apparatus at Rockabill was first permitted to revolve at its ordinary rate the action of its customary oil-light being observed, and subsequently the action of five different gas-burners of gradually augmenting power. The speed of revolution was then reduced so as to cause the axes of the rotating beams to succeed each other at intervals of one minute; but instead of allowing the beams to pass, as in the first experiments, steadily and unbroken before the observer's eye, they were resolved into a series of flashes by the regulated extinction and ignition of the gas. With the smaller burners the interval from flash to flash was one second, and this was found too rapid: the beam lost power as a signal light by this division. With the larger burners however the period of succession was augmented to two seconds and the flashes are here reported as "distinct and powerful." Observations were also carried out in which the gas was extinguished as each alternate panel of the revolving apparatus passed in front of the observer's eye.

But besides the period of one minute from flash to flash the wheel-work of Rockabill permitted of a period of forty five seconds. Partly for the purpose of control and partly for the sake of comparison I wished to witness the effect of this period also. It was likewise desirable to compare, in each case, the steady and unbroken beam with the flashing beam; it was also thought worth wile the to change the over-rapid period of one second for a period of two seconds in the case of the smaller burners. Finally, I desired before commencing observations on the 12th to assure myself that the oil-lamp at Rockabill yielded a proper flame. About 10 p.m. therefore an attempt was made to land upon the rock, but the surf and darkness defeated us. The following programme was subsequently carried out, the place of observation being about 5 miles distant from Rockabill

Up to 11.0. p.m. oil lamp; 45 seconds interval.
At 11.5 " " 28-jet burner; steady
" 11.10 " " 28-jet burner; flashing
" 11.15 " " 48-jet burner; steady
" 11.20 " " 48-jet burner; flashing
" 11.25 " " 68-jet burner; steady
" 11.30 " " 68-jet burner; flashing

The steady beams were here very fine, and unless distinction were of paramount importance I should be loth, especially as regards the smaller burners, to change them for the flashing beams.

At 11.35 p.m. we changed back to the 48-jet burner and to the old period of succession, namely 12 seconds, when a new series of observations, according to the method of extinction for every alternate panel, was carried out.

In the early part of Tuesday I had landed on Rockabill and inspected the apparatus. The oil lamp employed for the experiments was that always used in the Lighthouse. The pipe through which the wick was fed appearing rather narrow I thought it desirable to make a third series of experiments with a lamp which offered a freer passage to the oil. Such a lamp was found at Howth Baily, and on the Wednesday we steamed to Rockabill sufficiently early to set this lamp in full action before sunset. At 6 p.m. I inspected the lamp. It seemed to me that I had seen better flames in the experimental room of the Trinity House; but I have no reason to doubt the assurance both of the light-keeper[10] and of Mr Fitzsimon,[11] foreman to the inspector of Irish Lights, that the flame then exhibited was of the full average power employed in Ireland: at all events the oil-lamp was nursed up to its maximum strength. Having seen it in full action we steamed away to a distance of 4½ miles and observed the lights from this position.

At 7.30 we passed from the oil-lamp to the 28-jet burner and found, as before, a marked augmentation of the light. Owing to the quickness of rotation however (the ordinary rate at Rockabill) the beams, though brilliant while they lasted, passed rapidly over the observer's eye.

At 7.45. changed to the 48-jet burner. As before, the duration of each beam was found to be sensibly increased: in every respect indeed the light was a decided improvement on the last.

At 7.50 68-jet burner. The duration of the beam was still further augmented, and much more easy to be picked up by the mariner than the rapidly vanishing beams of the smallest light.

Up to this point the beams succeeded each other at intervals of 12 seconds, it being understood that the interval is reckoned from centre to centre of the beams. The time elapsing between the vanishing of one beam and the appearing of the next was less than 12 seconds, the interval of darkness diminishing and that of light augmenting as the diameter of the burner increased.

At 8.p.m. the rate of revolution was changed so as to permit the beams to succeed each other at intervals of a minute. Its previous performance caused me to confine my attention to the 68-jet burner, each beam of which, at the changed rate of rotation, occupied 16 seconds in passing. Judged by the naked eye the strength of the beam during this interval was sensibly uniform. It rose rapidly to a maximum after its first appearance, and on disappearing fell as rapidly to a minimum.

By the action of the machinery, this beam was converted into a flashing light, the time from flash-centre to flash-centre being 2 seconds. There were therefore 8 full flashes sent forth during every passage of the beam. The arrival of the full flashes was heralded and their disappearance was followed by a smaller flash. "A powerful and indeed splendid effect" are the words employed in my notes to describe this light. An interval of a minute between the beams seems to be the most suitable for flashes of this character.

When the 28-jet gas burner was employed, with the same rate of rotation, the number of flashes sank to 4. This result might have been predicted. The width of the 28-jet burner being exactly half that of the 68-jet one the number of flashes from the former ought to be only half of those from the latter.

Were its light fully invoked the 68-jet burner would be an expensive one, but by converting the beams into flashes the quantity of gas is reduced to little more than that consumed by 32 jets burning in the ordinary way. The cost of the 68-jet burner, thus applied, would probably not be more than that of the four-wick lamp hitherto at work at Rockabill.

Conclusion.

Weighing the observations here described, and others embodied in former reports, I think I am justified in recommending to the Board of Trade that the gas works recently erected at Rockabill be retained there, and that henceforth gas, instead of oil, be burned in the Lighthouse.

This recommendation is fortified by the consideration that in reference of the future of gas in Lighthouses it is highly important to have in the neighbourhood of Dublin both a fixed light and a revolving light where experiments on these respective modes of illumination may at all times be carried on. Such experiments would furnish sure guidance as to the best form of apparatus for other stations.

And should it be thought desirable to give a revolving light so distinctive a character as to render it perfectly unmistakeable "the group flashing light," as its inventor Mr. Wigham[12] calls it, secures this end. I have not been called upon to offer any recommendation as to its adoption, and would therefore merely refer to it as a light of unrivalled individuality, of great power, and, in Ireland at least, of moderate cost.

During my recent visit I also made experiments on the flashing of the fixed light at Howth Baily. The results assure me that with gas as a source of illumination an amount of variableness, & consequent distinctiveness, is attainable which is not attainable with any kind of oil. It would, I think, be easy to give to every lighthouse supplied by gas so marked a character that a mariner on nearing the light should know, with infallible certainty, its name. As stated in a former report,[13] I look in great part to the flexibility with which gas lends itself to the purposes of a signal light for its future usefulness. It may be beaten in point of cheapness by the mineral oils now coming into use, that is still to be proved, but in point of handiness, distinctiveness, and power of variability to meet the changes of the weather it will maintain its superiority over all oils. I therefore respectfully submit to the Board of Trade, that while withholding all countenance from extravagant or fanciful experiments, it will be wise to encourage, as hitherto, the gradual, economical, and consequent healthy expansion of the system of gas illumination in Ireland.

I have the honour to be | Sir | Your obedient Servant | John Tyndall

Cecil Trevor Esq[ui]re | & & & | Harbour Department | Board of Trade.

P.S. I am unwilling to close this report without reminding the Board of Irish Lights that our knowledge of the comparative powers of the larger and smaller burners, as regards the penetration of fogs, is not so accurate and complete as it might be. It is highly desirable that further experiments should be made on this point.

(x) The adoption of the method of alternate darkening would permit of the employment of a powerful burner as a moderate cost. A burner of 56-jets applied in this way would consume only the same amount of gas as one of 28 jets applied in the ordinary way. I have heard it urged that by reducing the speed of rotation at Rockabill to an interval of 26 seconds between the flashes duration may be secured even with the wall burner. This is true; but then the beams are not individually equal to those. By using the light, in short, for only half the time, we should be enabled, without any additional expense, materially to exalt its power.[14]

National Archives MT 10/131 H3953

---

1. *my reports*: see letter 3353, *Tyndall Correspondence*, vol. 11. In this volume, see letters 3400, 3417, 3475 and 3476.
2. *as illustrated at Howth Baily*: Baily is a lighthouse at Howth Head, Ireland. Here, Tyndall performed experiments on what kind of oil was best in lighthouses (see n. 1).
3. *in view of . . . Trade*: see letter 3475.
4. *I went . . . results*: see letters 3475, 3476, 3480 and 3483.
5. *I again visited Ireland*: see letter 3518, for example, where he told Bence Jones about his trip.
6. *former report*: letter 3475.
7. *island of Lambay*: the largest island off the east coast of Ireland.
8. *a former report*: Tyndall's report of 7 February 1871; see letter 3353, *Tyndall Correspondence*, vol. 11.
9. *Sir Leopold McClintock*: Francis Leopold McClintock (1819–1907), an Admiral and Irish explorer in the British Royal Navy. He entered the Navy when he was twelve, and when he

was nineteen he passed his exam and was made a mate. By 1845 he was made lieutenant. He was on a number of Arctic expeditions and gained experience surviving in extreme environments. He was promoted to Captain in 1854, to vice-admiral in 1877, and to Admiral in 1884. He was knighted in 1860.

10. *the lightkeeper*: not further identified.
11. *Mr Fitzsimon*: Fitzsimon (n.d.) was the foreman to the inspector of Irish lights; not otherwise identified.
12. *Mr. Wigham*: see letter 3475, n. 3.
13. *former report*: letter 3475.
14. *(x) . . . its power*: paragraph inserted by a notation.

To Henry Harrison Doty 19 September 1871 3522

Sub-enclosure 2, in No. 29.[1]

Royal Institution, 19 September 1871.

Sir,

On the 15th of August, you did me the honour of writing to me a letter,[2] in which you described the course which you consider experiment ought to take in the approaching trials of mineral oils for lighthouses.[3]

Immediately after my return from Switzerland you did me the favour of calling upon me, and of unfolding your views at considerable length upon the same question.

On Saturday morning last I was also favoured with a visit from you, during which you dwelt very fully upon the history and merits of your lamp, and on the mode of conducting the pending experiments which you consider to be the right one. I thought you then quitted me with the assurance that in the contemplated inquiry I had but one aim, namely, to elicit the truth.

Yesterday morning, however, I was favoured with another letter[4] from you, containing a reiteration of some of your views.

I have also learned that you have paid two visits to the gentleman whom I have appointed to assist me in the coming investigations—I mean Mr. Valentin.

I am sorry that you should have thought it necessary to take this course, and I think it especially objectionable that you should interfere in any way with Mr. Valentin. A single expression of your wishes was quite sufficient, and while it will give me pleasure to meet your views as far as it is expedient to do so, you must allow me to be guided in this inquiry simply and solely by what I consider right and fair towards all the parties concerned.

I have a very high opinion of Dr. Macadam,[5] and shall be glad to see you represented in the coming inquiry by so competent a man. I think the experiments are likely to proceed more smoothly if the principals, that is to say yourself and Mr. Douglass, were to hold yourselves aloof. You might come every morning and see your lamps lighted. But I would propose that each of you should then hand over his lamp to the care of a skilled assistant; and that during the experiments you should be represented by the gentleman whose presence you have expressed so strong a wish to secure.

I am, &c. | (signed) *John Tyndall.*

H. H. Doty, Esq., | &c. &c.

'Lighthouses (mineral oils). Copy of further correspondence relative to proposals to substitute mineral oils for colza oil in lighthouses (in continuation of parliamentary paper, no. 2, of session 1872)', HC 2 67,[6] Parliamentary Papers (1872), pp. 46–7
Typed Transcript Only

---

1. *Sub-enclosure 2, in No. 29*: In the Parliamentary Papers, Tyndall's letter, as well as letters 3540 and 3555, are enclosed in a letter from Trinity House to the Board of Trade. See Trinity House to Board of Trade, 19 December 1871, in 'Lighthouses (mineral oils). Copy of further correspondence relative to proposals to substitute mineral oils for colza oil in lighthouses (in continuation of parliamentary paper, no. 2, of session 1872)' HC 2 67, Parliamentary Papers (1872), p. 22.
2. *a letter*: letter missing.
3. *the approaching trials of mineral oils for lighthouses*: Tyndall spent much of the fall of 1871 on an investigation into competing designs for lighthouse lamps, part of his work as a scientific consultant to Trinity House. One of these lamps was Doty's; the other was designed by Douglass for Trinity House. See letter 3555 for the results of this investigation.
4. *another letter*: letter missing.
5. *Dr. Macadam*: Stevenson Macadam (1829–1901), a Scottish chemist whose research included the application of chemistry to manufacturing; he held the post of Lecturer on Chemistry at the University of Edinburgh. He was elected FRSE in 1855, and in 1877 was a founder of the Institute of Chemistry of Great Britain (now the Royal Institute of Chemistry).
6. *2 67*: at the time of writing, the ProQuest UK Parliamentary Papers database misnumbered this paper as 2 264.

To Cecil Trevor — 20 September 1871 — 3523

Dear Trevor.

Should the reader find any difficulty in deciphering the corrections in this Report[1] I shall be <gla>d to correct the printer's proof myself

John Tyndall | 20th Sept. 1871

National Archives MT 10/131 H3953

1. *this Report*: letter 3521.

To the Editor of *Nature*[1] 28 September 1871 3524

ON THE BENDING OF GLACIER ICE*

* The following is an extract from a note addressed to Prof. Hirst[2] and sent from Pontresina in the hope that it would reach Edinburgh in sufficient time to be communicated to Section A of the British Association. It was a few hours too late.—J. T.

[. . .][3]

I may add that various experiments were subsequently made, and a means discovered of rendering the bending very speedily visible. I hope before long to return to the subject.—J. T., September 28

J. Tyndall, 'On the Bending of Glacier Ice', *Nature*, 4 (5 October 1871), pp. 447–8
Typed Transcript Only

1. *Editor of Nature*: Norman Lockyer was the editor of *Nature* from 1869 until 1919.
2. *a note addressed to Prof. Hirst*: letter 3500.
3. *[. . .]*: Tyndall here included letter 3500.

From Emily Peel 28 September 1871 3525

Geneva | Sep[tembe]r. 28 / 71

My dear Professor,

I have been much disappointed that your rambles in Switzerland did not bring you this way—I heard of you however at Pontresina, having climbed up the Bernina[1] by way of a breather, therefore from that you must be in good health and fettle.[2]

A young friend of mine who was introduced to you at Pontresina is anxious to become a member of the English Alpine Club,[3] but being a German or rather an Austrian, he is not aware whether he could be received into the English section of that community. He is Baron Albert Rothschild,[4] the brother of our Swiss mutual friend[5] and I promised to ascertain the fact for him.

I believe in every other respect he is entitled to that distinction, having performed various ascensions & passes of importance, such as the Jungfrau,

Finsteraarhorn, Monte Rosa, Bernina Spitz, Weissthor, & I believe the Oberaarjoch.[6]

Would you be so very kind as to spare me one moment to tell me whether as a German he would be eligible as a member of the English Alpine Club.—He does not seem to care to belong to any other.—

I have read Whymper's last book[7] since I came here, & I admire more than I can say the dignified manner in which you have treated his false attacks[8] upon you.

Believe me | yours very kindly | Emily Peel

RI MS JT/1/P/55

1. *Bernina*: Piz Bernina, a peak in the Alps. See, for example, letters 3494, 3495, 3497 and 3498.
2. *fettle*: 'condition, state' (*OED*).
3. *English Alpine Club*: founded in 1857 in London as the world's first mountaineering club. See also letters 3500, n. 1 and 3531. On Tyndall as a member, see letter 3531, n. 1.
4. *Baron Albert Rothschild*: Albert Salomon Anselm Freiherr von Rothschild.
5. *Swiss mutual friend*: Rothschild had two brothers, so this could be either Nathaniel Meyer von Rothschild (1836–1905) or Ferdinand James Rothschild (1839–98). Ferdinand had moved to England and was an MP from 1885–98 so it was likely him. Otherwise not further identified.
6. *Jungfrau . . . Oberaarjoch*: various peaks and passes in the Alps.
7. *Whymper's last book*: *Scrambles amongst the Alps*.
8. *false attacks*: In *Scrambles amongst the Alps*, Whymper had implied that Tyndall's accounts of attempting to summit the Matterhorn were inaccurate; he was also critical of Tyndall's work on glaciers in the Alps. See p. 134 and pp. 317–31. Tyndall responded to Whymper briefly in *Hours of Exercise in the Alps*, concluding, 'touching Mr. Whymper's general tone towards myself, I do not feel called upon to make any observations' (p. 167).

To Cecil Trevor [September 1871][1] 3526

The substance of the foregoing Report[2] may be thus stated:—

1. A gas flame of the same size as an oil-flame may, in a revolving apparatus, be employed instead of the oil-flame. The flames are sensibly equal in luminous power and the experiments hitherto made in Ireland show that the gas there is cheaper than the oil.

2. At the Rockabill Lighthouse by making the interval of darkness 24 seconds instead of 12 seconds, an intermittent light might be produced with half the expenditure of gas necessary for a fixed light. This is to be understood as a statement of fact, not as a recommendation.

3. Should the more recent observations, and the judgement pronounced regarding them, be correct, every important form of intermittent light has been discovered.

Should the Board of Trade think the subject worthy of further attention I shall be happy, as soon as my duties here permit, to go again to Ireland and to subject the observations of M^{r}. Wigham[3] and Capt. Hawes[4] to strict personal examination.

I reserve for a future occasion such further remarks as might even now be made upon this subject. The Board may I think feel confident that before finally recommending any change in our system of Lighthouse illumination I shall assure myself by suitable experiments that the change would be an improvement optically or economically. That gas is capable of manifold application in Lighthouses is even now evident. We have to deal in fact with two elements, the quantity of the light and its distribution (in the former score gas, in Ireland, competes successfully with oil; but it is to the ease and promptness with which gas lends itself to various modes of distribution that the chief manifestations of its superiority are to be looked for. I anticipate great ameliorations and advances from its future use.

I have the honour to be Sir | Your obedient Servant | John Tyndall

Cecil Trevor Esq[ui]r[e]

National Archives MT 10/131 H648

---

1. *[September 1871]*: dated to sometime in September of 1871 because of letters 3521 and 3523.
2. *foregoing report*: see letters 3521 and 3523.
3. *Mr. Wigham*: see letter 3475, n. 3.
4. *Capt. Hawes*: see letter 3475, n. 4.

From Louis Pasteur 3 October 1871 3527

Paris le 3 8bre 1871

Monsieur et cher confrère,

Voici tout ce que j'ai pu réunir d'exemplaires à part de mes divers travaux sur le sujet qui vous intéresse présentement. C'est incomplet, mais les tables des comptes-rendus, table générale et tables partielles, vous mettront vite au courant.

Je vous le répète : le nœud de la difficulté touchant la question des générations spontanées se trouve délié dans ma note du B.56 des C^{ts} R^{dus} (1863). Cette note est le complément nécessaire, indispensable de mon mémoire des annales de chimie et de physique. Ultérieurement, M. Donné a reproduit avec les oeufs les expériences que j'avais faites sur le sang frais et l'urine fraîche. L'histoire des variations d'opinions de M. Donné et son aveu définitif qu'il s'était trompé peut-être précieux à enregistrer, si vous le rapprochez de la discussion que j'ai soutenue avec lui.

Je livre à tout votre mépris d'expérimentateur habile la sotte campagne et la retraite de mauvaise foi de Mm. Pouchet et Joly devant la commission de l'académie des sciences.

Je vous remets deux notes sur les fermentations et l'excellent travail de mon élève et ami Raulin parce que tout cela est lié au sujet que vous voulez traiter. Vous avez lu la preuve de la nécessité de germer, surtout si vous rapprochez les faits de l'absence absolue de formation d'organismes et de fermentation quand le germe est absent. Voilà le milieu minéral très propre à la vie, à la nutrition du germe et de ses descendances. *[Mais]*, ce germe, il le faut pour la manifestation de la vie et de ses effets. La première expérience de cette nature est celle relative à la [levure] de bière dans mon mémoire sur la ferme[ntati]on alcoolique des annales de Ch[imie] et de Physique.

Si j'en ai le temps, et si vous y tenez, je vous écrirai, *[1 word illeg]* lié et un peu plus complet, ce que je vous ai exposé de mes idées sur la vie, sur la dissymétrie moléculaire, sur les influences cosmiques dissymétriques en relation avec la vie et ses manifestations sur notre terre.

Je suis très heureux et très fier que vous ayez la pensée de consacrer quelques-uns de vos précieux instants de travail à l'étude de ces questions. Si vous les faites entrer dans le domaine de tous, gens du monde, médecins etc . . . avec *[l'exemplaire]* de votre rare génie, de votre brillante imagination, de votre incomparable talent d'exposition, vous rendrez un immense service et votre publication aura un succès extraordinaire, j'en ai la conviction.

Veuillez agréer, mon cher confrère, l'assurance de mes sentiments les plus distingués et les plus dévoués.

L. Pasteur

Paris, 3rd October[1] 1871

Sir, dear colleague,

Here are all the copies[2] I could gather of my various works on the subject that interests you at the moment.[3] It is incomplete, but the tables of reports, general table and partial tables will quickly get you up to speed.

I repeat: the central difficulty of the question of spontaneous generations is resolved in my memorandum of B.56 of the Comptes Rendus (1863).[4] That memorandum is the necessary, indispensable complement to my memoire in the Annals of Chemistry and Physics.[5] Subsequently, Mr. Donné[6] reproduced with eggs the experiments that I had done with fresh blood and urine. The history of the variations in Mr. Donné's opinion and his definitive confession that he was wrong may be valuable to record, if you connect it to the discussion that I had with him.

I leave you to judge, with your skilful experimenter's disdain, the foolish campaign and the retreat in bad faith of Messrs Pouchet[7] and Joly[8] before the commission of the Academy of Sciences.[9]

I give you two memoranda on the fermentations and excellent work of my student and friend Raulin[10] because all of this is related to the subject that interests you. You have read the evidence on the need to germinate, especially if you compare the absolute lack of organism formation to fermentation when the germ is absent. Here is the mineral environment that is suitable for life, for the nutrition of the germ and for its progeny. */But/*, that germ, it is needed for the manifestation of life and its effects. The first experiment of this kind is the one that deals with brewer's yeast in my memoir on alcoholic fermentation in the Annals of Chemistry and Physics.[11]

If I have the time and if you wish, I will write to you, *[1 word illeg]* related and a bit more complete, what I explained to you on my ideas about life, molecular asymmetry, the asymmetrical cosmic influences in relation to life and its manifestations on Earth.

I am very happy and very proud that you thought to devote some of your precious work time to the study of these questions. If you bring them within reach for all, people of the world, doctors, etc . . . with */the example/* of your rare genius, your brilliant imagination, your incomparable talent for explaining, you will provide a great service and your publication will be an immense success, I am sure.

Please accept, my dear colleague, the assurance of my most distinguished and devoted regards.

L. Pasteur

RI MS JT/1/P/4

---

1. *October*: in the French, 8bre is a shortened version for Octobre, or October.
2. *Here are all the copies*: copies missing.
3. *interests you at the moment*: Tyndall was working on dust and smoke and how these things in the air may cause illnesses. See *Ascent of John Tyndall*, pp. 263–80. His Friday Evening Discourse on this, delivered at the RI on 21 January 1870, also appeared in *Fragments of Science*, chapter XI, 'Dust and Disease', pp. 289–342.
4. *my memorandum of B.56 of the Comptes Rendus (1863)*: possibly L. Pasteur, 'Examen du rôle attributé au gaz oxygène atmospherique dans la destruction des matières animales et végétales après la mort', *Comptes rendus*, 56 (1863), pp. 734–40.
5. *my memoire in the Annals of Chemistry and Physics*: possibly L. Pasteur, 'Mémoire sur la fermentation alcoolique', *in Annales de chimie et de physique*, 58 (1860), pp. 323–426.
6. *Mr. Donné*: Alfred François Donné (1801–78), a French bacteriologist.

7. *Pouchet*: Félix-Archiméde Pouchet (1800–72), director of the Jardin des Plantes at Rouen and professor at the Rouen School of Medicine, and Pasteur's opponent in the debate about spontaneous generation.
8. *Joly*: Nicolas Joly (1812–85), professor at the University of Toulouse and Pouchet's supporter.
9. *retreat... Academy of Sciences*: Pasteur may be referring to the 1862 withdrawal of Pouchet from the competition organized by the French Académie des Sciences to prove spontaneous generation. See E. Duclaux, *Pasteur: The History of a Mind* (Philadelphia: W. B. Saunders, 1920), pp. 109–10.
10. *Raulin*: Jules Leonard Raulin (1836–96), professor at the Lycée de Caen.
11. *my memoir on alcoholic fermentation in the Annals of Chemistry and Physics*: possibly the same paper referenced in n. 5.

To Julius E. Hilgard[1] 4 October 1871 3528

Royal Institution of Great Britain | 4th Oct. 1871

My dear Sir,

Allow me to thank you cordially for the Report which you have been good enough to send me[2], and to say in addition that I still retain a very agreeable remembrance of your visit to London.[3]

faithfully yours | John <u>Tyndall</u>[4]

J. E. Hilgard Esq[ui]^{r}[e] | &c &c

Historical Society of Philidelphia, Gratz Collection

---

1. *Julius E. Hilgard*: Julius Erasmus Hilgard (1825–91), an American engineer and geodesist. From 1844 to 1860 he worked for the U.S. Coast Survey in the computing department and in 1867 was appointed the agency's assistant superintendent, a position in which he supervised the work of the Office of Weights and Measures and drafted legislation for the implementation of the metric system in the United States. Hilgard was elected president of the American Association for the Advancement of Science in 1877, the same year he won a gold medal for his design of an instrument to measure the density of seawater.
2. *Report... send me*: report missing.
3. *your visit to London*: in May 1869. See letters 2987 and 2996, *Tyndall Correspondence*, vol. 11.
4. *Tyndall*: On the back of the letter, in another hand, is written: 'Prof. John Tyndall | Eminent British scientist'.

To Thomas Henry Huxley 4 October 1871 3529

Royal Institution. | 4th. Oct. 1871.

My dear Huxley,

I was very glad to see your ugly old hand last night.[1] I am sound though hard pressed just now with trying and labourious work. I hope you and yours have returned from St. Andrews[2] in force.

Will you say to Mr Chamberlain[3] that it grieves me to decline such invitations as his. I am excessively busy, not only for the Royal Institution, but for the Board of Trade, for the Trinity House, and for other matters and masters. At the present moment I am dragged aside from an investigation[4] that I had quite given my heart to in order to decide upon the claims and merits of rival inventors of Lighthouse lamps.[5] My hands are in fact always full: So full that instead of adding to my labour by lecturing in the country, I am sometimes inclined to cut myself adrift from some of my present engagements. You know I quitted the School of Mines,[6] and thus did violence to all the affections of my heart simply to escape the labour of lecturing.

Say, therefore, in the kindest manner that your sweet nature is capable of to Mr Chamberlain that you feel it to be my duty to hold on to my work here, and not to give way to temptations which are only too strong, and which invitations like his throw in my way.

Yours always | John Tyndall

Take a Rowland for your Oliver[7] in this bit of writing.

By the way Thomson opened my eyes in Pontresina.[8] Seriously I wished that theory[9] unuttered.

IC HP 8.90
Typed Transcript Only

---

1. *old hand last night*: letter missing.
2. *St. Andrews*: a seaside town in Scotland.
3. *Mr. Chamberlain*: Joseph Chamberlain (1836–1914), an English industrialist and liberal politician. He was brought into industry by his father and uncle and apprenticed to his uncle's wooden screw-making business at the age of sixteen. He worked his way up through both industrial and political circles in Birmingham. He retired from business in 1874, and focused on politics and civic improvements. By 1880 he was the President of the Board of Trade, and friendly with Tyndall in later debates about lighthouses. In 1871, however, he was still a wealthy industrialist.
4. *an investigation*: having to do with dust, disease and spontaneous generation; see, for example, letter 3527.

5. *rival inventors of Lighthouse lamps*: Henry Harrison Doty and James Douglass. See letter 3522, n. 3.
6. *School of Mines*: Royal School of Mines. He was a professor of physics there until 1868, but he continued using their facilities for research as late as 1872 (see *Ascent of John Tyndall*, p. 299).
7. *Rowland for your Oliver*: Tyndall is referring to a story about two pages of Charlemagne's who were each so excellent that they were perfectly equal (*The Song of Roland*, Bodleian Library, MS Digby 23). We presume, and can attest, that Tyndall is saying that his handwriting in this letter is as bad as Huxley's.
8. *Thomson . . . Pontresina*: While in Pontresina, Tyndall read an account of Thomson's Edinburgh address to the BAAS. See letter 3503.
9. *that Theory*: the theory Thomson put forth in his address in Edinburgh, that because life comes only from other life, the Earth must have begun as 'moss-grown fragments of the ruins of another world' outside of Earth (*Ascent of John Tyndall*, p. 277).

To Margaret Ginty 7 October 1871 3530

Royal Institution | 7th. Oct. 1871.

My dear Mrs Ginty,

I have placed your dividend warrant in my bank, and send you in lieu of it a cheque for £85. This is the simplest solution of the matter.[1]

I am just now terribly pressed—otherwise I would write more.

Yours faithfully | John Tyndall.

RI MS JT/1/TYP/11/3667
LT Typescript Only

1. *the matter*: having to do with her husband William Ginty's estate. See letters 3496 and 3505.

To Emily Peel 7 October 1871 3531

7th Oct., 1871

My dear Lady Emily,

I am not a member of the English Alpine Club,[1] but immediately on receiving your letter[2] I wrote to the gentleman who was its first President, and who has immortalised himself by writing a Guide to the Alps[3]—I mean my friend John Ball. He does not see any difficulty in your friend the Baron being elected to the Alpine Club. He is quite prepared to propose him. But

I am almost ashamed to write down his preliminary condition, and his reasons for it. He does not place implicit trust in <u>your</u> account of the Baron's achievements, on the ground that ladies are usually inaccurate! And he therefore says that if the Baron will write to him giving him a statement of four or five ascents, sufficient to qualify for election, he, Ball, will do the rest. A letter addressed to

John Ball, Esq., | Athenaeum Club,[4] | London,

will reach him.

I trust this will meet your wishes. <u>Du reste</u>[5] I would give you four and twenty hours to convince the said Ball that you are as accurate as he is.

I am very glad that my way of meeting Mr Whymper[6] meets your approval. He has introduced a new phase into Alpine literature, and in my opinion a very contemptible one. But he is eaten up by conceit, which in his case is not toned down by the culture which renders it tolerable in some men.

Yours very faithfully, | John Tyndall.

RI MS JT/1/TYP/3/963
LT Typescript Only

---

1. *not a member of the English Alpine Club*: Tyndall had been elected to the Club in November 1858, and by 1861 he was Vice President of the Club. He resigned shortly after this, due possibly to the pressure of his work, or that he did not want the Club to have a claim over his writings about the Alps (see *Ascent of John Tyndall*, pp. 224–5). See also letters 1459, 1466 and 1478, *Tyndall Correspondence*, vol. 6; and letter 1849, *Tyndall Correspondence*, vol. 7.
2. *your letter*: letter 3525. See also letter 3533.
3. *A Guide to the Alps*: referring to John Ball's multi-part *Alpine Guide*, which included *A Guide to the Western Alps* (London: Longman, Green, Longman, Roberts, & Green, 1863); *The Central Alps, including the Bernese Oberland, and all Switzerland excepting the Neighbourhood of Monte Rosa and the Great St. Bernard* (London: Longman, Green, and Co., 1866); and *A Guide to the Eastern Alps* (London: Longman, Green, and Co., 1868).
4. *Athenaeum Club*: see letter 3372, n. 15.
5. <u>*Du reste*</u>: moreover, by the way.
6. *meeting Mr Whymper*: see letters 3462, 3464, 3470, 3471 and 3525.

To John Murray 8 October 1871 3532

Royal Institution of Great Britain | 8th Oct. 1871.

Dear Mr Murray,

I would like you to be aware that I have not interfered unwarrantably as regard your notice of the Bel Alp[1] and I therefore send you a couple of notes[2] bearing upon this point. Paris I daresay has already communicated to you a note from D[r] W[illia][m] Budd. One of those now enclosed[3] is also from him. If you could without trouble send the notes back to me I should be glad.

Yours faithfully | John Tyndall

NLS, John Murray Archive, Ms 41214

1. *your notice of the Bel Alp*: On the basis of this letter as well as 3535 we believe that this may refer to a bad review of a hotel given in one of Murray's handbooks for travelers (in which the firm specialized), but have not been able to identify to what exactly Tyndall is responding. In the 1871 edition of the *Handbook for Travellers in Switzerland*, for example, Murray gives a favorable account of an inn called the Belle Alp; see *A Handbook for Travellers in Switzerland, and the Alps of Savoy and Piedmont*, 14th edn (London: John Murray, 1871), pp. xliv, 111. Of course, this positive account does not fit with Tyndall's comments referred to in letter 3535, which suggest that Murray gave a negative review of the Hotel Belalp, built in the 1850s.
2. *a couple of notes*: notes missing.
3. *enclosed*: missing.

From Emily Peel 10 October 1871 3533

Oct: 10/ 71 | Villa Lammermoor | Sécheron | Genève

Dear Professor Tyndall

I hasten to thank you for your letter[1] and for the amiable manner in which you have conducted my little business.[2] It would be impossible to be presented for election to the Alpine Club under more flattering or more favourable auspices than proposed by Mr. John Ball, and I have written to Baron Rothschild, requesting him to comply with Mr. Ball's condition of giving him a statement of the extent of his mountaineering prowess. I am greatly amused at the discredit which Mr. Ball throws on my accuracy in general, and the accuracy in particular of my memory.[3] I do not expect that he will find my statement exaggerated, and to the one Pass I was not quite satisfied about (the Oberaarjoch,) I added the reservation, I <u>believe</u>. The others are all genuine.

The Jungfrau, Baron Rothschild escaladed after having talked with you on the subject this summer. He went up by moonlight, arriving on the summit just in time to see the sunrise, but he told me that your words in comparing it with the Bernina: "it is not more difficult, but you feel more helpless" constantly rung in his ear as he was ascending & descending the more precipitous passages.

I have the pleasure of a slight acquaintance with Mr. Ball, and altho' he may again fancy he has reason to doubt my accuracy, perhaps the mention of the detail may convince him that my memory is not altogether defective. I met him & was introduced to him at the Maison Italian a very few years ago—he had just crossed the Mont Cenis[4] on foot, notwithstanding that the Fell Railway[5] was open & working—I could even repeat the conversation we had together—but this is unnecessary. I hope you will chaff him, on my account next time you see him, in anticipation of my treatment of him when next I have the pleasure to meet him. We are having a very indifferent season as regards weather here, in fact I never remember so variable an autumn in point of weather as this one has been.

Let me thank you once more for your kindness in troubling yourself on my friend's account, and I remain yours | very sinc[ere]ly | Emily Peel

RI MS JT/1/P/56

1. *your letter*: letter 3531.
2. *my little business:* see letter 3525.
3. *my memory*: see letter 3531.
4. *Mont Cenis*: a pass in the French Alps, about six miles from the Italian border.
5. *Fell Railway*: a system of railway developed for steep inclines and declines. It used a central rail, in addition to the normal two-rail system, to provide more traction for movement and braking. The Fell system at Mont Cenis was in operation from 1868 to 1871, when a tunnel was built. See letter 3506, n. 3.

To Cecil Trevor 11 October [1871][1] 3534

Royal Institution of Great Britain | 11th Oct.

My dear Trevor.

I shall be always glad to correct my own proofs.

Your acceptance of my disposition proves that I made myself clear, which is a constant aim with me.

Yours faithfully | John Tyndall

National Archives MT 10/131 H3953

1. *[1871]*: dated to 1871 because it likely refers to the report in letter 3521.

From John Murray 11 October [1871][1] 3535

50 ALBEMARLE ST | Oct 11

Dear Prof[esso]r Tyndall

It needs not a word more from you respecting the Landlord of the Belle Alps[2] to insure the immediate cancel by me of the complaint against his house. But in justice to me your friends sh[oul][d] all be made aware that this is not a question of "the delicacies of the season" w[hi]ch no one in his senses can expect at such a place—but sheer absence of management & dirt—both of w[hi]ch can & ought to be corrected—& both of w[hi]ch have their origin doubtless as you & your friends know from the habits of the Landlord.

If you will induce some influential person (none less than yourself) to write & warn the man you may save him from ruin.

My dear sir | yours very sincerely | John Murray

I return your enclosures[3]

RI MS JT/1/M/154

1. *[1871]*: dated by reference to letter 3532, to which it responds.
2. *It needs not a word . . . Belle Alps*: see letter 3532.
3. *enclosures*: missing.

To Elizabeth Dawson Steuart 12 October 1871 3536

12th Oct 1871.

My dear Mrs. Steuart

I fear I have estranged some of my friends by my long silence. When I returned from Switzerland I found people lying in wait for me; new observations to be made, long and complicated contentions to be settled, in short a vast amount of laborious work to be accomplished. I went to Ireland but was forced to return when my work there was completed: I was not able to run down to Gorey to see Emma. There is a species of work rather new to me, and of a most disagreeable kind which I have been engaged upon early and late for the last three weeks: It is the decision of the claims of two rival inventors of lighthouses lamps;[1] and the enquiry has been preceded by a correspondence between the Board of Trade, the Trinity House, and the Commissioners of Northern Lighthouses,[2] which gives the work a special significance and renders it more searching and varied than it otherwise need be. A breathing time in the enquiry has just occurred, and I employ a portion of it to write to you.

I went this year to Pontresina in the Engadin. It is becoming as fashionable as Chamonix; and far more salubrious. It is about 6000 feet above the sea, and on the coachroad over the Bernina Pass—thus readily accessible. The Baths of St Moritz moreover are at hand, where of late years it has become the fashion for English physicians to send their patients, so that during the summer the nook which used to be so lovely years ago has become flooded with visitors. I had some good exercise with pleasant friends; and some harder exercise in which but few friends could join. There was one young man lost while I was there:[3] he had rashly gone up a mountain alone, and disappeared in ascending: I spent four or five days in the vain endeavour to recover his body.

There is a woman in Leighlin Bridge who writes to me from time to time, and is somewhat troublesome, but I suppose she is very miserable—Margaret Washington.[4] When I next write to you I will ask you to take charge of a trifle for her. I should like also if it could be managed to put W[illia]m. Tyndall[5] in a somewhat better position than his present one. I do not like his vocation, and if you could say he is a sober man I would endeavour to help him to something better. John Tyndall[6] is again troublesome—He sent to me a poor man some time ago of whom I know nothing, still he was in such low condition that I was forced to help him. This is somewhat vexatious—Now I have filled my sheet, and must again turn to my unpleasant work. Yours ever | John Tyndall

RI MS JT/1/T/1434
RI MS JT/1/TYP/10/3406–6a

1. *two rival inventors of lighthouse lamps*: Henry Harrison Doty and James Douglass. See letter 3522, n. 3.
2. *a correspondence*: see letter 3555, n. 3.
3. *lost while I was there*: Frederick Bodmer; see, for example, letters 3508 and 3512.
4. *Margaret Washington*: not further identified, but possibly related to Catherine Washington. See letter 3379.
5. *W[illia]m. Tyndall*: see letter 3379, n. 8.
6. *John Tyndall*: see letter 3507, n. 10.

To Henry Bence Jones 16 October 1871 3537

Royal Institution. | 16th, Oct, 1871.

My dear Friend

You express a hope that my experiments are going on well.[1] To my bitter regret I have been compelled to abandon them for the most uncomfortable inquiry I have ever entered into. Since I saw you I have been occupied in testing the claims of two rival inventors,[2] and I must say that if I had any notion of the consequent bother I should simply have sent in my resignation to the Trinity House. Do what you will—set your face like a flint with the determination to do even handed justice to both parties—you cannot avoid complaints and protests, and all manner of grumblings. It is simply disgusting, & I only wish my hands washed of it.

I will speak to Leibreich[3] and will also convey your wishes to Fergusson[4] probably to night.

These experiments have often kept me from 9 A.M. till 7. P.M. in a diabolical atmosphere. Sometimes even as late as 11.P.M. I only wonder that I have borne up so well against them. But I constantly have to think of the loss they have caused me by withdrawing me from dignified scientific work.

I had the foreman of the Firemen here today testing the respirator.[5] He pronounces it now perfect or nearly so—so I hope it will do good.

London is filling and I shall be right glad to see you here. The windows I think will need painting and if so the sooner it is done the better. They look rather shabby in contrast with the clean new coat. When you come we can decide the paint.[6]

Kind regards to all | Yours ever | John Tyndall

RI MS JT/1/TYP/3/803
LT Typescript Only

---

1. *You express . . . well*: This may have to do with his work with the fireman's respirator and his other work having to do with dust. See n. 5.
2. *two rival inventors*: Henry Harrison Doty and James Douglass. See letter 3522, n. 3.
3. *Liebreich*: Dr. Richard Liebreich (1830–1917), a German ophthalmologist, painter and sculptor; he moved to London in 1870 to practice medicine and became ophthalmic surgeon to St. Thomas's Hospital.
4. *Fergusson*: William Fergusson (1808–77), a Scottish surgeon. He was appointed professor of surgery at King's College in 1840 and published his *System of Practical Surgery* in 1842. In 1849 he became surgeon-in-ordinary to Albert, Prince Consort, and in 1867 sergeant-surgeon to Queen Victoria. He was one of the leading operating surgeons in London. He was elected FRS in 1848.

5. *the respirator*: Tyndall invented a fireman's respirator based on his theories of dust and disease. See letter 3411, n. 7.
6. *decide the paint*: see letter 3518, n. 4.

From Louis Pasteur 19 October 1871 3538

Paris, le 19 oct. 1871.

Mon cher confrère,

Vous aurez reçu, je l'espère, le paquet que je vous ai adressé à Londres, avec la lettre qui s'y trouvait incluse. Je mets à la poste de ce jour un travail d'un de mes élèves, extrait des Annales Scientifiques de l'Ecole Normale, par M. Duclaux. Il sera très utile que vous lisiez les pages 23, 24, 25. Vous y verrez que M. Duclaux a fait germer sous ses yeux des spores recueillies dans l'air atmosphérique ordinaire.

Votre bien affectionné | L. Pasteur

Le travail de M. Duclaux a été fait dans mon laboratoire. Il est très exact.

Mr Tyndall, à Londres.

Paris, 19 October 1871

My dear colleague,

You have received, I hope, the parcel that I sent to you in London, with the letter that was enclosed.[1] I am sending by today's mail a work,[2] extract from the Annales Scientifiques de l'Ecole Normale,[3] by one of my students, Mr. Duclaux.[4] It will be very useful for you to read pages 23, 24, 25. You will see there that Mr. Duclaux has germinated, before his very eyes, the spores collected in ordinary atmospheric air.

Affectionately yours | L. Pasteur

The work of Mr. Duclaux was done in my laboratory. It is very exact.

Mr. Tyndall, in London.

RI MS JT/1/P/5

---

1. *letter that was enclosed*: letter 3527.
2. *a work*: enclosure missing.
3. *Annales Scientifiques de l'Ecole Normale*: *Annales Scientifiques de l'École Normale Supérieure* is a journal founded in 1864 by Pasteur. The work itself cannot be further identified.
4. *Mr. Duclaux*: Émile Duclaux (1840–1904), a French chemist and bacteriologist. He was educated in Paris, and attended the Ecole Normale Supérieure where he studied with Pasteur. He worked with Pasteur in 1862 in his study on the disease of silkworms. In 1865, he moved to Tours, where he had been appointed professor of chemistry. In 1878, he was appointed professor of physics and meteorology at the Agronomic Institute, where he remained until his death; he took over the Pasteur Institute when Pasteur died.

From Edward Frankland 25 October 1871 3539

College of Chemistry | 25th October 1871.

Dear Tyndall

I am much gratified to learn that the claims of Mayer to the Copley Medal[1] are to be brought before the Council of the Royal Society this year. I have for many years wished and hoped that this highest honour which British science has to confer would be bestowed upon one so well worthy of it.[2]

I have read with care the chief of Mayer's publications and I do not remember any series of memoirs which so deeply impressed me with the profound knowledge and the clearness and accuracy of reasoning of their author,—and these papers were written at a time when anything like a clear conception of the conservation of force and its demonstration was scarcely beginning to dawn in the mind of any other man.[3]

When I consider the profound influence which this conception and its demonstration have exercised upon the subsequent development of experimental science, I cannot but regard Mayer as one of the very greatest of living philosophers and I certainly know of no one whose labours are better entitled to the recognition afforded by the Copley Medal.

Believe me | yours sincerely | E. Frankland.

RI MS JT/1/F/52

1. *Copley Medal*: the most prestigious award given by the RS.
2. *Mayer*: Tyndall, a member of the Council of the RS, proposed Mayer for the Copley Medal; he may have solicited the letter from Frankland in support of his efforts. Those efforts were successful and Mayer received the Copley Medal in 1871. See letters 3541, 3544, 3546 and 3625, as well as *Ascent of John Tyndall*, p. 278.
3. *these papers . . . any other man*: The science of energy saw numerous controversies over questions of priority in the nineteenth century. For Tyndall's involvement in these, see *Ascent of John Tyndall*, pp. 170–87.

From William Valentin 25 October 1871 3540

Sub-Enclosure 1, in No. 29.[1]

Royal College of Chemistry, | London, 25th October 1871.

Dear Sir,

I BEG to report to you on the results obtained by a chemical examination of a number of oils, used in the competitive trials conducted at this laboratory with two new 1st order lighthouse lamps.[2]

The following oils were submitted to me:—

1. Goddard's best petroleum oil,[3] marked 5a.

2. Young's paraffin oil,[4] No. 1, supplied by Captain Doty.

3. Ditto—ditto—No. 2-- ditto.

4. Young's best paraffin oil, marked F, procured from H. Povey's, oil and colourman, Market-place, Oxford Street.[5]

5. Ditto, sample marked G, procured from same source by Dr. Tyndall. | Crystal Oil.[6] | No. 2.

6. Young's paraffin oil, marked A, used during the experiments at the Trinity House[7] during the earlier part of the year.

7. Ditto, sample marked B.

8. Ditto, sample marked C, bought as Young's best oil on 30th September, and employed at the College for experiments.

9. Sample of Young's paraffin oil, marked D, bought at Messrs. Pinnell, Bros., wholesale oilmen, at 33, Minories, City.[8]

I determined the specific gravity[9] and the flashing point[10] of the several oils, and likewise submitted sample 5a, Young's oil Nos. 1 and 2, and Young's oil marked G, to fractional distillation.[11]

The results are subjoined in a tabular form:—

TABLE showing Specific Gravity, Flashing Point, and approximate Chemical Composition of

Petroleum and Paraffin Oils.

NAME of OIL.	Specific Gravity at 60° F.	Flashing Point.	1st Distillate upwards of 212° F.	2nd Distillate from 212–300° F.	3rd Distillate from 300° to 392° F.	4th Distillate from 392° to 482° F.	5th Distillate from 482° to 572° F.	6th Distillate above 572° F.	REMARKS.
			P.C.	P.C.	P.C.	P.C.	P.C.	P.C.	
Petroleum, 5a - -	·7869	126°–130° F.	0·6	21·7	60	18	- -	- -	Blackened slightly on distillation: no heavy oils present.
Young's Paraffin, No. 1	·8136	134°–138° F.	0·8	19·6	53·2	21·7	4·9	- -	Ditto - - ditto.
Ditto - - No. 2	·818	128°–134° F.	0·8	28·2	40·9	29·2	- -	- -	Ditto - - ditto.
Ditto - - F. -	·8122	134°–138° F.	—	—	—	—	—	—	—
Ditto - - G. -	·8267	150°–152° F.	0·2	0·8	13·2	26·0	40·8	19·0	Last portion of distillate semi-solid; blackened slightly; charred wicks.
Ditto - - A. -	·8141	146°–148° F.	- -	- -	- -	- -	- -	- -	Comp. Report of May 6th.
Ditto - - B. -	·8171	144° F.							
Ditto - - C. -	·8100	134°–136° F.							
Spurious sample of Young's Paraffin Oil, marked D.	·7966	84°–86° F.	- -	- -	- -	- -	- -	- -	Rejected as soon as the flashing point had been determined.
Petroleum, marked 4s, examined on former occasion.	·816	140° F.	Nil.	Nil.	Nil.	76·8		14·0 9·2	The last portion of this oil semi-solid; extinguished lights; most destructive to wicks.

The preceding results of the chemical analysis, when viewed in connection with the fact that certain oils charred the wicks of both lamps, that their burning was marked by a gradual decrease in the illuminating power, and they ultimately extinguished the lights altogether, demand most serious attention when balancing the advantages which mineral oils[12] possess over colza oil.[13]

It is evident that the oils which burnt best in the two lamps, viz., petroleum (5a), Young's paraffin (Nos. 1 and 2), and the oil marked F, were carefully distilled samples of mineral oils, obtained by rejecting both the lighter and the heavier portions of the natural oils, that the sample of Young's paraffin oil, marked G, contains more of the heavier oils than samples 1 and 2, and therefore affected the wicks unfavourably; also, that the sample of petroleum (4a) contained less of the lighter oils and somewhat more of the heaviest oils. This becomes apparent, likewise, by glancing at the fractionated portions of the oils which accompany this report. The different constitution of the several samples will readily become intelligible by a glance at the subjoined diagram, in which the shaded portion represents the range of temperature between which the oils distilled.

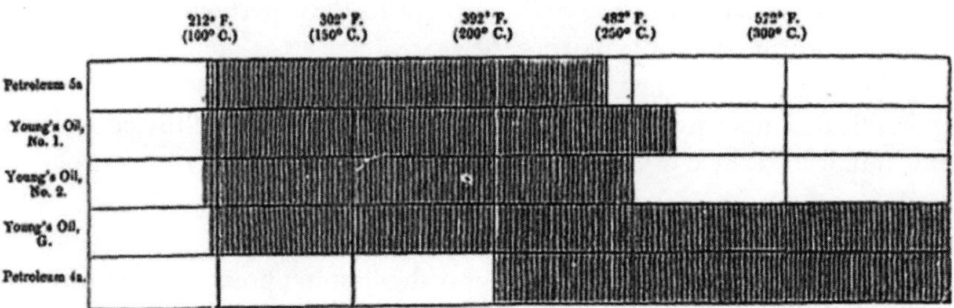

The observations made in my former report of May 6th[14] are thus fully borne out by the experience gained during the present trials, viz.:—

1. That the lamps as now constructed, and my remarks apply equally to the Doty and to the Trinity House lamp,[15] are only capable of burning one quality of oil with full advantage.

2. That it is unsafe to judge of the suitability of an oil by merely ascertaining its specific gravity and flashing point. (Comp. paraffin No. 2 and petroleum 4a.)

3. That oils of greater specific gravity, such as paraffin No. 2, sp. gr. 818, may even contain a greater proportion of comparatively lighter oils than oils of less specific gravity such as No. 1 paraffin.

4. That it would be unsafe to send out any oil without first submitting it to chemical analysis. That such analysis should not be confined to merely ascertaining the flashing point and specific gravity, but should include a fractional distillation as well.

N.B.—Although I am aware of the fact that mineral oils undergo a certain amount of change when distilled fractionally, that lighter oils "crack" to some extent into heavier oils and gaseous products, I would still recommend this mode of examination as perfectly adequate if conducted intelligently.

5. That no lamp should be sent out before it has been carefully tried with the oil,—no matter whether petroleum or paraffin, for I cannot discern any superiority of the one over the other,—which it is intended to burn, and which has been found by practical experience to be most suitable to the lamp, and to give the best results.

6. That further efforts should be made with a view of constructing 1st order lighthouse lamps capable of burning heavier oils, as such oils are known to produce a considerably increased illuminating power.

I have, &c. | (signed) *Wm. Valentin* | Principal Demonstrator of Practical Chemistry.

Professor J. Tyndall.

'Lighthouses (mineral oils). Copy of further correspondence relative to proposals to substitute mineral oils for colza oil in lighthouses (in continuation of parliamentary paper, no. 2, of session 1872)', HC 2 67,[16] Parliamentary Papers (1872), pp. 45–6
Typed Transcript Only

---

1. *Sub-Enclosure 1, in No. 29*: Valentin's report was enclosed in a report by Tyndall (letter 3555), which was in turn enclosed in a letter from Trinity House to the Board of Trade. See letter 3522, n. 1. These reports were part of an ongoing investigation into different designs for lighthouse lamps; see letter 3522, n. 3.
2. *1st order lighthouse lamps*: see letter 3422, n. 3.
3. *Goddard's best petroleum oil*: petroleum sold by S. A. Goddard and Co.
4. *Young's paraffin*: see letter 3422, n. 6.
5. *H. Povey's . . . Oxford Street*: Oxford Street is in Central London. We have not identified H. Povey's.
6. *Crystal Oil*: a refined petroleum oil. See F. A. Abel, 'On Accidental Explosions', *Roy. Inst. Proc.*, 7 (1873–75), pp. 390–429, on p. 403.
7. *Trinity House*: see letter 3400, n. 1.
8. *Pinnell . . . Minories, City*: Minories is a street in the City of London, a district of London. We have not identified Pinnell Bros.
9. *specific gravity*: 'the degree of relative heaviness characteristic of any kind or portion of matter; commonly expressed by the ratio of the weight of a given volume to that of an equal volume of some substance taken as a standard (viz. usually water for liquids and solids, and air for gases)' (*OED*).
10. *flashing-point*: 'the temperature at which the vapour given off from an oil or hydrocarbon will "flash" or ignite' (*OED*).
11. *fractional distillation*: 'designating a process in which different fractions of a mixture are separated in consequence of their differing physical properties' (*OED*).
12. *mineral oils*: derived from hydrocarbons.
13. *colza oil*: see letter 3422, n. 5.
14. *my report of May 6th*: see letter 3422.
15. *to the Doty and to the Trinity House lamp*: competing designs for lighthouse lamps; see letter 3522, n. 3.
16. *2 67*: At the time of writing, the ProQuest U.K. Parliamentary Papers database misnumbered this paper as 2 264.

To Henry Bence Jones 27 October [1871][1] 3541

Royal Institution | 27th, Oct.

My dear Friend

I wrote to Siemens[2] last night and hope for a favourable reply. Had your note[3] reached me a little earlier I might have spoken to him for we were both at the R[oyal].S[ociety]. Council meeting.

I have seen Liebreich[4] this morning. He will be ready whenever you call upon him.

His brother of chloral fame[5] is coming over here at Easter, and is willing to give an evening of his own subject. I do not know his skill in English. He would I believe read his lecture. Perhaps the intrinsic interest of the thing, as in the case of Max Müller[6] would carry the audience over this difficulty.

I shall be right glad to see you when you come. For my part my autumn has been so grievously broken that I have had thoughts of resigning my post at the Trinity House.[7]

We have a great conflict in the council about the Copley Medal.[8] It pains me to the quick to notice the attitude assumed by William Thomson towards Mayer. I hoped to live and die at peace with him, but there are now rumblings of war in the distance.

Yours ever | John Tyndall

Liebreich's brother is limited as to time, so that if you thought of securing him it must be within a space of 3 or 4 weeks at Easter.

The house looks exceedingly well.[9]

RI MS JT/1/TYP/3/802
LT Typescript Only

---

1. *[1871]*: dated by reference to the discussion of Mayer's candidacy for the Copley Medal. See letters 3539, 3544, 3546 and 3625.
2. *Siemens*: probably Charles William Siemens (1823–83), an electrical engineer and metallurgist; he and his brother Frederick Siemens invented the regenerative furnace. Siemens was born at Lenthe, Hanover and became a British citizen in 1859; he was elected FRS in 1862.
3. *your note*: letter missing.
4. *Liebreich*: see letter 3537, n. 3.
5. *His brother of chloral fame*: Oscar Liebreich (1839–1908), professor of pharmacology at the University of Berlin; Liebreich discovered that chloral hydrate could be used as an anaesthetic.
6. *Max Müller*: Friedrich Max Müller (1823–1900), a philologist and professor at Oxford,

who did significant early work in Vedic studies, comparative philology, comparative mythology and comparative religion. He also wrote and lectured for general audiences, including lecture courses at the RI on the 'science of language', delivered in 1861 and 1863. Müller was born in Dessau, Germany and emigrated to England in 1846.

7. *my autumn . . . House*: a reference to Tyndall's work on competing designs for lighthouse lamps. See letter 3522, n. 3.
8. *great conflict . . . Copley Medal*: see n. 1.
9. *The house . . . well*: see letters 3427, 3518 and 3537.

To Joseph Henry 31 October 1871 3542

London 31st October 1871

My dear Professor Henry

On my return from the Alps[1] I found your most friendly letter[2] here.

Since my return I have been entangled in an unexpectedly difficult and laborious enquiry.[3] Otherwise I should have given earlier expression to my deep sense of your kindness and of the uniform good-will of the American people.

I hoped, moreover, to be able to say to you, that under your wise guidance and advice I should endeavour to prepare some lectures expressly for America. But this I now clearly see will be out of my power

I am willing, however, to do what I can, and, if you desire it, to give at the Smithsonian Institution a short course of such lectures as I am in the habit of giving at the Royal Institution. You know the character of our lectures, and you can best decide whether they would be likely to serve any useful purpose in America.

With regard to lecturing at other places I should in the main be guided by the friendly counsel of yourself and Mr. Lesley,[4] and of my other scientific friends in the United States.

And with regard to the materialistic questions which cleave to lecturing I would say that my state of bachelorhood, unblessed as it otherwise may be, has this advantage:—it keeps me independent. It enables me to say that while American kindness has a strength sufficient draw me across the Atlantic, American money possesses no such power.

In the question of payment, therefore, you need not for a moment dwell:—it can never arise between us.

Believe me my dear Friend | Ever faithfully yours | John Tyndall

To Professor Joseph Henry

RI MS JT/1/T/509

1. *my return from the Alps*: on 3 September 1871.
2. *your most friendly letter here*: see letter 3510.
3. *an unexpectedly difficult and laborious enquiry*: into competing designs for lighthouse lamps. See letter 3522, n. 3.
4. *the friendly counsel of yourself and Mr. Lesley*: Tyndall wrote to Lesley on the same day; see letter 3543. There is no direct reply from Henry in this volume, but for more letters on this topic, see 3504, 3507, 3519, 3558, 3589, 3594, 3639, 3669, 3670, 3684, 3685, 3702 and 3703.

To Peter Lesley 31 October 1871 3543

London 31st October 1871.

My dear Mr. Lesley

In the autumn of 1872 we purpose making considerable alterations in our laboratories,[1] and while this work is going on I shall be prevented from pursuing the researches which usually occupy me during the Autumn months.

I purpose making use of my temporary freedom to visit the United States, and it would be satisfying to me indeed if, when there, I could in any adequate way respond to the invitation with which you have so highly honoured me.[2]

Was the necessary time at my disposal it would be a pride and a pleasure to me to meet the wish so kindly expressed by so many eminent persons, by preparing a special course of lectures for the United States. But the time is not at my disposal, and I falter through the consciousness that what I am able to offer falls far short of what you have a right to expect.

Your visit to England[3] will have made you acquainted with the character of the lectures at the Royal Institution. They are statements as plain and thorough as I can make them of the laws and phenomena of physical sciences, but they are extempore, and to literary finish they have no claim. They are usually illustrated by experiments, which, however, are kept entirely subordinate to the time of thought and exposition. This would certainly render my lectures tame to audiences accustomed, as I believe American audiences are, to large and brilliant effects.

I have already written to Professor Henry[4] expressing my willingness, if he desire it, to give a short course of such lectures at the Smithsonian Institution. As to lecturing elsewhere I should place myself entirely in his and your hands. If you think the cause of science or the public good can be promoted by such lectures in the United States I shall be willing to undertake to the small extent of my ability what you and Professor Henry advise. I wish you to take my words strictly and literally, and to allow no other consideration to influence you in this matter than the cause of Science and the public good. The work will be to me a labour of love.

Yours ever faithfully | John Tyndall
To Professor Lesley | &c. &c. &c.

RI MS JT/1/T/1037

1. *alterations in our laboratories*: see letter 3427, n. 3.
2. *the invitation*: see letters 3504 and 3510.
3. *Your visit to England*: Lesley made multiple visits to England; it is not clear to which this refers. See M. L. Ames (ed), *Life and Letters of Peter and Susan Lesley*, 2 vols (New York and London: G. P. Putnam's Sons, 1909).
4. *written to Professor Henry*: see letter 3542. Any response from Lesley to this letter is missing, but for more letters on the topic, see 3507, 3519, 3558, 3589, 3594, 3639, 3669, 3670, 3684, 3685, 3702 and 3703.

To Francis Galton[1] 1 November [1871][2] 3544

1st Nov.

My dear Galton,

If you want further data to convince you of the merit of D^r^. Mayer I can give them to you.

Even Sharpey, who proposes Claude Bernard,[3] says that he will take a first ballot, and if in a minority he will vote for Mayer.

All of us are ready to go heart & soul in for Helmholtz when his turn comes. I urge for Mayer what Sir W[illia]^m^ Thomson urged for Joule last year[4]—greatness of achievement & seniority of merit.

Yours ever | John Tyndall

GALTON/3/2/1/192 UCL Library Services, Special Collections

1. *Francis Galton*: Francis Galton (1822–1911), a biostatistician, human geneticist and eugenicist. His primary scientific work focused on heredity, which he used statistics to investigate. Galton's interest in the subject was animated by the question of whether mental traits could be inherited, and was probably in part inspired by his cousin, Charles Darwin; this research led Galton to coin the word 'eugenics' in 1883, and to argue that human breeding needed to be more carefully attended to. Galton was elected FRS in 1856 and he received the RS's Copley Medal in 1910; he was knighted in 1909. We know that Tyndall's correspondent in this letter is Francis rather than Douglas Galton because the letter comes from Francis Galton's archives.
2. *[1871]*: dated by reference to the controversy over Mayer and the Copley Medal. See letters 3539, 3541, 3546 and 3625.
3. *Claude Bernard*: Claude Bernard (1813–78), a French physiologist notable for his work

on the digestive and nervous systems, as well for pioneering methods of experimentation on animals. Bernard was made professor at the Sorbonne in 1854, and was elected to the Académie des Sciences the same year. In 1868 he moved to the Muséum d'Histoire Naturelle. He received the Légion d'Honneur in 1867 and became a member of the Académie Française in 1869. Bernard received the RS's Copley Medal in 1876.

4. *Joule last year*: Joule received the RS's Copley Medal in 1870.

To Hermann Helmholtz 2 November 1871 3545

Royal Institution | 2nd. November 1871.

My dear Helmholtz,

Will you permit me to introduce to you Dr Youmans, who is connected with the great publishing house of Appleton and Co of New York.

He has a project on foot which promises very important results[1]—nothing less, in fact, than a virtual international Copyright with America.

Huxley, Herbert Spencer, Bain,[2] Charles Darwin—all take a great interest in this project, and hope much from it. It is natural that Dr Youmans should seek to associate with it names so great as yours, and as I have entire confidence in Dr Youmans I beg to recommend him to your attention.

Yours ever faithfully | John Tyndall.

RI MS JT/1/T/488
LT Typescript Only

1. *a project . . . results*: In 1871, Youmans began an 'International Scientific Series', which aimed to produce short books for a general audience written by eminent scientific practitioners and which addressed problems of copyright; see letter 3484, n. 1. Helmholtz, busy with other matters, declined to become involved with the German-language series (D. Cahan, *Helmholtz: A Life in Science* (Chicago: University of Chicago Press, 2018), p. 511).
2. *Bain*: Alexander Bain (1818–1903), a Scottish psychologist; from 1860 he held the chair of logic at the University of Aberdeen. Seen as one of the founders of the discipline of psychology, Bain is considered notable for having shifted philosophical questions to an empirical framework.

To Julius Mayer 2 November 1871 3546

Royal Institution London | 2nd. November 1871

My dear Friend

It is one of the greatest pleasures of my life to announce to you that this day the Council of the Royal Society has awarded you the highest honour which it is in the power of the Society to bestow—namely the Copley Medal.[1]

Yours ever faithfully | John Tyndall

RI MS JT/1/T/1067
LT Typescript Only

1. *the Council . . . Copley Medal*: see letters 3539, 3541 and 3544 for discussion of how to award this medal.

From Mary Egerton 4 November [1871][1] 3547

Mountfield Court, | Robertsbridge, | Hawkhurst. | Nov[embe]r 4th

Dear Mr. Tyndall,

It was good of you to write,[2] for I am sure your object was to tell me you had not remained under the impression that I was quite such an idiot as that absurd begging letter story[3] made me out! I cannot say my friend showed much cleverness in the affair! Such a transparent piece of shows as his broken English, I never read!—May[4] is very angry at Mr Joddrell's[5] "putting a spoke in the wheels" as to your riding lessons—she declares you promised her a front place at the school to see you take the hurdles! However I sh[oul]d think the 'light of Nature', & a little more practice will give you as much skill as you are likely to require, unless you mean to *[1–2 words illeg]* a stud of horses, & a "box"[6] at Melton![7] It would be a more healthy & not much more unsatisfactory mode of spending time than that you are condemned to at present! I confess I grudge every hour that you have to descend from the higher regions of philosophic thought, to those of mere practical applications; but it is far worse when the matter is mixed up with the quarrels & jealousies that rival patentees imply![8]

When are you coming to walk in "a lovely park" in Sussex?[9] I assure you our little paddock did look lovely today. The autumn tints are positively dazzling. The brown and oranges of the bush combining into a sort of scarlet & bringing out in contrast the blue green of the Firs, and the hollies, with more berries that I ever saw, sparkling in the side gleam of the sun, setting under a heavy cloud, which cast deep blue shadows over the hollows. But I must end—I fancied (perhaps wrongly) from something in your letter, that you thought I might expect you to write. Never, never, think that;—of course it is a pleasure to me to hear from you, but I know quite well that I cannot be a friend to you in the same way that you are to me—a mutual friendship complete both in the inner & outer life can only exist between souls 'equal with diversity'[10] (like you and Mr Hirst for instance). But my mind is only a load of faint reflection of yours, so that I have nothing to give in return for all I gain from you, but sympathy, which is of no value to you who have plenty, but which I cannot help, because I care intensely for all the same things you do, from a fine day to the deep problems of physics & metaphysics, however little I may be able to understand them!

What a valuable possession the bust of that grand old Carlyle[11] will be to you. It was striking to see the obituary notice of Sir Rod[eric]k. Murchison[12] & poor old Mr Babbage[13] side by side in the paper the other day. How rapidly his contemporaries are following dear old Sir J. Herschel. Sir Henry Holland was over here the other day looking wonderfully fresh after his second visit to Iceland[14] at an interval of 60 years! Now I am afraid I have "punished" you, as you call it! But could I do otherwise than answer your letter? & could I answer it & not allude to its contents? I think I will send this to the Club, and then you will not be so cross as if it came in the middle of your morning business.

Y[ou]rs ever most truly | Mary F. Egerton

I wonder if y[ou]r. Manchester visit[15] was business or pleasure.

RI MS JT/1/E/57

---

1. *[1871]*: year established by the references to the deaths of Murchison and Babbage in October 1871.
2. *to write*: letter missing.
3. *absurd begging letter story*: The context here, and a possible letter, have not been identified.
4. *May*: see letter 3497, n. 11.
5. *Mr. Joddrell*: possibly Thomas Jodrell Phillips Jodrell (1807–89), a barrister, philanthropist and patron of science.
6. *box*: 'a small country-house; a residence for temporary use while following a particular sport, as a *hunting-box*' (*OED*).
7. *Melton*: Melton Mowbray, a small town in Leicestershire, in central England, was a centre for fox-hunting in the nineteenth century; many wealthy people had hunting boxes here.
8. *I grudge . . . imply*: a reference to Tyndall's work testing rival designs for lighthouse lamps; see letter 3522, n. 3.
9. *Sussex*: county in south-east England where the Egerton family home was located.
10. *equal with diversity*: reference unidentified.
11. *the bust of that grand old Carlyle*: a bust of Thomas Carlyle, sent to Tyndall by Carlyle himself. See letter 3550.
12. *Sir Rod[eric]k Murchison*: Roderick Impey Murchison (1792–1871), an English geologist who advocated catastrophist views and who, as director-general of the British Geological Survey, did much to shape Britain's overseas research and to tie the concerns of science

to those of empire. Murchison was made FRS in 1826, knighted in 1846, made Knight Commander of the Bath (KCB) in 1863 and baronet in 1866. He died in October 1871.

13. *M*[r]. *Babbage*: Charles Babbage (1791–1871), an English mathematician and engineer; Babbage invented a difference engine and an analytical engine that anticipated modern computers. He was made FRS in 1816. Babbage held the Lucasian chair of mathematics at Cambridge from 1828 until 1839. He died in October 1871.
14. *second visit to Iceland*: see letter 3518, n. 7.
15. *y[ou]*[r] *Manchester visit*: not identified.

From Emily Peel — 8 November 1871 — 3548

Whitehall | Nov[embe]r 8/71

Dear Professor Tyndall,

Could we persuade you to come to us at Drayton on Monday the 4th of Dec[embe][r]. or any other day that week to spend a few days with us.[1] I consider that you almost owe this little sacrifice to me; considering the shabby manner in which you have treated our poor little Geneva this year.[2]

Yours very sincerely, & with Sir Robert's best regards, Emily Peel

RI MS JT/1/P/57

1. *Could we . . . of Dec[embe]*[r]: Tyndall responded to this letter in 3549.
2. *our poor little Geneva this year*: see letter 3525.

To Emily Peel — 10 November 1871 — 3549

Royal Institution. | 10th Nov. /71

Dear Lady Emily,

If I can sufficiently lighten the heavy weight of work[1] now pressing upon me it will give me genuine pleasure to join you on the 4th of December. I would jump at your invitation[2] were I less entangled than I am.

I wonder did I send you a somewhat unorthodox book of mine, called "Fragments of Science," and if I have not, whether you would accept it from me?

I shall knock Ball off his pins when I next meet him. I would back your accuracy against his any day.[3]

With many kind remembrances to Sir Robert,

Believe me, | Most faithfully yours, | John Tyndall.

RI MS JT/1/TYP/3/964
LT Typescript Only

1. *the heavy weight of work*: Tyndall's investigation into rival designs for lighthouse lamps. See letter 3522, n. 3.
2. *your invitation*: see letters 3548 and 3553.
3. *I shall . . . Ball*: see letters 3531 and 3533.

To Thomas Carlyle 11 November 1871 3550

11th Nov. 1871

My dear Friend

You taught me when I was young to do the work nearest to me.[1] I have been doing so of late, and a very ugly piece of work it has been,[2] redeemed only by the thought that it ought to be done. I has prevented me from doing other and intrinsically better work, and it has also prevented me from doing in proper time and place my duty to you. You have been good enough to send me your bust:[3] on this head I will only say that no copy of any other features on the face of this earth could have the same value to me.

I will come down to see you soon though I should not be much surprised if you closed your door against me.

Yours ever | John Tyndall
Thos. Carlyle Esq[ui]re.

RI MS JT/1/T/150
RI MS JT/1/TYP/1/192

1. *You taught me when I was young to do the work nearest to me*: Carlyle's emphasis on hard work was present in much of his writing and influential throughout Victorian culture. See, for example, T. Carlyle, *Past and Present* (London: Chapman and Hall, 1843), p. 172. His ideas had a particular influence on Tyndall; see I. Hesketh, 'Technologies of the Scientific Self: John Tyndall and His Journal', *Isis*, 110:3 (2019), pp. 460–82.
2. *a very ugly piece of work it has been*: a reference to Tyndall's investigation into competing designs for lighthouse lamps. See letter 3522, n. 3.
3. *You have been good enough to send me your bust*: Any accompanying letter from Carlyle is missing.

From Oliver Lodge[1] 11 November 1871 3551

Hanley, 11. November 1871 | Staffordshire

Sir,

I hope you will excuse my writing to you, but would you allow me to suggest with diffidence that an answer might possibly be obtained to this question from Magnetic Storms.[2]

I believe that frequently an eruption or cataclysm of some sort has been observed in the telescope or on the spectroscope[3] to be taking place upon the solar surface and that the magnetic needles at Kew, Toronto,[4] and indeed all over the world, are found to have been suddenly & simultaneously affected.

It seems as if the distance between what I cannot help regarding as the cause & effect, is here so great as to give a good opportunity of measuring the intervening time. Supposing it granted that the solar disturbance be the cause, as it is the antecedent, of the terrestrial phenomenon; if the effect of the solar disturbance travelled with the same speed as light, the eye at the spectroscope & the magnetic needle at Kew, would indicate at the same instant; but if action at a distance were concerned, the needle would be 8 minutes beforehand, whereas I think I remember that it is somewhat behind.

From observations of this sort, might not the rate of propagation be ascertained.

At the best this could only shew that a something travelled at a certain rate, but then this something has an effect on the magnetic needle, and if not magnetic force, what is it? I suppose it would be premature to call the phenomenon observable on the sun, an electric discharge on a most gigantic scale, though this would almost agree with some of the observations such as Messrs. Carrington[5] & Hodgson's;[6] noticed at page 696 of Herschel's Astronomy.[7]

If the Electricity of the earth be grand what must that of the sun be.

With many apologies for my presumption in troubling you

I remain | Yours most respectfully | O. J. L.[8]

John Tyndall Esq FRS LSD

UCL, Lodge Papers, 109 in MS ADD 89

---

1. *Oliver Lodge*: Oliver Joseph Lodge (1851–1940), an English physicist. At the time of writing Lodge was working for his father, a merchant engaged in pottery manufacture, but his attendance at Tyndall's RI lectures in 1866–7 had inspired an interest in science that led to a distinguished career in physics and election as FRS in 1887. Lodge's work focused on electromagnetism, though his opposition to Einstein's relativity and advocacy of the existence of an ether diminished his standing.

2. *Magnetic Storms*: geomagnetic storm; 'a worldwide major disturbance of the earth's magnetic field caused by the interaction of the earth with high-speed charged particles ejected from the sun following a solar flare' (*OED*).
3. *spectroscope*: 'an instrument specially designed for the production and examination of spectra' (*OED*).
4. *Kew, Toronto*: Kew Observatory, in Richmond, London and the Toronto Magnetic and Meteorological Observatory were established as magnetic observatories by Edward Sabine. See L. Macdonald, *Kew Observatory and the Evolution of Victorian Science, 1840–1910* (Pittsburgh: University of Pittsburgh Press, 2018), pp. 26–8).
5. *Carrington*: Richard Christopher Carrington (1826–75), an amateur astronomer who maintained his own observatory in Reigate, south of London. Carrington, who did significant work on solar flares and sunspots, was elected FRAS in 1851 and FRS in 1860.
6. *Hodgson's*: Richard Hodgson (1804–73), a publisher and amateur astronomer. Carrington and Hodgson witnessed a significant solar flare in 1859.
7. *observations . . . Astronomy*: J. Herschel, *Outlines of Astronomy*, 11th edn (London: Longmans, Green, and Co., 1871), p. 696.
8. *O. J. L.*: Lodge's signature is faint, but Tyndall's response in letter 3552 establishes him as the writer.

To Oliver Lodge[1] 14 November [1871][2] 3552

14th Nov.

Sir,

You open a deep question,[3] which will be solved by and by, but which is not yet solved.

Yours truly | John Tyndall

Oliver J. Lodge Esq[uir]re

UCL, Lodge Papers, 109 in MS ADD 89

1. *Oliver Lodge*: see letter 3551, n. 1.
2. *[1871]*: This letter is a response to letter 3551, which is dated 1871.
3. *You open a deep question*: see letter 3551.

From Emily Peel 14 November 1871 3553

Drayton Manor, | Tamworth. | Nov[embe]r. 14/ 71

Dear Professor Tyndall

I have just received the Volume[1] which you so kindly sent me and I hasten to thank you for it. I shall value it very much, both for its intrinsic merit and as a gift from the author.—

I hope that you will manage to come down here for a couple of days.[2] I should have great pleasure in talking over with you your new prowesses of this year. Any day during the week of the 4th Dec[embe]r. would suit us.[3]

I remain yours | sincerely | Emily Peel

RI MS JT/1/P/58

---

1. *the Volume*: *Fragments of Science*; see letter 3549.
2. *I hope . . . days*: Tyndall did not visit the Peels because he injured his leg on 19 November; see letters 3566 and 3567.
3. *Any day . . . would suit us*: see letters 3548 and 3549.

From Edward Frankland 15 November 1871 3554

Royal College of Chemistry | Nov. 15/71

My dear Tyndall

I am threatened with trouble in connection with your Trinity Board experiments.[1] An Officer from the Board of Works[2] has called upon me & told me that I had no right to allow such experiments to be made at the College without special permission from Mr Ayrton & that I should shortly receive a communication from the Minister of Works[3] on this subject. I will therefore thank you to write me a note[4] describing the nature, importance &c. of the experiments, in order that I may defend myself as far as possible from what I fear will be considered an irregularity in our proceedings.

Yours sincerely | E. Frankland.

RI MS JT/1/F/53

---

1. *Trinity Board experiments*: the investigations Tyndall was pursuing into competing designs for lighthouse lamps. See letter 3522, n. 3, and letter 3555.
2. *Board of Works*: office of the British government concerned with public works.
3. *Minister of Works*: Acton Ayrton.
4. *a note*: Tyndall's response, if any, is missing.

To Robin Allen 16 November 1871 3555

Enclosure in No. 29.[1]

REPORT of Professor *Tyndall,* LLD., F.R.S., &c., to the Honourable Corporation of Trinity House, London, upon comparative Experiments made by him with "Trinity House" and "Doty" Lamps.[2]

Royal Institution, Albemarle-street, | 16 November 1871.

Sir,

I HAVE been recently engaged on comparative experiments with the Trinity House and Doty lamps, and for my own information have carefully read over the whole of the printed correspondence[3] relating to the employment of mineral oils in lighthouses. I think it will render the mastery of this subject more easy to the Elder Brethren and the Board of Trade, if I preface this Report by a brief analysis of the correspondence alluded to:—

1. On the 7th of September 1870, the Doty lamp was placed for trial in the lighthouse of Girdleness,[4] where it continued in action until the 5th of October. The Messrs. Stevenson were so satisfied with the performance of the lamp that they at once recommended it to the favourable consideration of the Commissioners of Northern Lighthouses.

2. Immediately after this, the Messrs. Stevenson, after conferring with Captain Doty, drew up a definite statement of terms, wherein it was agreed that for the use of his patent, Captain Doty should receive a sum equal to one year's saving in all the lighthouses in Scotland, which sum at that time would have amounted to 3,478*l*. 15 *s*. 7 *d*

3. On the 10th of November the Commissioners of Northern Lighthouses recommended that the Board of Trade should accede to the terms arranged between the Messrs. Stevenson and Captain Doty

4. Up to this date no photometric comparison[5] had been made between the colza and the paraffin flames. The price per gallon and the rates of consumption of the two oils were known; but the alleged relative cost of the two lights rested, not on photometric measurements, but solely on the inspection of the lights by the naked eye.

5. The relative atmospheric penetration of the lights also remained to be determined. To decide this the Messrs. Stevenson proposed to introduce paraffin into one of the two dioptric lighthouses[6] at Pentland Skerries,[7] and into one of the two catoptric lighthouses[8] at Pladda.[9] The colza and paraffin flames could thus be observed simultaneously. But it was while these observations were pending, and before they were executed, that the agreement with Captain Doty had been drawn up.

6. It is probable that the Commissioners, in recommending the foregoing agreement for adoption by the Board of Trade, took care not to commit themselves without reserve, to an entire change in their system of lighting, on data acknowledged to be incomplete. I say "incomplete," because, if the data were sufficient, the observations at Pentland Skerries and Pladda, and, indeed, the subsequent researches of Dr. Stevenson Macadam[10] would have been unnecessary. It is therefore to be assumed that the proposed agreement was made subject to the condition that future experiments should justify the opinion first formed of the paraffin light.

7. The recommendation of the Commissioners of Northern Lighthouses was forwarded to the Trinity House by the Board of Trade, and the remarks of the Elder Brethren upon the proposition were requested. The Elder Brethren stated in reply[11] "that the reports received from the M. Reynaud,[12] of the French Lighthouse Board, did not express the same confidence as those of the Messrs. Stevenson, but recognizing only a measure of success in the smaller lamps, did not appear to justify the expense of change in the first-class lights."[13]

8. The Elder Brethren stated further that they had themselves been occupied with experiments on Young's paraffin[14] in small burners, and had already decided to introduce it at the two new leading lights[15] on Great Castle Head;[16] and they proposed, "while establishing this permanently, and with no fear as to its results, to go on with experiments on lamps for first order dioptric lights." They trusted "eventually to produce a first-class burner, which will effect considerable saving in the cost of lighthouses illumination, provided the oil can be safely stored."[17] This communication is dated 18th November 1870.

9. On the 31st December a very able report, drawn up by Dr. Stevenson Macadam, the consulting chemist of the Northern Commissioners, was forwarded to the Messrs. Stevenson.[18] Dr. Macadam had instituted a thorough comparison between paraffin and colza oil for four orders of lights, and bad arrived at the following important conclusions:—

a. For first order lights the cost of light per hour per unit of paraffin is somewhat less than half the cost of colza.

b. For second order lights the cost of light per hour per unit is for paraffin about three-fifths of the cost of colza.

c. For third order lights the cost of light per hour per unit is less than one-third of the cost of colza.

d. For fourth order lights the cost of light per hour per unit is for paraffin about one-fourth of the cost of colza.[19]

10. The superiority in point of cost of the paraffin over the colza, strikingly brought out by these experiments, was proved to be greatest in the case of the smaller lights. From this fact Dr. Macadam drew the important practical inference, "that the full amount of light capable of being evolved from paraffin oil is not obtained in the largest sized paraffin lamps." This loss of light, he considered, might "be due to the escape of some of the paraffin vapours in an unconsumed state, owing to the great draught in the chimney; or it may be due to a want of sufficient air being admitted to the larger paraffin burners, or probably both of these causes may operate on producing the waste, which amounts to full one-seventh of the whole illuminating power."[20]

11. Dr. Stevenson Macadam winds up his report with the following words:—"After a full consideration of all the foregoing experimental

observations, I am of opinion that paraffin oil, consumed in Doty's lamps, alike from readiness in trimming[21] and lighting up, from steadiness of flame and from high photogenic power,[22] possesses decided advantages over colza oil for lighthouse illumination; whilst the quality of the best mineral oils now to be obtained in the market renders the employment of paraffin oil practically safe."[23]

I would, in passing, express the conviction that the matured opinion of Dr. Macadam ought to carry great weight with the Elder Brethren and the Board of Trade. In such portions of his work as have come under my inspection, his ability and his conscientiousness have been alike conspicuous.

12. Dr. Macadam's Report was forwarded to the Commissioners of Northern Lighthouses on the 7th of January 1871. It was accompanied by a letter from the Messrs. Stevenson, in which they are careful to say that "it must be kept in view that these photometric results obtained by viewing the light in a room must not, without further trial, be regarded as representing the relative penetrative power of the different lights as viewed from a distance in varying conditions of the atmosphere. This," they add, "remains to be determined by the observations to be made at Pladda and Pentland Skerries Lighthouses."[24]

13. Thus it would appear that after the important addition of Dr. Macadam's Report to the data, they are considered to be still incomplete. This confirms the surmise already expressed (6) that the agreement with Captain Doty could only be conditional. Otherwise it would have been hasty to commit the Board of Trade to a contract involving weighty pecuniary considerations, while various important points affecting the contract remained undetermined. Why it was deemed advisable to enter into an agreement at all before the experiments were completed does not appear.

14. On the 1st of March the Messrs. Stevenson communicated to the Northern Commissioners the observations made on the Pentland Skerries and Pladda Lights. They were uniformly in favour of the paraffin light.

15. Fortified by the report of Dr. Macadam and by the observations at Pentland Skerries and Pladda, the Commissioners of Northern Lighthouses, in a communication dated 9th of March 1871, reiterate their opinion, "that paraffin oil should be used in Northern Lighthouses, and would now wish to have the deliverance of the Board of Trade on the terms proposed by Captain Doty."[25]

16. Meantime, the experiments contemplated by the Elder Brethren, and referred to in their letter of the 18th of November, were being pushed forward by Mr. Douglass, the Engineer of the Trinity House. He began by testing the ordinary Trinity House burners, large and small, and found that paraffin consumed in the single-wick Argand burner[26] possessed an illuminating power considerably less than colza oil. The exact proportion was as 8·4 units to 13·9

units. With the four-wick burner a similar, though not so great, inferiority was manifested by the paraffin, the proportion here being as 210 to 209.

17. By altering the tips of the wick-case, introducing a perforated disk at the centre of the flame, and modifying the form of the glass chimney, Mr. Douglass succeeded in raising the performance of his Argand burner for paraffin from 8·4 units to 20·6 units, thus causing it to transcend considerably the performance of the colza lamp.

18. To economise the wick in the experiments on paraffin oil, the length of the Argand burner had been augmented.

19. Similar alterations were made in the four-wick burner, viz., "the tips of the wick-cases were closed so as to fit closely to the cotton wicks, and were bevelled in the same manner as the Argand lamp for the purpose of admitting the ascending currents of air freely into the lower part of each ring of flame, and the wick-cases were considerably lengthened for the purpose of economising the consumption of the cotton wick."[27]

20. A very important condition, though not involving any change of the burner itself, was this: to produce the maximum effect, it was necessary that the level of the paraffin oil in the wick-case should be three inches below the tip of the burner, instead of flowing over the tip, as in the colza lamp. This condition, rendered necessary by the greater volatility of the mineral oil, appears to have been discovered independently by several observers.

21. With the four-wick burner, thus altered and arranged, Mr. Douglass at once raised the illuminating power of the paraffin flame from 209 units to 280 units, thus causing it to transcend the colza flame of 269 units. Taking the lower price of paraffin into account, it was shown by these experiments that unit for unit the cost of the colza light is to that of the paraffin:—

For the small burner, as 55 to 15

For the large burner, as 38 to 15

It was also found that when the time of burning was extended to 16 hours, the colza wick suffered more from the long-continued action than the paraffin one.

22. "In conclusion," writes Mr. Douglass, "it may be stated as the result of the lengthened experiments which have been made at this house with paraffin as an illuminant for lighthouses:—

"1st. The cost of the light is 72·7 per cent. less when produced by the Argand, or single-wick lamp, and 60·5 per cent. less when produced in the first order, or four-wick lamp, than colza oil.

"2nd. The lamps burning paraffin will give a light of more uniform illuminating power throughout the night, without trimming, than the lamps burning colza oil.

"3rd. The lamps burning paraffin are more readily ignited; they burn with greater certainty, and require less attention than the lamps burning colza oil.

"4th. The lamps burning paraffin may be arranged for increasing the power of the light when the state of the weather requires it, as is now done with the electric light and coal gas.[28]

"5th. Paraffin can be stowed and used at lighthouses with safety, provided that ordinary care is used."[29]

23. These conclusions are substantially the same as those arrived at by Dr. Macadam, for the 1st and 4th order burners. Indeed, in the case of the four-wick burner, the coincidence of the two observers is surprising. With this burner, Dr. Macadam makes the cost of paraffin and colza per hour per unit of light ·014 *d.* and ·033 *d.* respectively; while with Mr. Douglass the figures are ·014 *d.* and ·033 *d.* respectively. With the single-wick burner the superiority of paraffin over colza is found somewhat greater by Dr. Macadam than by Mr. Douglass. Both experimentors, however, are agreed as to the greater gain in the case of the smaller lamps.

24. The agreement here pointed out furnished a strong guarantee that the recorded observations were substantially correct; nevertheless, the Elder Brethren thought it desirable to check them. This desire was communicated to me, and I, in consequence, invoked the aid of Mr. William Valentin, principal demonstrator in the Royal College of Chemistry, a practised photometric observer, in whose skill and care I had the greatest confidence. His results,[30] forwarded to me on the 6th of May 1871, confirm those previously obtained by Dr. Macadam and Mr. Douglass as to the superior economy and efficiency of the paraffin light. For the first-order lamp the cost of the paraffin was found by Mr. Valentin to be about half that of the colza, while for the single-wick lamp the cost of the paraffin was only one-fourth that of the colza. This conclusion is almost identical with that of Dr. Macadam already referred to *(see* clauses *a* and *d* of paragraph 9).

This consensus of evidence leaves no doubt upon the mind that, as regards cost and illuminating power, the paraffin light really possesses the advantages claimed for it.

25. In one of the tables of his first report, Dr. Macadam expresses in pounds of sperm[31] the total value of the light derived from one gallon of the respective oils, viz:—

	Paraffin	Colza
1st Order 4-wick Lamp - -	22·17 lbs. - -	21·56 lbs.
2nd Order 3-wick Lamp - -	26·46 " - -	19·21 "
3rd Order 2-wick Lamp - -	27·04 " - -	17·49 "
4th Order 1-wick Lamp - -	27·15 " - -	14·67 "

Here it will be observed that the value of the light derivable from a gallon of colza oil steadily augments with the size of the burner, the four-wick burner yielding almost 50 per cent. more light than the single wick. This indicates a defect of the small burner as regard the combustion of colza oil. With paraffin, on the contrary, the total value of the light diminishes as the size of the burner is increased, and this, according to Dr. Macadam (10), indicates a defect of the large Doty burner as regards the combustion of paraffin. This is precisely the conclusion of Mr. Valentin in reference to the Trinity House burner.

26. I have hitherto confined myself to a single report of Dr. Macadam, for when the experiments at the Trinity House were executed, this report only and that of Mr. Douglass were in manuscript before us. But the report of Dr. Macadam moved Captain Doty to attempt an improvement of his large burner. He changed his wick, and by this simple expedient abolished the discrepancy revealed in the foregoing table (25). "The result," says Dr. Macadam, in a second report, dated 10th of April 1871, "has been, that without the slightest structural alteration in the larger-sized lamps, but by a mere change in the thickness or body of the cottons, the waste of oil, which in the four-wick lamp was equal to one-seventh of the whole illuminating power, has been entirely done away with, and the full amount of light capable of being evolved from the paraffin oil has now been obtained in the larger paraffin lamps."[32]

27. The experiments, I think, hardly warrant the conclusion here drawn by Dr. Macadam. What he proves in his second report seems to be this, that by a certain arrangement of wicks, proportionality was established between the total value of the light and the quantity of oil consumed in the different Doty lamps.

28. I have thus endeavoured to place in a condensed and consecutive form the sum of our experience regarding the application of mineral oils to lighthouses, as recorded in the printed correspondence.

29. Captain Doty has directed my attention to one or two statements of the report[33] introduced at page 40 of the printed correspondence. That report was founded on experiments executed by Mr. Valentin at the Trinity House on the 22nd and the 29th of April. The first day's experiments yielded a cost per hour per unit of light considerably in excess of that found by Dr. Stevenson Macadam and Mr. Douglass. On the second day this excess was greatly diminished, but it did not entirely disappear. The cause of the discrepancy is that on the first day the action of the two lamps was so forced as to cause both colza and paraffin to pass unconsumed into the chimney, whereas on the second day this source of waste was in great part avoided. "In the first day's experiments," says Mr. Valentin, "I allowed the lights to be kept burning with the largest flame compatible with the non-production of smoke on the chimney-glasses, in harmony with the trials made in Scotland." This is the statement to which Captain Doty has drawn my attention, and it no doubt involves a misapprehension. Dr. Stevenson Macadam, it is true, states that

in his experiments "the lamps were kept burning with the largest flame compatible with the non-production of smoke, and the perfect combustion of the colza and paraffin oils." He does not, however, speak of smoke "on the chimney-glasses," but of smoke in any sense: the combustion, he says, was "perfect," whereas in Mr. Valentin's case, though no smoke sullied the glasses, "distinct clouds of smoke" were seen issuing into the air outside. Hence, though the experiments are useful as showing the constancy of the relation between colza and paraffin even when the lamps are unduly forced, they are not to be taken as the true expression either of the cost, or of the photogenic value of the two oils. In the printed table placed by the Trinity House at the end of Mr. Valentin's Report, these abnormal first day's experiments have been introduced. The table will therefore require correction.

30. Various communications not interwoven with the foregoing summary, because not necessary to its clearness, had passed between the Board of Trade, the Trinity House, and the Commissioners of Northern Lighthouses. One of their small lamps, for example, had been sent by the Elder Brethren to Edinburgh, and found, when there tested, to be markedly inferior to the corresponding Doty lamp. This result was ascribed by the Elder Brethren to the fact that the lamp had been sent away untested, and that it was structurally defective. Invitations to witness their respective experiments, or to test by direct comparison their respective lamps, had been given by both sides; but each side courted comparison on its own terms, and hence for a time the invitations came to nothing. At length it was resolved upon that the Trinity and Doty lamps should be tested together in London, the Board of Trade deciding, in reply to the recommendation of the Northern Commissioners, "that it would be desirable to postpone any general action until the entire case had received further and full consideration."

31. The subject accordingly was referred to me. It was thought desirable by the Elder Brethren to conduct the investigation away from the Trinity House, and after due consultation and discussion with Mr. Valentin, to whose practical knowledge and skill I have been so often indebted, it was decided to partition off and darken a portion of a large room in the Royal College of Chemistry in Oxford-street. I have to thank Dr. Frankland for his friendly consent to this arrangement. Flues were therefore set up, the Doty and Trinity House lamps mounted, and all things rendered ready for experiment for experiment on the 27th of September.

32. On the 15th of August, Captain Doty had written to me a letter[34] containing the following request:—"In view of the repeated searching tests of my burners in Scotland, the high illuminating power and economy accorded to them by the chemist to the Scotch Lighthouse Board, and in order to obviate the necessity of any further comparative trials in this country, I have asked the Board of Trade to move the Elder Brethren to grant that the high values of my

burners, as reported upon by Dr. Macadam, may be confirmed by him in your presence, and that the relative merits of the several burners to be tried, may be established by your united observations, or conjointly with such other experts as you may appoint to assist you."

33. Subsequently to the writing of this letter I had two visits from Captain Doty, and after these again, viz., on the 16th of September, he addressed to me the following note:[35]—"I beg to confirm my letter of the 15th ultimo, regarding the expediency of Dr. Macadam's presence at the proposed competitive trial of my lamps with those of the Trinity House Board. My conversation with you to day eliciting the fact of the near approach of this trial renders me anxious for your decision upon this point, which I cannot but consider one of great importance to my interests; as the presence of Dr. Macadam at the London trial would entirely obviate the necessity of a second series of experiments in the north."

34. I considered Captain Doty rather eager as regards visits and suggestions, and ventured to tell him so; but my opinion of Dr. Stevenson Macadam being such as I have already expressed (11), I could have no objection to his presence. Indeed, I hoped he might be permitted to take a leading part in the inquiry. Believing that it would promote the serenity of our work if the two principals could be induced to withdraw from it, and to commit their respective lamps and interests to the care of competent representatives, I proposed to Captain Doty that he should be represented by Dr. Macadam,* and to Mr. Douglass that he should be represented by Mr. Ayres,[36] Mr. Douglass agreed to this, but Captain Doty sent me the following reply:—

"Your suggestion, that both myself and Mr. Douglass should stand aloof during the progress of experimenting, would be cheerfully accepted by me, but for one consideration, which I submit to your candid judgment; I have no skilled assistant on whose management of my lamps I can sufficiently rely, whereas my rival has the advantage of a staff rendered perfectly *au fait*[37] therein by a long series of practical trials; I am therefore reluctantly compelled to be my own operator, pledging myself to offer no observations or suggestions whatever, relative to the conduct of the trials, and to maintain a perfectly neutral attitude."

I could have no objection to so reasonable a demand as this.

35. To my regret, however, when applied to by Captain Doty, Dr. Macadam expressed his inability to take part in the inquiry. I was thus brought into direct relations with the principals. Mr. Douglass was assisted by Mr. Taylor,[38] and Captain Doty by Mr. Easton, a skilled and highly-intelligent mechanic from the workshops of the makers of the Doty lamp.[39] During the earlier days of the inquiry Mr. Douglass was forced by other engagements to have his place taken by his assistant, Mr. Ayres.

36. An additional remark or two will bring us to the starting-point of the

inquiry. In his letter to me[40] (bottom note to page 42 of the printed correspondence on mineral oils), Mr. Valentin introduces a point of considerable practical importance, "viz., the suitability of mineral oils differing in specific gravity and in their chemical composition for one and the same lamp." At his desire Mr. Douglass furnished him with various samples of petroleum oil, and on page 43 of the correspondence will be found a "table showing the specific gravity, flashing point, and approximate chemical composition of petroleum and paraffin oils." Regarding one sample cited in this table Mr. Valentin makes the following remark:—"5*a* promises well, and should be tried in the new [Trinity House] lamp, but it is expensive, its price being two shillings a gallon."

37. In pursuance of this recommendation, Mr. Douglass had tried the oil 5*a*, and found it to yield in his lamp more light than the paraffin. It was this petroleum, therefore, and not Young's oil, that he proposed to burn on the 27th of September.

38. On my arrival at the College of Chemistry on the morning of that day, the work had already begun under the superintendence of Mr. Valentin. He reports thus:—"On starting the experiments on Wednesday morning it was understood that each party should choose whatever oil he preferred to burn. Mr. Douglass chose petroleum 5*a*.; Captain Doty chose to burn the same oil."

39. Thus we began, and thus for some time the experiments continued, "oil 5*a*," as reported by Mr. Valentin, "being burnt to the satisfaction of both parties." In a letter handed to me on the morning of the 27th, Captain Doty had suggested paraffin oil for the first trail, and also that the rival lamps should be placed at opposite ends of the photometer; the first suggestion had been waived before my arrival, the second it had always been my intention to carry out; but I agreed with Mr. Valentin that it would be desirable in the first instance to place the two lamps at the same end of the photometer under the same conditions of draught, and to determine, by the comparison of each with a standard sperm candle, not only the relative, but also the absolute illuminating power. The photometric readings of the first five days were taken by Mr. Valentin, with an occasional observation of my own, my attention being mainly directed to the flames. The subsequent readings were taken by my assistant, Mr. Cottrell, and myself.

Record of Experiments.

27th September, 9 A.M.—Position of lamps; side by side.

Charged Trinity lamp with four gallons of 5*a* Goddard's best petroleum oil.

Charged Doty lamp with four gallons of the same oil; reserved sample for chemical analysis.

Lighted both lamps at 9:40 A.M.

The illuminating power was taken by means of a Bunsen's Photometer

32 feet long, and best London sperm candles (Sugg and Co., Westminster). The candle was allowed invariably to burn for not less than 15 or 20 minutes before taking the photometric observation.

On this day the first quarter of every hour was devoted to the adjustment of the lamps by the respective operators; and without special permission to the contrary the lamps were expected to remain untouched for the rest of the hour. The order in which the lamps wens tested was reversed, every successive hour. In all cases, except when the contrary is formally stated, each series of experiments consisted of 10 photometric readings, the mean of which, squared and corrected with reference to the weight of sperm consumed by the burning candle, was taken to represent the candle power.

10.30 A.M.

Mean of 10 photometric readings, corrected for sperm consumed.

Candle Power.

Trinity lamp—-—250·9 | Doty lamp—-—- 154·25.

The Trinity flame was here too high, stretching upwards into the iron chimney which surmounted the lamp glass. The Doty lamp at the same time burnt without streaks, and was below its average height.

11.30 A.M.

Mean of 10 readings corrected for sperm consumed.

Candle Power.

Trinity lamp—-—204·49 | Doty lamp—-—- 186·59.

The Doty flame being raised, and the Trinity flame lowered, both were in a fairer condition than in the last experiment. The consequence is a fall in the photogenic power of the one, and a rise in that of the other.

12.30 P.M.

Candle Power.

Trinity lamp—-—257·92 | Doty lamp—-—- 75·82.

Here during the second half of the 10 photometric readings the Trinity lamp flared. At the outset it had been raised near its flaring point, long forks being present, and sometimes, but rather rarely, the column of flame ran into the iron chimney. When this was the case the photometer was affected by light which would be comparatively useless in a dioptric apparatus. Hence the intensity of the Trinity flame as here shown is greater than it ought to be.

1.30 P.M.

Candle Power.

Trinity lamp—-—235·93 | Doty lamp—-—- 245·55.

The Trinity flame was here remarkably steady, no forking being tolerated. It was below its average height. The Doty lamp also burnt with remarkable steadiness; though it forked slightly occasionally. The absence of flickering on the disk renders this a good and trustworthy experiment.

2.30 P.M.

Candle Power.

Trinity lamp—-—249·32 | Doty lamp—-—207·65.

The Doty lamp here burnt fairly, but it was on the verge of forking; it flared up occasionally. The Trinity flame was also on the verge of forking, but it was a good practicable lighthouse flame.

3.30 P.M.

Candle Power.

Trinity lamp—-—233·17 | Doty lamp—-—268·30.

The Trinity flame was here a very good one, and exhibited no forks. The Doty flame burnt with occasional forks, or was just on the verge of forking. The state of the candle during the first five of the 10 readings makes the Trinity power lower than it ought to be.

4.30 P.M.

Candle Power.

Trinity lamp—-—221·41 | Doty lamp—-—187·69.

Here the Doty lamp burnt without streaks, and was rather lower than in the former experiments. The Trinity lamp also burnt very fairly and without streaks.

With a view to ascertaining the consumption of oil, the lamps were permitted to continue burning till 9.40, no farther photometric readings being taken.

Temperature of Oil in Reservoirs at 10 P.M.

Doty lamp—-—- 40°C. | Trinity lamp—-—- 34°C.

Consumption of Oil during 12 *hours.*

Trinity lamp—-—- 2 gallons, 5 pints, 14 oz.

Doty lamp—-—- 2 gallons, 2 pints, 7 oz.

Or, as 100 : 84·56

Being a saving in oil of 15·44 per cent. in favour of the Doty lamp.

SUMMARY of Photometrical Results on Wednesday, 27th September.

	Trinity House Lamp	Doty Lamp.
1st Experiment—-—- -	250·90 sperm candles.	154·25 sperm candles 164*26 sperm candles.
2nd " —-—- -	204·49 "	186·59 "
3rd " —-—-—	257·92 "	175·82 "
4th " —-—-—	235·93 "	245·55 "
5th " —-—-—	249·32 "	207·65 "
6th " —-—-—	233·17 "	268·30 "
7th " —-—-—	221·41 "	187·69 "
Average Illuminating Power—-—	236·16	203·69

Or, as 100 : 86·25; being 14·75 per cent. in favour of Trinity lamp.

I do not attach much value to this day's results. The flames were not so steady as they ought to have been for photometric measurement. The Doty flame showed itself peculiarly sensitive to slight changes of draught: "Awfully sensitive" was the description applied to it by Captain Doty's assistant.[41]

Thursday, 28th September.—Second Day.

Position of lamps unaltered: the same oil (petroleum 5*a*) burnt in both lamps. Draught modified by the introduction of funnels, a portion of the day being consumed in making the alteration. Both lamps lighted at 3.30 P.M. Half an hour was allowed for adjustment: proceeded afterwards as on the previous day.

At 4.30 when the measurements should have commenced, the flames of both lamps were found too high, and after a few readings the measurements were discontinued.

4.40 P.M.

Candle Power.

Mean of 10 Readings.

Trinity lamp—-—222·01 | Doty lamp—-—- 218·45.

The Trinity flame was very steady during the whole time of its own photometric observation: the Doty lamp succeeded it, and by the time the measurements were concluded the Trinity lamp was flaring.

The Doty lamp burnt with fair steadiness. There were more forks than I liked; still the flame was not bad.

5.30 P.M.

Candle Power.

Trinity lamp—-—227·10 | Doty lamp—-—- 186·05.

The Doty lamp here burnt very steadily and without any flaring up. It was perhaps a little lower than it might have been. The Trinity flame was very good; no fault indeed to be found with either.

6.30 P.M.

Candle Power.

Trinity lamp—-—192·37 | Doty lamp—-—- 218·44.

The Trinity lamp here burnt with remarkable steadiness. The variations on the photometer[42] disk being almost reduced to *nil.* But the flame was decidedly below its average height.

7.30 P.M.

Candle Power.

Trinity lamp—-—219·63 | Doty lamp—-—- 202·21.

The Trinity lamp burnt here remarkably well, no flaring up, and scarcely any variations on the photometric disk. Previous to the measurements, Mr. Valentin insisted on the lowering of the flame, so that forking might be avoided. The Doty lamp burnt with remarkable steadiness.

The two measurements are excellent.

8.30 P.M.

Candle Power.

Trinity lamp—-—230·73 | Doty lamp—-—- 225·65.

The flame of the Doty lamp was here perfectly steady, showing no wavering, as in yesterday's experiments. The introduction of the funnels by Mr. Douglass, manifestly improved the draught for the Doty flame. It did not improve the draught for the Trinity flame. Both flames were, however, here faultless, and the result excellent.

9.30 P.M.

Candle Power.

Trinity lamp—-—249·64 | Doty lamp—-—- 222·90.

The steadiness of the Trinity flame was here admirable: the Doty flame was also very steady. It was smaller than its rival, but emitted a whiter light. I would sooner trust the three last measurements, as expressing the real relative powers of the two flames, than a far greater number of readings with the flames in a less satisfactory condition.

10.30 P.M.

Candle Power.

Trinity lamp—-—221·11 | Doty lamp—-—- 202·49.

Prior to this experiment the lamps had been readjusted at Captain Doty's request. I thought the altered Doty flame too forked, and said so. Captain Doty thought the flame right. He had his way, and at 10·25 the lamp had flared up and filled its glass chimney with a column of reddish flame. It was then lowered.

Immediately previous to the foregoing measurements Captain Doty raised his button. I have always noticed that a gradual increase which usually ends in flaring follows this act. This increase is plainly revealed in the 10 readings, which gradually augmented from 14·3 to 16·35. This corresponds to an increase of candle power in the ratio of 204 to 267. Long forks were thrown up, but as long as the measurements continued no actual flaring occurred. Soon afterwards, however, the glass chimney was filled with a column of reddish flame.

The Trinity lamp flared, though not so badly, soon afterwards.

The lamps were permitted to continue burning till 11.30 p.m., no photometric measurements being taken during the last hour.

Consumption of Oil during Eight Hours.

Trinity lamp—- - 1 gallon, 2 quarts, 37 oz.

Doty lamp—-— 1 gallon, 1 quart, 38 oz.

Or as 100 : 85·912.

Being equal to a saving of 14·08 percent, in favour of Doty.

SUMMARY of Photometric Results on Thursday 28th September.

——————-	Trinity House Lamp.	Doty Lamp.
1st Experiment -	222·01 sperm candles	218·45 sperm candles.
2nd " -	227·10 "	185·05 "
3rd " -	192·37 "	218·44 "
4th " -	219·63 "	202·21 "
5th " -	230·73 "	225·65 "
6th " -	249·64 "	222·90 "
7th " -	221·11 "	202·49 "
Mean—- -	223·23	210·88

Or as 100 : 94·46; being 5·54 percent, in favour of Trinity.

Friday, 29th September.—Third Day

The lamps were charged with Young's paraffin oil No. 2, which was submitted by Captain Doty. The position of the lamps (still side by side) remained the same. The wicks also remained unchanged. Half an hour was allowed for adjustment. Both series of experiments were preceded by a quarter of an hour, during which the lamps remained untouched.

At 11.30, when the first measurements were to have been taken, the Trinity lamp flared. The experiments were suspended for another half-hour, during the first quarter of which access was permitted to the lamps. The Trinity flame was lowered in this interval.

It had been lowered too much; and at 12 p.m. was so much below its usual height that Mr. Ayres objected to its being measured. The Doty lamp, judged at this time by the naked eye, seemed decidedly the most powerful.

After various efforts on the part of Mr. Ayers to get the Trinity lamp to perform satisfactorily, he retired from the contest at 2.45 p.m. Handled as it was on this day, the Trinity lamp was found practically incompetent to cope with paraffin No. 2.

The Doty lamp burnt on the whole fairly; still there was a good deal of flickering. Its performance ranged from a minimum of 190·44 candles to a maximum of 204·49 candles. The maximum was thus considerably less than that observed in the case of petroleum. "It is apparent," writes Mr. Valentin in the memoranda which he has handed me, "that the Doty lump did not relish oil No. 2 overmuch, as indicated by the constant flickering of the flame, and the lower illuminating power obtained; that Captain Doty was so ignorant about his own oils as to select a sample which gave a less illuminating power than oil 5a, and which evidently did not suit his burner perfectly."

It is here, however, to be borne in mind that Captain Doty did not introduce oil No. 2 with a view of obtaining greater illuminating power, but with the legitimate and subsequently avowed object of bringing his antagonists to a crucial test. He wished to show that his lamp could burn a heavier and more smoky oil than the Trinity lamp. Oil No. 2 had never been recommended for lighthouse purposes. Had it been of equal illuminating power with oil No. 1, Young's best paraffin, it, instead of No. 1, would have been used in the experiments in Scotland; for its price is only 1 *s*. 4 *d*. a gallon against 1 *s*. 6 *d*. a gallon, the cost of No. 1. The choice of No. 2 demonstrates, not the ignorance but the knowledge of Captain Doty. He relied upon the power of his lamp to burn this heavy oil, and upon the inability of the Trinity lamp to do so. This day's experience seemed fully to justify his confidence.*[43]

Saturday, 30th September.—Fourth Day.

The programme for this day was as follows:—

"The two lamps to be placed at the opposite ends of the photometer, the Trinity lamp at the north, and the Doty lamp at the south.

"Saturday to be devoted wholly to the removal and remounting of the lamps.

"Each inventor to have absolute control of the ventilation or draught of his own lamp, and over the oil to be used in his lamp."

This programme, which was signed by Captain Doty, Mr. Douglass, and myself, was duly carried out.

Monday, 2nd October.—Fifth Day.

The programme for this day was as follows:—

"Previous to each series of experiments, half an hour to be allowed to each inventor for the adjustment of his lamp; during the subsequent quarter of an hour prior to the measurements, the lamps to remain untouched.

"If, at the expiration of the last quarter of an hour, either or both of the lamps should be considered unfit for observation, the defect is to be recorded, and half an hour is to be again allowed for adjustment. This, as before, to be followed by a quarter of an hour, during which the lamps are to remain untouched.

"This process is to be continued until the measurements have been completed or abandoned.

"The experiments are to commence at 9 a.m., and the lamps are to continue burning for 16 hours."

At 9 a.m., the lamps were charged with petroleum 5a, and at 9.45 they were lighted. The measurements began at 10.30. The photometer being graduated from both ends, the disk was moved until the lights, falling upon its opposite sides, were equal. The distance of the disk from the two ends were then read off, 10 such readings being executed on each occasion. The squares of the distances give, of course, the relative intensities of the light.[44]

10.30 a.m.

Relative Intensities of Lights.

Trinity lamp—-—247·12 | Doty lamp—-—- 264·71

The Doty flame here was more forked than the Trinity flame, but there was no smoke. The Trinity flame appeared very solid and compact. It showed no forks whatever while the Doty flame towards the conclusion threw up six or eight inches high.

11.45 A.M.

Trinity Lamp—-—262·44 | Doty lamp—-—- 249·64

Doty flickering; Trinity very steady. No smoke, however, in Doty chimney. There was a natural desire on the part of each inventor to keep his flame at its maximum height, and one consequence of this was the flickering and tongues so often alluded to. These forks, in some cases very brilliant, threw their light upon the photometer, and gave the lamp from which they issued a corresponding advantage, whereas in a lighthouse, being out of focus, they would be of comparatively little use. I thought it fair to abolish their action by suspending in front of each lamp-glass a piece of sheet-iron six inches wide, and five inches high. The lower edge of the shade was four inches above the surface of the burner. The body of the flame was thus allowed free room to radiate, while the light of the comparatively useless forks was intercepted.

On this day, during the 15 minutes which preceded the next following series of observations, I observed from time to time the deportment of both flames.

1st Obs. No forks seen above Trinity shade, but they were seen above the Doty shade, sometimes reaching to the top of the glass-chimney.

2nd Obs. No fork over Trinity shade; forks over Doty shade sometimes stretching through the whole glass cylinder.

3rd Obs. About an inch of tongue showed itself above Trinity shade; Doty as before.

4th Obs. No flicker above Trinity shade; a tongue of flame about three inches long above Doty shade.

At 12.45, the Doty flame flickered so much and showed so many tongues that it was found impossible to proceed with the measurements. Half an hour was accordingly allowed; and the usual 15 minutes subsequently.

1.30 P.M.

Trinity Lamp—-—261·14 | Doty lamp—-—- 250·90.

These observations are perfectly good. During the experiments both lamps threw up short tongues from time to time above the sheet-iron shade; the Doty lamp most; but the defect was infinitesimal, and practically the same in both.

I again observed the flames during the 15 minutes which preceded the next series of experiments.

1st Obs. Doty, tongues; Trinity, none.

2nd Obs. Doty, one little tongue; Trinity, none.

3rd Obs. Doty, one tongue; Trinity, none.

4th Obs. Doty, long tongues; Trinity, short ones.

Trinity flame under its power; Doty flame full up to it. Hence the rise of the Doty and the fall of the Trinity in the subsequent series of observation.

2.30 P.M.

Trinity Lamp—-—237·47 | Doty lamp—-—275·23

All through this series, tongues of flame were seen above the Doty shade; but no smoke. Once during the observations the Trinity flame sent up a tongue. The Trinity wick was raised, and at 3.30 the measurements were commenced. At the seventh reading the Trinity lamp flared, and though some of the previous observations are marked "Good," I thought it fairest to suspend the measurements. In the readings thus marked, the Trinity flame exceeded the Doty in the proportion of 272 : 240.

4.30 P.M.

Trinity Lamp—-—246·49 | Doty lamp—-—- 265·69

These observations are very good, but the Trinity flame had been lowered a little too much. No trace of a tongue was observed over Trinity shade, but there was frequent shooting over the Doty shade.

From 5 to 6.40 p.m. readings were taken every half-hour by Mr. Valentin, Mr. Cottrell, and myself. During this time the flames were of nearly equal power. The mean of six closely concurrent readings gave 252·8 for the Trinity, and 259·2 for the Doty lamp; while a series of four readings taken every half-hour between 7 and 8.30 p.m., gave 246 for the Trinity, and 265 for the Doty lamp.

Throughout these observations the steadiness of the Trinity lamp was admirable: the Doty flame showed tongues, but no smoke; it was a good practical flame.

8.30 P.M.

Trinity Lamp—- 249·95 | Doty lamp—- 262·11

At this point the Doty lamp had already burnt for 7 1/2 hours, and the Trinity lamp for nearly four hours without being touched.

9.30 P.M.

Trinity Lamp—-—260·37 | Doty lamp—-—251·66

The Doty flame unsteady and showing long tongues, but no smoke; the Trinity flame steady and compact.

10.30 P.M.

Trinity Lamp—-—255·20 | Doty lamp—-—- 256·80

Extinguished lights at 1.45 a.m., the lamps having previously burnt for four and three-quarter hours without being touched.

Consumption of Oil during 16 *Hours' continuous burning.*

Trinity lamp -—- - 3 gallons, 3 quarts, 2 oz.

Doty lamp -—- - 3 gallons, 2 quarts, 2 oz.

Or as 100:93·35; being a saving of 6·65 per cent. in favour of Doty.

SUMMARY of the Photometrical Results of 2nd October.

———	Trinity Lamp.	Doty Lamp.
1st Experiment Square— -	247·12	264·71
2nd " " - -	262·44	249·64
3rd " " - -	261·14	250·00
4th " " - -	237·47	275·23
5th " " - -	246·49	265·69
7th " " - -	249·95	262·11
8th " " - -	200·37	251·66
9th " " - -	255·20	256·80
Average—- -	252·52	259·59

Or as 100: 102·79; being 2·79 per cent. in favour of Doty.

Tuesday, 3rd October.—Sixth Day.

The following programme was drawn out for this day:—

"The position of the two lamps to be reversed, the Doty lamp being placed at the north end and the Trinity lamp at the south end of the photometer.

"The lamps to be ready for the introduction of a measured quantity of oil at 12.45 p.m.

"Both lamps to be lighted at 1 p.m.

"At 1.45 the lamps to be handed over to the photometric observer, and to remain untouched for a quarter of an hour.

"At 2 p.m. the first series of readings to commence.

"Each succeeding series to be preceded by half an hour for adjustment, and a quarter of an hour during which the lamps shall remain untouched. These experiments being intended to check those of Monday, and eliminate all differences arising from the mere position of the lamps, the same oil is to be burnt as on Monday (5a petroleum), each inventor having the entire control of his own lamp.

"The lamp was to continue burning till 9 p.m."

The introduction of a bit of dirt into the feed-pipe of the Trinity lamp delayed the lighting for an hour.

Nothing calling for special remark occurred today; it is therefore needless to enter upon the details of the experiments. The consumption and relative illuminating power, deduced from the means of the observations, will suffice.

Consumption during Eight Hours.

Trinity lamp -—- 1 gallon, 3 quarts, 30 oz.

Doty lamp —- - 1 " 2 " 34 "

Or as 100 : 88·39; being a saving of 11·61 per cent, in favour of the Doty lamp.

Relative Illuminating Power.

Trinity lamp -—- 241·78 | Doty lamp —-—- 270·73

Or as 100 : 111·97; being a saving of 11·97 per cent. in favour of the Doty.

Both as regards consumption and illuminating power, the Trinity lamp suffered from its change of position. Indeed, it was obvious to the naked eye that the Trinity flame was below its average. The experiments on petroleum 5a were thus included.

Wednesday, 4th October.—Seventh Day.

This day was occupied in an undecisive comparison of the single-wick, or fourth order Doty and Trinity lamps. At the outset both inventors brought their lamps so near the flaring point that when the experiments began the lamps were unfit for measurement. A subsequent trial was contemplated, but this Captain Doty declined to carry out.

As far as I could judge, the two flames, before they flared, were of approximately equal intensity. There was certainly no marked superiority of the one over the other. In fact, the few correct measurements that were made demonstrated the practical equality of the two flames. Thursday was to be devoted to experiments with the four-wick lamps burning paraffin oil.

Towards the close of the day's work, however, while conversing with Mr. Douglass, I learned that he proposed to continue the experiments on Thursday, not with the actual burner previously employed, but with another burner of the same construction, containing a new wick. Seeing that Captain Doty could assure himself of the absolute identity of both burners, Mr. Douglass did not, he alleged, apprehend the least objection to this arrangement.

I could not share his confidence, and therefore asked Captain Doty to join us, permitting Mr. Douglass to state his own case. Captain Doty distinctly objected to the proposed change of burner.

Mr. Douglass remonstrated, declaring that he merely wanted to alter his wick. He urged that Captain Doty had, and knew that he had, a better wick than that of the Trinity lamp. Thereupon Captain Doty offered to abandon his wick, and to use in its stead that of the Trinity burner. This frank offer was declined by Mr. Douglass, who stated his determination not to risk comparative experiments before he had tested his wick. He was willing to proceed at 9 a.m. on Thursday, with the burner in which the wick was already fixed, but if compelled to continue with the old burner he would demand a day for preparation.

At the beginning of this inquiry, in which I knew forbearance and tact would be necessary, I had resolved that should differences arise, I would allow them to be arranged as far as possible by friendly discussion between the principals, thus restricting my own direct interference to a minimum. I was guided by this resolve on the present occasion. After a little time Captain Doty came to me and said, "I have another proposal to make; I will allow Mr. Douglass a day to change and test his wicks, provided no other alteration is made in the burner." The proposition was definitely placed before Mr. Douglass. He accepted it, and one article of the programme then drawn out and signed by Captain Doty, Mr. Douglass, and myself, was this:—"Mr. Douglass to prepare Trinity House four-wick lamp and photometer, but the preparation of the lamp not to involve any change in the metallic portion of the burner." I had previously had ground to notice Captain Doty's eagerness, and in some cases to question his reasonableness, but his conduct on this occasion appeared both liberal and courageous, and it materially exalted him in my estimation.

There was, it must be stated, no article in the programme drawn out on this occasion, restricting Mr. Douglass to the condition that the trial of his wick should be made at the College of Chemistry; but from our previous conversations I took it for granted that it was understood that the burner should not be removed. Mr. Douglass, however, must have thought otherwise. During the whole of Thursday the lamp was absent from the College of Chemistry, and it was first restored to its place on the Friday morning.

Serious imputations followed, which, had they been substantiated, or persisted in, would have certainly stopped the inquiry. They were, however, unreservedly withdrawn, without any interference on my part; and it was accepted by Captain Doty that beyond cleaning it no change had been made in the Trinity burner.

Friday, 6th October.—Ninth Day.

On the morning of this day I received a lengthy communication from Captain Doty[45] dated October 5th, in which, quoting largely from previous correspondence, he sought to show that the future experiments ought to be restricted to the heavier paraffin oils. The practical pith of the communication lies in this paragraph:—"As my burner has been subjected to four days' successful trial with petroleum, totally unfit for lighthouse purposes, I therefore most respectfully beg to request that all further serious comparative trials of my burners may be made *with safe paraffin oils* suitable for lighthouses." The italics are Captain Doty's. He goes on to "protest against the irregularities of the trials so far made," while recognizing my individual impartiality.

The protest immediately followed the altercation regarding the Trinity burner, and bears some evidence of the heat of altercation. Captain Doty was not aware at the time that the oil he denounces as "utterly unfit for lighthouse purposes,"[46] and which he insinuates to be unsafe, has a flashing point from 126° to 130° Fahr[enheit]., while the oil on which he most insisted viz., paraffin No 2, flashes from 128° to 134° Fahr. Further, the illuminating power of petroleum 5a, in Captain Doty's own lamp, is superior to that of paraffin oil No. 2. In fact, the former oil, which was recommended for trial by Mr. Valentin, yields a beautiful lighthouse flame, and the gravest, if not the only, obstacle to its employment is that it costs 2 *s*. a gallon.

It required, moreover, no alteration of my programme to meet substantially the request of Captain Doty. From the first the examination of paraffin oils had been resolved upon by me, and before the receipt of Captain Doty's letter it had been decided that the trial of these oils should commence this day.

Both lamps were charged with Young's paraffin oil, No. 1.

As on previous days, each series of ten readings was preceded by half-an-hour for adjustment of lamps, and a quarter of an hour during which they remained untouched.

At 11 a.m. the Trinity lamp was flaring; no measurements, therefore, were made.

11.45 A.M.

Trinity Lamp—-—229·93 | Doty lamp—-—- 283·50

Flames steady at the end of the readings.

12.45 P.M

Trinity Lamp—-—257·69 | Doty lamp—-—- 254·37

Left lamps for an hour untouched. A single observation at the end of this time made the Trinity lamp 272·25, and the Doty lamp 240·25.

The Doty lamp was then adjusted; Trinity untouched.

2.45 P.M.

Trinity Lamp—-—242·20 | Doty lamp—-—- 270·23

Ordered lamps to remain untouched till further notice. At 3.30 both were very steady, and both sides agreed that they were in a fair condition for experiments.

3.30 P.M.

Trinity Lamp—-—252·19 | Doty lamp—-—- 259·87

The flames were here very fairly matched, and the result perfectly trustworthy.

At 4.10 Doty lamp touched; Trinity untouched.

4.40 P.M.

Trinity Lamp—-—251·67 | Doty lamp—-—- 260·37

Both lamps very steady; an excellent series of observations.

6.25 P.M.

6.25 p.m. Doty lamp flaring high; adjusted; draught altered. At 6.30 and at 6.40 Doty lamp again adjusted.

7 P.M.

Trinity Lamp—-—- 289 | Doty lamp—-—- 225

7.30 P.M.

Trinity Lamp—-—284·6 | Doty lamp—-—- 229·0

At 7.35 the Doty flame was raised; it had been burning rather low for some time.

8 P.M.

Trinity Lamp—-—285·61 | Doty lamp—-—- 228·01

8:30 P.M.

Trinity Lamp—-—288·32 | Doty lamp—-—- 225·60

9 P.M.

Trinity Lamp—-—287·30 | Doty lamp—-—226·50

At 9.5[47] Captain Doty raised his flame.

9.30 P.M.

Trinity Lamp—-—280·90 | Doty lamp—-—- 232·26

From 9.36 to 9.56 Captain Doty was engaged in raising his flame and cleansing the spaces between the wicks by passing a wire between them.

10 P.M.

Trinity Lamp—-—260·82 | Doty lamp—-—- 251·22

At 10.12 Captain Doty adjusted his lamp.

10.30 P.M.

Trinity lamp—-—- 256 | Doty lamp—-—- 256

At 10.41 Captain Doty altered his draught; flame flaring. At 10.47 Captain Doty tilted his reservoir so as to raise the level of the oil in his burner.

11 P.M.

Trinity lamp—-—259·21 | Doty lamp—-—- 252·81

11.30 P.M.

Trinity lamp—-—262·44 | Doty lamp—-—249·64

12 MIDNIGHT

Trinity lamp—-—267·32 | Doty lamp—-—- 244·92

At 12.22 Captain Doty adjusted his lamp, the flame being low.

12.30 A.M.

Trinity lamp—-—259·21 | Doty lamp—-—- 252·81

1 A.M.

Trinity lamp—-—261·47 | Doty lamp—-—- 250·59

1.30 A.M.

Trinity lamp—-—266·34 | Doty lamp—-—- 245·86

2 A.M.

Trinity lamp—-—262·44 | Doty lamp—-—- 249·64

The lamps were then extinguished, having burned for 16 hours.

From 8.30 p.m. the lamps were under the direction of Mr. Cottrell, who, at my desire, took photometric observations every half-hour. All the changes made by Captain Doty were initiated by himself, Mr. Cottrell not interfering.

The Trinity flame burned from 12.45 p.m. to 2 a.m., that is, 13 1/4 hours, without any adjustment whatever. Up to my leaving the College of Chemistry at 8.30, the flame had been excellent. Mr. Cottrell's description of it, while under his superintendence, is this:—"General character of Trinity flame during the evening: A solid, compact, intensely brilliant, and occasionally forking flame; the forks sometimes stretching to the top of the glass chimney. The upper part of the chimney dulled with smoke." "General character of Doty flame: A solid, compact, intensely brilliant flame; very few forks; steady. Burnt low during the evening. Had to be raised and altered several times, and once cleansed during the evening; no smoke on chimney."

Consumption during 16 *Hours' burning.*

Trinity lamp—-—-—3 gallons, 1 quarts, 31 oz.

Doty lamp—-—-—- 2 " 3 " 17 "

Or as 100 : 83; being a saving of 17 per cent. in favour of the Doty lamp.

SUMMARY of Photometrical Results, 6th October, up to 4.40 p m.

————————-	Trinity Lamp.	Doty Lamp.
1st Experiment—- -	229·93	283·50
2nd "—- -	257·69	254·37
3rd "—- -	242·20	270·23
4th "—- -	252·19	259·87
5th "—- -	251·67	260·37
Mean—- -	246·73	265·67

Or as 100 : 107·67; being an excess of 7·67 per cent. in favour of Doty.

Means of Photometric Results from 6.25 *p.m. to* 2 *a.m. inclusive.*

Trinity lamp—-—271·33 | Doty lamp—-—- 241·32

Or as 100 : 88·93; being an excess of 11·07 per cent. in favour of the Trinity lamp.

The irregularity of the Doty lamp during the last eight hours of this day renders the observations valueless, save in so far as they demonstrate that even in the most practised hands a lamp burning paraffin may get out of order through causes which do not come into play in the case of colza. Captain Doty's explanation of the irregularities of his lamp will be given further on, and more correct results will, at the same time, be communicated.

Saturday, 7th October.—10*th Day.*

The task of drawing out, on Friday evening, a programme of this day's work, was by no means an easy one. Throughout the inquiry Captain Doty had urged with particular earnestness the trial of oil No. 2, affirming that his lamp had been specially constructed to burn such heavy oils. He more than once expressed his conviction to myself that the Trinity lamp could not burn the oil, and we have seen that, in the hands of Mr. Ayres, its performance seemed to justify the opinion of Captain Doty. Still Mr. Douglass affirmed that he had no doubt of the competence of his lamp to cope with the oil; the substance, however, was entirely new to him, and as Captain Doty had already prepared himself with experiments upon this oil, he, Mr. Douglass, demanded to be placed in some degree upon the same footing. He limited his demand to three hours' time, during which no photometric measurements were to be recorded. This time, moreover, he did not wish to abstract from the experimental day, but he proposed to make his trial so early in the morning, as to enable him to start with the experiments at the usual hour, viz. 9 a.m. He also pledged himself, whether his trial was successful or not, to submit his lamp to photometric comparison afterwards.

To this demand Captain Doty firmly objected. He urged that inasmuch as he had accepted Mr. Douglass's petroleum, though he regarded it as useless for lighthouse purposes, Mr. Douglass ought, in fairness, to accept his oil on the same terms. He contended that neither his rival nor himself came there to make experiments with their lamps, but to test lamps which ought to have been ready for experiments. On a former occasion I expressed a high opinion of Captain Doty's frankness, courage, and liberality. It is my duty to say that in the continuance of this discussion he did not exhibit the qualities for which I had previously given him credit, but perplexed and bewildered me extremely by the vague and often conflicting statements he made regarding Young's oils.* The unsatisfactory state of things here described was maintained for some time, when I at length ventured to advise Mr. Douglass to accept Captain Doty's terms. He did so under protest.

The lamps being first charged with oil No. 2, were both lighted at 10 a.m. The measurements began at

11 A.M.

Trinity lamp—-—165·45 | Doty lamp—-—- 366·18

The Trinity flame light and smokeless, but only an inch high, Mr. Douglass augmenting it cautiously.

At my request Mr. Cottrell wrote a description of the Doty flame:—"The lower part of flame remained a compact, brilliant mass, about two inches high, the whole being surmounted by innumerable spears of flame, some of them long and going up continuously into the iron chimney." It was, therefore, hardly fit for measurement.

12 NOON

Trinity flame—-—231·53 | Doty flame—-—281·71

Some smoke from outer wick of Trinity lamp; the flame itself brilliant and powerful. No smoke from Doty flame; occasional forks, but the flame practically good and powerful.

1 P.M.

Trinity lamp—-—229·74 | Doty lamp—-—- 283·69

Trinity flame much improved as regards smoke. Doty threw up

occasionally a long red tongue, which reached the iron chimney. The flame, however, was in fair condition, though it could not have been larger without flaring.

The lamps were left burning for an hour untouched. At 2.15 the Trinity flame had maintained its condition. The Doty flame was forking, but not prejudicially so.

3 P.M.

Trinity lamp—-—253·20 | Doty lamp—-—- 258·84

Neither flame objectionable, but Doty the steadier of the two. The observations were very satisfactory. Smoke not absolutely abolished from Trinity flame, but nearly so.

4 P.M.

Trinity lamp—-—250·34 | Doty lamp—-—261·73

Both flames steady and brilliant; in exceedingly good condition.

5 P.M.

Photometric readings commenced, but were stopped in consequence of the Doty flame streaming up into iron chimney. Colour of streaks smoky red.

5.20 P.M.

Trinity lamp—-—236·83 | Doty lamp—-—- 275·93

Doty was hardly in a condition for measurement, the forking was so vigorous. Sometimes we had a reddish smoky rush up the chimney, but this was rare and the smoke was not dense. The Trinity flame also showed momentary forks; but not so bad as the Doty flame.

6.30 P.M.

Trinity lamp—-—258·41 | Doty lamp—-—- 253·61

The flames here were much steadier than during the last readings, and it was much pleasanter to work at the photometer. Both flames were very brilliant.

At every half hour from 6.30 p.m. to 10 p.m., a photometric reading was taken. The lamps were almost alike in every case. The distances of the respective lamps from the photometric disk in these six observations were as follows:—

Trinity Lamp.	Doty Lamp.
15·95	16·05
16·00	16·00
15·85	16·15
15·87	16·13
16·00	16·00
15·80	16·20
Mean—- 15·91	16·09

So that for the last three hours and a half the illuminating powers of the two lamps were practically equal.

Consumption during 12 *hours.*

Trinity lamp:—2 gallons, 1 quart, 12 ounces.

Doty lamp:—2 gallons, 0 quarts, 33 ounces

Or as 100 : 94·89; being a saving of 5·11 per cent. in favour of Doty.

SUMMARY of Photometric Results, Saturday, October 7th.

	Trinity Lamp.	Doty Lamp.
1st Experiment	165·45	366·18
2nd "	231·53	281·71
3rd "	229·74	283·69
4th "	253·20	258·84
5th "	250·34	261·73
6th "	236·83	275·93
7th "	258·41	253·61
Mean	232·21	283·10

Or as 100 : 121·91;

Or, rejecting the first abnormal observation,

243·34—269·25

Or as 100 : 110·64;

Being in the one case an excess of 21·91, and in the other an excess of 10·64 per cent. in favour of Doty.

Throughout this day the Trinity lamp gained in power, and in some of the most accurate observations it came very close to the Doty lamp. The competence of the Trinity lamp to cope with oil No. 2 was certainly demonstrated by these experiments.

There can be no doubt that Captain Doty was not prepared for the performance of the Trinity lamp on this occasion; and soon afterwards, viz., on the 10th of October, he addressed to me another "protest:" "First," he says, "the iron pipes are not put up in such a manner as to comply with the conditions observed in lighthouses, leading as they do into a house chimney with irregular draughts of air, which affect the steadiness of the flame, and against which I have from the commencement of the trials objected.

"And that the fine white streaks of flame which you have frequently observed shooting up from the main body in my burner and on the appearance of which you have repeatedly compelled me to lower my light, thus lessening its photogenic record, are not prejudicial. These streaks will appear in every well-constructed mineral oil lighthouse burner. They frequently show themselves in lighthouses ascending first from one part of the burner, and then change to another, and generally, if left alone for a time, they will finally disappear altogether.

"This statement is based upon practical experience with mineral oils for a period of time more than three years, and while I affirm that I believe you conscientiously thought you were right in compelling me to lower my flame at the appearance of these points, it would be suffered to burn with greater strength were you acquainted with the peculiarities of this new light."

It is here to be remarked that the "conditions observed" in Scotland (where Captain Doty's lamp had been tried) in getting rid of the products of combustion are not quite the same as those observed in England. Now in our experiments neither inventor was forced to accept the conditions of the other. Each of them had absolute control over the draught of his own lamp. The "house chimney" referred to was occupied by them alone, such other flues as led into the chimney having been stopped during the experiments. The chimney was protected by a wind-guard outside. At the beginning of the inquiry Captain Doty spoke of having his lamp-chimney freely opening into the air of the room; if he chose not to do so the act was his own. I am of opinion that in all the experiments with the paraffin oils the Trinity lamp suffered more than the Doty from defective draught.

With regard to the "objections" stated to have been made from the commencement there is an exaggeration in the use of the word. Both Mr. Douglass and Captain Doty made observations from time to time regarding the draught, but no statement which could with propriety be called an *objection* was made to me.

Again, with regard to "the fine white streaks of flame," it is not to them that I objected. Captain Doty is credited with numerous readings where not only fine white streaks but thick white tongues were present. What I did object to is this: At an early stage of the inquiry I observed that by the careful raising of his button Captain Doty knew how to produce a very gradual augmentation of the strength of his flame. Taken at a certain stage of its development, when a brisk forking began to show itself, this growing flame exhibited a high photogenic power. But when it was permitted to remain untouched it infallibly ended in positive flaring; that is to say, it ended by sending a column of reddish smoky flame through the glass cylinder into the iron chimney above the cylinder. No lightkeeper would think of maintaining his lamp in this condition of unstable equilibrium. The Trinity flame, be it observed, was subjected to the same restrictions as the Doty flame, so that Captain Doty, in my opinion, lacks all just grounds of complaint or objection.

Friday, 13*th October,*—11*th Day.*

Paraffin oil is found to vary within certain limits through the accidents of manufacture. Two oils labelled alike do not always yield the same photogenic result. As Mr. Valentin first sagaciously pointed out,[48] it is of the first importance to ascertain how the lamps are affected by such variations of quality. If the burning range, to employ a new term, of one lamp could be proved greater than that of another, it would be a great point in favour of the former. I availed myself of an opportunity of testing the two lamps in this respect. At page 43 of the printed correspondence an oil called 4a is introduced, which was found by Mr. Valentin to rapidly char the wick of the Trinity lamp. If the Doty lamp could prove its competence to bum this oil it would establish its superiority on a very important point over its rival. A few days previously, moreover, Captain Doty had expressed a wish to have this oil placed in his lamp. Since the Trinity lamp had already been tested and found wanting, it is quite manifest that the chances were altogether in Captain Doty's favour.

At 3.30 p.m., Captain Doty drew my attention to the condition of his lamp, and strongly objected to the oil. At my request the following description was written down by my assistant from the dictation of Captain Doty himself:—"Lower end of glass chimney covered with brown spots, caused by the sputtering of the oil. The deposit of charred residuum of petroleum forming an incrustation upon the wicks produces an effect similar to the raising of the cottons, throwing out charred points and preventing the free circulation of the air at the point of ignition of the flame at the top of the burner; against which oil I protest."

Considering that the experiment had been arranged not only in the interest, but in accordance with the wish of Captain Doty, expressed a day or two previously, this "protest" seemed somewhat unreasonable. Prior to the experiments no objection had been made, and it is only when the result does not answer his expectations that Captain Doty thinks of protesting. The case is representative. Each of Captain Doty's protests follows a period of dissatisfaction *with the results*; which dissatisfaction by a process natural if not logical, is extended to *the arrangements*.

During this day's work the Trinity wick was also heavily charred, though not so much as the Doty. Various photometric measurements were taken, but as the oil proved unsuitable for both lamps it is unnecessary to dwell upon details. The illuminating power throughout the day was 9 per cent. in favour of Doty, while the consumption was 14·66 per cent. in favour of the Trinity. This leaves a clear balance of 5 1/2 percent, in favour of the latter.

To the verbal protest just recorded, Captain Doty added a written one,[49] dated the 14th of October. After quoting an expression contained in a previous letter of mine,[50] he goes on to say: "I now most respectfully beg to state, that from the commencement of these trials I have repeatedly expressed my wishes that they might be conducted in harmony with the published reports of the trials of the rival burners; and that the iron pipes surmounting the glass chimneys should be arranged in accordance with the practice adopted for lighthouses, instead of leading into the flues of a chimney with irregular draughts of air; and you will doubtless recall to mind that on the two first days of the trial I strongly protested against this arrangement. These two days were exhausted with petroleum experiments, when on the third day, the pipes were altered by and under the immediate direction of the engineer of the Trinity House. Both pipes were arranged exactly alike, while, on the third day, a trial was had with paraffin [No. 2], the result of which you are perfectly aware.

"Five days having been exhausted in trials with petroleum, for the combustion of which I have sufficiently demonstrated that my burner is perfectly adapted, I shall now be glad to have my burner put to the test with those oils referred to in the correspondence between the general lighthouse authority and the Board of Trade, as having been employed in their recent trials, and as being best suited for lighthouse purposes.

"Having spent too much time (five days) in experimenting with petroleum, I must respectfully decline any further trials except with those hydrocarbons referred to in the official reports, upon which the photometric values of the respective burners have been calculated."

I have already shown that Captain Doty's dislike to petroleum 5a was exaggerated; that he urged his objection to it before he was aware of its character. Four days had been devoted to this petroleum, and Captain Doty ought to have stated that the fifth day was spent on petroleum 4a in accordance with his own previously expressed desire. Captain Doty again recurs to the iron pipes and the chimney flue; to these points I have already referred, and I would only repeat now that the Doty lamp was, in my opinion, more removed from any evil arising from the arrangement of flues than the Trinity lamp. Captain Doty also reminds me of strong protests made during the first two days of the inquiry. I can only say, that had anything which could be interpreted as a protest been mentioned to me on either of the days referred to, I should not, in the face of it, have proceeded a single step with the investigation.

I could not indeed, at first, understand the meaning of this third protest, but I did get an inkling of it afterwards. Wishing to test paraffin sold by dealers at well as that sold at Young's office in Bucklersbury,[51] I had requested Mr. Douglass to send to some shop in the city for a sample of the same oil. He did so. Mr. Valentin afterwards gave himself the trouble of going to the city, and of purchasing from the same dealer a second sample of the same oil. In a subsequent table Mr. Valentin refers to it as "spurious paraffin." Now I have observed Captain Doty from time to time helping himself to specimens of the oils, and he probably had acquired a knowledge of this one when he wrote his protest. The protest was needless, for no "spurious" specimen could have escaped the rigid scrutiny to which all the oils were subjected. As soon as I ascertained the flashing-point of this oil, I condemned it as unfit for experiment.

With the view, however, of determining the range of the lamps, and the influence of variation in manufacture upon their performance, it was still thought desirable to test some of the oils sold by the customers of Young's paraffin company. The next two days were occupied with inquiries of this nature.

Saturday, 14th October.—12th Day.

Two specimens of Young's paraffin oil were purchased at Povey's,[52] Market-place, Oxford street, and examined this day. On inquiry afterwards at the office of Young's paraffin oil company in Bucklersbury, I learned that one of the oils was that described in the circulars of the company as "crystal oil;" the other being oil No. 2.

In the case of the crystal oil, the illuminating power of the Trinity lamp exceeded that of the Doty, but the consumption of the Doty fell considerably short of that of the Trinity. In the case of oil No. 2, the illuminating power of the Doty exceeded that of the Trinity, while the consumption of both lamps were the same.

As the crystal oil is not likely to be used in lighthouses, and as a lengthened series of experiments on Povey's oil No. 2 was executed on Tuesday, the 17th October, it is needless to refer further to this day's observations.

Tuesday, 17th October.—13th Day.

The experiments of to-day illustrate the variations sometimes observed in oils bearing the same label. The shopman at Povey's, whom I questioned myself upon the subject, was not acquainted with the specific names of the oils sold by him; and I went myself to Bucklersbury to determine this point. The heavier oil sold by Povey was found, by reference to the books at Bucklersbury, to be No. 2. But its action in the lamps was certainly different from the No. 2 oil introduced by Captain Doty. Throughout the day both wicks exhibited a gradual advance of charring, the Doty wick most, and at 10.45 p.m., when the lamps were extinguished, the amount of charred matter was considerable. It rested so loosely on the top of the wick that Captain Doty blew it off his burner. Throughout the day the Doty lamp manifested a greater vigour than its rival; the illuminating power deduced from a mean of 12 observations taken every hour, being:—

Trinity lamp-----225·71 | Doty lamp-----288·44

Or as 100:127·79; being an excess of 27·79 per cent. in favour of Doty.

While the consumption was—

For the Trinity lamp---------310 ounces.
" Doty lamp---------315 "

Being a saving of 12·90 per cent. in favour of the Trinity.

Tuesday, 31st October.—14th Day

The long interval between the last series of experiments and the present one, arises in part from the fact that Mr. Douglass had been called away to Scotland, and Mr. Ayres to Dungeness.[53]

The behaviour of the Doty lamp during eight hours of the 16 on the ninth day (Oct. 6th), was so irregular that no conclusion of any value could be drawn from the day's experiments. On the 7th Captain Doty wrote to me the following note, explaining the irregularity, and requesting a repetition of the experiments:—

"Yesterday, during the course of the trial of Young's Light Oil, I discovered that the outer wick of my lamp was too short to be properly nourished with fluid, to give the full effect of my burner, which must account for the numerous adjustments which it required, and from which cause I was compelled to raise the flow or level of the fluid. I will, therefore, feel obliged if you will allow the trial of yesterday to be repeated for any number of hours you may name."

This day was devoted to the repetition solicited by Captain Doty.

It is right, however, to remark that what is here called "light oil" was Young's best paraffin oil, commonly known as No. 1. It was brought to the College of Chemistry by Captain Doty himself, and accepted by Mr. Douglass on the faith of its being what the label represented it to be. It has, moreover, a flashing point from 134° to 138° Fa[ren]h[ei]t., whereas Captain Doty's favourite "heavy oil," No. 2, flashes from 128° to 134° Fa[ren]h[ei]t.

The lamps, charged with paraffin oil No. 1, purchased on Saturday, the 11th, at Young's office in Bucklersbury, were lighted at 9.20 a.m., and at 10.45 the first series of photometric observations was commenced.

Both flames were high at the time, and at the fifth reading the Doty flame because very unsteady. Soon afterwards it flared at times into the iron chimney. The Trinity flame remained tranquil, though it was quite at its maximum. The entire series of ten readings was taken, but from the fifth observation to the end the Doty flame was so unsteady as to render it desirable to cancel the series.

11.50 A.M.

Trinity lamp—-—265·05 | Doty lamp—-—- 247·13

Both flames tolerably steady; the Doty flame rather low.

Both lamps remained untouched till 12.45 pm., when the next series of 10 photometric readings was taken.

12.45 P.M.

Trinity lamp—-—258·61 | Doty lamp—-—- 253·49

This was a gusty day, and the flames rose and fell bodily at intervals, indicating variations of draught. The Trinity flame showed this action most. The Trinity lamp remained untouched until the next series. The Doty was adjusted.

1.50 P.M.

Trinity lamp—-—271·76 | Doty lamp—-—- 240·72

The Trinity lamp was here above its normal height; it forked briskly, shooting occasionally as high as the iron chimney. At my request the flame was lowered. The Doty flame small and steady.

3.15 P.M.

Trinity lamp—-—243·34 | Doty lamp—-—- 269·58

At the fourth observation the Trinity flame sank, through the variation of the draught the Doty flame began to throw up long tongues; at first occasionally, but it ended with continuous flaring into the iron chimney.

The augmentation of photogenic power, due to the growth of the Doty flame towards flaring, was here well illustrated. It began with a reading of 15·60, and ended with 16·85. These correspond respectively to the intensities 243·36 and 283·92.

Before the next readings the Doty flame was lowered and the Trinity flame raised. At 4.15, however, the Doty was again flaring; the disk was accordingly lowered, and at 4.30 the observations began.

4.30 P.M.

Trinity lamp—-—243·39 | Doty lamp—-—- 268·99

The lamps were permitted to remain untouched during the interval between this series and the next one.

5.10 P.M.

Trinity lamp—-—256·32 | Doty lamp—-—- 255·68

The lamps were here very fairly matched. The Trinity flame was largest; hence, volume for volume the Doty flame was most intense.

It was requested that, if possible, the flames were to remain untouched until the next observation, but at 5.55 the Doty flame was so high that it was lowered by Captain Doty himself.

6.30 P.M.

Trinity lamp—-—252·34 | Doty lamp—-—- 259·70

At 6.40 the photometric readings were concluded. Both flames burnt briskly at their maximum height, but remained steady throughout the entire series of observations.

Consumption of Oil in 12 *Hours.*

Trinity lamp:—351 ounces. Doty Lamp:—339 ounces.

Or as 100:96·57; being a saving of 3·43 per cent. in favour of the Doty.

SUMMARY of Photometric Results.

		Trinity Lamp.	Doty Lamp.
1st Experiment—-	—-	265·05	247·13
2nd "—-	—- -	258·61	258·49
3rd "—-	-—-	271·76	210·72
4th "—-	-—-	243·34	269·58
6th "—-	-—-	243·39	268·99
0th "—-	-—-	256·32	255·68
7th "—-	-—-	252·34	259·70
Mean	-—-	255·83	256·44

Or as 100 : 100·24; being a saving of 0·24 per cent. in favour of Doty.

Wednesday, 1st November.—15th Day.

From the 3rd of October up to the present date, the Doty lamp had stood at the north end and the Trinity lamp at the south end of the photometer. I believe, as regards draught, that this arrangement placed the Trinity lamp under a disadvantage throughout. At Mr. Douglass's request the lamps were now reversed; and several series of experiments, divided from each other by intervals equal to those of yesterday, were executed.

Lamps charged with paraffin oil (No. 1) were lighted at 10.40 a.m. The experiments commenced at—

	Trinity Lamp.	Doty Lamp.
12.5 P.M—-—- -	265·04	217·12
1 PM.—-—-—	249·96	262·12
1.55 P.M—-—- -	252·66	259·38
2.50 P.M.—-—- -	266·35	245·87
4.15 P.M.—-—- -	260·34	251·70
5.15 P.M.—-—- -	258·96	253·07
6.15 P.M.—-—- -	261·02	251·03
7.25 P.M.-—-—	266·18	246·02

The flames remained in excellent condition for one hour without being touched.

The lamps were extinguished at 10.40 p.m.

At 10.15 p.m., I made the following note regarding them:—Both lamps showed very fine flames; the Trinity flame largest, the Doty name whitest. They were well matched, and had remained steady for nearly four hours without being touched.

Consumption during twelve hours.

Trinity lamp—- 369 1/2 ounces. | Doty lamp—-—339 ounces.

Or as 100 : 91·74; being a saving of 8·26 per cent. in favour of Doty

SUMMARY of Photometric Results.

	Trinity Lamp.	Doty Lamp.
1st Experiment—-	265·04	247·12
2nd "—-—- -	249·96	262·12
3rd "—-—- -	252·66	259·38
4th "—-—- -	266·35	245·87
5th "—-—- -	260·34	251·70
6th "—-—- -	258·96	253·07
7th "—-—- -	261·02	251·03
8th "—-—- -	266·18	246·02
Mean—- -	260·06	252·04

Or as 100 : 96·92; being a saving of 3·08 in favour of Trinity.

Estimate of Results.

It will, I think, be most agreeable to the Elder Brethren and the Board of Trade if I now present, not a formal numerical summary of the foregoing results, but, employing my best judgment and plain language, as accurate an estimate of the comparative merits of the two lamps as the data enable me to form.

First, then, I will group together the four days occupied with petroleum 5a. Taking both consumption and illuminating power into account, the first day's experiments assign a gain of about 1·69 per cent. to the Doty lamp. The second day's experiments give a gain of 8·44 per cent.; the third a gain of 9·44 per cent., and the fourth a gain of 23·58 per cent. to the same lamp.

The results of the first and fourth day's experiments I hold to be abnormal, and not fairly representative of the action of the two lamps. The one gives the Doty lamp too little advantage, the other too much. The second and third days' experiments I hold to be fairly representative. Hence, as regards the employment of petroleum 5a, I should award a superiority of 9 per cent. to the Doty lamp.

Grouping also together the three days occupied with paraffin at No. 2, we find that on the first of these days, when Mr. Douglass was absent, the Trinity lamp showed itself to be practically incompetent to burn this oil; but that, on the second day, under the superintendence of Mr. Douglass, the competence of the lamp to burn oil No. 2 was demonstrated. Indeed, in some of the most accurate observations on this occasion the Trinity lamp showed an illuminating power but little short of that of the Doty lamp. The newness of the material and the previous experience of Mr. Ayres rendered Mr. Douglass cautious in the earlier part of the day; and the first series of experiments is manifestly not representative of the power of the Trinity lamp. Throwing this first series away, we have still an excess of 10·64 per cent. of illuminating power in favour of the Doty lamp, which added to the saving in consumption, would give the Doty a superiority of 15·75 per cent. On the third occasion where the oil used was Povey's No. 2 the Doty lamp had an advantage of 27·79 per cent. in illuminating power; the Trinity lamp, on the other hand, had a saving of 12·90 percent in consumption of oil. This leaves a clear balance of 14·89 per cent. in favour of the Doty.

It is to be remarked that No. 2 is not the oil which had been experimented with in Scotland by Dr. Macadam, and at the Trinity House by Mr. Douglass and Mr. Valentin; nor is it the oil which I should at present recommend for lighthouse purposes. Still, the experiments with it are valuable as demonstrating the superior burning power of the Doty lamp in the case of the heavier oils. There can be no doubt that, owing to the greater air spaces, to the artifices employed to cause the air to impinge upon the flame, or to both these causes combined, the Doty lamp maintains a more vigorous combustion than the Trinity lamp. This greater energy of combustion is indicated by the cleanliness of the interior surface of the Doty lamp-glass, and also by the whiteness of the Doty light. It comes most conspicuously into play in the case of the more smoky oils, and hence the desire manifested by Captain Doty throughout the inquiry to press the claims of oil No. 2.

But the advantage here claimed for the Doty lamp appears to be confined within narrow limits; for as regards the *charring* of the wicks by the heavier petroleum and paraffin oils, as illustrated by the oil 4a, and by Povey's paraffin No. 2, both lamps appear to suffer equally.

Grouping together the three days occupied with paraffin oil (No. 1), which is really the oil to be recommended under present circumstances for lighthouse use, we have first a day of sixteen hours, from which, owing to the derangement of the Doty lamp, no conclusion can be drawn, either as regards consumption or illuminating power. The day's work, however, demonstrates that great care is necessary in arranging the "flow"*· and in tying the wicks. On the other two days, however, valuable results were obtained. Taking both consumption and illuminating power into account, the first day gives the Doty lamp a superiority of 3·67 per cent.; while the second day gives the same lamp a superiority of 5·18 per cent. over its rival.

The merits of the lamps can now be stated in a few words. With oil No. 1, which appears to be more suitable for lighthouse purposes than any other, the Doty lamp shows a small, but distinct superiority. As regards this oil, the simple changes introduced by Mr. Douglass into the Trinity burner appear to be almost as effective as the more elaborate devices of a central button and an external jacket introduced by Captain Doty. But as regards the combustion of oils richer in carbon, and, therefore, more reliable to produce a smoky flame, the Doty lamp possesses, within certain limits, a distinct advantage. On the whole, therefore, I should pronounce the Doty lamp the more effective of the two.

Concluding Remarks.

The Doty lamp is patented, the Trinity lamp is unpatented, and *both* have been proved fit for lighthouse use. It is the difference between them that is to be taken into account in making an arrangement with Captain Doty. If his claims should be inordinate, the Elder Brethren have their own lamp to fall back upon; if his claims be moderate, I would venture to express the hope that they may be allowed.

For I cannot neglect dwelling upon the fact that Captain Doty was the first in this field; that from the time when he exhibited his first imperfect lamp at the Trinity House[54] he has steadfastly pursued this subject; that at a period when the Elder Brethren were merely proposing to extend their experiments to lamps of the first order, the Doty lamp had attained the high state of efficiency revealed by the experiments of Dr. Stevenson Macadam. While, therefore, the Board of Trade possesses an effectual check to all extravagant demands on the part of Captain Doty, I should learn with pleasure that a fair arrangement with him had been made.

The arrangement proposed by the Commissioners of Northern Lighthouses bears obvious marks of precipitation. The Board of Trade, I take it, has to look at this question from an imperial point of view. It has to take into account, not only Scotland, but England, Ireland, and the Colonies. Supposing the Doty lamp to possess every quality claimed for it in the north, and even still higher qualities, and that no other lamp of the kind existed, the Board of Trade would still have to consider what the terms recommended by the Northern Commissioners involved if applied to all the lighthouses over which the Board exercises control. Without such knowledge the adoption of these terms would be simply a leap in the dark.

Another point which ought to come under the consideration of the Board of Trade is this: Gas has been successfully applied in certain lighthouses in Ireland, and I think any practical mariner, reading the last report which I had the honour of submitting to the Board of Trade on this subject[55] must see that the special handiness with which gas lends itself to the purposes of a signal light, is a very strong point in its favour. Until we know what gas can do it would be unwise to make a sweeping alteration in favour of paraffin. Paraffin may, and probably will be proved cheaper than gas, but cheapness is only one element of the problem. If it can be shown that a gaslight, owing to its demonstrable "flexibility" can give the mariner more definite warning and more certain guidance, then it becomes a question whether these advantages might not be wisely purchased by a little additional outlay. The Elder Brethren are preparing to give gas a trial at Haisboro',[56] and until this trial is ended no contract, involving an entire change in our system of lighting, ought in my opinion, to be made.

I have thus done my best to give a full and fair account of an inquiry which, for many reasons, I should not like to repeat; and it will give me pleasure to learn that the results established have contributed to the final settlement of an entangled question. Were the combatants less eager, and Captain Doty less fond of "protests," the Report might be condensed to a statement of the results; but as the case stands, it is desirable to show upon the face of the Report that nothing was done or neglected which could justify complaint.[57] In conclusion, I wish to state some precautions which are essential in dealing with mineral oils. What we call paraffin oil is not a body of definite chemical composition, but a mixture of hydrocarbons differing widely from each other. Some are light and volatile, others heavy and non-volatile; some are even solid. Supposing, then, that we take two hydrocarbons from the extreme ends of the series and mix them together, we obtain in the mixture an oil of mean specific gravity. But suppose we choose our sample from the middle of the series, we also obtain an oil of mean specific gravity. But it is plain that these two oils, though agreeing in density, would be totally different from each other. One of them, by containing a light and volatile constituent, might be highly dangerous, while the other, through the absence of any such constituent, might be perfectly safe. Hence, specific gravity is no safe test of the quality of an oil. This point was brought out with perfect distinctness by Mr. Valentin in a previous report,[58] and I have now to direct attention to the conclusions enunciated by Mr. Valentin in the letter hereto appended.† [59] It is essential, especially in our colonial lighthouses, that a sufficient supply of the proper oil should always be present, and that they should never be dependent on the accidents of the market. The burners are so restricted in their powers of combustion, that an apparently slight variation in the quality of the oil might produce serious deterioration of the light, or even render the burner practically useless. By careful analysis, the proper qualities of oil can always be determined with ease, but careful analysis in the sense expressed by Mr. Valentin must in all cases precede the employment of paraffin oil.

I have, &c. | (signed) *John Tyndall.*

Robin Allen, Esq.

'* *See* Sub-Enclosure 2, on page 46.

° Mr. Douglass was absent from this day's experiments, and on his arrival expressed his discontent at being forced to burn an oil entirely new to him. It was an error on his part to suppose that any coercion had been employed. Mr. Ayres, who, it must be said, was avowedly far less conversant than Mr. Douglass with the Trinity lamp considered that lamp capable of burning anything, and he made no objection whatever to the trial of oil No. 2. That his other duties should have compelled Mr. Douglass to be absent was unfortunate, but he has, in my opinion, to complain of nothing unfair.

*Captain Doty informed me on the following day that he was bewildered and excited himself through the discovery that the Trinity burner had been tampered with. This charge of tampering has been already referred to as withdrawn.

*That is, the level of the oil in the burner, which, as already stated, is about three inches below the tips of the wicks.

†*See* Sub-Enclosure 1, at p.45

'Lighthouses (Mineral Oils). Copy of Further Correspondence relative to Proposals to Substitute Mineral Oils for Colza Oil in Lighthouses (in Continuation of Parliamentary Paper, no. 2, of Session 1872)', HC 2 67,[60] Parliamentary Papers (1872), pp. 23–47
Typed Transcript Only

---

1. *Enclosure in No. 29*: see letter 3522, n. 1. The typescript of the present report, our only source, has a number of errors; many of these have been silently corrected.
2. *Trinity . . . lamps*: Since 1867 Tyndall had been involved in a dispute between Trinity House and the Commissioners of the Northern Lighthouses (the comparable body for Scotland) concerning competing lamp designs. This report concerns a lamp used by Trinity House, designed by Douglass, and one used by the Scots, designed by Doty. See *Ascent of John Tyndall*, pp. 284–5.
3. *the whole of the printed correspondence*: Tyndall was probably referring to the previous Parliamentary Paper on the subject. See 'Correspondence Between the General Lighthouse Authorities and the Board of Trade, relative to Proposals to Substitute Mineral Oils for Colza Oil in Lighthouses', HC 318, Parliamentary Papers (1870).
4. *lighthouse of Girdleness*: or Girdle Ness; a lighthouse south of Aberdeen's harbour, in the north of Scotland.
5. *photometric comparison*: comparison of the intensity of two sources of light.
6. *dioptric lighthouses*: a reference to the type of optical apparatus used in lighthouses. Dioptric, or refractive, lighting used glass to concentrate light into narrow beams ('Lighthouses', in *Encyclopedia Britannica*, 11th edn, 11 vols (New York: Encyclopaedia Britannica, 1911), vol. 16, pp. 627–50, on p. 633).
7. *Pentland Skerries*: a lighthouse in the north of Scotland.
8. *catoptric lighthouses*: catoptric, or reflective, lighting used reflective surfaces to concentrate light into narrow beams ('Lighthouses', *Encyclopaedia Britannica*, vol. 16, p. 633).
9. *Pladda*: a lighthouse in the west of Scotland, off the island of Arran.
10. *Dr. Stevenson Macadam*: see letter 3522, n. 5.
11. *The Elder Brethren stated in reply*: see 'Correspondence Between the General Lighthouse Authorities and the Board of Trade', HC 318, p. 10.

12. *M. Reynaud*: Léonce Reynaud (1803–80), a French engineer who directed the French lighthouse service from 1849 until 1878.
13. *first-class lights*: probably first-order lights. French physicist and engineer Augustin Fresnel, who designed the widely used Fresnel lens, divided his lens into different orders based on size. First-order lights were about six feet across; second-order lights about four feet across; third-order lights about three feet. Fourth- and fifth-order lights were approximately nineteen and fourteen inches across, respectively, and were used for local hazards. See T. Levitt, *A Short, Bright Flash: Augustin Fresnel and the Birth of the Modern Lighthouse* (New York: W.W. Norton, 2013), pp. 83, 168.
14. *Young's Paraffin*: see letter 3422, n. 6.
15. *leading lights*: lights used to lead ships through narrow channels or provide direction. See 'Lighthouses', *Encyclopaedia Britannica*, vol. 16, p. 635.
16. *Great Castle Head*: a lighthouse in Wales.
17. *they proposed . . . safely stored*: 'Correspondence Between the General Lighthouse Authorities and the Board of Trade', HC 318, p. 10.
18. *On the . . . Stevenson*: see 'Correspondence between the General Lighthouse Authorities and the Board of Trade', HC 318, pp. 13–8.
19. *For first order lights . . . cost of colza*: for the different orders of lights, see n. 13.
20. *that the full amount . . . illuminating power*: see 'Correspondence between the General Lighthouse Authorities and the Board of Trade', HC 318, p. 18.
21. *trimming*: to trim a lamp is to put it 'in proper order for burning' (*OED*).
22. *photogenic*: 'producing or emitting light' (*OED*).
23. *After a full . . . practically safe*: see 'Correspondence between the General Lighthouse Authorities and the Board of Trade', HC 318, p. 18.
24. *it must be . . . lighthouses*: see 'Correspondence between the General Lighthouse Authorities and the Board of Trade', HC 318, p. 18.
25. *that paraffin . . . Doty*: see 'Correspondence between the General Lighthouse Authorities and the Board of Trade', HC 318, p. 19.
26. *Argand burner*: lamp invented in 1780 by François-Pierre-Ami Argand (1750–1803), a Swiss physicist.
27. *the tips . . . cotton wick*: see 'Correspondence between the General Lighthouse Authorities and the Board of Trade', HC 318, p. 37.
28. *electric light and coal gas*: Electric light was first used for a lighthouse in England in 1858; coal gas was first used in 1837; see 'Lighthouses', *Encyclopaedia Britannica*, vol. 16, p. 641, p. 640. Tyndall had investigated the question of coal gas as part of his work for the Board of Trade; see *Ascent of John Tyndall*, pp. 284–5.
29. *In conclusion . . . used*: see 'Correspondence between the General Lighthouse Authorities and the Board of Trade', HC 318, pp. 38–9.
30. *his results*: see letter 3422.
31. *Sperm*: oil from sperm whales, used in English lighthouses until the middle of the nineteenth century. See 'Lighthouses', *Encyclopaedia Britannica*, vol. 16, p. 640.
32. *The result . . . lamps*: see 'Correspondence between the General Lighthouse Authorities and the Board of Trade', HC 318, p. 25.

33. *the report*: letter 3422.
34. *On the 15th . . . letter*: letter missing.
35. *on the 16th . . . note*: letter missing.
36. *Mr. Ayres*: Arthur Ayres (1830–1907) served as Chief Assistant Engineer at Trinity House from 1865–92; he worked with Tyndall on fog-signalling as well on the use of oil, gas, and electricity in lighthouses.
37. *au fait*: expert or skilled.
38. *Mr. Taylor*: Taylor (n.d.) was described in a later report by Douglass as a 'foreman lamp-maker at the Blackwall workshops'. See J. N. Douglass to Robin Allen, in 'Copies of Further Correspondence relative to Proposals to Substitute Mineral Oils for Colza Oils in Lighthouses', HC 3 378, Parliamentary Papers (1873), pp. 45–8, on p. 48.
39. *Mr. Easton . . . lamp*: not further identified.
40. *his letter to me*: letter 3422.
41. *Captain Doty's assistant*: Mr. Easton; see n. 39.
42. *photometer*: instrument 'for comparing the intensities of light from different sources' (*OED*).
43. *This day's experience seemed fully to justify his confidence.**: Although an asterisk is used within the text to mark a footnote of Tyndall's, the footnote itself is marked with °.
44. *The squares . . . of the light*: makes reference to the inverse square law stating that 'the intensity of a particular force or phenomenon at a given point is inversely proportional to the square of the distance between that point and the source of the force or phenomenon in question' (*OED*).
45. *a lengthy communication from Captain Doty*: letter missing.
46. *utterly unfit for lighthouse purposes*: In the paragraph above, Tyndall has 'totally unfit' but here has changed it to 'utterly unfit'.
47. *9.5*: presumably an error in the original, but we are unable to reconstruct the correct time.
48. *As Mr. Valentin first sagaciously pointed out*: see letter 3540.
49. *a written one*: letter missing.
50. *letter of mine*: possibly letter 3522.
51. *Bucklersbury*: a street in London.
52. *Povey's*: not identified.
53. *Dungeness*: on the coast of Kent in the southeast of England.
54. *the time when he exhibited his first imperfect lamp at the Trinity House*: In 1868 Tyndall examined the lamp and counselled Trinity House not to adopt it. See T. A. Tag, 'The Doty Dilemma: Technological Advancements in Lighthouse Lamps and Subsequent Patent Infringement', U.S. Lighthouse Society *Keeper's Log*, 16, no. 4 (Summer 2000), pp. 28–33.
55. *Gas . . . this subject*: Tyndall had at various times reported favorably on gas in Irish lighthouses. See letters 3475 and 3521 (the 'last report' referenced here), as well as *Ascent of John Tyndall*, pp. 284–5.
56. *Haisboro'*: sandbank off the coast of Norfolk.
57. *it is desirable . . . complaint*: Notwithstanding Tyndall's caution, Doty articulated several further complaints; see letter 3568.
58. *in a previous report*: see letter 3422.

59. *the letter hereto appended*: see letter 3540.
60. *2 67*: At the time of writing, the ProQuest U.K. Parliamentary Papers database misnumbered this paper as 2 264.

To Jane Barnard 16 November [1871][1] 3556

[Royal Institution][2] | 16th Nov.

My dear Miss Barnard.

Friday, tomorrow and Tuesday next week are days on which I am in bonds: Let us postpone it[3] until a glut of engagements which now pester me are over.[4]

I am right glad to hear good accounts of you.

Yours ever | John Tyndall

RI MS JT/1/T/125
RI MS JT/1/TYP/12/4201

1. *1871*: LT dates this letter to 1871, which makes sense given how busy Tyndall was at this time with research into competing lighthouse lamp designs and given that in letter 3579 Barnard's aunt, Sarah Faraday, referred to not having seen Tyndall for some time.
2. *Royal Institution*: LT annotation.
3. *Let us postpone it*: Tyndall seems to be responding to an invitation in a letter that is missing.
4. *a glut of engagements*: possibly a reference to Tyndall's experiments testing rival designs for lighthouse lamps. See letter 3522, n. 3.

From Henri Deville[1] 17 November 1871 3557

Laboratoire de Chimie | Ecole Normale Supérieure | Paris, le 17 Nov. 1871

Cher ami—Je viens de recevoir votre ouvrage sur la physique destinée aux ignorants, race bien heureuse qui se *[préfère]* aux *[demi-]*savants & à laquelle vous faites bien de vous adresser. Pour moi qui ignore tant de choses je vais me mettre à vous étudier & je suis sûr d'y trouver à faire mon profit, comme cela m'est arrivé bien souvent avec vos ouvrages. J'aurai peut-être recours à vos renseignements & à vos conseils parce que plusieurs de mes amis & moi nous désirons établir en France votre Institution royale. Je compte que vous voudrez bien me dire vos idées à la page 8.

M. Jules Simon, notre ministre de l'instruction publique m'a dit que vous lui aviez écrit au sujet d'un de mes élèves M. Croullebois.

Sachez que c'est un jeune homme fort peu estimé de ses maîtres & de ses camarades. Sa conduite privée laisse beaucoup à désirer & nous savons pertinemment que ses travaux sont faits sans confiance. Il rapporte des expériences qu'il n'a jamais faites & publie des nombres qu'il n'a jamais obtenus. C'est du moins l'impression générale & les réclamations qu'il a soulevées partout *[valent]* donner parfaitement raison. J'ai été obligé d'intervenir auprès du ministre pour qu'on ne le nomme qu'après une enquête à la place qu'il sollicite & cette enquête ne lui est pas favorable. Ainsi méfiez vous à son sujet. Je suis bien désolé, cher ami, de ne vous avoir pas vu en Suisse cette année. Nous avons été ici si troublés dans nos *[instituts]*, dans nos *[facultés]*, dans nos travaux que je suis bien triste & bien découragé.

Votre ami dévoué | H Sainteclair Deville

Chemistry Laboratory | Ecole Normale Supérieure[2] | Paris | 17 Nov. 1871

My dear friend—I just received your book on physics for the uneducated,[3] that happy race so *[preferable]* to the *[half-]*educated, which you do well to address. As for me, knowing so little, I will start to study your work & I am sure I will find something profitable in it, as I often have in your books. Perhaps I will be able to draw on your expertise & your advice, as many of my friends & I wish to establish your Royal Institute in France. I hope that you will kindly tell me about your ideas on page 8.[4]

Mr. Jules Simon,[5] our minister of public instruction, told me that you have written to him[6] about one of my students, Mr. Croullebois.[7]

Know that he is a young man who is not held in high esteem by his teachers & comrades. His private conduct leaves much to be desired & we know for a fact that his work is unreliable. He reports on experiments he has never done & publishes numbers he never obtained. At least, that is the general impression & the complaints that arose wherever he went make it worth believing. I was forced to intervene with the minister so that Croullebois would only be appointed to the position he requested after an investigation, & that investigation has not been in his favour. So, do not trust him. I am very sorry, dear friend, not to have seen you in Switzerland this year. We were so troubled here in our *[institutes]*, in our *[faculties]*, in our work, that I am very saddened & discouraged.

Your devoted friend | H Sainteclair Deville

RI MS JT/1/D/132

---

1. *Henri Deville*: Etienne Henri Sainte-Claire Deville (1818–81), a French chemist. Deville was a professor at the Sorbonne from 1859–81; his most significant research focused on inorganic and thermal chemistry, including the isolation of compounds and the theory of reversible reactions.
2. *Ecole Normale Supérieure*: elite graduate school in Paris.
3. *your book on physics for the uneducated*: *Fragments of Science*.
4. *your ideas on page 8*: possibly a reference to a discussion of molecular motion. Tyndall

writes, 'We on the earth's surface live night and day in the midst of aethereal commotion' (Tyndall, *Fragments of Science*, p. 8).

5. *Jules Simon*: see letter 3467, n. 5.
6. *Mr Jules Simon . . . to him*: letter missing, but the correspondence is referred to in letters 3467 and 3469.
7. *Mr. Croullebois*: Marcel Croullebois.

To Hector Tyndale 18 November 1871 3558

[Royal Institution][1] | 18th Nov 1871

My dear Hector

Long before this reaches you, in fact before it is written, you will have received my letter.[2] At the same time letters were dispatched to Prof. Henry and to Prof. Lesley.[3] I must say a word regarding the cause of the delay in replying to you. Firstly, I remained in Switzerland much longer than I intended, and secondly after I came home I was seized upon by the Board of Trade and the Trinity House and had to throw myself into what was to me a most unpleasant investigation.[4] It quite made shipwreck of my autumn's scientific work. Before I replied I wished to see my course clear; for I shrank from the idea of going to America to lecture without being able to introduce something new and good. Pressed as I have been I shall not have time to prepare such lectures as I should deem most suited to my American audiences, and I shall therefore be obliged to offer them poorer fare than I should willingly offer if I had the time to prepare it. I trust that this will explain my tardiness—I am now committed to the thing—and must get through with it as best I can; consoling myself by the thought that I shall at all events make a first acquaintance with American audiences, and that at some future day I may make atonement for the shortcomings of the next year.

Mr. Secretary Fish[5] telegraphed some time ago to General Schenk[6] making enquiries about the matter & Mr Moran[7] had the great kindness to call upon me here. I was absent, but went down to the embassy afterwards and explained all.

I am just on the point of quitting London for a day or two—your letter[8] arrived last night, & I did not like to go away without first sending you this hasty reply.[9]

My kindest regards to your wife.[10]

Yours affectionately | John Tyndall

RI MS JT/1/T/1451
RI MS JT/1/TYP/4/1692

1. *Royal Institution*: LT annotation.
2. *my letter*: This letter is missing, but seems to have been a response to 3507. The present letter responds to 3519.
3. *letters were dispatched to Prof. Henry and to Prof. Lesley*: see letters 3542 and 3543.
4. *a most unpleasant investigation*: Tyndall's investigation into competing designs for lighthouse lamps; see letter 3522, n. 3.
5. *Mr. Secretary Fish*: Hamilton Fish (1808–93), an American lawyer and politician. He served as Secretary of State from 1869 to 1877. In this position he helped to negotiate the Treaty of Washington of 1871, for which see letter 3616, n. 4.
6. *General Schenk*: Robert Cumming Schenk (1809–90), an American lawyer and politician. Schenk served as a member of Congress and a diplomat; Abraham Lincoln appointed him as a general in the American Civil War. From 1870 to 1876 he was the American minister to Great Britain.
7. *Mr Moran*: Benjamin Moran (1820–86), an American diplomat who held the post of secretary of legation in London from 1855 to 1874. In this position he did much to promote amicable Anglo-American relations and helped to avoid a rupture between the two countries during the American Civil War.
8. *your letter*: probably letter 3519.
9. *this hasty reply*: Tyndale responded to the present letter and another letter from Tyndall, apparently missing, in letter 3594. For additional letters related to this topic, see 3504, 3510, 3589, 3639, 3669, 3670, 3684, 3685, 3702 and 3703.
10. *your wife*: Julia Tyndale.

To Thomas Archer Hirst 24 November 1871 3559

Friday | Nov. 24th 1871

Dear Tom

Something might arise to render the expression of an opinion from me desirable: I therefore send you the accompany note.[1] Do what you please with it.

J. T.

I have been quite laid up.[2]

RI MS JT/1/T/688
RI MS JT/1/HTYP/584

1. *the accompany note*: note missing; presumably Tyndall meant to write 'accompanying'. Hirst noted on this letter that it was sent 'in re extraordinary general meeting of Athenaeum Club'. According to Hirst's Journal, the general meeting of 24 November was 'convened to consider the late charges in the tariff for table and Billiards'. Hirst reports

reading Tyndall's letter, which opposed the 'table money and produced an effect' (Brock and Macleod, *Hirst Journals*, p. 1921). We get some sense of this effect from letters 3561 and 3562.

2. *I have been quite laid up*: Tyndall had injured his right calf on 19 November 1871; see letters 3564 and 3567.

To unidentified 24 November [1871][1] 3560

Royal Institution | 24th Nov.

My dear Sir

Dr Bence Jones has just left me, and his stern commands are that I must not leave my sofa tomorrow.

I have strained my leg, and am quite a prisoner.[2]

This I regret very much as I had many pleasurable thoughts regarding my visit to Cambridge[3]

Yours faithfully | John Tyndall

RI MS DPM/1/I

---

1. *[1871]*: dated by reference to Tyndall's injured leg; see n. 2.
2. *I have . . . prisoner*: see letter 3559, n. 2.
3. *[Cambridge]*: We propose Cambridge as a possible reading here, but have no further information about who Tyndall might have been planning to visit there.

From Abraham Hayward[1] 24 November [1871][2] 3561

Copy | Athenaeum[3] Nov. 24

Dear Professor Tyndall

A letter from you was read at the meeting this day containing a proposition which (I own) struck me with surprise. It was to the effect that this Club[4] owed its principal distinction to a class of members to whom small economies are an object: meaning (it was supposed) poor or struggling men of science—

Now it strikes me that it owes its distinction to the eminent & successful men of all professions & vocations—

In the successful clergy, headed by the bishops—the visiting lawyers[5] with the judges—the leading politicians, including ex & actual members of the Cabinet[6]—the officials and ex-officials well-salaried or pensioned: the city magnates headed by L[or]d Overstone[7] & Baring:[8] the great nobles, like the Dukes of Devonshire,[9] Somerset,[10] Cleveland,[11] Argyll[12] &c—the leading physicians and surgeons—the artists, who are Academicians[13] to a man: the military and naval notabilities who are almost all generals and admirals—the literary men of established reputation, like Lytton,[14] Houghton,[15] Stirling-Maxwell,[16] Tennyson,[17] Carlyle, Froude, Browning,[18] Kinglake[19] &c. &c and the men of science headed by Professors like yourself.

I looked round the crowded meeting today and could not discover more than four members under forty, and not one to whom to the best of my knowledge, the 3$^{\underline{d}}$ in dispute could be an object

The great majority are men who <u>have risen</u>, or who hold a position implying at least a competency—

It strikes me that the class to which you allude must be a <u>very</u> small minority and that they do not form the principal distinction of the Club. But I differ with hesitation and regret from any opinion so formally put forward by you—

Ever faithfully yours | A. Hayward

HWD 10.107 (COPY), Hayward Correspondence, The Carl H. Pforzheimer Collection of Shelley and His Circle, New York Public Library, Astor, Lenox, and Tilden Foundations

---

1. *Abraham Hayward*: Abraham Hayward (1801–84), an essayist and translator. He wrote for major periodicals including the *Morning Post*, the *Edinburgh Review* and the *Quarterly Review* on topics including politics, literature and history-writing. Hayward was elected to the Athenaeum Club in 1835 and, as a bachelor, passed much of his time there.
2. *[1871]*: Hayward responded to a letter of Tyndall's that Hirst read at a general meeting of the Athenaeum Club on 24 November 1871. See letter 3559.
3. *Athenaeum*: the Athenaeum Club; see letter 3372, n. 14.
4. *Club*: the Athenaeum Club.
5. *visiting lawyers*: Visiting lawyers have the privilege of practicing law in other jurisdictions.
6. *the Cabinet*: The body that, with the Prime Minister, leads the British government.
7. *Overstone*: Samuel Jones Loyd, Baron Overstone (1796–1883), a banker with Jones, Loyd, and Co., which had been founded by his father and uncles. He became well known in the first half of the nineteenth century for his writings on banking and finance, which were politically influential. Loyd was made Baron Overstone in 1849. He was a member of the BAAS.
8. *Baring*: possibly Thomas Baring (1799–1873), a banker and politician. With American merchant Joshua Bates, Baring was a partner in Baring and Co., a firm that dealt in trade and merchant financing, and in sterling bond issues. Baring first entered the House of Commons as a Tory in 1835.
9. *Devonshire*: William Cavendish; see letter 3468, n. 5.
10. *Somerset*: Edward Adolphus St Maur (1804–85), twelfth duke of Somerset and politician.

St Maur first entered the House of Commons as a Whig in 1830; he entered the House of Lords in 1855 upon his father's death and became first lord of the Admiralty in 1859. He was made a knight of the Garter in 1862, and Earl St Maur of Berry Pomeroy on 17 June 1863.

11. *Cleveland*: Harry George Powlett (1803–91), fourth duke of Cleveland; his name was Harry Vane until 1864. He entered the House of Commons in 1841, and the House of Lords in 1864 upon his elevation to the dukedom. He was made a Knight of the Garter in 1865.
12. *Argyll*: George Campbell.
13. *Academicians*: members of the Royal Academy of Art, which was founded in 1768 as a society to promote the fine arts.
14. *Lytton*: Edward George Earle Lytton Bulwer Lytton (1803–73), the first Baron Lytton, an author and politician. He entered the House of Commons as a radical in 1831, and returned in 1852 as a Tory. His literary works, which included novels, non-fiction and plays, were extraordinary successful, although today he is much less well known than many of his fellow novelists.
15. *Houghton*: Richard Monckton Milnes (1809–85), first Baron Houghton, a politician and author. He entered the House of Commons in 1837, and the House of Lords in 1863, when he was made first Baron Houghton. His *Life, Letters, and Literary Remains of John Keats* (1848) recuperated Keats's poetry.
16. *Stirling-Maxwell*: William Stirling Maxwell (1818–78), an art historian and collector of books and art; he was particularly interested in Spanish art. He entered the House of Commons as a Conservative in 1852. Born William Stirling, he assumed the name Stirling-Maxwell when he was allowed to succeed his maternal grandfather's baronetcy in 1865.
17. *Tennyson*: Alfred Tennyson (1809–92), a major Victorian poet. Tennyson is the author of such works as *In Memoriam* (1850), *Maud* (1855) and the Arthurian romance *The Idylls of the King* (1859–85). Queen Victoria appointed him Poet Laureate in 1850 and made him Baron Tennyson in 1883.
18. *Browning*: Robert Browning (1812–89), another major Victorian poet, was notable especially for his dramatic monologues. His most famous works include *Men and Women* (1855), *Dramatis Personae* (1864) and *The Ring and the Book* (1868–69). Browning was married to another important Victorian poet, Elizabeth Barrett Browning (1806–61).
19. *Kinglake*: Alexander Kinglake (1809–91), a travel writer, historian, and politician; he composed the eight volume work *The Invasion of the Crimea* (1863–87). He was a Liberal member of the House of Commons from 1857–69.

To Abraham Hayward[1] 25 November [1871][2] 3562

Royal Institution of Great Britain | 25th Nov.

My dear Hayward.

I regard your note as a compliment, and thank you for it accordingly.

And I entirely concur in all that you say regarding the distinction conferred upon the Athenaeum[3] by the high personages & classes you have mentioned.

Looking at the great Clubs of London we find some features common to them all. The Athenaeum resembles the Carlton[4] and the Reform[5] in possessing the magnates to which you refer. They confer <u>distinction</u> on all three clubs, but they do not <u>differentiate</u> the one from the others. Is there any basis of differentiation? Obviously yes, the Carlton and the Reform are also distinguished from each other through the one being conservative and the other liberal. The Athenaeum is distinguished from both by its strong infusion of Literature, Art and Science.

Literature, Art and Science, then, give the Athenaeum—whether 'distinction' or not I will not say, for I never mentioned the word—but its "distinctive character".

I might take some of the very highest names that you have addressed as illustrating my point that it is desirable, for the sake of such, to keep the charges of our club within moderate limits. They are now old, and happily raised above all considerations of the kind, but in the eyes of competent judges they were masters long before their works obtained the market value which they now enjoy. I should expect you <u>above</u> most others to lend a hand in */preserving/* a broad approach for such men to the Athenaeum.

Pray banish the idea that I was thinking of science or scientific men when I wrote; and give me credit for an extension of sympathy beyond the narrow margin of my own studies. In fact had I had full liberty when young those studies would have taken a totally different direction. For though I loved Science well, I loved your friend Byron[6] far better.

Now be at peace with me & believe me | Yours faithfully | John Tyndall | A Hayward Esq[ui]^{r}[e] | &c &c

HWD 10.106, Hayward Correspondence, The Carl H. Pforzheimer Collection of Shelley and His Circle, New York Public Library, Astor, Lenox, and Tilden Foundations

---

1. *Abraham Hayward*: see letter 3561, n. 1.
2. *[1871]*: written on the first page in pencil and on the fourth page, vertically on the left side in red pen, in unknown hand. It is confirmed by the letter's continuity from 3559 and 3561.

3. *Athenaeum*: see letter 3372, n. 14.
4. *the Carlton*: founded in 1832, as the Conservative Party regrouped after failing to prevent the passage of the democratising First Reform Act of 1832.
5. *the Reform*: founded in 1836, as a Whig and Radical counterpart to the Carlton Club.
6. *your friend Byron*: George Gordon Noel Byron (1788–1824), the sixth Baron Byron, a major Romantic poet. Famous in his lifetime both for his poetry (including *Childe Harold's Pilgrimage* (1812–18) and *Don Juan* (1819–23)) and for his sensational personal life, he had enormous posthumous influence on the literature of the nineteenth century.

From Jacques-Louis Soret[1] 25 November [1871][2] 3563

Genève 25 Nov. *[1871]*

Mon cher Tyndall

Je vous ai fait expédier 10 vues stéréoscopiques de glaciers; c'est ce qui m'a paru le mieux dans ce que j'ai trouvé à Genève en ce moment. Si cela ne vous suffit pas écrivez-moi ce que vous désirez de plus je tacherai de me le procurer; il y a une collection de vues de l'Oberland qui ne se trouve pas actuellement ici, mais qu'il serait facile de faire venir & où l'on trouverait sans doute de bonnes choses.

Il y a quelques années j'ai fait un cours sur les Glaciers & j'avais fait usage comme vous comptez le faire de projection de photographies. Pour la plus grande partie j'avais employé des vues stéréoscopiques mais pour quelques points spéciaux, je n'avais pas trouvé ce qu'il fallait. Alors j'ai simplement fait photographies sur verre des vues d'après de grandes photographies de Bisson. Cela n'avait pas trop mal réussi. Je vous envoie avec les 10 vues stéréoscopiques, deux échantillons de vues qui me sont restés. L'un représente assez bien les crevasses transversales et les moraines je n'ai rien trouvé de ce genre en stéréoscopie.

En tout cas disposez de moi si vous avez besoin d'autre chose; si quelques unes des vues que l'on vous a expédiées ne vous conviennent pas vous pouvez me les renvoyer pour les changer.

Le prix de ces vues est de 5fr. la pièce, et le *[*brute*]* je crois deux francs. Quant au mode de paiement ce sera comme vous voudrez, le marchand peut *[*faire*]* traite sur vous ou si vous le préférez il vous indiquera l'adresse de son correspondance à Londres.

Mme Soret vous remercie de votre aimable souvenir. Il est toujours assez *[*souffrante*]*. J'ai fait un compliment à M. de la Rive et à M. Prevost qui est passablement quoique un peu affaibli.

Rec[eve]z mon cher Tyndall l'expression de mes sentiments les plus affectueux.

L. Soret

Geneva, 25 Nov. *[1871]*

My dear Tyndall,

I sent you 10 stereoscopic views of glaciers; the ones that currently seemed to be the best from those that I found in Geneva.[3] If this is not sufficient, write to tell me what else you would like and I will try to acquire it. There is a collection of Oberland[4] views that are not currently here, but which would be easy to acquire & where we could certainly find some good things.

A few years ago, I did a course on Glaciers & I made use of projected photographs, as you wish to do. For the most part I used stereoscopic views, but for certain special points I did not find what I needed. Thus, I simply made photographs on glass of views, like the great photographs of Bisson.[5] That did not turn out too badly. I am sending you, with the 10 stereoscopic views, two samples of views that I still have. One shows very well the transverse crevasses and the moraines, which I have not found in stereoscopy.

In any case, I am at your disposal if you need anything else; if some of the views that I have sent you do not satisfy you, you can send them back to me to change them.

The price of these views is 5fr. each, and the *[*raw images*]* I believe two francs. As for the method of payment, it will be as you wish; the dealer could *[*make*]* a deal for you, or if you prefer, he will give you the address of his correspondent in London.

Mrs Soret[6] thanks you for your friendly remembrance. *[*She*]* is still *[*suffering*]*. I paid a compliment to Mr de la Rive and Mr Prevost[7] who is passably well, although a bit weakened.

Please accept, my dear Tyndall, the expression of my most affectionate sentiments.

L. Soret

RI MS JT/1/S/91

---

1. *Jacques-Louis Soret*: see letter 3438, n. 4.
2. *[1871]*: Soret's handwriting is not clear, and the last letter of the date may be a '1' or a '6'. We have decided on the former on the basis of the reference to Prevost (see n. 7) and the probable reference to Tyndall's Christmas lectures (see n. 3).
3. *I sent . . . Geneva*: Tyndall was probably using these images in his Christmas lectures at the RI on 'Ice, Water, Vapour, and Air'. The fifth of these lectures, delivered on 6 January 1872, focused on glaciers and 'was illustrated by many very beautiful photographs on glass, which were magnified and projected upon the screen, by means of the electric lantern' ('Ice, Water, Vapour, and Air', *Alnwick Mercury*, 13 January 1872, p. 2). On the lectures, see letter 3565, n. 2.
4. *Oberland*: mountainous region in the Swiss canton of Bern.

5. *Bisson*: probably Auguste-Rosalie Bisson (1826–1900), a French photographer. Auguste-Rosalie was in business for some time with his brother, Louis-Auguste Bisson (1814–76). In the 1850s Auguste-Rosalie took some successful photographs of the Alps, particularly of glaciers, after which the brothers pursued a number of Alpine photography expeditions.
6. *Mme Soret*: see letter 3438, n. 1.
7. *Mr Prevost*: Alexandre-Pierre Prevost (1821–73). Interested during his early years in animal physiology, Prevost turned to a career in banking in London before resuming his scientific studies, in part under the influence of de la Rive, his father-in-law. He died in 1873 following a long illness.

To Mary Egerton 29 November 1871 3564

Royal Institution | 29th. Nov. 1871

My dear Lady Mary.

You are good and kind and constant! and I would come to you on Sunday if I could. But I had promised more than a week ago to spend at all events a portion of the day with Mr. Lefevre[1] at Ascot;[2] and this promise I fear was unavailing, for I shall probably not be able to go. I do not suppose that any man in England has subjected his muscles to greater and more various strains than I have. Judge then my astonishment to find that in scampering leisurely over English fields and crossing a few English fences and walking afterwards briskly 12 or 13 miles, that I should have strained or ruptured the fibres of the calf of my right leg so as to confine me now for a week to the sofa![3] I cannot comprehend it, and I don't know anybody that can. It is the most paltry and ignoble break down that I can call to mind. I was at first very patient, and worked as well as I could upon the sofa but I am now getting irate.

As I had written so far my friend Busk made his appearance, and he examined my leg. He is going to wrap something round it tomorrow which will help me to keep it quiet. Then Bence Jones came in and we had a learned talk over this and other matters. Were it not for this interruption I should have posted this note to you to day. I had four or five engagements out last week; but this abhorred accident swept them all away.

But what a ridiculous thing it is to waste your time with my poor private sufferings. There is no limitation to people coming here at Christmas. But if Hugh[4] wishes to have the thing without hearing it—he shall have a copy of my notes, which are models of clearness and concentration![5]

Lady Emily Peel wants me to go down to Drayton,[6] and she was so good to me in my hour of need[7] that I should like to go—I fear however I cannot.

To day I attended a meeting of the Governors of Harrow School.[8] I remember when a child hearing of Harrow as a place existing at almost Olympian heights. I remember my poor father telling me of the great men who were educated there; and here I am a Governour of Harrow. Were I to write a simple account of my life it would be better than many romances. I wonder what effect it would have upon you and "May!"[9]

Grace and benediction! | Always Yours | John Tyndall

RI MS JT/1/TYP/1/395–6
LT Typescript Only

---

1. *Mr. Lefevre*: possibly George John Shaw-Lefevre (1831–1928), a Radical politician first elected to Parliament in 1863. Shaw-Lefevre was involved in efforts to preserve open space and rights of public access, and he was instrumental to the passage of the Married Women's Property Act of 1870. He became secretary to the Board of Trade in 1868. Shaw-Lefevre built a house at Ascot in 1865.
2. *Ascot*: small town in East Berkshire, west of London.
3. *I should . . . to the sofa*: see letter 3567 for another discussion of this injury.
4. *Hugh*: Hugh Egerton; see letter 3405, n. 4.
5. *There . . . concentration*: probably a reference to Tyndall's Christmas lectures at the RI on 'Ice, Water, Vapour, and Air'. See letter 3565, n. 2.
6. *Lady Emily Peel . . . Drayton*: see letters 3548 and 3567.
7. *my hour of need*: see letter 3431, n. 6.
8. *Harrow School*: renowned public school founded in 1572 and located in Harrow, London. Tyndall had been involved with Harrow since the mid-1860s. His appointment as one of the school's ten governors followed the Public Schools Act of 1868, which, although it gave governors oversight of the curriculum at elite schools, also entrenched their associations with privilege and wealth. Tyndall served as governor for twenty years. See *Ascent of John Tyndall*, pp. 290–2.
9. *May*: Mary Egerton.

To John Murray 29 November [1871][1] 3565

29th Nov.

My Dear Sir,

I am going to talk to the boys at Christmas about glaciers.[2]

I am writing notes & the lectures for the little fellows: it would add much to their plainness to have a few illustrations. Could you let me have the blocks[3] of the "Glaciers of the Alps"[4] so as to enable me to select from them the particular ones suitable for my purpose?

Yours very faithfully | John Tyndall
John Murray Esq.

NLS, John Murray Archive, Ms 41214

1. *[1871]*: dated by reference to Tyndall's Christmas lectures; see n. 2.
2. *I am going to talk to the boys at Christmas about glaciers*: Tyndall delivered the Christmas lectures at the RI on 'Ice, Water, Vapour, and Air' on 28 and 31 December 1871, and on 2, 4, 6 and 9 January 1872; they are described as 'adapted to a Juvenile Auditory' (*Roy. Inst. Proc.*, 6 (1870–72), p. 384). The Christmas lectures were begun by Faraday in 1825, with the aim of educating young people about science; they continue into the present.
3. *blocks*: Wooden blocks, with images in relief, were used to produce illustrations through the nineteenth century (M. Twyman, 'The Illustration Revolution', in D. McKitterick (ed) *Cambridge History of the Book in Britain* (Cambridge: Cambridge University Press, 2010), pp. 117–43, on p. 122).
4. *Glaciers of the Alps*: Tyndall's 1860 book *Glaciers of the Alps*.

From Emily Peel 1 December 1871 3566

Drayton Manor, | Tamworth. | Dec[embe]r. 1/71

My dear Professor,

I hope you are not going to forget us next week.

I only write these lines as a reminder in case your kind promise might escape your memory.[1]

Yours sincerely | Emily Peel

RI MS JT/1/P/59

1. *your kind promise*: see letters 3549 and 3553. Tyndall responded in letter 3567.

To Emily Peel 2 December 1871 3567

Royal Institution, | 2nd Dec., 1871

My dear Lady Emily,

Will you graciously listen to a tale of "prowess"[1] and form your own judgment as to whether the word applies to me?

Last Sunday week[2] I was down at Berkhampstead.[3] Before breakfast I was scampering about, climbing all the eminences I could find, which were not many. After breakfast I took a walk of two hours with Mr John Ball, crossing fields and fences in the quietest manner possible. After luncheon I took a scamper of about 13 miles to Ashridge,[4] and returned I thought with an amount of vigour in my limbs that would carry me over at least thirty miles more. That night I felt a tingle in my leg. I neglected it for three days, but on the Wednesday, at the command of Dr Bence Jones, I was obliged to take to the sofa, where I have been ever since.[5] They say I ruptured a small pipe[6]—but anything so contemptible, and indeed so unaccountable never occurred to me before. I limped to the Royal Society dinner on Thursday because I had to speak; having been concerned in obtaining the Copley Medal for an eminent foreigner;[7] but it did me no good.

My friend Busk, an eminent surgeon, was with me this morning—He says leaving London on Monday would be madness; but he thinks I might be able to escape at the end of next week. If so, and if you would then receive me, I would gladly go down to Drayton.[8]

Or shall you not be there in January? If so, it would give me extreme pleasure to join you then.

Had the thing occurred at Drayton Manor instead of Berkhampstead you might have had the trouble of nursing me once more, and for a cause still more contemptible than the last.[9]

I am right sorry that it has occurred at this particular juncture.

Yours ever faithfully, | John Tyndall.

RI MS JT/1/TYP/3/965
LT Typescript Only

---

1. *prowess*: see letter 3553. Tyndall had accepted an invitation from Peel to spend a few days at her home in Drayton Manor; see also letters 3548, 3549 and 3566.
2. *Last Sunday week*: 19 November 1871.
3. *Berkhampsted*: small town in the Chiltern hills northwest of London.
4. *Ashridge*: country estate in the Chilterns.
5. *I was obliged . . . ever since*: For further discussion of this injury, see letters 3560 and 3564.
6. *small pipe*: We have not been able to locate a use of the word 'pipe' that fits with Tyndall's discussion of his injured calf.
7. *obtaining the Copley Medal for an eminent foreigner*: Julius Mayer, who received the RS's Copley Medal in 1871; see letters 3539, 3541, 3544 and 3546.
8. *he thinks . . . Drayton*: Peel responded in letter 3569.
9. *you . . . last*: see letter 3431, n. 6.

From Henry Harrison Doty 2 December 1871 3568

15, Bury-street, St James's, London, | 2 December 1871.

Dear Sir,

FOLLOWING up my personal application to you,[1] several days since, for a copy of the paper signed by Mr. Douglass and myself, which was put forth by you as a "programme" for our guidance on the 5th and 6th October ultimo,[2] in conducting the competitive trials, under your direction, of our hydrocarbon lighthouse burners, I will feel greatly obliged if you will permit me to take a copy of the said paper, or if you will kindly furnish me with a copy of it.[3]

I am, &c. | (signed) *H. H. Doty.*

Professor John Tyndall, | &c. &c. &c., | Royal Institution.

'Lighthouses (Mineral Oils). Copy of Further Correspondence relative to Proposals to Substitute Mineral Oils for Colza Oil in Lighthouses (in Continuation of Parliamentary Paper, no. 2, of Session 1872)', HC 2 67,[4] Parliamentary Papers (1872), p. 50
Typed Transcript Only

---

1. *my personal application to you*: letter missing. The present letter, along with letter 3578, appeared in the Parliamentary Papers as enclosures in a letter from Doty to the Board of Trade, in which Doty demanded that the Board furnish him with a copy of the programme to which he refers here. See Captain Doty to Board of Trade, 29 December 1871, 'Lighthouses (Mineral Oils). Copy of Further Correspondence relative to Proposals to Substitute Mineral Oils for Colza Oil in Lighthouses (in Continuation of Parliamentary Paper, no. 2, of Session 1872)', HC 2 67, Parliamentary Papers (1872), p. 50. For the Board's refusal, see Board of Trade to Captain Doty, 1 January 1872, "Lighthouses (Mineral Oils). Copy of Further Correspondence relative to Proposals to Substitute Mineral Oils for Colza Oil in Lighthouses (in Continuation of Parliamentary Paper, no. 2, of Session 1872)', HC 2 67, p. 51.
2. *a "programme" for our guidance on the 5th and 6th October ultimo*: see letter 3555.
3. *I will feel . . . a copy of it*: Tyndall delegated his response to John Cottrell, ostensibly because of his injury. See Cottrell to Captain Doty, 'Lighthouses (Mineral Oils). Copy of Further Correspondence relative to Proposals to Substitute Mineral Oils for Colza Oil in Lighthouses (in Continuation of Parliamentary Paper, no. 2, of Session 1872)', HC 2 67, p. 51.
4. *2 67*: At the time of writing, the ProQuest U.K. Parliamentary Papers database misnumbered this paper as 2 264.

From Emily Peel 3 December 1871 3569

Drayton Manor, | Tamworth. | Dec[embe]r. 3/71

Dear Professor Tyndall,

You are indeed too unlucky,[1] and we are also to be pitied as suffering from your misfortune—but any day you like to come Thursday Friday or Saturday we shall be charmed to receive you, & you can pass a "healthy" Sunday here preparatory to your week's labours.—

I hope you informed M^{r}. Ball of my indignant protest against his suspicions of any inaccuracy; and that my young friend has a chance of being elected to the E[nglish]. Alpine Club.—[2]

Make haste & get well—& let us hear that you are coming to us soon.[3]

Yours sincerely | Emily Peel

The 3 ocl[ock]: P.M. from Euston Square[4] is a very good train as is also the 11 ocl[ock]: A.M. if that is not too early.

RI MS JT/1/P/60

1. *You are indeed too unlucky*: see letter 3567.
2. *I . . . Club*: see letter 3533.
3. *let us hear that you are coming to us soon*: Tyndall responded in letter 3576.
4. *Euston Square*: railroad station in London.

To Margaret Ginty 4 December [1871][1] 3570

Royal Institution | 4th. Dec.

My dear Mrs Ginty,

I have been on the sofa for the last ten or twelve days through a strained muscle.[2] I regret very much that I could not call upon you. I also regret not being free when you came to day—Any other day between now and Friday at 2 P.M. would find me at home.

Yours faithfully | John Tyndall.

RI MS JT/1/TYP/11/3717
LT Typescript Only

1. *[1871]*: year established by reference to Tyndall's calf injury.
2. *I . . . muscle*: Tyndall injured his calf in November 1871. See letters 3564 and 3567.

To Elizabeth Dawson Steuart 4 December [1871][1] 3571

Royal Institution | 4th Dec.

My dear Mrs Steuart,

I had a note[2] a few days ago from the Garrison[3] asking me to subscribe to the Leighlin Clothing fund,[4] and I replied that I was about to send some money to you. I now do so £5 for the fund and £1. for Catharine Washington.[5]

I am sorry that good people cannot work always together.

For 10 days or so I have been on the sofa with a strained muscle.[6]
ever Yours faithfully | John Tyndall

RI MS JT/1/TYP/10/3342
LT Typescript Only

---

1. *[1871]*: LT dated this letter to 1861, but the reference to Tyndall's injury puts it in 1871.
2. *a note*: letter missing.
3. *the Garrison*: presumably Garrison House in Leighlinbridge, though it is unclear who lived there in 1871.
4. *Leighlin Clothing fund*: the Leighlinbridge clothing fund, to which Tyndall donated regularly (*Ascent of John Tyndall*, p. 179). Tyndall was born in Leighlinbridge and maintained charitable ties to it throughout his life.
5. *Catharine Washington*: see letter 3379. For a continuation of the present letter, see letter 3572.
6. *For . . . muscle*: Tyndall injured his calf in November 1871. See letters 3564 and 3567.

To Elizabeth Dawson Steuart — 5 December [1871][1] — 3572

Royal Institution | 5th. Dec.

My dear Mrs Steuart,

I should like Andrew McAssey[2] to have one pound out of the five[3]—or at least 10s/–

Yours ever | John Tyndall

RI MS JT/1/TYP/10/3348
LT Typescript Only

---

1. *[1871]*: LT dated this to 1861, as she did letter 3571, but continuity with 3571 places it in 1871.
2. *Andrew McAssey*: Tyndall's mother's maiden name was McAssey (sometimes spelled Macassey or Macasey), and he had a number of cousins by this name. Andrew McAssey is presumably one of these relatives but has not been further identified.
3. *one . . . five*: see letter 3571.

From William Carpenter 5 December 1871 3573

University of London, Burlington Gardens, W., | December 5th, 1871.

My dear Tyndall,—

If I correctly apprehended what you said at the Dinner of the Royal Society in regard to Dr. Mayer,[1] you repeated what you had previously stated in your Lecture at the Royal Institution in 1863,[2] as to the entire ignorance of Mayer's work which prevailed in this country until you brought it into notice on that occasion.

Now, I very distinctly remember that a few days previously to that Lecture, I mentioned to you that as far back as 1851 I had become acquainted, through the late Dr. Baly,[3] with one of Dr. Mayer's earlier publications;[4] and that, in bringing before the readers of the *British and Foreign Medical Review*[5] (of which I was then the Editor) the 'Correlation' doctrine, as developed in Physics by Grove,[6] and in Physiology by myself, I had stated that we had both been to a great extent anticipated by Mayer—as I should have shown much more fully if the pamphlet had earlier come into my hands.[7]

I also most distinctly remember that, as you stated in that Lecture, no one in this country—'not even Sir Henry Holland, who knows everything'[8]—had ever heard of Mayer, I spoke to you again on the subject a few days afterwards; and that you then expressed your regret at having entirely forgotten what had previously passed between us on the subject.

As it would seem that this second mention of the matter has also passed from your mind, I shall be obliged by your looking at the passages I have marked in pp. 227 and 237 of the accompanying volume,[9] from which I think that you will be satisfied that I had at that date correctly apprehended Mayer's fundamental idea, and that I have done the best to put it before the public that I could under the circumstances—the article having been in type and ready for press before his pamphlet came into my hands.

Since, in thus bringing forward Mayer, I spontaneously abdicated the position to which I had previously believed myself entitled, of having been the first to put forward the idea that all the manifestations of Force exhibited by a living organism have their source *ab extra*,[10] and not—as taught by physiologists up to that time—*ab intra*,[11] I venture to hope that you will do me the justice of stating the real facts of the case in a short communication either to the *Athenæum*[12] or to NATURE.[13]—I remain, my dear Tyndall, yours faithfully, | William B. Carpenter

Prof. Tyndall.

Nature, 5 (21 December 1871), p. 143

---

1. *Dr. Mayer*: Tyndall spoke about Mayer during the toast to the latter as the year's Copley medalist; see letter 3584. For Mayer's receipt of the medal, see letters 3539, 3541, 3544 and 3546.
2. *your Lecture at the Royal Institution in 1863*: Tyndall's lecture 'On Force' was delivered at the RI as a Friday Evening Discourse on 6 June 1862, not 1863 (*Roy. Inst. Proc.*, 3 (1858–62), pp. 527–36). See also *Ascent of John Tyndall*, p. 170.
3. *Dr. Baly*: William Baly (1814–61), a physician who was appointed to report on conditions at Millbank prison in London, where his research on dysentery and cholera helped to stem an outbreak in the 1840s. Baly was elected FRS in 1847. In 1859 he retired from his work at the prison and became physician-extraordinary to Queen Victoria. Baly translated the work of German physiologist Johannes Müller between 1837 and 1842.
4. *Dr. Mayer's earlier publications*: J. R. Mayer, *Die organische Bewegung in ihrem Zusemmenhange mit dem Stoffwechsel. Ein Beitrag zur Naturkunde* (Heilbron, 1845).
5. *British and Foreign Medical Review*: Titled *The British and Foreign Medical Review* from its initial publication in 1836, the journal's name changed to *The British and Foreign Medico-Chirurgical Review* in 1848.
6. *Grove*: see letter 3428, n. 1.
7. *in bringing before . . . my hands*: The article to which Carpenter refers is [W. Carpenter], 'Grove, Carpenter, &c., on the Correlation of Forces, Physical and Vital', *The British and Foreign Medico-Chirurgical Review*, 8:15 (July 1851), pp. 206–37; Carpenter briefly discusses Mayer's publication on the final page of his article (p. 237).
8. *'not even Sir Henry Holland, who knows everything'*: This line is not present in the published version of Tyndall's lecture ('On Force', *Roy. Inst. Proc.*, 3 (1858–62), pp. 527–36). Holland was president of the RI from 1865 to 1873.
9. *the accompanying volume*: presumably volume 8 of the *British and Foreign Medico-Chirurgical Review*. The enclosure is missing.
10. *ab extra*: 'from without' (*OED*).
11. *ab intra*: 'from within' (*OED*).
12. *the Athenæum*: The *Athenaeum, London Literary and Critical Journal* was a weekly that ran from 1828 to 1921 and was one of the most influential publications of the nineteenth century, though its standing varied under different editors (M. Demoor, 'Athenaeum', in L. Brake and M. Demoor (eds), *Dictionary of Nineteenth-Century Journalism in Great Britain and Ireland* (London: The British Library, 2009), pp. 26–8, on p. 26).
13. *or in Nature*: Tyndall responded to Carpenter's request by forwarding his letter to *Nature*, along with a short covering letter. Both appeared under the title 'Dr. Carpenter and Dr. Mayer'. For the covering letter and full reference, see letter 3584. For other letters relevant to this incident, see 3575, 3581, 3582, 3586, 3587 and 3598.

To Jane Barnard 7 December [1871][1] 3574

[Royal Institution][2] | 7th Dec.

My dear Miss Barnard

Three tickets are allotted to you for the Juveniles[3] if you care to use them.

Yours ever | John Tyndall

Kindest regards to your aunt.[4] I have thought of her of late while attempting to get up stairs.[5]

RI MS JT/1/T/126
RI MS JT/1/TYP/12/4202

---

1. *[1871]*: year assigned by the reference to Tyndall's Christmas lectures (see n. 3) and the probable reference to his injury (see n. 5).
2. *[Royal Institution]*: LT annotation.
3. *for the Juveniles*: Tyndall's Christmas lectures at the RI; see letter 3565, n. 2.
4. *your aunt*: Sarah Faraday.
5. *while attempting to get up stairs*: probably a reference to Tyndall's injured leg. See letters 3564 and 3567.

To Thomas Archer Hirst [7 December 1871][1] 3575

[Royal Institution][2] | *[7th Dec.]*[3]

Dear Tom.

I wish very much to use, should I think it right to do so, the account that I have given of Joule & Mayer before the Council of the Royal Society.[4] Please consult Sharpey[5] about it today, for I would not willingly take any step which could in the slightest degree be objected to by the Council. Of course both documents are historic statements merely I cannot for my own part see any objection to the publication of these two documents; if needed in any way for public enlightenment.

I shall be very glad indeed should Sharpey see no harm in this proceeding.

Yours aff[ectionatel]y. | John.

The attitude of Carpenter[6] & others renders it very desirable that what I did say on both these occasions should be known.

RI MS JT/1/T/946
RI MS JT/1/HTYP/584/2

1. *[7 December 1871]*: dated by the reference to Tyndall's conflict with Carpenter about Mayer; see n. 4 below, as well as letters 3573, 3581, 3582, 3584, 3586, 3587 and 3598.
2. *[Royal Institution]*: LT annotation.
3. *[7th Dec.]*: This date may not be in Tyndall's hand, but seems reasonable given the ongoing controversy relating to the RS's Copley Medal; see n. 4.
4. *the account that I have given of Joule & Mayer before the Council of the Royal Society*: This account was probably given during discussions of the Copley Medal, which was awarded to Mayer in 1871 and had been awarded to Joule in 1870. See letters 3539, 3541, 3544 and 3546. Tyndall may have wanted this account as the basis for two publications that appeared in the following weeks: J. Tyndall, 'The Copley Medallist of 1871', *Nature*, 5 (14 December 1871), pp. 117–20, which discussed Mayer; and J. Tyndall, 'The Copley Medallist of 1870', *Nature*, 5 (21 December 1871), pp. 137–8, which discussed Joule.
5. *Sharpey*: William Sharpey was secretary of the RS from 1854 to 1872.
6. *The attitude of Carpenter*: see letters 3573 and 3584.

To Emily Peel 8 December 1871 3576

Royal Institution. | 8th Dec., 1871

My dear Lady Emily,

I am in hopes of being able to go down to-morrow;[1] but I am not certain. Yesterday I drove to the British Museum[2] to see a book; and the walking up the steps and through the galleries brought on aching.[3] I will send you a telegram to-morrow morning and I hope it will be to say that I am well enough to join you[4]—for it would be a great pleasure to me.

Most faithfully yours, | John Tyndall.

RI MS JT/1/TYP/3/966
LT Typescript Only

1. *go down to-morrow*: see letter 3569.
2. *British Museum*: Located in the Bloomsbury neighborhood of London, the British Museum houses antiquities, books, and manuscripts.
3. *brought on aching*: Tyndall injured his calf in November 1871; see letters 3564 and 3567.
4. *I will . . . to join you*: Any telegram is missing, but Tyndall seems not to have joined the Peels; see letter 3577.

From Emily Peel 12 December 1871 3577

Drayton Manor, | Tamworth. | Dec[embe]r 12/71

Dear Professor Tyndall,

We all regretted your absence.[1] Mrs Norton[2] begged me to give you many pretty messages which my inelegant style prevents my transmitting in the charming form in which she delivered them to me, and we moreover deeply sympathised with your ill luck in breaking down so ignominiously over English fields and country lanes. Decidedly the highest pinnacle of the Alpine range is where you are most at home.

Many thanks for your most interesting notes.[3] I have already looked through them & endeavoured to understand them and indeed I should do so for they are beautifully clear & simple.

Sir Robert[4] begs to be kindly remembered to you & to express his regret at y[ou]r not having been able to keep y[ou]r promise with us. Young Stonor[5] is now at College studying for his examinations & kept rather hard at work.

Yours sincerely | Emily Peel

RI MS JT/1/P/61

---

1. *your absence*: see letters 3569 and 3576.
2. *Mrs Norton*: not identified.
3. *your most interesting notes*: possibly notes for his Christmas lectures at the RI on 'Ice, Water, Vapour, and Air'; see letter 3565, n. 2.
4. *Sir Robert*: Robert Peel.
5. *Stonor*: see letter 3431, n. 5.

From Henry Harrison Doty 13 December 1871 3578

15, Bury-street, St. James's, London, | 13 December 1871.

Dear Sir,

I BEG to acknowledge receipt of your favour of the 5th instant,[1] supplying me with an extract from "programme" for 6th October.[2] I have also to acknowledge receipt of an *extract* from programme for the 5th October, received on the 2nd instant;[3] but I am sorry to have to trouble you again, as you appear to have misunderstood what it is I desire, and I shall therefore feel obliged if you will be good enough to furnish me with a *complete* copy of the programme drawn up on the evening of the 4th October, and signed by Mr.

Douglass, yourself, and myself, and which regulated our experiments on the 5th and 6th October.

I am, &c. | (signed) *H. H. Doty*.

To Professor Tyndall, | &c. &c. &c.

'Lighthouses (mineral oils). Copy of further correspondence relative to proposals to substitute mineral oils for colza oil in lighthouses (in continuation of parliamentary paper, no. 2, of session 1872)', HC 2 67,[4] Parliamentary Papers (1872), p. 51

1. *5th instant*: the fifth day of the present month, thus 5 December.
2. *your favour… October*: Doty treats a letter from Cottrell as a letter from Tyndall; see letter 3568, n. 3.
3. *receipt of… the 2nd instant*: letter missing.
4. *2 67*: At the time of writing, the ProQuest U.K. Parliamentary Papers database misnumbered this paper as 2 264.

From Sarah Faraday 13 December [1871][1] 3579

Barnsbury Villa, | 320, Liverpool Road, N. | 13th. Dec.

Dear Dr. Tyndall

It seems so long since I had any direct communication with you that I think I must write a few lines to tell you that tho' I cannot hope to see you at present (I suppose) as you now must be quite taken up preparing for the juveniles.[2] Yet I do look forward to a pleasant hour or two, when you are at leisure, or at least not so busy; I have been keeping well considering the severe weather—but we seem to have much around us to call for our sympathy. Dr. Gladstone[3] called here yesterday but I did not see him, it was during my afternoon rest when I am hardly able, and I could not have helped him if I had. I fear he will be disappointed as to his Biography.[4]

Believe me dear Dr. Tyndall | Yours most truly | S.Faraday

We have just had a lively visit from the Miss Moores[5] who told us that they called to see whether you would take a drive with them to see me, I am very sorry to hear you are so lame[6] but think when you can spare the time you will take a cab and let us have you all to ourselves.

RI MS JT/1/TYP/12/4185
LT Typescript Only

1. *[1871]*: year established by reference to Tyndall's injury; see letter 3564 and 3567.
2. *the juveniles*: Tyndall's Christmas lectures at the RI; see letter 3565, n. 2 and letter 3574.

3. *Dr. Gladstone*: John Hall Gladstone (1827–1902), an English physical chemist and friend of Michael Faraday. After his education at University College, London and Giessen (PhD 1848), he became FRS in 1853 and Fullerian Professor of Chemistry at the RI from 1874–7. He was a leader in using spectroscopy in chemistry for which he won the RS's Davy Medal in 1897.
4. *his Biography*: J. H. Gladstone, *Michael Faraday* (London: Macmillan, 1872).
5. *the Miss Moores*: Julia and Harriet Moore; see letter 3518, n. 3. Their family had close ties to the RI.
6. *I am very sorry to hear you are so lame*: see n. 1.

To Elizabeth Dawson Steuart 14 December [1871][1] 3580

Royal Institution, | 14th. Dec.

My dear Mrs Steuart,

My notion was to give a pound to each of the two persons mentioned, by degrees, and four pounds to the Clothing-fund[2]

The two persons mentioned[3] would thus be withdrawn from the list of those to be relieved

ever Yours | John Tyndall.

The leg is much better:[4] but my work is weighty.

I hope most earnestly the Prince will pull through.[5] I have a nice letter that he wrote to me some years ago.[6]

RI MS JT/1/TYP/10/3344
LT Typescript Only

---

1. *[1871]*: dated by reference to Tyndall's leg injury in November 1871 (see n. 4, below) and continuity from letters 3571 and 3572.
2. *My notion . . . fund*: see letters 3571 and 3572.
3. *two persons mentioned*: Catharine Washington (see letters 3379 and 3571) and Andrew McAssey (see letter 3572, n. 2).
4. *The leg is much better*: Tyndall injured his leg in November 1871; see letters 3564 and 3567.
5. *I hope most earnestly the Prince will pull through*: Albert Edward, the Prince of Wales (see letter 3398, n. 6), had fallen ill of typhoid fever in October 1871; he was not expected to live, but recovered in mid-December. He had been rather unpopular, partly as a result of his tense relationship with his mother, Queen Victoria, and his consequent exclusion from public affairs, but his illness and recovery helped to redeem him in the eyes of the public.
6. *I hope . . . ago*: for Albert Edward's 1868 letter to Tyndall, see letter 2788, *Tyndall Correspondence*, vol. 10.

To [Norman Lockyer][1] 18 December [1871][2] 3581

18th Dec

My dear Sir

You are really kind. I dare say you know that in former years I have had to meet some brisk criticism regarding Dr. Joule:[3] this makes me anxious that no ground should exist that could intimate that I rated him lower than Dr. Mayer. Were this not the case I should not for a moment think of asking you either to alter the heading or to give the article the same place as that accorded to the one on Mayer.[4]

If you wish it pray postpone the article until the next week, making a note that press of matter compelled you to do so.

I hardly think however that it would make the least difference to Dr. Sanderson[5] whether his article comes first or second.[6]

"Dr. Carpenter & Dr. Mayer"[7] may come any where—but I think Joule ought to have the same position as Mayer.

Excuse all the bother | Yours Faithfully | John Tyndall

RI MS JT/1/OTHER/2
RI MS JT/8/3/3

---

1. *[Norman Lockyer]*: Lockyer was the editor of *Nature* from 1869 until 1919; Tyndall's letter is presumably addressed to him, as the letter makes reference to publications by Tyndall that appeared in *Nature* in December 1871. See n. 2, n. 6 and n. 7.
2. *[1871]*: year assigned by reference to the publication of Tyndall's articles about Mayer and Joule in *Nature*: for these articles, see letter 3575, n. 4. Tyndall was prompted to write them by a conflict with Carpenter, discussed in letters 3573, 3575, 3582, 3584, 3586, 3587 and 3598.
3. *I have had to meet some brisk criticism regarding Dr. Joule*: For example, in 1862 William Thomson and P. G. Tait criticized Tyndall's championing of Mayer. See W. Thomson and P. G. Tait, 'Energy', *Good Words*, 3 (December 1862), pp. 601–7, on p. 603.
4. *I should not . . . Mayer*: Tyndall's article on Mayer appeared in *Nature* on 14 December 1871; his article on Joule appeared in *Nature* on 21 December 1871. Each came first in that week's issue.
5. *Dr. Sanderson*: John Scott Burdon Sanderson (1828–1905), a pathologist and physiologist who did research on the transmission of infectious disease and helped to advance the germ theory of disease; he also made significant contributions to establishing the field of experimental physiology. In 1872 Sanderson was appointed first professor-superintendent of the Brown Institute at the University of London, which was devoted to research on veterinary diseases. He was elected FRS in 1867 and made a baronet in 1899.

6. *his article comes first or second*: Sanderson's unsigned article appeared second: [J. Sanderson], 'The Brown Institute', *Nature*, 5 (21 December 1871), pp. 138–40.
7. *Dr. Carpenter and Dr. Mayer*: see letters 3584 and 3573.

To [Norman Lockyer][1] 18 December [1871][2] 3582

18th Dec.

My Dear Sir

Many thanks for the proof;[3] it shall be sent to Bedford Street,[4] tomorrow morning in good time.

Faithfully yours | John Tyndall

RI MS JT/1/OTHER/1–3
RI MS JT/8/3/3

1. *[Norman Lockyer]*: see letter 3581, n. 1.
2. *[1871]*: dated by continuity with letter 3581.
3. *the proof*: presumably of Tyndall's forthcoming article in *Nature* on Joule. See letter 3575, n. 4. This exchange is part of a larger conversation involving Tyndall and Carpenter; see letters 3573, 3575, 3581, 3584, 3586, 3587 and 3598.
4. *Bedford Street*: *Nature*'s offices were located at 29 Bedford Street, in London.

To the editor of *The Index*[1] 21 December 1871 3583

21st December 1871

Dear Sir,—I have been away from London, laid up with a hurt, and also excessively occupied under these unfavorable conditions.[2] My reply to your friendly letter[3] has been thus retarded.

It would give me sincere pleasure to lend you the aid you require,[4] if it were in my power to do so. But unhappily it is not.

To do any work of permanent value in this modern Babylon,[5] I find it absolutely necessary to shut my eyes against the numberless temptations that offer themselves to cause me to swerve from my proper work.

I suppose I have already had this year fifty such requests as yours.[6] And they come to me backed by inducements which are usually found available; but I do not undertake any work of the kind.

I hope these considerations will acquit me in your eyes of all unkindness, or want of sympathy with the earnest work which you have taken in hand.

I am, dear sir, faithfully yours, | JOHN TYNDALL

The Index, 3 (20 January 1872), p. 20

1. *the editor of* The Index: Francis Ellingwood Abbot (1836–1903), an American philosopher and theologian who endeavored to put religion on a scientific basis. Abbot established *The Index*, subtitled *A Weekly Paper, Devoted to Free Religion*, in 1870; he was the editor and wrote much of the content.
2. *I have . . . conditions*: It is not clear to what trip Tyndall is referring. The 'hurt' he mentions is the injury to his leg in November 1871, for which see letters 3564 and 3567. His 'excessive occupation' may be his work testing rival designs for lighthouse lamps; see letter 3522, n. 3.
3. *your friendly letter*: letter missing. Tyndall's response, and an editorial piece that follows, are printed under the title 'Prof. John Tyndall'.
4. *the aid you require*: An editorial piece that follows the letter comments, 'It would have been especially gratifying, if the brilliant author of "Fragments of Science for Unscientific People" could have accepted our invitation to express in THE INDEX his independent views on science and religion' ('Prof. John Tyndall', *The Index*, 3:108 (20 January 1872), p. 20).
5. *modern Babylon*: London (*OED*).
6. *fifty such requests as yours*: No other letters in our volume correspond to such requests.

To the Editor of *Nature*[1] 21 December 1871[2] 3584

AT the Anniversary Dinner of the Royal Society on November 30, I was honoured by a request from the President[3] to say a few words in acknowledgement of the toast to the Copley Medalist.[4] I did so, stating briefly the origin of my acquaintance with Dr. Mayer's writings. Though Dr. Carpenter at the time was within sight of me, it did not occur to me to introduce his name into my remarks. A few days afterwards I was favoured by a letter from Dr. Carpenter, in which he reminds me somewhat sharply of this and other lapses as regards himself, and requests me to rectify the omission by a brief communication to the *Athenæum*[5] or to NATURE. It will be fairer to Dr. Carpenter, and more agreeable to me, if he would state his own case *in extenso*.[6] Here is his letter:—[7]

Nature, 5 (21 December 1871), p. 143

1. *editor of* Nature: Norman Lockyer.
2. *21 December 1871*: the date of the letter's publication; the date of composition is not known.
3. *the President*: George Biddell Airy, president of the RS from 1871 until 1873.

4. *the Copley Medalist*: Julius Mayer. See letters 3539, 3541, 3544 and 3546.
5. *the Athenæum*: see letter 3573, n. 12.
6. *in extenso*: 'at full length' (*OED*).
7. *his letter*: In *Nature*, letter 3573 follows here. Both Tyndall's letter and Carpenter's were published under the title 'Dr. Carpenter and Dr. Mayer'. For other letters relevant to this issue, see 3573, 3575, 3581, 3582, 3586, 3587 and 3598.

To Thomas Archer Hirst 21 December 1871 3585

21st Dec 1871

My dear Tom

On turning into the open air from the Athenaeum[1] last night the matter we were speaking of[2] thus presented itself to me:—

You accommodate your eye to the small illuminated aperture: On your retina therefore you have an image of that aperture.

You now interpose a pin. The portion of the circle occupied by the pin is cut away. You see a circle minus that particular area of light. Consequently that area will appear to you of the form of the pin cut out of the bright ground.

It is not an image of the pin that you see, but an image of the aperture minus the rays intercepted by the pin.

And I think you will see that under the circumstances the outline of the pin will be erect upon the retina, and that this will correspond to an apparent inversion of the pin outside. In fact the pin close to the pupil acts as if it were actually laid against the retina, or as if a portion of the retina corresponding to the area of the pin were blind.

Yours aff[ectionate]ly. | John

RI MS JT/1/T/689
RI MS JT/1/HTYP/584

1. *Athenaeum*: see letter 3372, n. 14.
2. *the matter we were speaking of*: Hirst appended the following note to this letter: 'Mem. The fact of which the above is an explanation is this: If a sheet of paper in which a small aperture has been made with a pin be held between an illuminated surface and the eye, and at about a foot from the latter; then when the pin is held vertically with its head close to the eye an inverted image of it is seen apparently beyond the aperture in the paper'.

From Thomas Henry Huxley — 22 December 1871 — 3586

Jermyn St. | Dec. 22nd. 1871.

My dear Johnny,

You are certainly improving. As a practitioner in the use of cold steel myself, I have read your letter[1] in to-day's "Nature" "mit Ehrfurcht und Bewunderung".[2] And the best evidence of the greatness of your achievement is that it extracts this expression of admiration from a poor devil whose brains and body are in a colloide[3] state and who is off to Brighton[4] for a day or two this afternoon.

God be with thee my son and strengthen the contents of thy gall-bladder!

Ever thine | T. H. Huxley.

P.S. Seriously I am glad that at last a protest has been raised against that process of anonymous self-praise to which our friend is given. I spoke to Smith[5] the other day about that dose of it in the "Quarterly" article on Spirit rapping.[6]

IC HP 8.91
Typed Transcript Only

---

1. *your letter*: Tyndall's letter to *Nature* forwarded William Carpenter's letter complaining that Tyndall had not given Carpenter sufficient credit for his recognition of Joule's work. See letters 3573 and 3584. For Tyndall's response to Huxley, see letter 3587. For other letters relevant to this issue, see 3575, 3581, 3582 and 3598.
2. *mit Ehrfurcht und Bewunderung*: with awe and admiration (German).
3. *colloide*: colloid; 'a homogeneous or slightly granular gelatinous substance into which the cells are changed in certain forms of degeneration of tissue' (*OED*).
4. *Brighton*: sea-side town in the southeast of England.
5. *Smith*: not identified.
6. *the "Quarterly" article on Spirit rapping*: [W. Carpenter], 'Spiritualism and its Recent Converts', *Quarterly Review*, 131:262 (October 1871), pp. 301–53.

To Thomas Henry Huxley — 24 December [1871] — 3587

Ascot, 24th. Dec.

My dear Hal,

It certainly gives me great pleasure to learn that the arch-master of the "cut and thrust" approves of my little prod.[2] The man[3] may cut up rough, and then I think I must push the steel a little deeper: he is becoming intolerable.

And I am right sorry to hear about the brain: that colloidal state is so well known to me. In fact for some years no other state was known to me. But the crystallization is again growing firm: and I shall by and by be all right.[4]

You will be right in a jiffy, but do not test the colloide too much.

I came out here on Friday night, and was joined last night by Hirst and Debus.[5] To day we are to have a walk together.

I wish you were with us. | Love to all around you | Ever | John Tyndall.

IC HP 8.92
Typed Transcript Only

1. *[1871]*: dated by continuity with letter 3586.
2. *It certainly . . . little prod*: see letter 3586 for Huxley's approval. The 'little prod' is Tyndall's publication, in *Nature*, of a letter in which William Carpenter complained that Tyndall had not given Carpenter sufficient credit for his recognition of Joule's work. See letters 3573 and 3584. For other letters relevant to this issue, see 3575, 3581, 3582 and 3598.
3. *The man*: Carpenter.
4. *But the . . . be all right*: In letter 3586 Huxley used the word 'colloide' in a physiological sense; Tyndall here plays on its use in chemistry 'to describe a peculiar state of aggregation in which substances exist; opposed to *crystalloid*' (*OED*).
5. *Hirst and Debus*: Hirst noted in his journal of 25 December 1871 that he, Tyndall and Debus walked from Ascot to Windsor. See Brock and MacLeod, *Hirst Journals*, p. 1923.

From Mary Coxe[1] 24 December [1871][2] 3588

Kinellan[3] | Sunday 24th Dec.

A merry Xmas and a happy New Year and very many of them to you, my dear friend—and a nice wife before another comes round. That is the best wish I can wish you!

I am sorry to hear of "hurts"[4]—You do not say of what nature, but I conclude they were the result of scientific experiments. I remember Sir David Brewster[5] very nearly putting out his eyes.[6] Yes—we were grievously disappointed at your non appearance I had always a lingering hope that you might cast up, albeit such a Will o'the wisp—and once in Switzerland it is a case of "lasciate ogni speranza"[7] of your return till the very last moment.

We had a very jolly time at the Association,[8] (22 in the house—up stairs and down) The Rose Innes's[9] and two girlies, Elma (Mrs Stewart when you saw her) and her boy[10]—&c, Among others, the author of the two best works I know—"Robin Gray" and "For lack of gold"[11] This last Lord Lorne[12] told me when here the other day—he and "his wife"[13] had read aloud to each other in order to enjoy it together; it was at the Queen's desire they had got them

You remember meeting—dining with us in London—and once calling also—Ian Sobieski Stuart[14]—the elder of "the Princes". I am sorry to say, he is dying—and I fear rapidly—a complication of maladies. The most hopeless of which is a gradual closing of the throat. More dreadful even than the prisoner, who daily saw the walls of his cell approximate till at last he was crushed between them.[15]

We are all three[16] well—and all desire kindest memories to you.

Ever believe me, & in all sincerity | Your affectionate friend | Mary Anne Coxe.

RI MS JT/1/TYP/1/296
LT Typescript Only

---

1. *Mary Coxe*: see letter 3406, n. 2.
2. *[1871]*: year suggested by LT and confirmed by reference to the BAAS meeting at Edinburgh and to Tyndall's leg injury; see n. 4 and n. 8 below.
3. *Kinellan*: the Coxe's home near Edinburgh.
4. *hurts*: Tyndall was probably referring to the injury to his leg in November 1871; see letters 3564 and 3567.
5. *David Brewster*: David Brewster (1781–1868), a Scottish natural philosopher, author, and university administrator; his career included appointments at the University of St Andrews and the University of Edinburgh. His scientific work focused on optics and vision and included an interest in instrumentation, which led to, among other things, the invention of the kaleidoscope. He was elected FRS in 1815 and received the RS's Copley Medal the same year. Brewster also authored an important biography of Isaac Newton, and was among the editors of the *Phil. Mag.*, from 1832–68.
6. *I remember . . . his eyes*: In 1831 a chemical explosion nearly cost him his eyesight (M. M. Gordon, *The Home Life of Sir David Brewster* (Edinburgh: Edmonston and Douglas, 1869), p. 151).
7. *lasciate ogni speranza*: abandon all hope (Dante Alighieri, *Divina Commedia*, III.9).
8. *the Association*: the BAAS, which met in Edinburgh in August 1871. For Tyndall's non-attendance, see letter 3503.
9. *The Rose Innes's*: James Rose-Innes (1824–1906), a South African magistrate; he and his wife Mary Ann (née Fleischer) had nine children, but we are not sure if this is the family to whom Coxe refers.
10. *Elma . . . and her boy*: not identified.
11. *the author . . . gold*: The author is Scottish novelist Charles Gibbon (1843–90). The novels are *Robin Gray* (London: Blackie and Son, 1869) and *For Lack of Gold* (London: Blackie and Son, 1871).
12. *Lord Lorne*: John George Edward Henry Douglas Sutherland Campbell (1845–1914), marquess of Lorne and MP for Argyll from 1868 to 1878. Campbell served as governor-general of Canada from 1878 until 1883. He became Duke of Argyll in 1900 upon the death of his father, George Campbell.

13. *"his wife"*: Louise Caroline Alberta (1848–1939), the fourth daughter of Queen Victoria; she married Lorne in March 1871. Louise was artistically and politically inclined; she exhibited sculptures and watercolors, and she supported progressive causes including the women's movement.
14. *Ian Sobieski Stuart*: John Sobieski Stuart (real name John Carter Allen) (*c.* 1795–1872); he and his brother Charles Edward Stuart (real name Charles Manning Allen) (*c.* 1799–1880) pretended to be the heirs of Charles Edward Stuart (Bonnie Prince Charlie), who unsuccessfully claimed the British throne after 1766.
15. *the prisoner . . . them*: An early version of this story was 'The Iron Shroud', by William Mudford (*Blackwood's Edinburgh Magazine*, 28:170 (August 1830), pp. 364–71), but the device is used elsewhere, including in Edgar Allen Poe's story 'The Pit and the Pendulum' (E. A. Poe, 'The Pit and the Pendulum', *The Gift: A Christmas and New Year's Present* (Philadelphia: Carey and Hart, 1843), pp. 133–51).
16. *all three*: The other two are probably James Coxe (see letter 3406, n. 1) and William Cumming (see letter 3406, n. 7).

To Peter Lesley 27 December 1871 3589

Royal Institution of Great Britain | London 27th Dec. 1871.

My dear Professor Lesley

Your excellent letter[1] has just reached me and I cannot sufficiently thank you for it and for the capital sketch of Philadelphia which accompanies it.

I hope you will allow me to jot down from time to time such memoranda as may be conducive to the success of the lectures. Such jottings may render arrangements possible which if proposed suddenly might create difficulty.

One arrangement which I variably adopt here, and which I should be disposed to regard as indispensable, is that I should have the room in which the lecture is to be given at my disposal during the day previous to the lecture. The preparation of the experiments would I think render this arrangement absolutely necessary.

I think the plan you propose an excellent one. To the Smithsonian Institution,[2] the Peabody Institute,[3] and the Lowell Institution[4] I am in the first instance pledged.[5] In fact Mr. Lowell[6] was the first to invite me to go to America,[7] and the Director of the Peabody Institute[8] has also a claim upon me. It will soon appear whether the lectures are suitable to American audiences, so that the other places you mention may then act with their eyes open.

Accept my very best thanks for your offer of hospitality. I know I can ask you to grant me the tolerance which is so often granted to me by my friends at home. In fact a considerable residue of the primeval savage still adheres to me and this causes me in certain moods to enjoy the barbarous liberty of a hotel. Should the savage mood predominate when I arrive I hope you will be good natured enough to humour me and help me to find a quiet room in a hotel.

It will however give me exceeding pleasure to make the acquaintance of M[rs]. and the Misses Lesley.[9] I wish I could talk to the young lady about art; but though I have a kind of instinct toward it I lack the educated judgement.

I heartily wish you could come over to London. The doors of the Athenaeum Club[10] would be open to you.

Ever yours faithfully | John Tyndall

Mss.B.L56.Letter 11. American Philosophical Society

---

1. *Your excellent letter*: letter missing; perhaps a response to letter 3543, in which Tyndall discussed plans to visit and lecture in the United States. For additional letters related to this trip, see 3504, 3507, 3510, 3519, 3542, 3558, 3594, 3639, 3669, 3670, 3684, 3685, 3702 and 3703.
2. *Smithsonian Institution*: see letter 3504, n. 2.
3. *Peabody Institute*: see letter 3510, n. 4.
4. *Lowell Institution*: founded by a bequest from John Lowell (1799–1836) and opened in 1839 with the intent of presenting popular lectures and instruction in a range of subjects (*ANB*).
5. *I am . . . pledged*: see n. 1.
6. *Mr. Lowell*: John Amory Lowell (1798–1881), an American businessman and philanthropist. He served as the first trustee of the Lowell Institute, which had been endowed by his cousin. See n. 4.
7. *the first to invite me to America*: Lowell invited Tyndall to the United States in November 1864. See *Ascent of John Tyndall*, p. 300; and letter 2160, *Tyndall Correspondence*, vol. 8.
8. *the Director of the Peabody Institute*: Nathanial Holmes Morison; see letter 3510, n. 4.
9. *the Mrs. and Misses Lesley*: Lesley's wife, Susan Lesley (née Lyman, 1823–1904), and two daughters, Mary Lesley (n.d.) and Margaret Lesley (b. 1857).
10. *Athenaeum Club*: see letter 3372, n. 14.

From Joseph Dalton Hooker — 27 December 1871 — 3590

Kew[1] | Dec. 27th/1871.

Dear Tyndall

The Emperor of Brazil[2] has written asking me to procure for him two copies of your photograph, with your signature on both. I am thus obliged to trespass on your good-nature. I promised to get them before he left England, but did not, and he has sent over from France for them! so he must be in earnest.

Ever your aff[ectionately] | J. D. Hooker.

RI MS JT/1/TYP/8/2601
LT Typescript Only

---

1. *Kew*: the Royal Botanic Gardens at Kew, in London. Joseph Hooker's father, William Hooker, was director from 1841 until his death in 1865, at which point Joseph succeeded him. The gardens were open to public visitors and were also used for scientific purposes; they were extensively imbricated with Britain's imperial enterprise. On Joseph Hooker and Kew, see Endersby, *Imperial Nature*, and MacLeod, "Ayrton Incident." On the general history of the Royal Botanic Gardens at Kew, see L. H. Brockway, *Science and Colonial Expansion: The Role of the British Royal Botanic Gardens* (New Haven, CT: Yale University Press, 2002); R. Desmond, *The History of the Royal Botanic Gardens Kew* (London: Royal Botanic Gardens, Kew, 2007); R. Drayton, *Nature's Government: Science, Imperial Britain, and the "Improvement of the World"* (New Haven, CT: Yale University Press, 2000); and K. Teltscher, *Palace of Palms: Tropical Dreams and the Making of Kew* (London: Picador Books, 2020).
2. *Emperor of Brazil*: Pedro d'Alcantara.

From Rudolf Clausius 30 December 1871 3591

Bonn, 30/12 71.

Lieber Tyndall,

Du hast nicht nur dem Johnny, sondern uns allen eine grosse Freude gemacht durch Deinen freundlichen Brief.

Wenn Du darin sagst ich habe Dich vergessen, so ist das natürlich ein Scherz. Du weisst recht gut, dass, wenn ich auch selten schreibe, wenn nicht eine besondere Veranlassung vorliegt da ich in meinem Amte viel zu thun habe, ich darum doch meine Freunde nicht vergesse, und am allerwenigsten Dich.

Diesen Sommer war ich meines leidenden Knies wegen im Seebade auf der Nordsee-Insel Borkhum bei Emden. Da traf ich mit Jenkin zusammen, der bei der Legung eines Kabels betheiligt war, welches von dort nach England geführt ist. Ich habe die Seefahrt, auf welcher die erste Hälfte des Kabels gelegt wurde, mitgemacht, was mir sehr interessant war. Auch freute ich mich die Bekanntschaft von Jenkin zu machen, der sehr gefällig war.

Mein Knie ist jetzt soweit hergestellt, dass ich in meinen Geschäften nicht gehindert bin, und auch täglich etwas spaziren gehen kann; aber viel darf ich ihm noch immer nicht bieten. Hoffentlich wird es mit der Zeit wieder kräftiger werden.

Vor Kurzem ist ein Buch von Maxwell erschienen "Theory of Heat" welches mich in das grösste Erstaunen versetzt hat. Ich habe nie geglaubt dass ein Mann, der wissenschaftlich so hoch steht, so ungerecht sein und den historischen Sachverhalt wissentlich so falsch darstellen konnte. Mit Ausnahme der Stellen wo von der Molecularconstitution der Körper die Rede ist, kommt mein Name nur einmal vor, indem gesagt wird ich habe den Namen Entropy eingeführt, wobei aber hinzugefügt wird, die Theorie der Entropy habe Thomson schon vor mir gegeben. Alles, was ich gethan habe, wird Thomson und Rankine zugeschrieben. Mayers Name kommt im ganzen Buche nicht vor.

Du erinnerst Dich vielleicht noch, dass Du mir bei der Besteigung des Drachenfels, als wir von dem Historical Sketch of Thermodynamics von Tait sprachen, riethest, etwas dagegen zu schreiben, ich aber sagte, solche persönlichen Erörterungen seien mir zu unangenehm. Jetzt habe ich mich aber doch entschlossen eine kurze Darstellung des Sachverhaltes, soweit er mich betrifft, zu schreiben. Ich kann glücklicherweise durch ganz bestimmte Daten, und durch Aussprüche von Rankine und Thomson selbst meine Rechte so klar nachweisen, dass gar kein Zweifel darüber bestehen kann. Würde ich noch länger schweigen, so könnte dieses als eine Zustimmung zu den falschen Darstellungen betrachtet werden, und mein Name würde allmälig ganz aus der mechanischen Wärmetheorie verdrängt werden.

Ich bin gerade damit beschäftigt den Artikel zu schreiben, und werde ihn, wenn er fertig ist, an Francis schicken mit der Bitte ihn in das Phil. Mag. aufzunehmen.

Hoffentlich geht es Dir und Hirst gut. Ich wünsche Euch beiden von Herzen Glück zum neuen Jahre. Johnny schreibt selbst, um seinen Dank für den Brief auszusprechen, auf den er sehr stolz ist. Es ist der erste Brief, den er durch die Post bekommen hat, und nun gar ein Brief aus England. Ich hatte den Kindern zu Weihnachten eine grosse Landkarte von Europa geschenkt. Da war Johnny's erstes, England und London und den Weg dahin aufzusuchen, was ihm ungeheure Freude machte.

Mit herzlichen Grüssen von meiner Frau und mir, | Dein | Clausius.

Bonn, 30/12 71.

Dear Tyndall,

You[1] have given not only Johnny,[2] but all of us great pleasure through your kind letter.[3]

When you say in it that I have forgotten you, then that is of course a joke. You know quite well that, even if I write only rarely, unless there is a special reason for it, as I am busy in my post, I am therefore not forgetting my friends, and least of all you.

This summer, because of my sore knee,[4] I visited the seaside resort on the North Sea island of Borkum, near Emden.[5] There I met Jenkin,[6] who was involved with the laying of a cable which is taken from there to England. I went along on the sea trip on which the first half of the cable was laid, which I found very interesting. I was also pleased to make the acquaintance of Jenkin, who was very obliging.

My knee has now recovered to the point that I am not hampered in my daily business, and can even go on short daily walks; however, I am still not allowed to demand much from it. Hopefully it will get stronger again with time.

A book by Maxwell has appeared recently, "Theory of Heat",[7] which utterly astonished me. I never believed that a man who stands so highly in the scientific world could be so unjust and knowingly represent historical facts so falsely. With the exception of the passages where the molecular constitution of bodies is discussed, my name occurs only once, where it is said that I introduced the term entropy,[8] but at which point it is added that Thomson had already formulated the theory of entropy before me. Everything that I have done is ascribed to Thomson and Rankine. Mayer's name does not occur in the entire book.

You will possibly still remember that, at the time we climbed the Drachenfels,[9] as we were talking about the Historical Sketch of Thermodynamics by Tait,[10] you advised me to write something against it, but I said that personal debates of that kind were too unpleasant for me. Now, however, I have decided to write a short representation of the facts, so far as they concern me. Fortunately, I can prove my claims so clearly, through quite specific dates, and through remarks from Rankine and Thomson themselves, that absolutely no doubt about them can exist. Were I to remain silent any longer, then this could be viewed as agreeing with the false representations, and my name would gradually be displaced from the mechanical theory of heat entirely.

I am just now occupied with writing the article, and shall send it, when it is ready, to Francis,[11] asking him to accept it in the Phil. Mag.[12]

Hopefully you and Hirst are well. I sincerely wish you both good fortune for the New Year. Johnny will write himself to express his thanks for the letter, of which he is very proud. It is the first letter that he has received through the mail, and what's more a letter from England. I had given the children[13] a large map of Europe for Christmas. The first thing Johnny did was to search out England and London, and the way there, which gave him immense pleasure.

With warm regards from my wife[14] and me, | Your | Clausius.

RI MS JT/1/TYP/7/2323–4

1. *You*: Clausius uses 'du', the informal German 'you'.
2. *Johnny*: Rudolf John Clausius; see letter 3444, n. 7.
3. *your kind letter*: letter missing.
4. *my sore knee*: Clausius was injured during the Franco-Prussian war; see letter 3444.
5. *the North Sea island of Borkum, near Emden*: in northwest Germany; Emden is a port city.

6. *Jenkin*: Henry Charles Fleeming Jenkin (1833–85), a British electrical engineer, professor of engineering at the University of Edinburgh and collaborator with William Thomson. Jenkin's work focused primarily on marine telegraph engineering, on which he held thirty-five patents by the end of his life. He was elected FRS at age thirty-two.
7. *"Theory of Heat"*: J. C. Maxwell, *Theory of Heat* (London: Longmans, Green, and Co., 1871).
8. *my name . . . entropy*: see Maxwell, *Theory of Heat*, pp. 186–8.
9. *Drachenfels*: hill near Bonn, Germany.
10. *Historical Sketch of Thermodynamics by Tait*: P. G. Tait, *Sketch of Thermodynamics* (Edinburgh: Edmonston and Douglas, 1868).
11. *Francis*: William Francis (1817–1904) was the editor, with his father Richard Taylor (1781–1858), of the *Phil. Mag.* Francis was friendly with Tyndall and helped to support him early in his career. See *Ascent of John Tyndall*, pp. 61–86; and the lengthy correspondence between them in R. Barton, J. Rankin, and M. S. Reidy (eds), *The Correspondence of John Tyndall, Volume 3: The Correspondence, September 1843—December 1849* (Pittsburgh: University of Pittsburgh Press, 2017).
12. *I am . . . Phil. Mag.*: R. Clausius, 'A Contribution to the History of the Mechanical Theory of Heat', *Phil. Mag.,* 42:284 (1872), pp. 106–15.
13. *the children*: Clausius had six children by 1875.
14. *my wife*: Adelheid Clausius.

From Mary Egerton [1871][1] 3592

Thoresby Park, | Ollerton[2]

French.[3]

Have you forgiven our unfortunate game at cross purposes? It only struck me afterwards, that, to make the imbroglio more complete, you probably thought Wallace[4] was a friend of mine, & that my vexation was in the interest of his reputation—whereas I know nothing whatever of him, but his books on natural history & his noble generous conduct regarding Darwin.

It is an old story now perhaps, but I cannot resist saying how much good I believe your "Discourse"[5] at Liverpool[6] has done. I have seen it in many allusions, from the Church Congress[7] (!) downwards. A generous tone on one side induces the same on the other, and you have shown that men of science are not all indiscriminate Iconoclasts, & that the sincere seekers after Truth need not be at enmity, even though they should differ as to the roads open for reaching it. I hope you don't mind my saying this? I don't mean to be presumptuous—but it has been so often in my thoughts.

We are living here in a corner of the magnificent new house[8] my brother[9] is building, with the poor old home of my childhood all dismantled staring me in the face. But the melancholy recollections[10] of my visit last year reconcile me in some degree to the wrench of leaving it. We go from hence to London on Monday next, and shall be at home all the evening from 6 o'clock upwards, in case you were inclined to repeat the proposal, you made us last time, of a visit. I humbly beg pardon for filling a second sheet,[11] but the post goes early & I had not time to consider what ought to be left out! What a miserables.

RI MS JT/1/E/99p

---

1. *[1871]*: Egerton's lax use of time periods means this letter can only be roughly dated. References to Tyndall's Liverpool discourse (see n. 5) and the Church Congress (see n. 7) place it in late October 1870 at the earliest. Egerton's mention of the death of her husband (see n. 10) and the building of the new house (see n. 8) places her letter in late 1870 or early 1871, when the house was finished. We place it here, as a letter that was written sometime in 1871, in accordance with our editorial practice.
2. *Ollerton*: a small town in Nottinghamshire, England. Thoresby Park is the home of Mary Egerton's brother.
3. *French*: This letter is a fragment, so the word 'French' ends a sentence from the first page. See n. 11.
4. *Wallace*: Alfred Russel Wallace (1823–1913), a British naturalist, geographer, and evolutionary theorist. He was well-known for his work in the Amazon and Malay Archipelago and his research in biogeography. He was the co-discoverer of the theory of evolution by natural selection, and his work spurred Darwin to publish his theory after decades of work. It is unclear to what Egerton is referring in this paragraph.
5. *your "Discourse"*: Tyndall's discourse 'On the Scientific Use of the Imagination'; see letter 3370, n. 5.
6. *at Liverpool*: The BAAS met in Liverpool from 14–21 September 1870.
7. *Church Congress*: annual meeting of the members of the Church of England, used to discuss religious, moral, and social issues. The meeting in 1870 took place in Southampton on 11–14 October.
8. *magnificent new house*: Thoresby Hall, Ollerton, Nottinghamshire, England, was the Egerton family home. It was rebuilt between 1864 and 1871.
9. *my brother*: Sydney William Herbert Pierrepont (1825–1900), Egerton's brother and the Third Earl Manvers. He attended Christ Church College, Oxford, earning his BA in 1846. He became Viscount Newark in 1850, and Third Earl Manvers in 1860. He was a Conservative MP for South Nottinghamshire from 1852 to 1860.
10. *the melancholy recollections*: Mary Egerton's husband, Edward Christopher Egerton (1816–69) died unexpectedly in 1869. He was MP for Macclesfield from 1852–68, and for East Cheshire from 1868–9. He was Under-Secretary for Foreign Affairs from 1866–9 under both the Earl of Derby and Benjamin Disraeli.
11. *second sheet*: The letter transcribed here is only one sheet of stationery, meaning this is the second half of her letter.

From Herbert Spencer [1871][1] 3593

White Hart Hotel | Salisbury | Thursday

My dear Tyndall,

My conscience has been pricking me all the week about my rudeness to you the other night.[2] It was partly real partly apparent; for what I intended as illustration you apparently construed into a proposition. However there was quite enough of impropriety in it to call for an expression of regret.

I am, as you see in the controversy[3] with Hutton,[4] moderate enough in feeling and expression when I have time to think. But my excitable brain, apt to become suddenly congested, carries me away before I am aware of it, on these occasions when I get into argument. Pray do not measure me by such manifestations.

ever yours sincerely | Herbert Spencer

RI MS JT/1/S/185
RI MS JT/1/TYP/3/1173

---

1. *[1871]*: dated to sometime in 1871 because of the controversy with Richard Hutton; see n. 3 and n. 4.
2. *rudeness to you the other night*: It is not clear to what this refers.
3. *the controversy*: In 1869, Hutton had given a paper at the Metaphysical Society critiquing Spencer's theory of evolution and the development of morals. In July, Hutton published a version of this paper as an article, 'A Questionable Parentage for Morals', *Macmillan's Magazine,* 20 (July 1869), pp. 266–73. Spencer did not seem to have read it, or he did not react to it, until he published a rebuttal in April 1871, 'Morals and Moral Sentiments', *Fortnightly Review*, 52 (April 1871), pp. 419–32. In response to Spencer's having published words about him that Hutton considered 'as a little harsh', Hutton issued a rebuttal: R. H. Hutton, 'Mr. Herbert Spencer on Moral Intuitions and Moral Sentiments', *The Contemporary Review*, 17 (April 1871), pp. 463–72. It is unclear how long the two were embroiled in this controversy. We assume it was much of the spring and summer of 1871, but cannot place it clearly. For more on this, see I. Hesketh, 'Evolution, Ethics, and the Metaphysical Society, 1869–1875', in C. Marshall, B. Lightman, and R. England (eds), *The Metaphysical Society (1869–1880): Intellectual Life in Mid-Victorian England* (Oxford: Oxford University Press, 2019), pp. 185–203.
4. *Hutton*: Richard Holt Hutton (1826–97), an English journalist and theologian. He attended University College London where he studied philosophy and mathematics. From 1850–3, Hutton was the vice-principal and then principal of University Hall, Gordon Square, London, which was a mostly Unitarian congregation. Hutton became a coeditor of the weekly *Spectator* in 1861, a position he held until 1897, despite his embrace

of controversial positions; for instance, Hutton was anti-slavery, and his support for the North in the American Civil War caused the *Spectator*'s subscriptions to decline. He was an original member of the Metaphysical Society and was known to attack materialism in the works of Darwin, Spencer and Tyndall.

1872

From Hector Tyndale 2 January 1872 3594

Philadelphia Jan[uar]'y 2d 1872

My dear John

Your two letters with acknowledgement of my previous letters—and Prof:[1] Lesley's &c—came duly.[2]

To day Rev[eren]'d Mr: Furness[3] D.D. &c (Unitarian Clergyman of this City) sent me a copy of a print of Emerson and another for you—he seems to think much of you. The painter of the original was a son of Doctor Furness.[4] He died, a young man, about five or six years ago and it is to him he alludes when he says "tell him (you) (what I cannot) that the blessed artist is no more and what a gentle and true artist he was &c". As you may meet Doctor Furness when you come here, I enclose his note to me,[5] so that he can tell you himself of this and of his "little book 'Jesus'"[6] which he has sent you and of the arrival of which he would be glad to know. His son, the painter, was a very conscientious, upright self-expressing man, whom I knew and esteemed very greatly. No man here or in Boston—he is a Boston man I think—stands higher than Dr: Furness. He was a classmate of Emerson's and is one of his nearest friends. He is an earnest, hopeful, modest and able man of great culture and faculty who lives a pure and upright life.

By to day's mail I have sent a roll with the copy of the print of Emerson on it which I trust may arrive safely. It is a private print of Dr. Furness' and therefore otherwise unattainable—and only twenty copies made.

Some days ago I wrote to Cousin Emma[7] and acknowledged the receipt of some kind Xmas Souvenirs for Mrs Tyndale and myself from Kathleen, Annie & Georgina—our Cousins of Gorey.[8]

They are very kind & lovable people. I have a great regard for them and for Cousin Emma.

Please remember me kindly to Messrs Hirst and Debus, who, I am glad to know, think sometimes of me.

Mrs: Tyndale thanks you for your remembrances and desires hers and the compliments of the season, if it is not too late, to you. Some days ago Lesley told me he had written to you again.[9]

Affectionately yours | Hector Tyndale
Prof. John Tyndall &c.

RI MS JT/1/T/65

---

1. *Prof:*: for this punctuation, see letter 3507, n. 11.
2. *Your . . . duly*: One of these letters is 3558; the other is probably a missing letter. See letter 3558, n. 2.
3. *Rev[eren]'d Mr: Furness*: William Henry Furness (1802–96), minister of the First Unitarian Church of Philadelphia from 1825 to 1875. Furness wrote numerous studies of the New Testament, his approach to which was influenced by German Biblical criticism, as well as by Emerson and American Transcendentalism more broadly.
4. *a son of Doctor Furness*: William Henry Furness (1827–67), a portrait painter.
5. *his note to me*: letter missing.
6. *his little book 'Jesus'*: W. H. Furness, *Jesus* (Philadelphia: B. Lippincott, 1871).
7. *Cousin Emma*: probably Tyndall's sister, Emily, or Emma, Tyndall.
8. *Kathleen, Annie & Georgina—our Cousins of Gorey*: possibly the family of John Tyndall, a cousin in Gorey. See letter 3507; and letter 2095, *Tyndall Correspondence*, vol. 8.
9. *Lesley told me he had written to you again*: probably the missing letter to which Tyndall replied in 3589.

From James Coxe[1] — 3 January 1872 — 3595

General Board of Lunacy | Edinburgh, Jan 3rd 1872.

My dear Dr Tyndall,

Do you know what it means to feel black affronted?[2] Well, I feel so at your repeated recognitions of my existence.[3] But the pleasure which you thus bestow upon me will I trust be to some extent reflected back upon yourself, by the assurance that your pearls are not cast before swine[4]—or if before swine by swine who can and do appreciate in a very high degree the mind and heart that have their local habitation within the earthly tabernacle of John Tyndall.

Remember that precious to your country is your life. So beware of over work in the laboratory or on the mountain. And strive so to moderate your zeal that fifty years hence you will be receiving from friends and admirers as hearty good wishes for your health and happiness during the coming year as I now send you for 1872.

Yours most truly | J. Coxe.

RI MS JT/1/TYP/1/297
LT Typescript Only

1. *James Coxe*: see letter 3406, n. 1.
2. *black affronted*: Scottish expression meaning deeply embarrassed (*OED*).
3. *your repeated recognitions of my existence*: possibly a reference to a missing letter from Tyndall, perhaps a response to Mary Coxe's letter of 24 December 1871, letter 3588.
4. *your pearls are not cast before swine*: Matthew 7:6.

From Robin Allen 4 January 1872 3596

TRINITAS IN UNITATE[1] | TRINITY HOUSE, |

LONDON. E.C. | 4·1·72

My dear D[r] Tyndall

The recollection of Pygmalion[2] is an especially pleasant one to me *[*causing*]* as I think it does your kind *[*permission*]* for me to send you the accompanying copy of Orion[3]—the best Horne[4]-book on the philosophy of noble Love that I know.

Believe me | v[er]y faithfully y[ou][r] serv[an][t] | Robin Allen

While we are thus engaged there Chamois look down from the heights ab[o]v[e][5]

RI MS JT/1/A/83

1. *TRINITAS IN UNITATE*: this letter is on Trinity House letterhead featuring the House's coat of arms, with the words 'TRINITAS IN UNITATE', meaning 'Three in One', a reference to the Holy Trinity.
2. *The recollection of Pygmalion*: W. S. Gilbert's play, *Pygmalion and Galatea*, premiered on 9 December 1871 ('Dramatic Gossip', *Athenaeum*, 2302 (9 December 1871), pp. 765–6, on p. 765).
3. *the accompanying copy of Orion*: a book-length poem by Richard Horne (see n. 4), first published in 1843. In December 1871 it was published in a 'Library Edition' (R. H. Horne, *Orion* (London: Ellis and Green, 1871)). *The Examiner* listed it among their 'Christmas Books' ('Christmas Books', *Examiner*, 3334 (23 December 1871), pp. 1272–3).
4. *Horne*: Richard Hengist Horne (1802–84), an author of poetry, drama and fiction; *Orion* (see n. 3) was his best-known work. Horne briefly worked as a sub-editor of *Household Words* in the 1850s, but relocated to Australia in 1852. He remained there until 1869, when he returned to London and re-entered the literary scene.
5. *While we . . . ab[o]v[e]*: This sentence appears on the recto of the letter. We have been unable to trace its referent or source.

To [John] Sedgwick[1] 5 January 1872 3597

5th. Jan 1872

My dear Sir

The icicles are perfectly beautiful; but I am terribly hampered by the ground glass.[2] Mr. Ladd states that he can receive it and restore it perfectly. Dare I ask you to permit me to remove the glass for tomorrow's lecture?

faithfully yours | John Tyndall

J B Sedgwick Esq

RI MS JT/1/T/1274

1. *[John] Sedgwick*: A John Bell Sedgwick (d. 1895) appears in the list of visitors to the RI. Sedgwick may also have been a member of the Council of the Photographic Society of London.
2. *The icicles . . . glass*: probably a reference to a demonstration Tyndall performed during his Christmas lectures at the RI. See letter 3565, n. 2.

From James Prescott Joule 6 January 1872 3598

5 Cliff Point Broughton | Manchester 6/1/72

My dear Tyndall

I saw the articles in "Nature"[1] from your pen and thank you most heartily for the kind expressions you have made respecting myself. I have not yet quite done my part as regards the mechanical equivalent and am about making fresh experiments with a view to further accuracy and if I work now slower; than I once did, I hope to make a better experiment as far as absolute accuracy is concerned.[2]

Wishing you the blessings of the season believe me

Yours most truly | James P. Joule

RI MS JT/1/J/141

1. *articles in "Nature"*: see letter 3575, n. 4.
2. *I have not . . . concerned*: In his article about Joule in *Nature* Tyndall had emphasized Joule's contributions as an experimenter to the mechanical theory of heat ('The Copley Medalist of 1870', *Nature*, 5 (21 December 1871), pp. 137–8, on p. 138).

From Lyell Adams[1] 7 January 1872 3599

United States Consulate, | Malta, January 7th 1872.

Dear Sir,

If you will take the trouble to present the enclosed receipted freight bill to Messrs. P. and W. Smith No. 2 Crosby Square, you will receive a small case of Malta oranges, Egg and Mandarin. I tried to arrange for their delivery at the Royal Institution, but the red tape of the shipping agents is insuperable.

Lest the oranges should seem unaccountable to you I ought to explain that they have reference to the Sundry doses of Chloral[2] administered last September at Pontresina—forgotten very likely by you, but quite memorable to me for the good effect they had. I say nothing of the other debts contracted in late years, for if one were to acknowledge all obligations of that sort I should be shipping oranges to Mr. Carlyle—which indeed I should be very much pleased to do—and Mr. Emerson and many more. It is plainly necessary to draw the line somewhere, so I draw it at eminent Authors to whom I have been indebted for Chloral.

I hope the oranges will reach you in good condition—although I can't promise that they will—and do as much good as the Chloral did to.

Yours very sincerely | Lyell T. Adams.

Professor Tyndall

I see by some numbers of the Atlantic Monthly[3] which I have just received that Mr. Clarence King having scaled a great Snow-peak Sierra Nevada named it Mount Tyndall.[4] This seems to me a very cold compliment as compared with the tropical warmth of mine. Moreover it can't be repeated, for to make the christening effective it is indispensible that there shall be no more Mount Tyndalls. Whereas, if you are fond of oranges it will always be easy and pleasant to send you some.

L. T. A.

RI MS JT/TYP/1/6
LT Typescript Only

---

1. *Lyell Adams*: Lyell Thompson Adams (1837–92), the American consul at Malta. In his journal, Tyndall recorded the receipt of the oranges on 24 January 1872, noting that he had met Adams at Pontresina (RI MS JT/2/13c/1385).
2. *chloral*: chloral hydrate, used as an anesthetic in the nineteenth century. Tyndall died of an overdose of chloral mistakenly administered to him by his wife Louisa Tyndall in 1893.
3. *Atlantic Monthly*: leading American journal of literature and culture, founded in 1857.

4. *I see . . . Mount Tyndall*: King published a series of articles about the California geological survey he was a part of with Josiah Dwight Whitney, titled 'Mountaineering in the Sierra Nevada' in the *Atlantic Monthly* in 1872. Adams is referring to C. King, 'Mountaineering in the Sierra Nevada. III. The Ascent of Mount Tyndall.', *Atlantic Monthly*, 28:165 (July 1871), pp. 64–76, on p. 76. The survey climbed the peak King named Mount Tyndall, which rises to just over 14,000 ft., on 6 July 1864. King recounts the ascent and descent of Mount Tyndall in chapters 3 and 4 in his *Mountaineering in the Sierra Nevada* (Boston: James Osgood & Co., 1872), pp. 49–93. See also letter 2224 in Michael D. Barton, Iwan Rhys Morus, and James Ungureanu (eds), *The Correspondence of John Tyndall, Volume 9: The Correspondence, February 1865–December 1866* (Pittsburgh: University of Pittsburgh Press, 2022).

From Archibald Hamilton 12 January 1872 3600

Trillick Omagh | Jan. 12/72

Dear Sir

I received your kind letter[1] of the 2nd Inst[ant].[2] for which I am much obliged—

My letter consisted of (I think) 5 sheets (20 pages) of paper similar to this, rolled up into a tight roll & directed to you to the Royal Institution London[3] }

Though begun in June I think it was not posted till August; I have no exact note of the time.

The substance of it was as follows:

1. I assert the possibility of conveying to the Deaf Mute something of the profit & (even) something of the pleasure which is produced in those who are gifted with the sense of Hearing, through the agency of the Ear & the operations of the Mind unconsciously connected therewith.

2. For this purpose, after preliminary remarks on what I may call the <u>intellectual</u> department connected with every Sense I proceed boldly <u>to classify</u> the Senses under the 3 primary Intuitions of SPACE, TIME & KIND, giving

Sight to Space,
Hearing to Time
& Taste Smell to Kind }

reserving Touch for the present or keeping it in view only as a kind of Supplementary Sense— Each of these 3 Classes of Sense has its corresponding Science:—

Space has Geometry

Time has Algebra (considered as the Science of Pure Time as my father the late Sir W. R. Hamilton[4] considered it)—

Kind or Quality has Chemistry.

A great assistance is derived from the mutual correlation of the Senses thus suggested; analogies being carried from one Sense to another—

3. I begin with the visible point in Geometry, & endeavour to draw out in an interesting & popular style the possibilities or capabilities which it suggests to the geometer—(suppose you pick up a paper in the Study of Sir I. Newton or Sir John Herschel, with just one point marked on it, after his death)—

I then endeavour to draw out the analogy between the visible point & what I may call the audible one, or the least perceptible portion of the Sound wave, (an idea I have never seen in books; but I do not see any absurdity in it). The analogies of Chemistry should also be glanced at here—

I would thus go back & forwards continually from the Senses my pupil has to the Senses he has not, & from the Sciences connected with them to those which are connected with the others, proceeding gradually & interweaving the known & the unknown & to speak in a popular & entertaining manner if I could.

I set down here a few theses which I would maintain.—

1. A priori considerations are absolutely essential to a firm or an exalted view of such subjects—

2. The analogies I point out & the Classification I suggest are not arbitrary;

(3. I even venture to think that they are indicated in the formation of the bodies of man & of the brute creation—)

4. They help greatly to the object I have in view, ie. to convey to the Deaf Mute some of the profit & even of the pleasure which he is naturally deprived of.

5. The system I propose suggests a priori a much subtler analysis of sound than I have ever seen attempted.

6. It suggests that all musical (or probably all articulate) sounds are only an infinitely particular case of Sound. The unmusical have not received the attention they deserve. $D=o^5$ proposed expression for musical sound.

7. suggestions as to new Instruments & new Experiments to follow.

8. As an example of its power as a Calculus a priori! (to use grandiose language)

[Space] i.e. Vision |

The SPACE=requires a field of view or Extension in SPACE }
The TIME=sense——————————————- } Extension in TIME
The Quality=senses ought to require Extension in Quality }

by a verbal substitution. What is the meaning of this latter phrase? To me it seems to lead up to the time=honoured idea of the Philosopher's Stone; an idea which I believe Faraday and Herschel would not ridicule.

Finally, while asserting for the Ear a power of judging of small vibrations independent of that "Harp of many strings"[6] which Anatomists assert to exist inside that organ, I maintain also in full the claims of ordinary Music upon our attention, & endeavour to draw out some of those prerogatives which it enjoys—

9. The actual vibration of a Pendulum is not perfectly isochronous.[7] May we infer from this that every Musical note rises slightly in the scale as it dies away, the vibrations quickening?

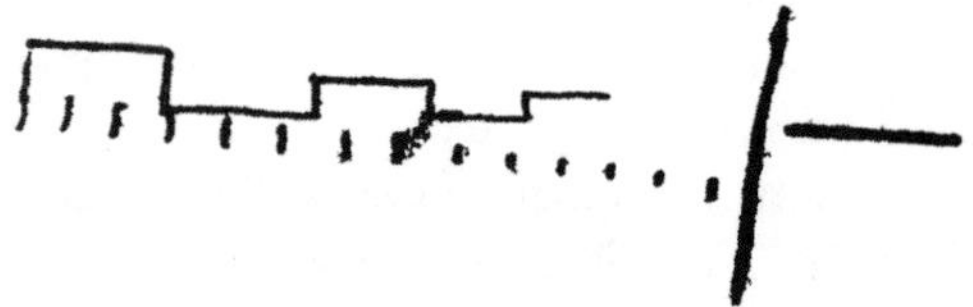

isochronous actual? Quere. what curve?

10. The air we breathe contains a record of every undulation ever impressed on it? This would require high skill to *[power]*, from the interference & mutual collision of different sets of undulations.

These are a few of the points I would maintain—

I want you to be so good as to say whether you think this attempt at a combined Theory of the Senses altogether too venturous, or on the other hand, not new. Also whether there is any doubt in your own mind that the ear possesses a distinct independent power of registering unmusical notes, or whether *[1 word illeg]* any possible property of elastic bodies *[1–2 words illeg]*[8] reduced to a series of short musical notes rapidly ascending or descending (the inclined plane breaking into a flight of short steps)—

I also asked if you could direct me to any commentary on Sir I. Newton's treatment of the matter in the Principia—[9]

Kindly excuse me for so far trespassing on your very valuable time & with best wishes both for time & for Eternity

Believe me | yours faithfully in the Lord | A. H. Hamilton Clk.[10]

RI MS JT/1/H/13

1. *your kind letter*: letting missing.
2. *Inst[ant].*: 'of the current calendar month' (*OED*). Hamilton thus makes reference to a letter of Tyndall's dated 2 January 1872.
3. *My letter . . . London*: either letter 3456 or 3499.
4. *W. R. Hamilton*: William Rowan Hamilton (1805–65), an Irish mathematician known for his work in algebra, including his discovery of quaternions, which were used in the development of linear algebra as well as in the physics of the early twentieth century. Hamilton was Andrews professor of astronomy at Trinity College, Dublin, and in conjunction with this served as Royal Astronomer of Ireland.
5. *D=o*: It is unclear what Hamilton means by this notation.
6. *"Harp of many strings"*: see letter 3456, n. 5.
7. *isochronous*: 'equal in duration, or in intervals of occurrence, as the vibrations of a pendulum' (*OED*).
8. *[1 word illeg] . . . [1–2 words illeg]*: An inkblot makes it difficult to discern whether or not there are illegible words on this page.
9. *Principia*: I. Newton, *Philosophiæ Naturalis Principia Mathematica* (London: Royal Society, 1687).
10. *Clk*: see letter 3456, n. 6.

To Mary Lyell[1] 21 January [1872][2] 3601

21st Jan[uar][y]

My dear M[rs]. Lyell

Most gladly would I join you but up to February the 2nd I am a hermit: working day and night to accomplish a piece of work which has been unexpectedly thrown upon me.[3]

Yours ever faithfully | John Tyndall.

My heart smote me when I saw Rosamond[4]—I forgot to send her a ticket

RI MS JT/1/T/1057
RI MS JT/1/TYP/3/841

1. *Mary Lyell*: Mary Lyell (née Horner, 1808–83), a British geologist and conchologist; she was married to geologist Charles Lyell and collaborated in his research.
2. *[1872]*: year established by the reference to Tyndall's Friday Evening Discourse at the RI on 2 February 1872; see n. 3.
3. *a piece . . . me*: Bence Jones asked Tyndall at the last minute to give a Friday Evening Discourse on 2 February 1872. See *Ascent of John Tyndall*, p. 279, and J. Tyndall, 'On the Identity of Light and Radiant Heat', *Roy. Inst. Proc.*, 6 (1870–72), pp. 417–21.
4. *Rosamond*: Rosamund Lyell (1856–1904), daughter of Mary Lyell's sister Katherine Lyell and Charles Lyell's younger brother Henry Lyell.

From Gustav Wiedemann 21 January 1872 3602

Leipzig 21 Jan 72

Mein lieber Tyndall!

Beifolgend erhalten Sie eine lange Arbeit von mir; den ersten Theil von Untersuchungen über Funkenentladungen; die vielleicht etwas zur Aufhellung mancher Räthsel beitragen können. Experimentell hatte die Arbeit ihre ganz besonderen Schwierigkeiten, wie Sie aus der Abhandlung ersehen werden.

Ihr, jeden Besucher der Schweiz erfreuendes Werk „hours of exercise ist jetzt von meiner Frau fast vollständig übersetzt und der Druck soll demnächst bei Vieweg beginnen. Hoffentlich sind Sie auch mit dieser Uebersetzung zufrieden. Uns hat sie sehr grosse Freude gemacht.—Nur über den deutschen Titel bin ich etwas im Zweifel „Arbeitsstunden oder Uebungsstunden in den Alpen“ klingt mir nicht gut Deutsch: fast würde ich „Ferienarbeiten in den Alpen sagen. oder etwas ähnliches.

Wo haben Sie die Thermosäulen für Ihre Vorlesungen über die Wärme her bezogen und welchen Preis haben sie? Prof Czermak, der jetzt hier öffentlich physiologische Vorträge halten will, möchte es gern wissen.

Im August ist hier die deutsche Naturforscherversammlung. Wollen Sie nicht herkommen und bei uns wohnen? Sie würden uns dadurch eine sehr grosse Freude machen.

Mit besten Grüssen von meiner Frau | Ihr | treu ergebener | G. Wiedemann

Leipzig 21 Jan 72

My dear Tyndall!

Enclosed you will receive a long paper of mine;[1] the first part of investigations on the discharges of sparks; which will possibly be able to contribute something to clarifying a good many riddles. In an experimental sense, the paper had its quite special difficulties, as you will see from the treatment.

Your work, “Hours of Exercise, which will please every visitor to Switzerland, has now been almost completely translated by my wife,[2] and printing should begin soon with Vieweg.[3] Hopefully you will be satisfied with this translation too. It has given us very great pleasure.—It is only about the German title that I am somewhat in doubt. „Arbeitsstunden or Uebungsstunden in den Alpen“[4] does not sound like good German to me: I would almost say „Ferienarbeiten in den Alpen.[5] Or something similar.

Where did you obtain the thermopiles[6] for your lectures on heat,[7] and what is their price? Prof. Czermak,[8] who wants to hold public lectures on physiology here now, would like to know.

The meeting of German natural scientists will be held here in August.[9] Won't you come here and stay with us? You would bring us very great pleasure by doing so.

With best regards from my wife | Your | loyally devoted | G. Wiedemann

RI MS JT/1/W/50

1. *a long paper of mine*: possibly G. Wiedemann and R. Rühlmann, 'Ueber den Durchgang der Elektricität durch Gase', *Annalen der Physik*, 145:2 (12 March 1872), pp. 235–58.
2. *my wife*: Clara Wiedemann.
3. *Your work . . . Vieweg*: see letter 3451.
4. *„Arbeitsstunden or Uebungsstunden in den Alpen"*: "Hours of Work or Exercise in the Alps"; that is, Wiedemann is proposing different possible titles. The German translation was ultimately published under the title *In den Alpen*; see letter 3640.
5. *Ferienarbeiten in den Alpen*: Holiday Work in the Alps. For further discussion of how to translate this title, see letters 3615, 3618, 3632, 3640 and 3649.
6. *thermopiles*: 'a thermo-electric battery, used in connection with a galvanometer, for measuring minute quantities of radiant heat' (*OED*).
7. *lectures on heat*: the lectures on heat Tyndall delivered at RI in 1862, published as *Heat Considered as a Mode of Motion* (London: Longmans, Green, and Co., 1863).
8. *Prof. Czermak*: Johann Nepomuk Czermak (1828–73), a Prague-born physiologist who developed a medical procedure for examining the larynx and throat, known as laryngoscopy.
9. *meeting of German . . . scientists*: The *Gesellschaft Deutscher Naturforscher und Ärzte*, the Society of German Natural Scientists and Physicians, was founded in 1822 and is the oldest German scientific association.

To [George Campbell][1] 22 January 1872 3603

[Royal Institution][2] | 22nd Jan[uar]y 1872.

My dear Duke

I send you an uncorrected proof of a brief article on the "Physical Basis of Solar Chemistry" written many years ago,[3] which refers to and defines some of the points raised in your letter.[4]

I know very well that the word "Identity" is much abused—absurdly so by those who base the affirmation of identity on microscopic examination—You will not I think find me neglecting to limit duly the extension of the word.

Faithfully yours | John Tyndall

The Lecture is on Friday week, not Friday next[5]

RI MS JT/1/T/96
RI MS JT/1/TYP/5/71

---

1. *[George Campbell]*: A note in LT's handwriting on this letter reads ' "I have not seen the Proof of the article. | W.H.B." is written outside this letter'. Although the initials do not align, we believe that Tyndall's correspondent is George Campbell, the Duke of Argyll. In his journal for 22 January 1872, Tyndall wrote 'A letter from the Duke of Argyll about the <*1 word missing*> of Light and Radiant Heat. He is a clearheaded, downright and courageous little fellow' (RI MS JT/2/13c/1385).
2. *[Royal Institution]*: LT annotation.
3. *a brief article on the "Physical Basis of Solar Chemistry"*: J. Tyndall, 'On the Physical Basis of Solar Chemistry', *Roy. Inst. Proc.*, 3 (1858–62), on pp. 387–96.
4. *your letter*: letter missing.
5. *The Lecture is on Friday week, not Friday next*: Tyndall's Friday Evening Discourse at the RI on 2 February 1872. See letter 3601, n. 3.

From Jessie Huxley[1] [22 January 1872][2] 3604

4 Marlborough Place | Abbey Road N.W.

Dear Dr. Tyndall,

I think that I should hardly have liked writing to you, about the birthday present, you so kindly asked me to choose, but now that you yourself have written[3] to reproach me for my unwillingness to do so, I suppose I must decide in some measure.

I think I should like some ornaments, as it will never wear out, but not a broach and not earrings, which I never wear.

I should like you to choose the kind of present, as I shall then have more pleasure in wearing it.

Moo[4] sends her love and so do all of us,

Believe me | yours affectionately | Jessie[5] Huxley.

RI MS JT/1/TYP/9/3054
LT Typescript Only

---

1. *Jessie Huxley*: Jessie Oriana Huxley (1858–1927), second child and oldest daughter of Thomas and Henrietta Huxley.
2. *22 January 1872*: The date is handwritten by LT, with the year marked by one square bracket: [1872. However, the letter seems to fit with letter 3608, which references a note to Jessie, and with Jessie's birthday, in mid-February.
3. *you yourself have written*: letter missing.
4. *Moo*: the Huxley children's nickname for their mother, Henrietta Huxley. See Desmond, *Huxley*, p. 502.
5. *Jessie*: LT's transcript of the letter reads 'Jessiot', but we have not been able to confirm that this was a nickname; it seems more likely to have been an error on LT's part.

From James Kenward[1] 27 January 1872 3605

Harborne,[2] near | Birmingham | 27 Jan[uar][y] 1872

My dear Sir,

I have been reading your "Holiday Scenes in the Alps"[3]—with what pleasure I need hardly say. Unfortunately I can judge of no continental mountain from personal knowledge—but the life and eloquence of your book I can quite well appreciate. There is one descriptive passage surpassing anything I have seen since Ruskin's[4] summing up of Verona.[5]

But—'je vous en veux'[6]—as the French say, for one portion of the work. I have a light velleity[7] of quarrel against you on account of the article on Snowdon.[8] I compare it with the original[9] in the Saturday Review of 5 Jan[uar]y. 1861, and I find a curious lowering of the enthusiastic tone which gave a real charm to the latter. You have qualified, changed, or suppressed epithets and phrases that signified admiration of the Welsh scenes. In particular take the following in the Review:[10]—

"Hardly on Mont Blanc,[11] hardly on Monte Rosa,[12] hardly on the Görner-Grat,[13] hardly even on the Mer de Glace[14] in winter, had we seen anything to excel in beauty this scene from the top of Snowdon. Full of the vigour of the glorious region, we rushed down the mountain to Llanberis."[15]

This is attenuated in the volume[16] for

"The scene would bear comparison with the splendours of the Alps themselves" (strong though this be)

You know that Snowdon is an older brother of Mont Blanc, and probably in the days of his geologic youth he soared quite as high to heaven. May he not justly say to you when your praise mounts higher with each higher young Alp—Bless me also—hast thou reserved no blessing for me?[17]

I am bound to believe that you felt what you wrote in 1861.[18] Can ten years touch the question? With Snowdon most certainly—"tempora non mutantur." I do not want to think of you "mutaris in illis."[19]

Yours very truly | J. Kenward

Prof. Tyndall

RI MS JT/1/K/5

---

1. *James Kenward*: James Kenward (b. 1828), a poet and author. He lived near Birmingham, in the midlands of England, and published poetry and prose about Wales under the name Elfynydd.
2. *Harborne*: suburb to the southwest of Birmingham.
3. *Holiday Scenes in the Alps*: presumably *Hours of Exercise in the Alps*.
4. *Ruskin*: John Ruskin (1819–1900), an English art critic, social critic, and aesthetician. He

championed the work of contemporary artists including William Turner and the Pre-Raphaelites, and also celebrated Gothic art and architecture. Ruskin was critical of contemporary science, which he saw as clashing with his religious values. His criticisms extended to Tyndall; see *Ascent of John Tyndall*, pp. 254–5.

5. *Ruskin's summing up of Verona*: Kenward may have in mind a Friday Evening Discourse on Verona that Ruskin delivered at the RI on 4 February 1870; see J. Ruskin, 'A Talk Respecting Verona and its Rivers', *Roy. Inst. Proc.*, 6 (1870–72), pp. 55–61. As the lecture was not printed in its entirety until the 1890s, it is possible that Kenward had attended in person.
6. *je vous en veux*: I am annoyed with you (French).
7. *velleity*: 'the fact or quality of merely willing, wishing, or desiring, without any effort or advance towards action or realization' (*OED*).
8. *the article on Snowdon*: 'Snowdon in Winter', in *Hours of Exercise in the Alps*, pp. 421–8. Snowdon, a mountain in Wales, is the highest point in the British Isles outside of Scotland.
9. *the original*: [J. Tyndall], 'Snowdon in Winter', *Saturday Review*, 11:271 (5 January 1861), pp. 11–12.
10. *the following in the Review*: see 'Snowdon in Winter', p. 12.
11. *Mont Blanc*: highest mountain in the Alps, on the border of France and Italy.
12. *Monte Rosa*: second highest mountain in the Alps, on the border of Italy and Switzerland.
13. *Görner-Grat*: ridge in the Alps near Zermatt.
14. *Mer de Glace*: glacier on the northern slopes of Mont Blanc.
15. *Llanberis*: village in Wales at the foot of Snowdon.
16. *in the volume*: see *Hours of Exercise in the Alps*, p. 427.
17. *Bless . . . for me?*: Genesis 27:36.
18. *what you wrote in 1861*: that is, in the *Saturday Review*, the first passage that Kenward quotes.
19. *"tempora . . . illis"*: a reference to the saying 'tempora mutantur et nos mutamur in illis': 'the times change, and we change with them' (*OED*).

From Paul Niemeyer[1] 28 January 1872 3606

Magdeburg, 28. Januar | 1872.

Hochverehrter Herr Profeßor!

Sie haben mich durch Ihre werthvolle Zusendung, welcher heute das freundliche Schreiben folgte, ebenso geehrt als erfreut und ich sage Ihnen meinen wärmsten Dank.

Was [den] Respirator betrifft, so darf ich mir die Bemerkung erlauben, daß das Instrument des Mr. Carrick mir insofern unvollkommen erscheint als es *[nur]* den Mund bedeckt. Sowohl Herr M'Cormak als Mr. Catlin („shut your mouth") haben neuerdings dargethan, daß die Nase der natürliche Respirator sei u.s.w. Daß Ihr fire-man's respirator auch die Nasenöffnung schüzt, obgleich es im Texte nicht ausdrücklich bemerkt ist und aus der Zeichnung nicht ersichtlich ist, glaube ich aus dem Umstande schließen zu */dürfen/*, daß Sie */und/* Herrn Captain Shaw's Leute so lange */damit/* im Rauche aushielten. Es wäre mir wichtig, dies ausdrücklich zu erfahren, um so mehr, als unsere Polemik gegen den Jeffrey'schen Respirator gerade auf diesen Umstand Gewicht legt. Daß ich Ihnen, Hochverehrter Herr Profeßor, für die gütige Mittheilung */einer/* „sketch of the respirator in his most */recent/* form" sehr dankbar sein würde, brauche ich wohl kaum ausdrücklich zu sagen.

Ich darf mir bei dieser Gelegenheit erlauben, Ihnen meine, zum Theil aus Ihren Vorlesungen geschöpfte neue Theorie der Auscultations-Zeichen vorzulegen, wenngleich Sie sich gegen das „statement" verwahren, daß Sie „are quitting your own métier" etc. Wie Sie aus p. 42 ersehen, erkläre ich die Schallzeichen sowohl im Blutgefäß- als im Luftröhren-System aus der Bewegung von Flüßigkeit, also aus einem identischen Vorgange; ich gelange dabei zu dem ebenso einfachen als überraschenden Resultate, daß z.B. das breathing murmur of the air cells (vesicular inspiratory murmur) identisch ist mit dem placentarium murmur rising in the */vessels/* of the placenta during pregnancy. Both these murmurs are the result of the unisono of so many small constrictions, */provocating/* the stream of air or of blood to form what Mr. Salter calls */œdema/*, Savart more correctly: veine fluide.

Ich habe [dieses] Buch bereits der New-Sydenham-Society vorgelegt, jedoch keine Antwort erhalten. In Deutschland ist es in über 3000 Exemplaren (seit Juli) verbreitet, hat auch eine italienische und rußische Übersezung [erfahren] und soll noch in's Polnische und Spanische übersezt werden. Von meinem großen Handbuche wird bald eine zweite Auflage erscheinen.–

Ich bin, Hochverehrter Herr Profeßor, mit vorzüglicher Hochachtung

Ihr | ergebenster Diener | Dr. P. Niemeyer.

Anbei unter Kreuzband mein „Grundriß", den ich leider geheftet */nicht zur/* Hand habe.

Magdeburg, 28. January | 1872.

Esteemed Professor!

You have honoured me just as much as you have delighted me by your valuable delivery, which was followed by your kind letter[2] today, and I give you my warmest thanks.

As far as [the] respirator is concerned, I may permit myself to remark that the instrument of Mr Carrick[3] appears imperfect to me, insofar as it covers */only/* the mouth.[4] Both Mr */M'Cormak/*[5] and Mr Catlin ("shut your mouth")[6] have recently set forth that the nose is the natural respirator, etc. That your fire-man's respirator also protects the nasal opening, even though this is not expressly noted in the text and is not visible from the drawing, is something I believe I */may/* conclude from the circumstance that you */and/* Captain Shaw's[7] people held out for so long in the smoke */with it/*.[8] It would be important for me to learn this expressly, and all the more considering that our polemics against the Jeffrey respirator[9] set great store precisely on this circumstance. That I would be very grateful to you, esteemed Professor, for kindly communicating [a] "sketch of the respirator in its most [recent] form", is something I probably hardly need to state expressly.

I may permit myself on this occasion to present to you my new theory of auscultation signals,[10] which is drawn in part from your lectures, even though you protest against the "statement" that you "are quitting your own métier", etc. As you see from p. 42, I explain the acoustic signals both in the blood vessel and in the tracheal system from the motion of liquid; that is, from an identical process; I arrive in doing so at the simple as well as surprising result that, e.g., the breathing murmur of the air cells (vesicular inspiratory murmur) is identical to the placentarium murmur rising in the [vessels] of the placenta during pregnancy. Both these murmurs are the result of the unisono[11] of so many small constrictions, provocating[12] the stream of air or of blood to form what Mr. Salter[13] calls [œdema],[14] Savart[15] more correctly: veine fluide.[16]

I have already presented [this] book to the New-Sydenham-Society,[17] but have not received a reply. In Germany, it has had a circulation of more than 3000 copies (since July), has also seen an Italian and Russian translation, and is still supposed to be translated into Polish and Spanish. A second edition of my large handbook[18] is going to appear soon.

I am, esteemed Professor, respectfully,

Your | humble servant | Dr. P. Niemeyer.

Enclosed under separate cover my "Grundriss", which I unfortunately do [not] have [at] hand as a bound version.

RI MS JT/1/N/15

1. *Paul Niemeyer*: Paul Niemeyer (1832–90), a German physician and author of scientific medical texts as well as popular works about hygiene.
2. *your kind letter*: letter missing.
3. *Mr Carrick*: James Carrick (n.d.), an inventor and hotel proprietor in Glasgow.
4. *the instrument . . . mouth*: Tyndall discussed the respirator invented by Carrick in his Friday Evening Discourse at the RI on 9 June 1871; his interest in the subject was connected to his advocacy of the germ theory of disease. See 'On Dust and Smoke', *Roy. Inst. Proc.*, 6 (1870–72), pp. 365–76, on p. 371; and *Ascent of John Tyndall*, pp. 274–5.
5. *Mr. [M'Cormak]*: not identified.
6. *Mr Catlin ("shut your mouth")*: George Catlin (1796–1872), an American artist and author who had studied and painted Native Americans. He argued for the benefits to health of breathing through the nose in an 1869 work titled *Shut Your Mouth and Save Your Life*, which claimed to draw on what Catlin called 'Ethnographic labours' among indigenous

people in North and South America. See G. Catlin, *Shut Your Mouth and Save Your Life* (London: N. Trübner, 1869), pp. 1–2.

7. *Captain Shaw*: Eyre Massey Shaw (1828–1908), chief officer of London's Metropolitan Fire Brigade from 1866 to 1891. Shaw expanded the use of fire engines and telegraph communication; he was something of a celebrity, though also considered not as knowledgeable about fires as the previous chief officer, James Braidwood (1800–61), who died in a fire on 22 June 1861.
8. *From the circumstance . . . [with it]*: described in 'On Dust and Smoke', pp. 374–5. .
9. *the Jeffrey respirator*: invented by British surgeon Julius Jeffreys (1800–77). Niemeyer misspells his name 'Jeffrey'.
10. *my new theory of auscultation signals*: P. Niemeyer, *Grundriss der Percussion und Auscultation* (Erlangen: F. Enke, 1871). Auscultation is the medical practice of listening to a patient's heart, lungs, or other organs (*OED*).
11. *unisono*: unison (German).
12. *provocating*: Niemeyer presumably meant 'provoking'.
13. *Mr. Salter*: Hyde Salter (1823–71), a physician whose research focused on asthma, from which Salter himself suffered, and other diseases of the chest.
14. *[œdema]*: also edema: 'the localized or generalized accumulation of excessive fluid in tissues or body cavities' (*OED*).
15. *Savart*: Felix Savart (1791–1841), a French physician and physicist whose research focused on acoustics and vibration. He was a professor of experimental physics at the Collège de France.
16. *murmur . . . veine fluide*: This portion of the original letter is written in English, not German. It is possible that Niemeyer was quoting from Tyndall's missing letter.
17. *New-Sydenham-Society*: British publishing enterprise focused on translating European medical works.
18. *my large handbook*: P. Niemeyer, *Handbuch der theoretischen und clinischen Percussion und Auscultation vom historischen und clinischen Standpuncte* (Erlangen: F. Enke, 1868–71).

From Adam Sedgwick[1] 29 January [1872][2] 3607

Cambridge Jan. 29 | 187[2].

My dear Professor

I write to thank you for the little book upon the Glaciers of the Alps[3] you had the kindness to send me, & for the instruction & delight its perusal gave me. It is in fact a concise & very luminous synopsis of your enormous Alpine labours & discoveries; & it is of its kind I think the most perfect work I ever read—Every page marks the singular power you possess of putting your works, in the clear bright colours of day light, before the readers eye & makes him feel as if he were your happy companion & fellow labourer.

But why have I been so long in telling you this?—Because the infirmities of old age make writing at all times painful to me. The posture of writing often makes me giddy & a sad chronic malaise (which I shall never shake off) produces such a lassitude & weariness of spirit that I am very ill able to go my[4] daily little tasks & duties. Hence, I generally relieve myself & spare my dim & very irritated old eyes by dictating my letters to my servant or to any present friend who may be willing to help me—This is a very dark day: so that with spectacles on now with a large lens ("a Reader" I think now called) in my left hand can hardly read the ugly scrawls my pen is tracing.[5] But I could find in my heart[6] to dictate this note of thanks to my servant.

I wish I could discuss cleavage plains,[7] with you in my Museum.[8] In any case they are demonstrably produced by pressure. Are they always so produced? I will notice some difficulties—(1) It is not rare to see the traces (sometimes quite distinctly) of a second cleavage (2). I saw mountain chains of beds highly inclined & contorted (e.g. the Scotch chain running across the Island from St Abbs Head[9] to the Mull of Galloway[10]) the cleavage planes of the numerous slate quarries are coincident with the beds, yet there must have been enormous lateral fissures & acting in various directions—3. With slate quarries under Ingleborough[11] in the North of England, the cleavage planes are beds; or surfaces parallel to *[them]*—yet there are also transverse cleavage plains of structure—and we have <u>striped slates</u> of which the cleavages are parallel to the bedding; & the stripes show the *[direction]* of the cleavage plains intersecting the surfaces of the slates—(4) Among the contorted slates of South Wales not unusual to see good cleavage planes in a position nearly horizontal—no easy matter to conceive a pressure capable of doing this—(5). There are quarries & tracts of *[carting]* in Westmorland[12] well impressed by plains of cleavage—yet abounding in fossils that are not distorted or compressed by by[13] oblique plains. &c &c &c.—My poor head is swimming on my shoulders—I must conclude. I remain

Truly & gratefully yours | <u>A Sedgwick</u>

RI MS JT 2/10/476–7a
RI MS JT/2/13c/1387–8

---

1. *Adam Sedgwick*: Adam Sedgwick (1785–1873), a geologist and Woodwardian professor at the University of Cambridge from 1818 until his death. Partly in collaboration with Roderick Murchison, Sedgwick proposed the Cambrian and Devonian geological periods; he was also Darwin's teacher at Cambridge, but opposed the theory of evolution. He was elected FRS in 1820 and won the RS's Copley Medal in 1863.
2. *187[2]*: Sedgwick's handwriting is unclear, but Tyndall entered this letter in his journal on 30 January 1872, noting 'I prize this letter very much' (RS MS JT/2/13c/1387–8).
3. *the little book upon the Glaciers of the Alps*: Tyndall's *Glaciers of the Alps* was published more

than a decade earlier. Sedgwick was likely referring to *Hours of Exercise in the Alps*; the chapter 'Structure of Glaciers' could have prompted his discussion of planes of cleavage. See *Hours of Exercise in the Alps*, pp. 369–78.

4. *go my*: Sedgwick presumably meant 'go about my' or something similar.
5. *the ugly scrawls my pen is tracing*: Sedgwick's handwriting in this letter is difficult to read; in places, we have used the version of his letter reproduced in Tyndall's journal to decipher hard-to-read words.
6. *I could . . . heart*: Sedgwick presumably meant to write, 'I could not find it in my heart'.
7. *cleavage plains*: Sedgwick apparently wrote 'plains' for 'planes' at various points in this letter.
8. *Museum*: the Woodwardian museum at the University of Cambridge, established in 1728. The museum was renamed in Sedgwick's honor after his death and is now the Sedgwick Museum of Earth Sciences.
9. *St Abbs Head*: promontory in Berwickshire, Scotland.
10. *Mull of Galloway*: Scotland's most southerly point; 'mull' is a Scottish word for promontory or headland (*OED*).
11. *Ingleborough*: the second highest peak in the Yorkshire Dales.
12. *Westmorland*: in the north-west of England.
13. *by*: word repeated by Sedgwick.

From Henrietta Huxley [30 January 1872][1] 3608

26 Abbey Place | Tuesday

Dear Brother John

I know you will be pleased to hear that I had a famous long letter from Hal[2] yesterday morning from Tangier[3] and afterward from Gibraltar.[4] The weather had been fearful, so wet that it was impossible to get to Tetuan.[5] Hal is better but has rheumatism in one of his knees. He is enchanted with the Eastern appearance of the people and the town and gave a vivid description of a story teller and his audience, and the water-carriers, and the two women grinding at the mill just as in the time of Abraham.[6] I had nearly written Hercules![7] Sir John Hay,[8] at Tangier had been very civil and kind to Hal, and Dr Hooker sent me a note from the former saying that "Professor Huxley" had arrived there and they found him "charming"—Naturally!

May I have two tickets for Friday evening?[9]

Ever with love (2nd best you know) | Sincerely Yours | Nettie Huxley

Behold your note[10] to Jess[11] has just come.

RI MS JT/1/TYP/9/2952
LT Typescript Only

1. *[30 January 1872]*: Huxley's letter is only dated 'Tuesday'. We propose the date of 30 January 1872 based on Thomas's travels and Henrietta's request for 'tickets for Friday evening'. We presume here that she is referring to Tyndall's Friday Evening Discourse at the RI, for which see letter 3601, n. 3. Huxley had left England in early January, ill as a result of overwork and exhaustion, to recover while traveling in the Mediterranean. See Desmond, *Huxley*, pp. 409–10. Further letters in this volume related to this trip include 3616, 3623, 3631, 3635 and 3644.
2. *Hal*: her nickname for Thomas Huxley.
3. *Tangier*: Moroccan port on the Strait of Gibraltar; Huxley was there around 23 January 1872 (Desmond, *Huxley*, p. 411).
4. *Gibraltar*: British territory at the southernmost point of Spain; Huxley was there on 16 January 1872 (Desmond, *Huxley*, p. 411).
5. *Tetuan*: Tetuán or Tétouan; city in Northern Morocco, close to the Strait of Gibraltar.
6. *two women . . . Abraham*: possibly an allusion to Matthew 24:41: 'Two women shall be grinding at the mill; the one shall be taken and the other left'; a similar verse appears in Luke 17:35.
7. *Hercules*: hero of Greek and Roman myth, renowned for his strength.
8. *Sir John Hay*: John Hay Drummond-Hay (1816–93), diplomat stationed in Morocco from 1844 until 1886; he was minister resident during Huxley's visit. Huxley had a connection to him via Joseph Hooker (Desmond, *Huxley*, p. 412).
9. *two tickets for Friday evening*: see n. 1.
10. *your note*: letter missing, but possibly Tyndall's response to letter 3604.
11. *Jess*: Jessie Huxley; see letter 3604, n. 1.

From Henry Roscoe[1] and Balfour Stewart[2] — 30 January 1872 — 3609

Owens College[3] | Manchester | Jan. 30. 1872

Dear Sir

We feel that the pre-eminent scientific merits of our friend Dr Joule have, partly through his own retiring disposition, not been sufficiently appreciated by the prominent men in this district and it seems to us that a proper recognition of his great services ought if possible to take place during his life time.[4]

We have in contemplation a scheme for achieving this in a way agreeable to his own feelings but before taking any steps it would be desirable to obtain an expression of the opinion entertained by a few prominent scientific men respecting the great value of his researches and the position which they have gained for him in the world of science.[5]

We shall therefore be obliged by a short statement which can be shown to our Manchester Friends.

H. E. Roscoe B Stewart

Prof. Tyndall

RI MS JT/1/R/38
RI MS JT/1/TYP/2/666

1. *Henry Roscoe*: see letter 3387, n. 21.
2. *Balfour Stewart*: Balfour Stewart (1828–87), a Scottish physicist and meteorologist whose most important research concerned radiant heat; he was elected FRS in 1863. Stewart directed the Kew observatory from 1859 until 1871 and became a professor at Owens College in 1870.
3. *Owens College*: Founded in 1851 by industrialist John Owens as a non-sectarian college, Owens became a major center for research.
4. *We feel . . . life time*: Roscoe and Stewart were attempting to secure a pension for Joule; though they were unsuccessful in 1872, Joule was granted a pension of £200 a year in 1878 (D. S. L. Cardwell, *James Joule: A Biography* (Manchester: Manchester University Press, 1989), pp. 241–50).
5. *the great . . . science*: Joule's claims to recognition for his scientific work had recently involved Tyndall in controversy; see letters 3573, 3575, 3581, 3582 and 3584.

To Jane Barnard [1 February 1872][1] 3610

My dear Miss Barnard

In my stress and pressure I had nearly forgotten these:[2] I suppose it would not much matter for Dr Bence Jones said to me some time ago that you had a supply from him.

Yours ever | John Tyndall

RI MS JT/1/T/137
RI MS JT/1/TYP/2/4202

1. *[1 February 1872]*: The date is supplied by LT; it seems likely to be correct, given the date of Tyndall's Friday Evening Discourse at the RI on 2 February 1872. See letter 3601, n. 3.
2. *I had nearly forgotten these*: presumably tickets to Tyndall's discourse; see n. 1.

To unidentified[1] 7 February [1872][2] 3611

Royal Institution of Great Britain | 7th Feb[ruar]y.

Dear Sir

My purpose was to add one or two illustrations to a small book on Ice and Glaciers which I am about to publish.[3] Should you have the least scruple about the matter pray do not hesitate to tell me.

Yours truly | John Tyndall

I should be quite content with one or two illustrations—not including your frontispiece. But I shall be also content to forego all

Tyndall, John. Papers. MSS 001486A. Dibner Library of the History of Science and Technology, Smithsonian Libraries and Archives

---

1. *unidentified*: In a pencilled note on the original letter, the recipient is identified as Alfred Russel Wallace (for whom see letter 3592, n. 4). We believe that this is a misidentification, as none of Wallace's published books would have had suitable illustrations for *Forms of Water* (see n. 3). The addressee may be an artist whose work had previously illustrated one of Tyndall's books, possibly Whymper, whose work appeared in *Hours of Exercise in the Alps*, or wood-engraver James Davis Cooper (1823–1904), whose work appeared in Tyndall's *Glaciers of the Alps*.
2. *1872*: dated by reference to the publication of *Forms of Water* (see n. 3). For Tyndall's work on this volume, see letter 3639.
3. *a small book on Ice and Glaciers*: *Forms of Water*, published in the fall of 1872. *Forms of Water* was the first volume of the International Scientific Series, for which see letter 3484, n. 1.

From Alfred Mayer 10 February 1872 3612

Stevens Institute of Technology,[1] | Department of Engineering, | Hoboken,[2] New Jersey, U.S.A. | February 10th 1872.
To | Professor John Tyndall F.R.S. D.C.L.

My dear Sir;

I trust that you have not construed my long silence into a want of appreciation of your marked kindness in having published so handsomely in the Phil. Mag., my paper on the "magnetic spectra,"[3] and in having also given a specimen plate to adorn your valuable "Fragments".[4] In fact, a change of residence from the Lehigh University[5] to the well endowed Technological Institute of Hoboken (just across the Hudson River from N[ew]. York) with the constant press of the business of a new organization upon me, has deprived me of the leisure to attend properly to my correspondence.

Alas, I wished to wait until one of the several original researches I have now on hand had reached sufficient perfection to be laid before you. This privilege has also been denied me and with the exception of the enclosed piece[6] of sketchy work—conceived and accomplished on the same day—nothing else has advanced sufficiently for publication.

With all allowance for that paternal feeling which we always will have for our intellectual offspring I cannot but regard what I now send you as having considerable theoretic beauty embodied in experiments of some finesse—At best the results were to me charming and the means of attained[7] them certainly ridiculously simple.

You will oblige me greatly, my dear Sir, if you will have this short paper published in the Philosophical Magazine, either as embodied in a letter to you or as a direct contribution from me to the editors.[8] These gentlemen I am unknown to & this is the reason I take the liberty of again intruding on your kind intervention; well knowing, from experience, how difficult it is for a stranger at such a distance to have such matters properly attended to.

There has been some talk among us American men of Science of having soon the pleasure of seeing you among us.[9] I do not know what foundation, if any, there may be for such a report, but, I, at least, can be confident in assuring you a most cordial and enthusiastic reception—for you are as well known and as much read among us as you are in England.

I trust also that you will not suppose all American public exposition of science (so called) to consist of fiz! flash!! and bang!! I own that this has indeed been the case with two or three "popular lectures" or as the papers style them lectures on popular science. What species of science this is I do not know—probably a hybrid—half pyrotechnist half prestidigitation.—

The fact is that few of our real workers ever appear before the public, but when they do they meet with a much more critical audience than the popular men, and never fail to gratify those who are alone worth the endeavour to please—

With highest consideration of regard I remain | Yours respectfully & truly | Alfred M. Mayer

RI MS JT/1/M/73

---

1. *Stevens Institute of Technology*: Founded in 1870 by a bequest from railroad executive Edwin Augustus Stevens (1795–1868), the Stevens Institute offered a curriculum in engineering.
2. *Hoboken*: city in New Jersey immediately west of Manhattan.
3. *my paper on the "magnetic spectra"*: A. M. Mayer, 'On a method of fixing, photographing, and exhibiting the magnetic spectra', *Phil. Mag.*, 41:275 (June 1871), pp. 476–80.
4. *a specimen plate to adorn your valuable "Fragments"*: probably the image discussed in letter 3377; see n. 2 in that letter.
5. *Lehigh University*: university in Bethlehem, Pennsylvania, founded in 1865.
6. *enclosed piece*: The enclosure is missing, but may be the paper later published as A. M. Mayer, 'Acoustical experiments showing that the translation of a vibrating body causes it to give a wave-length differing from that produced by the same vibrating body when stationary', *Phil. Mag.*, 43:286 (April 1872), pp. 278–81.

7. *attained*: Mayer presumably meant to write 'attaining'.
8. *either as . . . the editors*: Mayer's paper was noted in the *Phil. Mag.* as 'Communicated by the Author'. See A. M. Mayer, 'Acoustical Experiments', p. 278.
9. *There has . . . among us*: Tyndall was planning a trip to America; see especially letter 3510.

To unidentified[1] 11 February 1872 3613

My dear Sir, Were you to see my private account of disbursements, and the number of things to which I "lend a hand," I hardly think you would have any high opinion of my prudence.

Still I should pause before declining to comply with your request if I felt any lively sympathy with our latter day movements regarding "the rights of Women."

But there are many of us who would go any length with you in the endeavours to right the real wrongs of women, and who, in common with what they regard as the true womanhood of England, consider much of the present movement as leading to mischievous results.

In fact I sometimes think that is because the <u>men</u> of the age have become to some extent women, that a few of the women aspire to be men.

Faithfully yours | John Tyndall.

RI MS JT 2/13c/1389–90
LT Typescript Only

---

1. *unidentified*: In the typescript of the journal, this letter is prefaced with the remark, 'Mr <name missing> wrote to ask me to subscribe for and support a paper called "Woman" I wrote this reply'. While we have not identified Tyndall's correspondent, we believe the 'paper' in question was the short-lived periodical *Woman; a weekly journal embodying female interests from an educational, social, and domestic point of view*, which ran from January to July 1872 and was edited by Amelia Lewis; see M. E. Tusan, *Women Making News: Gender and Journalism in Modern Britain* (Urbana: University of Illinois Press, 2005), p. 86. While it seems logical that Lewis would have been Tyndall's correspondent, that is hard to fit with the male form of address that Tyndall used (as well as his closing insult, directed to a correspondent presumed to be male). It is possible that Lewis's signature was ambiguous, or that the correspondent was one of *Woman*'s male contributors. Tyndall's attitude in this letter is of a piece with his attitude towards women more broadly. See *Ascent of John Tyndall*, pp. 286–7.

To [Charles Tomlinson][1] 11 February [1872][2] 3614

11.th Feb[ruar][y].

My dear Sir

I have hardly altered a word of what you have written;[3] but I have added a few lines which you will be good enough to deal with as you please.

I think you have embraced in a very short space to say all that the public would care to know about me. Of course you are responsible for the flattering judgement which you have pressed upon me and my work.

The notion of my "personality" rose, I think in this way:[4] In writing out these lectures on sound and heat I wished to make the picture before the readers mind as vivid as that before the hearer's eye, and was thereby betrayed into a very copious use of the personal pronoun I. A criticism was once shown to me wherein it was noticed that I had used eight or ten I's in a very short space—Where I /learned to then/ place the I's consisted of statements such is this—"I pull this wire aside" I shake the centre of the wire" "I draw my /rubber/ over this rod" and so forth. These are blemishes certainly as regards style, but they naturally arose out of the desire above expressed. I do not think you will find this gratuitous to an obtrusive degree in the papers I have communicated to the Royal Society. In your concluding paragraph there I think you have interpreted me aright.

I have asked the Mess[rs]. Longman to forward you a copy of "Fragments of Science" & of "Faraday as a Discovererer".[5] and I shall be glad if you would accept them from me.

Yours faithfully | John Tyndall

IET SC MSS 003/B/2/100

1. *[Charles Tomlinson]*: Charles Tomlinson (1808–97), a science teacher and author of popular articles and books about science. He was elected FRS in 1867. 'To Tomlinson' is written at the top of this letter in LT's hand; this identification is confirmed by Tyndall's journal entry for 10 February 1872, in which he noted 'Tomlinson has sent me an article to look over' (RS MS JT/2/13c/1389).
2. *[1872]*: year established by Tyndall's journal; see n. 1.
3. *what you have written*: Tomlinson's entry for Tyndall in Charles Knight (ed.), *Biography, or Third Division of the "English Cyclopedia"* (London: Bradbury, Evans, 1872), Supplement, pp. 1157–8. In his journal, Tyndall commented, 'I had a mind to strike out four things, but on second thought concluded that he was expressing opinions which I had no right to change' (RS MS JT/2/13c/1389).

4. *The notion . . . in this way*: Tomlinson's entry concluded, 'In all Tyndall's labours, whether as a writer or a lecturer, there is an unmistakable personality, so that you cannot forget the man in his work' (Tomlinson, p. 1158).
5. *"Faraday as a Discoverer."*: J. Tyndall, *Faraday as a Discover* (London: Longmans, Green, and Co., 1868).

From Friedrich Vieweg & Sohn 12 February 1872 3615

Braunschweig 12 Febr. 1872.

Hochgeehrter Herr!

Eine neue Auflage der Übersetzung Ihres Werkes „on Sound" ist augenblicklich noch nicht nöthig, möglicherweise können wir aber in den Fall kommen, im Laufe des nächsten Sommers Vorbereitungen dazu treffen zu müßen, und alsdann müßen wir jedenfalls die Veränderungen berücksichtigen, welche das Original inzwischen erfahren hat, wenn dieselben auch nur gering sind. Es wäre uns sehr lieb, wenn Sie uns in einem Expemplare der neuen engl. Auflage die vorgenommenen Veränderungen anstreichen wollten.

Von der Übersetzung der „Fragments of science" */hat/* uns Herr Helmholtz vor einigen Tagen die erste Hälfte des Manuscriptes zugesandt, und wir sind grade im Begriff, den Druck zu beginnen. Jedenfalls müßen */wir/* die Veränderungen der neuen Auflage des Originals noch berücksichtigen; wir dürfen Sie wohl bitten, daß Sie Ihren Verleger veranlaßen, uns sogleich ein Exemplar zu übersenden.

Die Übersetzung der „hours of exercise" ist noch nicht so weit gediehen, daß der Druck beginnen könnte, doch hat uns Herr Wiedemann mitgetheilt, daß wir in einigen Wochen darauf rechnen könnten. Sollten für dieses Werk ebenfalls Veränderungen in Aussicht stehen, so bitten wir um Mittheilung derselben. Herr Wiedemann war in Zweifel darüber, wie der Titel des Buches am besten im Deutschen wiederzugeben sei; er schlug vor, ihn „Stunden der Arbeit in den Alpen" zu übersetzen. Das Wort „exercise" ist in dem Sinne, wie es dort angewendet ist, nicht leicht wiederzugeben, denn es bezieht sich eben sowohl auf die körperliche Übung des Bergsteigens als auf die geistige Arbeit der dabei gemachten wißenschaftlichen Beobachtungen. Es wäre uns lieb, wenn Sie selbst den Ausschlag geben wollten.

Wir hatten geglaubt, daß Ihen Herr Lederer über unsere Correspondenz mit ihm im Betreff des „Diamagnetism" berichtet hätte, und es deshalb unterlaßen, Ihnen darüber zu schreiben. Nach unserer eigenen sowie nach der Ansicht sachverständigen Freunde dürfte dieses Buch in Deutschland nicht auf so große Theilnahme zu rechnen haben, wie Ihre früheren Werke, da schon früher ein großer Theil dieser Arbeiten in Poggendorff's Annalen ausführlich mitgetheilt worden ist und sich sonach in den Händen jedes Gelehrten befindet, der ein Intereße daran nimmt; auch durch Wiedemann's „Lehre vom Galvanismus" ist vieles, wenigstens im Auszuge, bereits bekannt geworden.

Ein großer Theil des Buches bringt Historisches, was für das größere Publicum in Deutschland ebenso wie die Briefe, weniger Intereße hat, als für die gleichen Kreise in England. Wir haben uns diesen Erwägungen gegenüber, ebenso aber auch, weil wir Herrn Lederer als Übersetzer gar nicht kennen, und es nicht möglich war, die früheren bewährten Übersetzer Ihrer Arbeiten dafür zu gewinnen, zu der Herausgabe von Herrn Lederers Übersetzung nicht entschließen können.

Sollten Sie aber besonderen Werth darauf legen, das Buch in Deutschland publiciert zu sehen, so findet sich wohl binnen einiger Zeit eine Gelegenheit, die Übersetzung ganz bewährten Händen anzuvertrauen.

Mit bekannter ausgezeichneter Hochachtung | Ihre | ganz ergebenen | Friedr. Vieweg & Sohn

Braunschweig[1] 12 Feb. 1872.

Esteemed Sir!

A new edition of the translation of your work "on Sound"[2] is still not necessary at the moment, but we can possibly get to the point of having to make preparations for this in the course of next summer, and we shall then, at any rate, have to take into consideration the changes which the original has undergone in the meantime, even if these are only minor.[3] We should be very glad if you would mark for us in a copy of the new Engl[ish]. edition the changes that have been made.

Herr Helmholtz has sent us the first half of the manuscript of the translation of "Fragments of science" some days ago, and we are just about to commence printing.[4] At any rate we shall still have to take into consideration the changes to the new edition of the original;[5] we think we can probably ask you that you get your publisher to send us a copy straight away.

The translation of "Hours of Exercise" has still not reached the stage that printing could commence, but Herr Wiedemann has informed us that we could expect doing so in some weeks' time. Should any changes be expected for this work also, then we ask you to inform us about them. Herr Wiedemann was in some doubt about how the title of the book was best reproduced in German; he suggested translating it as "Stunden der Arbeit in den Alpen".[6] The word "exercise" is, in the sense it is used there, not easily reproduced, for it refers to both the physical exercise of climbing mountains and the mental work involved in the scientific observations made in doing so. We should be glad if you would give a definitive ruling yourself.

We had believed that Herr Lederer[7] had informed you about our correspondence with him regarding "Diamagnetism"[8] and therefore neglected to write to you about it. In our own view and that of expert friends, this book might not expect such great interest in Germany as your previous works, as a large part of these papers has previously been communicated at length in Poggendorff's Annalen[9] and consequently can be found in the hands of every scholar who takes an interest in them; through Wiedemann's "Lehre vom Galvanismus"[10] also, much, at least in abstract, has already become public knowledge.

A large part of the book provides historical material, which, like the letters, has less interest for the wider audience in Germany than for the same circles in England. In the face of these considerations, but also because we do not know Herr Lederer as a translator at all and it was not possible to obtain the previous tried and tested translators of your papers for this purpose, we have not been able to make a decision on publishing Herr Lederer's translation.

Should, however, you place special importance on seeing the book published in Germany, then there will probably be an opportunity within a short time of entrusting the translation to some entirely tried and tested hands.

With familiar excellent esteem | Your | most humble servants | Friedr[ich]. Vieweg & Son

RI MS JT/1/V/10

1. *Braunschweig*: city in Lower Saxony, in the northern part of Germany.
2. *"on Sound"*: *Sound* (1867). The second German edition appeared in 1874 as J. Tyndall, *Der Schall: Acht Vorlesungen Gehalten in der Royal Institution von Grossbritannien*, trans. H. Helmholtz and G. Wiedemann, 2nd edn (Braunschweig: Vieweg, 1874). It had in fact been translated by Anna Helmholtz and Clara Wiedemann.
3. *A new . . . minor*: The second English edition of *Sound* appeared in 1869 (J. Tyndall, *Sound: A Course of Eight Lectures Delivered at the Royal Institution of Great Britain*, 2nd edn (London: Longmans, Green, and Co., 1869)).
4. *Herr . . . printing*: on the translation of *Fragments of Science* into German, see letter 3432. The book was published in German as J. Tyndall, *Fragmente aus den Naturwissenschaften. Vorlesungen und Aufsätze*, trans. A. Helmholtz (Braunschweig: Vieweg, 1874).
5. *the new edition of the original*: *Fragments of Science* had gone into a third edition by the end of 1871.
6. *Stunden der Arbeit in den Alpen*: 'Hours of Work in the Alps'; for Wiedemann's discussion of the title, see letter 3602. For Tyndall's response, see letter 3618. For additional discussion of this title, see letters 3632, 3640 and 3649.
7. *Herr Lederer*: not identified.
8. *Diamagnetism*: J. Tyndall, *Researches on Diamagnetism and Magne-crystallic Action* (London: Longmans, Green, and Co., 1870). This book seems not to have been translated into German.

9. *Poggendorff's Annalen*: *Annalen der Physik und Chemie*, edited by Johann Christian Poggendorff, was the leading German journal for the publication of research in physics.
10. *Wiedemann's "Lehre vom Galvanismus"*: Wiedemann's 'Theory of Galvanism'; G. Wiedemann, *Die Lehre von Galvanismus und Elektromagnetismus* (Braunschweig: Vieweg, 1861).

To Thomas Henry Huxley 14 February 1872 3616

Royal Institution. | 14th. February, 1872

My dear Hal,

From time to time I look in at 26 Abbey Place to get some intelligence of your condition. Your first letters rejoiced Mrs Huxley much—they were "famous" letters;[1] but I did not know what to make of them as they, I was told, contained nothing about yourself. They were purely objective, dealing with savages and their habits without including yourself. But the last letter was a comfort, for it explained the past and brightened the future.[2]

At the suggestion of your good wife I had a long chat with Bence Jones about you on Sunday night. He will write to you himself—and he has written to Lord Ripon requesting an extension of your leave[3]—for he deems it of great importance that your health should be thoroughly consolidated before your return.

We are here in the midst of political rows. The American treaty[4] for the moment is snuffed out by the terrible intelligence from India.[5] By the way we hear much of classical training as bearing upon English composition. But the achievements of our Statesmen in this matter of the treaty show that classical knowledge and good English by no means illustrate the principle of indissoluble association.[6] I fear, however, that the defects of the treaty are defects of character rather than of style: either our Commissioners were unable to see clearly what they themselves meant and wished for, or seeing it they were afraid to express it—It is a bad business.

The children[7] seemed very charming on Sunday. Leonard[8] was in the act of ringing the bell when I reached the door, and a sweet gentlemanly little fellow he appeared to be. And within they were all so affectionate and lovable that it did me good to see them. What an immense motive force you must derive from those children. They may be a responsibility, but assuredly they must be a comfort and a joy. If the devil should ever tempt you to go in for the coarser forms of radicalism the thought of the purity and refinement of the little ones at home ought to restrain you. There is a homily for you.

I have not been at the Metaphysical[9] since you left: Carpenter, who is on a lecturing tour in the north, had his paper upon Common Sense,[10] and

Gregg,[11] last night, made a communication about the beauty of ruins.[12] I am told the discussion last night was very agreeable and interesting. We had a lecture here on Friday on Sleep by Humphrey of Cambridge:[13] he is an easy fluent speaker but there seemed singularly little in the lecture: and certainly part of it was wrong.[14]

I heartily wish I was with you—I long to see those regions, but my holiday time does not answer for regions regarding which the question has been asked "Oh God why did you make Hell, when the deserts of the East were already made to your hand."[15]

Yours ever affectionately | J. Tyndall.

IC HP 8.94
Typed Transcript Only

---

1. *they were "famous" letters*: see letter 3608. For Huxley's trip, see letter 3608, n. 1; related letters include 3623, 3631, 3635 and 3644.
2. *the last letter . . . future*: In his journal for 11 February 1872, Tyndall noted a visit to Henrietta Huxley: 'Hal writes from Malta to say the cloud is lifting—how like my own experience!' (RS MS JT/2/13c/1390).
3. *an extension of your leave*: Henrietta Huxley had Bence Jones request an additional month's leave for Huxley. See Desmond, *Huxley*, p. 414. For further discussion of this subject, see letters 3631 and 3635.
4. *The American treaty*: The Treaty of Washington, signed 8 May 1871, dealt with disputes between the United States and Great Britain that emerged out of the American Civil War (1861–5). These included the U.S.'s pursuit of damages done to Northern ships by British-built vessels that had been sold to the Confederacy, including the *Alabama*. Although the treaty was developed by a joint high commission, it remained controversial in Britain after its signing.
5. *the terrible intelligence from India*: the assassination of Richard Southwell Bourke (1822–72), Governor-General of India, on Thursday, 8 February 1872, by a convict at Port Blair, in the Andaman Islands, an archipelago in the Bay of Bengal. Tyndall notes his death in his journal on 12 February 1872, writing 'Calamities are heaped upon us' (RS MS JT/2/13c/1390).
6. *But the achievements . . . association*: The language of the Treaty of Washington had come in for commentary and critique; see the leader, *Times*, 29 January 1872, p. 9; leader, *Times*, 12 February 1872, p. 12, and S., 'The Treaty of Washington', letter to the editor, *Times*, 13 February 1872, p. 12. Tyndall joined this debate; see letter 3617.
7. *the children*: see letter 3411, n. 2.
8. *Leonard*: Leonard Huxley (1860–1933), who became an author, writing a biography of his father and one of Joseph Hooker.
9. *the Metaphysical*: The Metaphysical Society, founded in 1869, was an association of élite

male British intellectuals and politicians whose aim was to discuss religious, moral and philosophical issues. The Society met monthly while Parliament was in session to discuss a previously circulated paper by one member (see C. Marshall, B. Lightman, and R. England (eds), *The Metaphysical Society (1869–1880): Intellectual Life in Mid-Victorian England* (Oxford: Oxford University Press, 2019), pp. 185–203). Tyndall participated but never wrote a paper (see *Ascent of John Tyndall*, pp. 217–8).

10. *his paper upon Common Sense*: Carpenter's paper was the subject of discussion on 17 January 1872. See W. B. Carpenter, 'What is Common Sense?', in C. Marshall, B. Lightman and R. England (eds), *The Papers of the Metaphysical Society, 1869–1880* (Oxford: Oxford University Press, 2015), pp. 319–32.
11. *Gregg*: William Rathbone Greg (1809–81), a civil servant and essayist who wrote for the periodical press and published works of social and political commentary.
12. *a communication about the beauty of ruins*: Greg's paper was the subject of discussion on 13 February 1872. See W. R. Greg, 'Wherein Consists the Special Beauty of Imperfection and Decay?', in *The Papers of the Metaphysical Society*, pp. 333–9.
13. *Humphrey of Cambridge*: George Murray Humphry (1820–96), a surgeon and professor of surgery at the University of Cambridge. Humphry did important research in human anatomy and was elected FRS in 1859.
14. *We had . . . wrong*: Humphry delivered the Friday Evening Discourse at the RI on 9 February 1872. See G. M. H., 'On Sleep', *Roy. Inst. Proc.*, 6 (1870–72), pp. 424–5. Tyndall made a similarly critical comment in his journal of 9 February 1872 (RS MS JT/2/13c/1389).
15. *Oh God . . . to your hand*: quotation not identified.

To the Editor of the *Times*[1] 14 February 1872[2] 3617

Ambiguity of the English Language[3] | To the Editor of the Times

Sir,—

Will you permit me to add to the excellent remarks on English composition published in The Times of Monday[4] a reference to the prevalent delusion that it is easy to write good English? It was once remarked to Jacobi,[5] of Berlin, who was, I believe, the greatest mathematician of his age, that the ease and clearness of his style proved him to write without effort. "You have little idea," he replied, "of the labour expended in attaining the perspicuity you are good enough to admire. I sometimes copy my papers five times over before what I write expresses what I wish to write." The best writers in our own day and language labour quite as hard as Jacobi to attain clearness and precision, of expression. Mr Mill,[6] I am told, does so; and I know that Mr Huxley, who has a gift of off-hand expression possessed by few, does not find himself absolved from a labour equal to that of Jacobi in reaching the clearness and concentration for which his writings are so remarkable. The improvement of style alone by no means expresses the result and value of such labour. By trying to clothe thought in the best possible language we give precision to thought itself; and the mere act of rewriting, with intent to improve, is a wonderful solvent of the haze which usually accompanies first conceptions.

But the submergence of the individual in the committee, which finds so much favour under Constitutional Governments, is a serious drawback as regards good composition. It is necessary, of course, that every member of a committee should have an opportunity of stating his views or offering his suggestions, and that the committee as a body should decide on the substance of its reports. But in difficult cases the work done under the eyes of the committee resembles raw sugar, and must undergo refinement in the individual mind. Had Sir Stafford Northcote[7] trusted himself more and the High Commission less, the penetration and power of correct expression which he has been known to display, even on the committee of this Club would, I am persuaded, have saved us from our present deplorable complications.

Athenaeum[8] | J. T.

RI MS JT/2/13c/1391
Times, 14 February 1872, p. 4

---

1. *The Editor of the Times*: Delane; see letter 3403, n. 1.
2. *14 February 1872*: This is the date of the letter's publication; the date of composition is not known.
3. *Ambiguity of the English Language*: This title appears in both the *Times* and in Tyndall's journal.
4. *the excellent . . . Monday*: In a leader, the *Times* criticized the ambiguous language of the Treaty of Washington (*Times*, 12 February 1872, p. 12). See letter 3616, n. 4 and n. 6.
5. *Jacobi*: Carl Jacobi (1804–51), a German mathematician whose work linked different fields of mathematics and included number theory and the theory of elliptic functions.
6. *Mr Mill*: John Stuart Mill (1806–73), well-known utilitarian philosopher, MP from 1865 until 1868 and author of books including *On Liberty* (1859) and *The Subjection of Women* (1869). Tyndall seems to have had a mixed relationship to him politically. See *Ascent of John Tyndall*, pp. 201 and 244.
7. *Stafford Northcote*: Stafford Henry Northcote (1818–87), a English politician. He is known for the 1853 report he issued, with Charles Trevelyan, recommending that civil service appointments be made by examination, and for his position as a financial advisor to Benjamin Disraeli. Northcote joined the commission that examined the *Alabama* claims in February 1871; for those claims, see letter 3616, n. 4.
8. *Athenaeum*: see letter 3372, n. 14.

To Friedrich Vieweg & Sohn 15 February [1872][1] 3618

15th. Feb[ruar]y.

Gentlemen,

I have desired the Messrs. Longman to send you a copy of the second edition of my book "on Sound";[2] but I regret to say that it is not possible to mark the places where the style of the book has been altered.

Were I to go over the book again I daresay I should make further alterations, but I do not think that, save in an aesthetic point of view, they would be of any importance.

My wish to make the impression of the experiments as vivid as possible caused me in the first edition to make too frequent use of the pronoun I,[3] but this is not a fault of any serious gravity.

I have also directed the Messrs Longman to send you a copy of the new edition of "Fragments".[4] Some pages of the Essay "on the scientific Use of the Imagination" have been struck out; because though proper in the spoken discourse they may be dispensed with in the book. The short essay on the Physical Basis of Solar Chemistry; on Force; The Copley Medalist of 1870; the Copley Medalist of 1871 have been added.

I have already correspondend with Prof. Wiedemann regarding the title of "Hours of Exercise";[5] but you seem to mistake the character of the work. It is, as stated in its preface, for the most part a record of bodily action.

I am not at all particular about diamagnetism.[6] It has had a very far circulation in England, and will come to a second edition.[7] It is exceedingly convienent to scientific workers and readers to have memoirs thus connected together. I gave myself some trouble to make the papers intelligible, and hence I dare say their success in England.

Yours very faithfully | John Tyndall

TU Braunschweig/Universitätsbibliothek/UABS V 1 T : 63

---

1. *[1872]*: The date is written on the back of this letter, probably by Vieweg, and is confirmed by the letter's relationship to letter 3615, to which it responds point by point.
2. *"on Sound"*: see letter 3615, n. 2.
3. *My wish . . . pronoun I*: Tyndall also discussed this issue in letter 3614.
4. *new edition of "Fragments"*: see letter 3615, n. 5.
5. *I have . . . Exercise"*: see letter 3602. The title of this book is also discussed in letters 3632, 3640 and 3649.

6. *diamagnetism*: Tyndall's book *Diamagnetism*, for which see letter 3615, n. 8.
7. *a second edition*: In fact, *Diamagnetism* did not reach a second edition until 1888. See J. Tyndall, *Researches on Diamagnetism and Magne-crystallic Action*, 2nd edn (London: Longmans, Green, and Co., 1888) and *Ascent of John Tyndall*, p. 438.

From Anna Helmholtz 17 February 1872 3619

Berlin 45. Königin | Augusta Strasse | 17/2. 72.

Dear Sir

You know perhaps, that I am translating your „Fragments of Science"[1] and very difficult I find them. Would you be kind enough to tell me where the verses by Goethe pag 70 the motto to Matter and Force are to be found?[2] Both my husband and I looked over all the volumes without finding the passage. I hope that I am not giving you a great deal of trouble—We thought the verses were taken out of the 2d part of Faust,[3] which alas is not as present to our minds and memories as it ought to be. So pray pardon my troubling you.

With my husbands[4] | herzlichsten Grüßen[5] | Yours very truly | Anna Helmholtz

RI MS JT/1/H/34

1. *I am translating your „Fragments of Science"*: For discussion of this translation, see letters 3432, 3602, 3615, 3618, 3619, 3622, 3625, 3632, 3640 and 3649.
2. *the verses . . . to be found*: Chapter 4, 'Matter and Force', begins with a poem attributed to Goethe. See *Fragments of Science*, p. 68. The same poem appears at the end of Book Two of Carlyle's *Past and Present*. See T. Carlyle, *Past and Present* (London: Chapman and Hall, 1843), p. 116. For the full reference, see letter 3622, n. 3 as well as letter 3625. For additional discussion of Tyndall's use of Goethe in *Fragments of Science* see letter 3397, n. 3 and n. 4.
3. *the 2d part of Faust*: The first part of Goethe's masterwork Faust was published in 1808; a revised version appeared in 1828–9. The second part was published in 1832, after Goethe's death.
4. *my husbands*: Hermann Helmholtz.
5. *herzlichsten Grüßen*: warmest greetings.

From James Gibbs[1] 20 February 1872 3620

Brompton Barracks | Chatham | 20. 2. 72 |
Royal School of Military Engineering[2]

Dear Sir,

I have been engaged for some time in a series of experiments with the respirator you invented for the use of firemen.[3] The two papers I send you with this letter[4] will shew[5] you, how I first conceived the idea of using such a respirator in the more dangerous operations of military mining. Having proved experimentally the chemical efficacy of your respirator, I have proceeded to make a model of what I believe to be the most suitable shape for practical use.

Mr Ladd, of Beak Street,[6] kindly lent me a mouthpiece that he had designed, which I dare say you have seen; but he remarked that he did not know what he should do next, as he did not know how to connect it with a box for chemicals.

With the aid of his mouthpiece I have succeeded in hitting upon what I believe will be at the same time effective and comfortable. I have received permission from the Commandant of the School of Military Engineering[7] to place myself in communication with you on the subject, as it is your invention, in order with your advice to perfect it as much as possible. Thus we may obtain a form that will be useful both to us in military mining,[8] and to firemen according to your original idea. If you wish to see my model, I shall be delighted to bring it to you, either Friday or Saturday this week, or any day after (except a Monday or a Thursday) in the afternoon.

I remain, Sir, | Y[ou]^r obedient servant | J. E. Gibbs

RI MS JT 2/13c/1393

---

1. *James Gibbs*: James Edward Gibbs (n.d.), an officer of the Royal Engineers. By 1873 he was Assistant Superintendent of the Great Trigonometrical Survey of India.
2. *Brompton . . . Engineering*: The Royal School of Military Engineering, located at Brompton Barracks, Chatham, in south-east England, is the training school for the Royal Engineers, the engineering division of the British army.
3. *the respirator you invented for the use of firemen*: for Tyndall's interest in respirators, see letter 3367, n. 4.
4. *The two . . . letter*: The enclosures are missing, but the following month *Nature* published Gibbs's account of a series of experiments with the respirator. See J. Tyndall and J. E. G., 'Foul Air in Mines and How to Live in It', *Nature*, 5 (7 March 1872), pp. 365–6.
5. *shew*: alternative spelling of 'show', now obsolete.

6. *Beak Street*: street in Soho, London.
7. *Commandant of the School of Military Engineering*: Thomas Lionel John Gallwey (1821–1905); the Commandant was the director of the school. Gallwey held the position from 1868 until 1875.
8. *military mining*: in warfare, the construction of subterranean passages with the object of exploding something about them. See O. H. Ernst, *A Manual of Practical Military Engineering* (New York: D. Van Nostrand, 1873), p. 13.

To Frances Colvile 23 February [1872][1] 3621

23rd Feb

My dear Lady Colvile

I am conditionally bound to another engagement for Saturday evening; but if the condition should not hold I will join you, and thank you for letting me do so.[2]

I was so pleased to see you last night[3]—Indeed I had begun to think that I should never see you again and that you had erased me from your list of friends.

Yours ever faithfully | John Tyndall

BL Add MS 60634, ff. 128–9

1. *[1872]*: The year appears on the stamped envelope.
2. *I am . . . do so*: Possibly a white lie, as in his journal entries of 23 and 24 February 1872 Tyndall wrote about resolving to avoid dining out for the sake of his health (RS MS JT/2/13c/1394).
3. *last night*: Tyndall recorded dining at the Busks on 22 February (RS MS JT/2/13c/1394).

From Thomas Carlyle 23 February 1872 3622

5, Cheyne Row, Chelsea. | 23rd Feb. 1872.

Dear Tyndall,

My brother[1] answers punctually by return of post; and here is a copy of the poem itself, as well as perfect indication of its place in Goethe's Werke. It appears we had it in our hands that night;[2] but in our hurry did not know.

Yours ever truly, | Signed. T. Carlyle.

Symbolum[3]

Des Maurers Wandeln
Es gleicht dem Leben
Und sein Bestreben
Es gleicht dem Handeln
Der Menschen auf Erden.

Die Zukunft decket
Schmerzen und Glücke
Schrittweis' dem Blicke
Doch ungeschrecket
Dringen wir vorwärts.

Und schwer und schwerer
Hängt eine Hülle
Mit Ehrfurcht Stille
Ruhn oben die Sterne
Und unten die Gräber.

Betracht' sie genauer
Und siehe, so melden
Im Busen der Helden
Sich wandelnde Schauer
Und ernste Gefühle.

Doch rufen von drüben
Die Stimmen der Geister
Die Stimmen der Meister
Versäumt nicht zu üben
Die Kräfte des Guten

Hier winden sich Kronen
In ewiger Stille.
Die sollen mit Fülle
Die Thätigen lohnen!
Wir heissen euch hoffen.[4]

Goethe's Werke, Dritter Band, Seite 69. Ausgabe letzter Hand 1828. Unter "love" im Verzeichniss des Inhalts.[5]

RI MS JT 2/13c/1394
LT Typescript Only

---

1. *My brother*: possibly John Aitken Carlyle (1801–79), a physician and writer. John and Thomas had a close if sometimes tense relationship.

2. *that night*: 20 February 1872. Tyndall recorded in his journal, 'Dined at the Athenaeum and spent an earnest hour with Carlyle afterwards' (RS MS JT/2/13c/1393). The present letter, copied in his journal minus the poem, is prefaced with the remark, 'The prophet sought to find it for me, but failed. He said however that he would get it for me within the week, True as a law of Nature he has done so' (RS MS JT/2/13c/1394). Tyndall was looking for the poem in response to Anna Helmholtz's request; see letters 3619 and 3625.
3. *Symbolum*: 'Symbolum', *Goethe's Werke*, 55 vols (Stuttgart and Lubingen: J. G. Cotta, 1827–33), vol. 3 (1828), pp. 69–70.
4. *Des Maurers . . . hoffen*: The mason's trade | Resembles life, | With all its strife,—| Is like the stir made | By man on earth's face. | Through weal and woe | The future may hide, | Unterrified | We onward go | In ne'er changing race. | A veil of dread | Hangs heavier still. Deep slumbers fill | The stars overhead, | And the foot-trodden grave. | Observe them well, | And watch them revealing | How solemn feeling | And wonderment swell | The hearts of the brave. | The voice of the blest, | And of spirits on high | Seems loudly to cry: To do what is best, | Unceasing endeavor! | In silence eterne | Here chaplets are twined, | That each noble mind | Its guerdon may earn,—| Then hope ye forever! (see 'A Symbol', *The Complete Work of Johann Wolfgang von Goethe*, 10 vols (New York: P. F. Collier, 1839), vol. 5, pp. 241–2).
5. *Goethe's Werke . . . Inhalts*: Goethe's Works, Third Volume, p. 69. Last edition printed in his lifetime 1828. Under "love" in the table of contents.

From Henrietta Huxley 24 February 1872 3623

26 Abbey Place | St John's Wood | 24th Feb 1872.

Dear Brother John,

Thank you very much for your ready compliance with my request—and help in my difficulty.[1]

Are you going to St. Pauls on Tuesday—or to the Athenaeum?[2] I have been stupid in not asking in time for a place anywhere.[3] Mr Darwin has promised me his ticket at the Athenaeum but as it has not turned up, I think that as many are given as there is room for—After all I don't care to see the procession—One has done that sort of thing before but I should think that the music in St Pauls would be worth hearing. Did I tell you that Dr Bence Jones called and left me thoroughly depressed? He said that he didn't think the Nile[4] quite the right place for Hal[5] to have visited—that he had warned him of rheumatic fever[6] and the general diseases which people who went there were subject to that it would be a bad thing if he were laid up there, and so on—but that he Dr Jones didn't see any thing better to advise.

Why can't people know the value of silence. What was the use of looking at the "possibilities" after he had advised Hal to go there. I should like to have screamed—instead of which I looked calm and indifferent and held myself tight—and yet he was so kind that I forgave him

Ever | Yours most sincerely | Henrietta Huxley.

RI MS JT/1/TYP/9/2950
LT Typescript Only

---

1. *Thank you . . . difficulty*: possibly a reference to Tyndall's assistance when additional money was needed for the house the Huxleys were expanding. See Desmond, *Huxley*, p. 414.
2. *Are you going to St. Pauls on Tuesday—or to the Athenaeum?*: On 27 February 1872 a Royal Thanksgiving was held in St Paul's Cathedral, London, for the recovery of Prince Albert of Wales from typhoid fever. The procession of the royal family from Buckingham Palace to St Paul's passed by the Athenaeum Club on Pall Mall, a street in central London. See 'The National Day of Thanksgiving', *Times*, 28 February 1872, pp. 5–9, on p. 5. On the Prince's illness, see letter 3580, n. 5.
3. *I have . . . anywhere*: In his journal Tyndall recorded 'sending the two tickets with which the Lord Chamberlain had favoured me to Mrs Huxley' (RS MS JT/2/13c/1395). Those present in St. Paul's included '400 holders of "Lord Chamberlain' tickets"' ('National Day', p. 6).
4. *the Nile*: Huxley began traveling up the Nile on 15 February. See Desmond, *Huxley*, p. 414, as well as letters 3631 and 3635. For Huxley's trip more generally, see letters 3608, n. 1, 3616 and 3644.
5. *Hal*: nickname for Thomas Huxley.
6. *rheumatic fever*: The term was originally used to mean 'any fever accompanied by pain in the joints or muscles' (*OED*). We believe that it is used in this sense here.

From Hervé Mangon[1] 24 February 1872 3624

N°. 69 rue S^t^. Dominique | Paris le 24 février 1872

Monsieur,

J'ai l'honneur de vous adresser, par le courrier de ce jour, un exemplaire de mes titres publiés à l'occasion de ma nomination à l'académie des Sciences de l'Institut de France—Je vous prie de recevoir cette brochure en témoignage de ma sincère admiration pour vos beaux travaux et en souvenir de votre gracieux accueil lorsque j'ai accompagné à Londres mon beaupère Mr. Dumas, il y a quelques années.

Recevez, je vous prie, | Monsieur, | l'assurance de mes sentiments les plus entièrement dévoués | Hervé Mangon

M^r^. Tyndall.

N°. 69, Rue S^t^. Dominique | Paris, 24 February 1872

Sir,

I have the honour of sending you, by today's mail, a sample of my titles[2] published on the occasion of my appointment to the Académie des Sciences of the Institute of France.[3] I ask you to receive this brochure as a token of my sincere admiration for your outstanding work, and in memory of your gracious welcome when I came to London with my father-in-law Mr. Dumas[4] a few years ago.

Please accept, Sir, the assurance of my most devoted sentiments | Hervé Mangon

M^r^. Tyndall.

RI MS JT/1/M/45

---

1. *Hervé Mangon*: Charles François Hervé Mangon (1821–88), a French engineer who specialized in agricultural work; he was named commander of the Legion of Honor in 1878.
2. *a sample of my titles*: missing.
3. *Académie des Sciences of the Institute of France*: established in 1666 for scientific research and discussion.
4. *Mr. Dumas*: see letter 3502, n. 1.

From Anna Helmholtz 27 February 1872 3625

27th. February 72. | 45. Königin Augusta | Str.

Dear Professor Tyndall

I am so sorry to have given so much trouble not only to you but to Mr Carlyle too—but nevertheless I am very glad of your answer.[1] Certainly I should never have found the verses without your help. And I got your letter[2] and all that explanation about the Copley medal—though I had never dreamed of thinking my husband ought to have had it.[3] To Dr Meyer[4] it has surely been not more precious perhaps but certainly of more importance for his life is I am told not a very bright one.

We saw him 2 years ago at Insbruck[5] and I was very much struck—painfully so—with his confused way of speaking—and had some pains in believing that this was the man I had heard so much of.

Will you not come over and see us here in Berlin? We have got two empty rooms which is a rare thing in a Berlin appartment, and there are so many people who would be happy to see you—dear Professor Tyndall.

I wish they would build a decent laboratory for my husband. They promised it should be ready in 75 but the ground is not even bought. Whatever is not military is most miserable here.[6] It is a great shame and I am very often

very unhappy to see my husband come home from the university *[quite]* exhausted and having lost precious hours with stupid mechanical works—carrying instruments to and from places where they can't remain etc.[7]
Well, and life is short—as we all know—except this abominable *[2–4 words illeg]* once more

Many thanks | Yours very truly | Anna Helmholtz

RI MS JT/1/H/35

---

1. *I am . . . answer*: Anna had written to Tyndall asking for the source of a Goethe quotation in *Fragments*, which she was translating. See letters 3619 and 3622. Tyndall's letter is missing. For additional discussion of this translation, see letters 3432, 3602, 3615, 3618, 3632, 3640 and 3649.
2. *your letter*: If this is a different letter, it is also missing.
3. *the Copley . . . had it*: In a letter of 13 November 1879 Tyndall told Rudolf Clausius that William Thomson and others had wanted Helmholtz to receive the Copley Medal in 1871, when the medal went to Mayer; see *Ascent of John Tyndall*, p. 278. On controversies over the medal and Tyndall's involvement, see letters 3546 and 3584. See also letter 3544.
4. *Dr Meyer*: Anna misspelt 'Mayer', for Julius Robert Mayer.
5. *Insbruck*: Innsbruck, Austrian city in the Alps.
6. *Whatever is . . . here*: Helmholtz's comment may register the mood in Germany in the wake of the Franco-Prussian War (for which see letter 3370, n. 3).
7. *carrying instruments . . . etc*: The Berlin Physics Institute, where Helmholtz worked, was at this time located in very small quarters; a much larger building was opened in 1878. See D. Cahan, *Helmholtz: A Life in Science* (Chicago: University of Chicago Press, 2018), pp. 429–66.

To William Spottiswoode 28 February 1872 3626

28th Feb / 1872

My dear Spottiswoode,

It is a great relief to me to learn that Joule has been selected by the Council[1] for the next president.[2]

It would be a reproach to me not to have mentioned him had I not been under the impression that he had already been asked to preside & had declined, I know I spoke to Hirst about him long ago.

Whether Joule accepts or not, (and if he accepts I will go all the way to Bradford to support him) you and my other friends must cease to think of me for some years to come;[3] and you will act the part of what you have always been, a true friend, if you would not only make no attempt yourself, but prevent any member of the Council from making the attempt, to alter this resolution—I am loaded with labours, and this presidency far from being any thing desirable, would be a serious drawback to me.

Yours truly | John Tyndall

RI MS JT/1/T/1301

---

1. *Council*: of the BAAS.
2. *next president*: for the BAAS's 43rd meeting, in 1873, in Bradford, a town in Yorkshire. In the event, Joule withdrew from the position due to illness and Tyndall was invited to serve, though he declined; see letters 3627 and 3628. Alexander William Williamson (1824–1904), a chemist, served as president.
3. *you . . . to come*: Tyndall was BAAS president in 1874.

From Douglas Galton — 4 March 1872 — 3627

4 March 1872

My dear Tyndall

The council of the British Association have desired me to convey to you[1] their request that you will do them the honour to allow them to nominate you for the Office of President of the Association for the meeting at Bradford in 1873.[2]

It is with sincere pleasure that I convey this request to you in the earnest hope that you will answer to their request.

Believe me yours very truly | Douglas Galton

J. Tyndall, Esq. L.L.D. F.R.S.

RI MS JT 2/13c/1397a

---

1. *The council . . . to you*: Galton was a general secretary of the BAAS from 1871 until 1895.
2. *the Office of President of the Association for the meeting at Bradford in 1873*: see letters 3626 and 3628.

To Douglas Galton — 7 March 1872 — 3628

7th. March 1872

My dear Galton,

I have duly considered the request which you have been good enough to convey to me from the Council of the British Association.[1]

Need I say that I highly appreciate this honouring mark of the Council's kindness and recognition? If any thing could add to my gratification it would be the kind, and even cordial manner, in which the request has been conveyed.

Flattering however as the invitation is, I learn from it with regret that we are not to have the pleasure of supporting Mr Joule at Bradford.[2]

This regret is heightened by my own inability to undertake the duty. For,—though Bradford would possess more than ordinary interest for me, and though my presiding there would certainly be the greatest public distinction of my life, the nature of my engagements compel me to forego the distinction.

To the Council I offer my grateful acknowledgements, and to you, the friendly medium of their wishes, my warmest thanks

Believe me | very faithfully yours | John Tyndall

Capt. Douglas Galton | Sec[retary]. Brit[ish]. Assoc[iation].

RI MS JT 2/13c/1397a
LT Typescript Only

1. *the request . . . Association*: the invitation to be president of the BAAS. See letter 3627, as well as letter 3626.
2. *we are not . . . at Bradford*: see letter 3626, n. 2.

From Elizabeth Rutherford[1] 8 March 1872 3629

13 Shandwick Place | Edinburgh | 8th March. 1872

My dear Sir,

A fit of shyness took possession of me, the beginning of the year, and of remorse at having occupied any portion of your time. I don't know that I should ever have recovered from it, had I not seen "Tyndall's Attempt on the Matterhorn" and on the perusal of Mr E. Whymper's letter of explanation,[2] I was filled with such a mixed feeling of contempt & vanity at having rightly judged the said gentlemans character; that my courage was restored to me.—

I sincerely hope you have quite recovered from your accident,[3] and when I am in London for a few days in April, I mean to come to ask if you are able, & again going to make an attempt, and when?

You will surely laugh at the very unsentimental offering I shall bring you from Scotland consisting of 2 shirts of Shetland wool[4]—"Tell it not in Gath Whisper it not in Askelon"[5] but after all it is not so very wanting in a romantic feeling—as it is a remembrance of the Weisshorn,[6] Orion, Castor & Pollux, gazing at you, lying under the ledge of the rock[7]—A child I dearly love, on receiving a Shetland vest and describing it to a friend said "It looks terribly thin and holey, but it feels as if warmth came out of its holes."

This has been a very sad and dreary Winter. Fever of different sorts, Scarlet, Typhoid, Smallpox etc. has carried off a very unusual number. We have been as ill off in Edinburgh as either London or Dublin—Friends & relatives, both, I have lost many of, this year, but at my age one's mind is made up to that—: unfortunately one rarely makes a new friend.

Believe me, My dear Sir | With great esteem, | Very Sincerely Yours, | Elizabeth Rutherford

RI MS JT/1/R/85

1. *Elizabeth Rutherford*: possibly Elizabeth Rutherford (née Bunyan, n.d.), the mother of physiologist William Rutherford.
2. *"Tyndall's . . . letter of explanation*: see letter 3525, n. 8.
3. *your accident*: possibly a reference to Tyndall's injury to his leg in November 1871, for which see letters 3564 and 3567.
4. *Shetland wool*: wool from sheep native to the Shetland Isles, in Scotland.
5. *Tell it not in Gath Whisper it not in Askelon*: 2 Samuel 1:20.
6. *Weisshorn*: peak in the Swiss Alps which Tyndall was the first to summit, on 19 August 1861 (see *Hours of Exercise in the Alps*, pp. 91–113, and J. Tyndall, *Mountaineering in 1861: A Vacation Tour* (London: Longmans, Green, and Co., 1862), pp. 41–58).
7. *a remembrance . . . rock*: an allusion to a passage in the chapter 'The Weisshorn' in *Hours of Exercise in the Alps* (p. 96): Tyndall describes sheltering under a ledge and watching Orion (a winter constellation) rise above the mountains. Castor and Pollux are twin peaks on the Swiss-Italian border.

From Joseph Dalton Hooker 12 March 1872 3630

Royal Gardens | Kew. | March 12th 1872.

Dear Friend,

I have had an explanation with D. Galton, which though utterly unsatisfactory to my powers of reasoning, I am bound as a gentleman to accept.[1] I will show you the very explicit correspondence we had one day soon.

Ever aff[ec]t[ionat]ely yours | J. D. Hooker

I am summoned to a committee of the Cabinet at Lord Ripon's house tomorrow upon this wretched Ayrton affair.[2]

RI MS JT/1/TYP/8/2606
LT Typescript Only

1. *I have . . . accept*: Douglas Galton was appointed director of public works in 1870 by Ayrton, in his position as First Commissioner of Works under Gladstone. In the position, Galton had oversight over hothouses under construction at Kew, which Hooker resented as an incursion on his authority. This incident was part of a long conflict between Hooker and Ayrton. See Endersby, *Imperial Nature*, p. 284, and MacLeod, "Ayrton Incident," p. 54. The next letters about this conflict in this volume are 3633, 3634 and 3638.
2. *I am . . . affair*: Hooker, at this meeting, asked the Cabinet to delineate the responsibilities of Kew and the Board of Works, and to restore his full powers. See Endersby, *Imperial Nature*, p. 285, and MacLeod, "Ayrton Incident," p. 58.

From Henrietta Huxley 12 March [1872][1] 3631

26 Abbey Place | March 12th

Dear Brother John

Good news from Egypt! from Thebes[2] Hal[3] writes that he is feeling better than he has done—since he left England—and intends extending his holiday (he has not heard of the leave)[4] so that he will not return before the first week in April. He says he did it with hesitation but thought it was the right thing to do. Really he has more sense than I gave him credit for. The travelling in the drowsy Dahabieh[5] has been most soothing and the air though cold is as pure as mountain air—and invigorates him wonderfully—There was a ridiculous description of his and Mr Ellis's[6] descent into a mummy pit chiefly of crocodiles.

Well, I need hardly add that I am very happy and I know well that you—his dear old friend, will rejoice with me at the good news

Ever | Yours affectionately | Nettie Huxley

RI MS JT/1/TYP/9/2953
LT Typescript Only

1. *[1872]*: dated by reference to Huxley's trip to the Mediterranean; see letter 3608, n. 1. Related letters include 3616, 3623, 3635 and 3644.
2. *Thebes*: ancient city along the Nile in southern Egypt. Huxley was there on 23 February 1872 (Desmond, *Huxley*, p. 414).
3. *Hal*: her nickname for Thomas Huxley.
4. *the leave*: see letter 3616, n. 3 and letter 3635.
5. *Dahabieh*: a sailing vessel used on the Nile (*OED*). Huxley traveled on one with Frederick Ouvry (1814–81), the President of the Society of Antiquaries, and Charles Ellis, for whom see n. 6. See Desmond, *Huxley*, p. 414.
6. *Mr Ellis's*: probably Charles Arthur Ellis (1839–1906), son of Charles Augustus Ellis (1799–1868), sixth Baron Howard de Walden, a diplomat stationed in Portugal and Belgium.

From Gustav Wiedemann 17 March 1872 3632

Leipzig d 17 März 1872

Mein lieber Tyndall!

Ich habe Vieweg beauftragt, Ihnen ein Exemplar der ersten Lieferung der II Auflage meines Galvanismus zu übersenden. Nehmen Sie dasselbe als ein Zeichen meiner steten freundschaftlichen Gesinnung und wissenschaftlichen Hochachtung an. Die neue Auflage hat mir wieder manche Mühe gemacht, da das Buch, wie Sie wissen, keine Compilation, sondern eine gründliche Verarbeitung des gesammten vorliegenden Materials sein soll. Einiges Neue, zB. die Theorie der Kette §11 und namentlich §34, so wie eine Menge kleinerer Notizen werden Sie wohl darin finden. Hoffentlich sehen Sie einen Fortschritt gegen die I te Auflage, die ich gänzlich umgearbeitet habe. Möchte doch das Buch nun auch dazu dienen, die vielen Wiederholungen älterer Beobachtungen, wie sie uns namentlich Frankreich so oft in fast unveränderter *[Form]* ins *[1 word illeg]* liefert, nach und nach zu beseitigen. Es kann dadurch viel Arbeit erspart werden.

Ihre hours werden fleißig gedruckt. Sie haben nun doch den Namen Stunden der Arbeit erhalten, nachdem sich meine Frau im Manuscript in der Ueberschrift *[1 word illeg]* komischer Weise geirrt und Stunden der Andacht (Sie kennen das etwas süß fromme Buch von Zschokke) geschrieben hatte. Arbeit haben Sie doch einmal körperlich und geistig in den Alpen vollbracht, und diese darf dem Publicum nicht vorenthalten werden.

Herzlichste Grüße meiner Frau, die sehr bedauert, Sie im Herbst nicht hier sehen zu sollen. Indeß lockt die Naturforscherversammlung nicht sehr; auch ich würde ihr gerne entfliehen.

In treuer Ergebenheit | Ihr G. Wiedemann

Leipzig, 17 March 1872

My dear Tyndall!

I have instructed Vieweg to send you a copy of the first instalment of the 2nd edition of my Galvanism.[1] Accept it as a token of my continued friendly disposition and scientific respect. The new edition has again caused me quite some bother, as the book, as you know, is not supposed to be a compilation, but a thorough processing of the entire material currently available. You will probably find some new material in it, e.g. the theory of the circuit in §11 and in particular in §34, as well as a quantity of shorter notes. Hopefully you will see some improvement on the 1st edition, which I have completely revised. May the book also serve now to gradually eliminate the many repetitions of older observations as France in particular so often provides them to us *[1 word illeg]* in almost unchanged *[form]*. Much work can thereby be saved.

Your "Hours"[2] are being busily printed. They have now received the title "Stunden der Arbeit"[3] however, after my wife had made a *[1 word illeg]* amusing error in the title in the manuscript and had written "Stunden der Andacht"[4] (you are familiar with the somewhat sweetly-pious book[5] by Zschokke).[6] You did perform work physically and intellectually once while in the Alps, and this should not be withheld from the readership.

Kindest regards from my wife,[7] who very much regrets that she shall not see you here this autumn. However, the Meeting of natural scientists is not very enticing; I, too, would gladly flee from it.[8]

In loyal devotion | Your G. Wiedemann

RI MS JT/1/W/51

---

1. *my Galvanism*: see letter 3615, n. 10.
2. *"Hours"*: *Hours of Exercise in the Alps*.
3. *"Stunden der Arbeit"*: for discussion of how to translate the title 'Hours of Exercise', see letters 3602, 3615, 3618 and 3649. For Tyndall's response to this letter, see 3640.
4. *Stunden der Andacht*: 'Hours of Devotion'.
5. *the somewhat sweetly-pious book*: J. Zschokke, *Stunden der Andacht zur Beförderung wahren Christenthums und häuslicher Gottesverehrung*, 8 vols (Aarau, Switzerland: H. R. Sauerländer, 1816).
6. *by Zschokke*: Johann Heinrich Daniel Zschokke (1771–1848) was German-born but relocated to Switzerland in 1796; there, he was an educator and became involved in Swiss politics. His writings included political books, volumes of tales and the *Studen der Andacht* (see n. 5), which ran to twenty-seven editions in his lifetime.
7. *my wife*: Clara Wiedemann.
8. *the Meeting . . . from it*: see letter 3602.

From Edward Stanley 27 March 1872 3633

(Private) | 23 St James's Square | March 27th 1872.

Dear Professor Tyndall,

You were speaking at The Club[1] last night about Dr Hooker and the treatment which he had received from (I suppose) Mr Ayrton.[2]

The time and place made further enquiry impossible, but I know Dr Hooker to be a man deserving of all possible support, on the general ground of his services to science and though entirely ignorant of the ground of difference in this particular case, I should think him unlikely to put himself in the wrong; while Ayrton's habit of harassing and ill-using his subordinates is well known.

If the case is of a nature in which a question or a motion for production of papers[3] would be of service, it would give me pleasure to assist Dr Hooker in that way—always supposing, first that he wished anything of the kind to be done, and next that on looking into the facts I could satisfy myself that he had been right.

I need hardly say that my sole feeling in the matter is a wish that an eminent scientific man, whose position makes it difficult for him to defend himself, should not suffer injustice.

If it does not seem to you that any help can be given in the manner suggested, there is no harm done: we are as we were: if otherwise, I would gladly look into the case and see if it admits of action being taken.[4]

Tell me what you think and believe me | Very faithfully yours | Derby.

RI MS JT/1/TYP/8/2608
LT Typescript Only

1. *The Club*: see letter 3381, n. 4.
2. *Dr Hooker . . . Mr Ayrton*: Hooker and Ayrton came into conflict over the management of Kew Gardens. See letter 3630, n. 1.
3. *a question or a motion for production of papers*: a motion in Parliament for the printing of correspondence prior to a Parliamentary inquiry.
4. *If it . . . action being taken*: Tyndall welcomed Derby's assistance; see letters 3634 and 3638.

To Joseph Dalton Hooker [29 March 1872][1] 3634

29th. March. | /72

My dear Hooker
I forward you a letter (from Lord Derby)[2] which will interest you much. Last Tuesday at the dinner of 'The Club'[3] a question arose regarding the relationship of official *[1 word illeg]* to Science, and I then and there took occasion to express somewhat emphatically what I thought of Ayrton's proceeding. Lord Derby was beside me, and indeed what I said was mainly meant for him, though I believe all at the table, including the Chancellor of the Exchequer,[4] could hear me. This morning I received the letter now enclosed to you. I shall write to Lord Derby to thank him & to say that I shall see you immediately after my return to town on Monday.[5]

Yours ever | John Tyndall

Shall you be in town on Monday? we have a managers meeting[6] from 1 to 2. or thereabouts—Could you not come dine with me at the Athenaeum?[7] Or if you cannot I will go to you. I should do this without hesitation but I am desperately occupied[8]

RBG, Kew Papers Relating to Kew 1867–1872, Ayrton Controversy, volume 1, f. 137

---

1. *[29 March 1872]*: The date on this letter appears to be in a different ink; we are not sure whether it is in Tyndall's hand or Hooker's. For further issues with the dating of letters on the topic of Hooker's controversy with Ayrton, see 3682, 3683, 3690, and 3694. The date of 29 March 1872 makes sense in the timeline of this controversy.
2. *I forward you a letter (from Lord Derby)*: letter 3633.
3. *'The Club'*: see letter 3381, n. 4.
4. *Chancellor of the Exchequer*: Robert Lowe (1811–92), a British politician. Lowe had entered Parliament as a Liberal in 1852, and was appointed chancellor of the exchequer by Gladstone in 1868. Lowe and Huxley were friendly (Desmond, *Huxley*, p. 422) and Lowe was sympathetic to science (MacLeod, "Ayrton Incident," p. 56).
5. *I shall write . . . on Monday*: We believe that this is letter 3638, and thus that Tyndall did not write to Derby until after he had seen Hooker. It is not clear where Tyndall was while writing the present letter; letter 3637 seems to reference this trip, but indicates that Tyndall returned on Saturday.
6. *manager's meeting*: at the RI.
7. *Athenaeum*: see letter 3372, n. 14.
8. *I am desperately occupied*: probably with work on the books *Molecular Physics* and *Forms of Water*; see letters 3637 and 3639.

From Thomas Henry Huxley 31 March 1872 3635

Hotel de Grande Bretagne | Naples[1] | March 31st. 1872.

My dear Tyndall,

Your very welcome letter[2] did not reach me until the 18th of March when I returned to Cairo from my expedition to Assouan.[3] Like Johnny Gilpin I "little thought, when I set out, of running such a rig."[4] But while at Cairo I fell in with Ouvry[5] of the Athenaeum[6] and a very pleasant fellow, Charles Ellis,[7] who had taken a dahabieh and were about to start up the Nile.[8] They invited me to take possession of a vacant third cabin and I accepted their hospitality, with the intention of going as far as Thebes and returning on my own hook. But when we got to Thebes I found that there was no getting away again, without much more exposure and fatigue than I felt justified in facing just then, and as my friends showed no disposition to be rid of me, I stuck to the boat, and only left them on the return voyage at Roda,[9] which is the terminus of the railway, about 150 miles from Cairo.

We had an unusually quick journey, as I was little more than a month away from Cairo, and as my companions made themselves very agreeable it was very pleasant. I was not particularly well at first, but by degrees the utter rest of this "always afternoon" sort of life did its work and I am as well and vigorous now as ever I was in my life.

I should have been home within a fortnight of the time I had originally fixed. This would have been ample time to enable me to fulfil all the engagements I had made before starting and Donnelly[10] had given me to understand that "My Lords" would not trouble their heads about my stretching my official leave.[11] Nevertheless I was very glad to find the official extension (which was the effect of my wife's and your's and Bence Jones's friendly conspiracy)[12] awaiting me at Cairo. A rapid journey home via Brindisi[13] might have rattled my brains back into the colloid state[14] in which they were when I left England. Looking through the past six months I begin to see that I have had a narrow escape from a bad break-down and I am full of good resolutions.

As the first-fruit of these you see that I have given up the school-board,[15] and I mean to keep clear of all that semi-political work hereafter. I see that Sandon[16] (whom I met at Alexandria) and Miller[17] have followed my example and that Lord Lawrence[18] is likely to go. What a skedaddle!

It seems very hard to escape however. Since my arrival here on taking up the 'Times' I saw a paragraph about the Lord Rectorship of St. Andrews. After enumerating a lot of candidates for that honour, the paragraph concluded, "But we understand that at present Professor Huxley has the best chance"[19] It is really too bad if anyone has been making use of my name without my permission. But I don't know what to do about it. I had half a mind to write to Tulloch[20] to tell him that I can't and won't take any such office, but I should look rather foolish, if he replied that it was a mere newspaper report and that nobody intended to put me up.

Egypt interested me profoundly but I must reserve the tale of all I did and saw there for word of mouth. From Alexandria[21] I went to Messina,[22] and thence made an excursion along the lovely Sicilian coast to Catania[23] and Etna.[24] The old giant was half covered with snow and this fact, which would have tempted you to go to the top, stopped me. But I went to the Val del Bove[25] whence all the great lava streams have flowed for the last two centuries, and feasted my eyes with its rugged grandeur. From Messina I came on here and had the great good fortune to find Vesuvius[26] in eruption. Before this fact the vision of good Bence Jones forbidding much exertion vanished into thin air, and on Thursday up I went in company with Ray Lankester[27] and my friend Dohrn's father,[28] Dohrn himself being unluckily away. We had a glorious day and did not descend till late at night. The great crater was not very active and contented itself with throwing out great clouds of steam and volleys of red-hot stones now and then. These were thrown towards the south-west side of the cone, so that it was practicable to walk all round the northern and eastern lip, and look down into the Hell-Gate. I wished you were there to enjoy the sight as much as I did. No lava was issuing from the great crater, but on the north side of this, a little way below the top, an independent cone had established itself as the most charming little pocket-volcano imaginable. It could not have been more than 100 feet across. Out of this, with a noise exactly resembling a blast furnace[29] and a slowly-working high pressure steam engine combined, issued a violent torrent of steam and fragments of semi-fluid lava as big as one's fist and sometimes bigger. These shot up sometimes as much as 100 feet and then fell down on the sides of the little crater, which could be approached within 50 ft., without any danger. As darkness set in, the spectacle was most strange. The fiery stream found a lurid reflection in the slowly drifting steam cloud, which overhung it, while the red hot stones which shot through the cloud shone strangely beside the quiet stars in a moonless sky.

Not from the top of this cinder cone but from its side, a couple of hundred feet down, a stream of lava issued. At first it was not more than a couple of feet wide, but whether from receiving accessions merely from the different form of slope, it got wider on its journey down to the Atrio del Cavallo,[30] a thousand feet below. The slope immediately below the exit must have been near 50 but the lava did not flow quicker than very thick treacle would do under like circumstances. And there were plenty of freshly cooled lava streams about, inclined at angles far greater than those which that learned Academican Elie de Beaumont[31] declared to be possible. Naturally I was ashamed of these impertinent lava currents and felt inclined to call them "Laves mousseuses"[32]

Courage my friend, behold land! I know you love my handwriting. I am off to Rome to-day and this day week if all goes well I shall be under, my own roof-tree again. In fact I hope to reach London on Saturday evening. It will be jolly to see your face again.[33]

Ever yours faithfully | T. H. Huxley.

My best remembrances to Hirst if you see him before I do.

IC HP 8.95
Typed Transcript Only

---

1. *Naples*: city on the coast of southern Italy.
2. *your welcome letter*: letter 3616.
3. *Assouan*: city in southern Egypt, now spelled Aswan. See letters 3623 and 3631 for Huxley's travels on the Nile. For Huxley's trip more generally, see letter 3608, n. 1 and letters 3616 and 3644.
4. *Johnny . . . rig*: Huxley alludes to 'The Entertaining and Facetious History of John Gilpin', a ballad by poet William Cowper (1731–1800), in which the eponymous hero's horse runs away with him. See W. Cowper, 'The Entertaining and Facetious History of John Gilpin, Showing How He Went Farther than He Intended, and Came Home Safe at Last', *Public Advertiser*, 15122 (14 November 1782), n.p.

5. *Ouvry*: see letter 3631, n. 5.
6. *Athenaeum*: see letter 3372, n. 14.
7. *Ellis*: see letter 3631, n. 6.
8. *But while . . . Nile*: see letter 3631, n. 5.
9. *Roda*: Rodu or Roāu, also called Turah, is the location of a quarry on the east bank of the Nile. See F. C. H. Wendel, *A History of Egypt* (New York: D. Appleton, 1890), p. 37.
10. *Donnelly*: John Fretcheville Dykes Donnelly (1834–1902), a British army officer who helped build the center for science at South Kensington and who remained involved with it for the rest of his career.
11. *my official leave*: from his teaching duties at the School of Mines. See Desmond, *Huxley*, p. 410.
12. *the official . . . conspiracy*: see letter 3616, n. 3.
13. *Brindisi*: port on the Adriatic Sea, in the south of Italy.
14. *the colloid state*: see letter 3586.
15. *school-board*: see letter 3411, n. 5.
16. *Sandon*: Dudley Ryder (1831–1900), known as Viscount Sandon until he became Earl of Harrowby upon his father's death in 1882. Ryder was a politician and MP who served on the first London school board.
17. *Miller*: John Cale Miller (1814–80), Church of England clergyman, was the rector of St Martin's Church, Birmingham, from 1846–66, where he was active in philanthropy and social outreach. In 1866 he became vicar of St Alfege, Greenwich, a post he held until 1873.
18. *Lord Lawrence*: John Laird Mair Lawrence (1811–79), a politician and Indian administrator. Lawrence became a hero in Britain after his role in violently suppressing the Indian Rebellion of 1857, and he held the position of viceroy of India from 1864 until 1869. Lawrence was the first chair of the school board; in this position, he tried to refuse Huxley's resignation (Desmond, *Huxley*, p. 410).
19. *After . . . best chance*: see 'St. Andrews', *Times*, 26 March 1872, p. 9.
20. *Tulloch*: John Tulloch (1823–86), professor of theology at St. Andrew's, educational reformer and essayist, with interests in Germany and German biblical criticism.
21. *Alexandria*: Egyptian city on the Mediterranean.
22. *Messina*: harbor city in Sicily.
23. *Catania*: province in Sicily.
24. *Etna*: active volcano in Catania.
25. *Val del Bove*: valley carved into the side of Mount Etna.
26. *Vesuvius*: active volcano on the bay of Naples.
27. *Ray Lankester*: Edwin Ray Lankester (1847–1929), a zoologist and supporter of Darwin's theory of evolution. His work included phylogeny and his book *Degeneration: A Chapter in Darwinism* (1880) argued that less stimulating conditions could lead species to degenerate into simpler forms; he was elected FRS in 1875. At the time of Huxley's visit Lankester was studying with Anton Dohrn (see n. 28). He worked as a demonstrator for Huxley in the summer of 1871.

28. *my friend Dohrn's father*: entomologist Carl August Dohrn (1806–92). His son was Felix Anton Dohrn (1840–1909), a German zoologist who worked on the phylogeny of arthropods and on the relations between arthropods and vertebrates. In 1874 he founded the Naples Zoological Station, a model for research institutes that followed (see C. Groeben, 'Marine Biology Studies at Naples: The Stazione Zoologica Anton Dohrn', in K. S. Matlin, J. Maienschein, and R. A. Ankeny (eds), Why Study Biology by the Sea? (Chicago: University of Chicago Press, 2020), pp. 29–67).
29. *blast furnace*: 'a furnace in which a blast of air is used; *spec.* the common furnace for iron-smelting' (*OED*).
30. *Atrio del Cavallo*: valley on the northern side of Vesuvius, between Vesuvius and Mount Somma. See G. Forbes, 'The Observatory on Mount Vesuvius', *Nature*, 6 (20 June 1871), pp. 145–8, on p. 145.
31. *Elie de Beaumont*: Jean-Baptiste Armand-Louis-Léonce Élie de Beaumont (1798–1874), French geologist, professor at the Collège de France and member of the Académie Française de Sciences. De Beaumont helped to create a geological map of France in the first half of the nineteenth century. His geological theories included the argument that lava can only spread widely across a horizontal surface.
32. *Laves Mousseuses*: foamy lava.
33. *this day week . . . your face again*: see letter 3644 for Tyndall's response.

To Charles d'Almeida[1] [March 1872][2] 3636

Cher Monsieur,

J'ai reçu votre nouveau *Journal de Physique*, et je vous en remercie ; il satisfait un besoin qui est senti en Angleterre aussi bien qu'en France. Les noms des collaborateurs sont une garantie de la pureté des doctrines qu'il va propager. Je désire son succès ([1]).

En fait d'enseignement, j'ai essayé récemment de montrer à un nombreux auditoire la rotation du plan de polarisation des rayons obscurs. Pour cela, j'ai employé deux prismes de Nicol assez gros, et j'ai adopté la disposition de MM. De la Provostaye et Desains, qui placent les sections principales des prismes non pas à angle droit, mais sous un angle de 45 degrés. Je me suis servi de la lampe électrique et j'ai fait usage de l'iode dissous dans le sulfure de carbone pour intercepter la lumière. J'ai pu obtenir avec la chaleur obscure une déviation qui s'élevait à 150 divisions de mon galvanomètre, lorsque je faisais agir l'électro-aimant sur le verre pesant traversé par les rayons de chaleur. Je pense que MM. de la Provostaye et Desains employaient la chaleur solaire lumineuse dans leurs expériences, et que l'effet produit s'élevait à 2 ou 3 degrés de leur galvanomètre.

Je regarde comme erronés les résultats affirmatifs publies avant les expériences de MM. de la Provostaye et Desains. Par les moyens employés et avec les sections principales des Nicols à angle droit l'un de l'autre, l'action était certainement trop faible pour être observée.

Votre etc. | John Tyndall

Dear Sir,

I received your new *Journal de Physique*,[3] and I thank you for it; it satisfies a need that is keenly felt in England as much as in France. Seeing the names of the colleagues involved is enough to guarantee the precision of the theories it disseminates. I hope for its success ([1]).[4] I wanted to inform you that I recently attempted to show a large audience the rotation of the plane of polarization of infrared light.[5] To do so, I used two fairly large Nicol prisms[6] and adopted the placement of Messrs. de la Provostaye[7] and Desains,[8] who align the main sections of the prisms not a right angle, but at an angle of 45 degrees. I relied on an electric lamp and used iodine dissolved in carbon disulfide to intercept the light. With infrared light, I was able to obtain a deviation that measured up to 150 on my galvanometer,[9] when I used an electromagnet on the leaded glass through which the infrared passed. I think that Messrs. de la Provostaye and Desains used visible light from the sun in their experiments, and that the resulting effect measured up to 2 or 3 degrees on their galvanometer.

I consider the affirmative results published before the experiments of Messrs. de la Provostaye and Desains to be erroneous. Considering the methods that were used and having the main sections of the Nicol prisms at right angles to one another, the effect would certainly be too small to be observed.

Yours, etc. | John Tyndall

Journal de Physique Théorique et Appliquée, 1 (1872), pp. 101–2

---

1. *Charles d'Almeida*: Joseph-Charles d'Almeida (1822–80), a French physicist. D'Almeida was the author of a popular textbook, the *Cours élémentaire de Physique* (1862), and the founder of the *Journal de Physique Théorique et Appliquée* in 1872 as well as the Societé de Physique in 1873.
2. *[March 1872]*: On 16 February 1872, Tyndall recorded in his journal, 'A new Physical Journal has been started in Paris by M. D'Almeida' (RS MS JT/2/13c/1392). On 20 April d'Almeida wrote to Tyndall to thank him for his letter, which was published in the third issue; see letter 3662. We thus propose that this letter was written in March and published in April, although the exact date of publication of this number of the *Journal de Physique* is unknown because surviving copies exist as bound volumes from which information about individual parts has been excised. The letter was published under the title 'ROTATION DU PLAN DE POLARISATION DES RAYONS DE CHALEUR OBSCURE | (Lettre de M. John Tyndall)', or 'THE ROTATION OF THE PLANE OF POLARIZATION OF INFRARED LIGHT | Letter from Mr. John Tyndall)'.

3. *your new* Journal de Physique: possibly the first number of the journal; see n. 2.
4. *I hope for its success*: In the original, this sentence is annotated with the following footnote: 'Nous avons traduit tout entière la lettre que l'éminent physicien a bien voulu nous adresser. Nous tenons à lui témoigner publiquement notre reconnaissance des vœux qu'il fait pour le succès de ce Journal. Sa collaboration y contribuera certainement. (C. d'A.)' ('We have translated the entire letter that this eminent physicist was kind enough to send us. We would like to publicly state our gratitude for his commitment to the success of this Journal, to which his contributions will certainly contribute. (C. d'A.)).
5. *I recently . . . infrared light*: in his Friday Evening Discourse at the RI on 2 February 1872; see J. Tyndall, 'On the Identity of Light and Radiant Heat', *Roy. Inst. Proc.*, 6 (1870–72), pp. 417–21.
6. *Nicol prisms*: designed by British geologist and lecturer William Nicol (1770–1851) to filter polarized light.
7. *de la Provostaye*: Joseph Prudent Frédéric Hervé de La Provostaye (1812–63), a French physicist who worked on chemistry and crystallography, as well as heat and light in collaboration with Desains (see n. 8).
8. *Desains*: Paul-Quentin Desains (1817–85), a French physicist who worked with de La Provostaye on light and heat.
9. *galvanometer*: used to detect the direction and intensity of a galvanic current (*OED*).

To Thomas Archer Hirst 2 April 1872 3637

2nd. April, 1872.

My dear Tom

Would you glance at the analysis of Mem I. & II. and at the historic remarks between them?[1]

I came home on Saturday[2] after two days of wet weather found 5 packets of proofs here waiting for me and was altogether so oppressed by the quantity of work weighing on my mind that I resolved not to stir till I had broken the heart of it.

Yours affectionately | John

RI MS JT/1/HTYP/585
LT Typescript Only

---

1. *the analysis of Mem I. & II. and at the historic remarks between them*: 'Analysis of Memoir I.', 'Analysis of Memoir II.' and 'Historic Remarks on Memoir I.', in *Molecular Physics*, pp. 2–5, 59–64, 66–8.
2. *I came home Saturday*: It is not clear where Tyndall had traveled.

To Edward Stanley 2 April 1872 3638

Royal Institution, | 2d April 1872.

Dear Lord Derby

On my return to town I called to see Hooker and I had previously made him acquainted with the kind interest that you took in his case.[1] He is very desirous indeed that you should be adequately informed of the facts, though as matters are still pending, and as Sir John Lubbock has undertaken to do in the House of Commons what you have so kindly offered to do in the House of Lords, no action can, I suppose, be at the present moment taken.

Hooker has drawn out a clear and consecutive account of the whole transaction and this I now beg to place in your hands.[2] It seems to my mind the statement of a very hard case. In 1840 the late Sir William Hooker became director of Kew. It had been a Royal garden nine acres in area, and had been actually advertised for sale. In its present form it is virtually the creation of Sir William Hooker and his son. Dr Hooker began his travels in 1839, and throughout those travels his collections were made with reference to the wants of Kew.[3] Sir William Hooker at the time of his appointment was the possessor of a fine herbarium without which nothing could have been done, and for five and twenty years this herbarium was employed in the development of Kew. True at his death, as he did not think his son so able as himself to disregard pecuniary matters, he recommended that it should be valued by different competent persons and sold to the public at the lowest valuation, which was done.

There is a most interesting and instructive museum at Kew.[4] With his own private collections of drugs, dyestuffs, textile fabrics, and other things illustrative of the relation of the vegetable kingdom to the useful arts, Sir William founded this museum, and never received a farthing for his contributions to it. As regards the flora of Asia, Africa, and America, the herbarium at Kew had been unrivalled, but Europe was scantily represented. Three years ago a collection embracing the very Flora needed for the completion of Kew was offered for sale in Paris. Dr Hooker purchased the collection for £400 of his own money and presented it to the Kew herbarium. His income as Director is £800 a year, and here is half of it at once devoted to the interests of the public and of science.

The real character of Kew is only too likely to escape the superficial observer. The place has been made so beautiful and so attractive to the public that its immense scientific importance is likely to be overlooked.[5] That it is of the very highest importance to botanical science and to the application of that science in India and the Colonies might be readily demonstrated.

In regard to Dr Hooker's personal qualities I hardly know how to speak of them. A man of exalted natural endowments which, as regards his own great department of science, have been cultivated to the very highest degree. With one exception, if it be one, I do not suppose that England ever possessed a botanist like Hooker, nor do I suppose that Europe at the present moment has his equal. And associated with this knowledge and culture is a character of the very finest fibre, a sense of duty of the very highest cast, which would render the slightest neglect of public interests simply impossible to him. It is a man of this stamp that is placed under the thumb of Mr Ayrton; to be worried, first out of his peace of mind, and afterwards out of a position which his father[6] and he have so long held with such unparalleled success. There is not a scientific botanist in Europe who would not deplore such a consummation as a calamity to his science. With regard to Mr Ayrton himself, there is nothing to be done. His nature is too low to permit of appeal to it. But I am told on all hands that the sin of this transaction lies at the door of Mr Gladstone, who backs up this dog in office. Hooker's communication is "confidential", but I desire no privacy for these sentiments of mine. I should be only too happy of the opportunity of expressing them in the presence of the Prime Minister himself.

One word of personal thanks I must add for your timely and most excellent letter.[7] I know you will act[8] in this matter with justice and wisdom. Many of the best intellects in Europe will feel themselves to be your debtors if you can do anything to prevent this inversion of equity, this placing of knowledge under the control of ignorance, this subjection of a high and refined nature to the tyranny of an Ayrton.

I am, my lord, | Yours most faithfully | John Tyndall.

RI MS JT/1/TYP/8/2610–1
LT Typescript Only

---

1. *the kind interest you took in his case*: see letters 3633 and 3634; the latter discussed Tyndall's plan to see and confer with Hooker. The present letter is a response to letter 3633.
2. *Hooker . . . in your hands*: enclosure missing; Derby seems to respond to this account in letter 3659.
3. *Dr Hooker . . . Kew*: Hooker traveled in the southern hemisphere, on the *Erebus*, from 1839 until 1843, and in the Himalayas from 1847 until 1849.
4. *museum at Kew*: the Museum of Economic Botany, opened by William Hooker, Joseph's father, in 1848. See Endersby, *Imperial Nature*, p. 11.
5. *The real . . . overlooked*: on the 'hybrid' character of Kew, as both a center for research and a garden for the public, see Endersby, *Imperial Nature*, pp. 277–8, 281.
6. *his father*: William Hooker.

7. *timely and most excellent letter*: letter 3633.
8. *I know you will act*: for Derby's response, see letters 3643 and 3654.

To Hector Tyndale 2 April 1872 3639

[Royal Institution][1] | 2nd April 1872.

My dear Hector.

A newspaper bearing your well-known direction[2] reached my hands ten minutes ago and brought to a determinate issue thoughts which have been long flying through my brain. I have had two books on hand,[3] both of which shall be sent to you in due time. My lectures were in front of me[4] and I must get the books finished before the commencement of the lectures or postpone them for a year, to the detriment of them & me. I tried to get Dean Stanley[5] or Froude to take two or three of the first lectures off my hands but failed. And there they are now looming upon me for the 11th of April. The heart of my biggest book[6] is broken; but it is not finished yet; and my little one leaves much to be desired.

Under these circumstances I have been living for the last 6 weeks the life of a hermit; steadily declining all invitations, and I fear giving umbrage to some of my friends. It was that cause, and this only, that prevented me from writing to you.

And to add to my calamities I have not yet been able to give a single minute's thought to my American lectures.[7] My main preparation for them must be made during my vacation. I have begun to gather in pieces of suitable apparatus, and on the whole though I shall lag at an immense distance behind my ideal, I am not without a hope of being able to gather together materials for two or three short courses.

This is the arrangement that I would propose. I am quite willing to have the expenses of my assistants paid, and the cost of such apparatus as may be bought expressly for America. I am even willing to permit my own travelling and Hotel expenses to be paid; but beyond this I should not be willing to accept a farthing. My notion is to lecture in the cause of Chicago,[8]—and to devote the proceeds, such as they may be, to the founding or maintenance of some scientific or other institution in that city. This I trust is but a preliminary visit of mine to America, and if my friends humour me this time, I promise, should I ever feel the need of money, to replenish my purse by giving at some future period another and I trust better course of lectures in America.

Tom[9] & Debus, were they here, would send their love to you.

Give my kindest regards to your friend Leslie—I hope he does not consider me an unsocial brute.[10]

With best wishes to your wife[11] | I am dear Hector | Yours affectionally | John.

RI MS JT/1/T/1452
RI MS JT/1/TYP/4/1694–5

---

1. *[Royal Institution]*: LT annotation.
2. *A newspaper bearing your well-known direction*: correspondence missing.
3. *two books on hand*: *Molecular Physics* and *Forms of Water*.
4. *my lectures were in front of me*: Tyndall gave a series of nine weekly lectures on 'Heat and Light' at the RI on Thursdays from 11 April to 6 June 1872 (*Roy. Inst. Proc.*, 6 (1870–72), p. 384). See *Ascent of John Tyndall*, p. 296.
5. *Dean Stanley*: see letter 3419, n. 4.
6. *my biggest book*: probably *Molecular Physics*; see letter 3637.
7. *my American lectures*: for discussion of Tyndall's American trip, see letters 3504, 3507, 3510, 3519, 3542, 3543, 3558, 3589, 3594, 3669, 3670, 3684, 3685, 3702 and 3703.
8. *to lecture in the cause of Chicago*: The Great Chicago Fire burned from 8–10 October 1871, destroying several square miles of the city and costing hundreds of lives. The disaster attracted significant national and international attention and assistance.
9. *Tom*: Thomas Hirst.
10. *Give my . . . brute*: see letter 3589, in which Tyndall declined Peter Lesley's invitation to stay with him.
11. *your wife*: Julia Tyndale.

To Gustav Wiedemann 3 April 1872 3640

3rd April 1872 | Royal Institution of Great Britain

My dear Wiedemann

The book which you have been good enough to send me has not yet arrived,[1] but when it does arrive I expect it to illustrate those high qualities which have hitherto characterised your labours—in which profound knowledge and thoroughness of workmanship have been combined.

I have thought over the title of the little book,[2] and talked with Debus over it. I think the very best title you could adopt would be simply

In den Alpen[3] | by J. Tyndall.

This would leave the precise character of the work undetermined, and would not lead any body to imagine that the book is a scientific one.

I have been working very hard of late putting my separate memoirs on Radiant Heat[4] together, analysing the contents of each of them, and showing how they are connected together. I hope in a month or five weeks to be able to send you a copy. I have resumed the consideration of the differences between Magnus[5] and myself, and though I have studiously avoided every unnecessary word I can hardly prevent what I write from appearing somewhat controversial in character. If Magnus were alive I should feel myself much more free in this discussion; but I shrink from criticising a dead man's labours. Still there is great error afloat on this subject, and I think it ought not to be allowed to remain.

Clausius has just sent me a small controversial paper from Poggendorf.[6] I think it is most deplorable that questions of nationality should be thus introduced into science; but our Scotch friends[7] are to blame for this—the sooner it is discontinued the better.

This spirit of nationality coupled with a good deal of personal bitterness, directed in part against myself, appears in a book which was sent up to me last night, written by your colleague Zöllner.[8] Certainly I never expected that my little cometary hypothesis—expressed, I may say, in the midst of astronomers,[9] and regarding which I have never been censured by an astronomer, could have provoked such animosity as it appears to have done in the breast of Zöllner. In his notices of me he is certainly making all unscientific use of his Imagination.[10]

Give my kindest regards to Mrs Wiedemann[11] and believe me always yours
John Tyndall

EH FAM123

---

1. *The book which you have been good enough to send me has not yet arrived*: see letter 3632. The book is G. Wiedemann, *Die Lehre von Galvanismus und Elektromagnetismus*, 2nd edn (Braunschweig: Vieweg, 1872).
2. *the title of the little book*: see letters 3602, 3615, 3618, 3632 and 3649. The book under discussion is *Hours of Exercise in the Alps*.
3. *In den Alpen*: 'In the Alps'.
4. *memoirs on Radiant Heat*: *Molecular Physics*; for Tyndall's work on this volume, see letters 3637 and 3639.
5. *Magnus*: see letter 3450, n. 14.
6. *a small controversial paper from Poggendorf*: possibly J. C. Poggendorf, 'Versuch einer Theorie der Elektro-Doppelmaschine', *Annalen der Physik und Chemie*, 221:1 (24 February 1872), pp. 1–24.
7. *Scotch friends*: probably a reference to the 'North British' group of physicists and engineers including Thomson, Tait, and others. This group came into conflict with Tyndall and other scientific naturalists. See C. Smith, *The Science of Energy: A Cultural History of Energy Physics in Victorian Britain* (Chicago: University of Chicago Press, 1998).

8. *a book . . . Zöllner*: The book is J. Zöllner, Über *die Natur der Cometen, Beiträge zur Geschichte und Theorie der Erkenntniss* (Leipzig: Engelman, 1872), in which Zöllner attacked not only Tyndall but numerous other scientific practitioners (*CDSB*). Zöllner criticized both Tyndall's ideas about comets and his popular lectures. See *Ascent of John Tyndall*, p. 277. For Wiedemann's response about this issue, and to this letter more broadly, see letter 3649. For other letters about Zöllner, see letters 3642, 3652, 3655, 3695 and 3697.
9. *my little cometary . . . astronomers*: Tyndall articulated his theory at the Cambridge Philosophical Society (see letter 3652) and published it as a letter to the *Phil. Mag.* See J. Tyndall, 'On Cometary Theory', *Phil. Mag.*, 37:249 (April 1869), pp. 241–5.
10. *all unscientific use of his imagination*: a reference to Tyndall's lecture 'On the Scientific Use of the Imagination', for which see letter 3370, n. 5.
11. *Mrs Wiedemann*: Clara Wiedemann.

From Joseph Dalton Hooker [3 April 1872][1] 3641

Royal Gardens | Kew | Wed[nesda]y.

Dear Friend

I do not know how to express myself to you in the matter of the enclosed.[2] I cannot keep it till Thursday's X;[3] it makes my old cheeks burn so hotly. Mrs Hooker[4] has however taken the liberty of keeping a copy of it. Ever your affectionate &. . . . . | J. D. Hooker.

RI MS JT/1/TYP/8/2609
LT Typescript Only

1. *[3 April 1872]*: We propose this date based on Hooker's reference to a meeting of the X Club on Thursday; according to Hirst's journal, the X Club met on 4 April 1872 (Brock and MacLeod, *Hirst Journals*, p. 1939). The X Club was a group of nine prominent men of science who met regularly, beginning in 1864, and which became highly influential. Members included Hooker, Tyndall, Hirst, and Huxley. See R. Barton, *The X Club: Power and Authority in Victorian Science* (Chicago: University of Chicago Press, 2018).
2. *the enclosed*: enclosure missing. It is possible Tyndall had sent Hooker his letter of 2 April; see letter 3638, whose praise for Tyndall might correspond to Hooker's expression of embarrassment.
3. *Thursday's X*: see n. 1.
4. *Mrs Hooker*: Frances Harriet Hooker (née Henslow, 1825–74), the oldest daughter of botanist John Stevens Henslow (1796–1861) and Harriet Jenyns. In 1851 she married Joseph Hooker, with whom she had four sons and two daughters. She was fluent in French and

German, and in addition to making private translations for her husband she published a translation of *A General System of Botany*, by Emmanuel le Maout (London: Longman, 1873). Although no letters between them survive for the period of this volume, she corresponded with Tyndall and shared literary references, poetry, and novels. She was a skilled editor and occasionally read over manuscripts for him.

From Rudolf Clausius 4 April 1872 3642

Bonn, 4/4 72.

Lieber Tyndall,

Ich danke Dir sehr für Deinen freundlichen Brief den ich soeben erhalten habe. Du kannst mir glauben dass es mir sehr schwer geworden ist einen Artikel der Art, wie den gegen Maxwell, zu schreiben. Ich hatte solche Auseinandersetzungen so lange es ging, vermieden, und hatte daher auf das Buch von Tait kein Wort entgegnet. Das Buch von Maxwell zwang mich aber zu einer Vertheidigung meiner Rechte. Es versteht sich von selbst, dass meine Bemerkung am Anfange des Artikels sich nur auf eine bestimmte Anzahl englischer Physiker beziehen soll, nämlich die Gruppe, welche sich um Thomson schaart.

Da Du in Deinem Briefe das Buch von Zöllner erwahnst so kann ich nicht unterlassen, Dir zu schreiben, dass ich über dieses Buch auf das Äusserste entrüstet bin, und ich kann Dir auch sagen, dass alle, mit denen ich darüber gesprochen habe, die darin vorkommenden persönlichen Angriffe entschieden missbilligen und verdammen. Die Art und Weise wie er spricht, ist nicht blos für ein wissenschaftliches Werk unerhört, sondern sie ist auch sonst so unerklärlich und sonderbar dass man gar nicht begreift, wie er dazu gekommen ist. Es sind dadurch sogar Gerüchte über seinen Geisteszustand hervorgerufen, die für ihn wenn sie sich bewahrheiten sollten, sehr traurig sein würden. Ich möchte Dich aber bitten von dieser letzten Mittheilung keinen weiteren Gebrauch zu machen, da es eben nur ganz unbestimmte Gerüchte sind, die durch die Sonderbarkeit seines Buches entstanden sind.

Lass Dich aber durch diese Angelegenheit nicht zu sehr verstimmen. Du musst bedenken, dass gerade diejenigen, die sich eine hervorragende Stellung errungen haben, und von denen daher viel gesprochen wird, am meisten derartigen Angriffen ausgesetzt sind.

Mit herzlichen Grüssen von meiner Frau und mir an Dich und Hirst,

Dein | Clausius.

Bonn, 4/4 72.

Dear Tyndall,

I thank you[1] very much for your kind letter,[2] which I have just received. You can believe me that it has become very difficult for me to write an article of the kind like that against Maxwell.[3] I had avoided such arguments for as long as possible, and had therefore not replied one word to the book by Tait.[4] The book by Maxwell,[5] however, forced me to defend my rights. It goes without saying that my remark at the beginning of the article should refer only to a specific number of English physicists, namely the group which is gathering around Thomson.

As you mention the book by Zöllner[6] in your letter, I cannot refrain from writing to you that I am extremely indignant about this book, and I can also tell you that everyone whom I have I have spoken to about it firmly disapproves of and condemns the personal attacks that occur in it. The way in which he speaks is not only outrageous for a scientific work, but it is also otherwise so inexplicable and strange that one does not comprehend at all how he has come to it. There have even been rumours aroused by it about his mental condition, which, should they prove to be true, would be very sad for him. I would, however, like to ask you to make no further use of this latter information, as it is simply only quite vague rumours that have arisen through the strangeness of his book.

But don't let yourself be annoyed too much by this matter. You have to remember that precisely those who have achieved a prominent position, and about whom much is therefore spoken, are exposed most of all to attacks of this sort.

With kind regards from my wife[7] and me to you and Hirst,

Your | Clausius.

RI MS JT/1/TYP/7/2325–6

---

1. *I thank you*: Clausius uses the familiar German 'you' ('Du').
2. *your kind letter*: letter missing.
3. *an article of the kind like that against Maxwell*: see letter 3591, n. 12.
4. *the book by Tait*: see letter 3591, n. 10.
5. *The book by Maxwell*: see letter 3591, n. 7.
6. *book by Zöllner*: see letter 3640, n. 8. For additional discussion of Zöllner, see letters 3649, 3652, 3655, 3695 and 3697.
7. *my wife*: Adelheid Clausius.

From Edward Stanley 7 April 1872 3643

Knowsley, | Prescot,[1] | April 7th 1872

Dear Professor Tyndall

I send this line only as an acknowledgement of your letter of the 2d.[2] I shall be in town by the end of this week and ready to go into the case.[3]

Very faithfully yours | Derby.

RI MS JT/1/TYP/8/2616
LT Typescript Only

---

1. *Knowsley, | Prescot*: the Derby family seat in Lancashire, near Liverpool in the north of England.
2. *your letter of the 2d*: letter 3638.
3. *I shall be . . . the case*: see letter 3654.

To Thomas Henry Huxley 8 April [1872][1] 3644

Royal Institution. | 8th. April.

Dear old Hal,

A thousand welcomes back to your friends.[2] Your capital letter[3] prepared me for your arrival, and Spencer tells me you are in first rate trim. I should have run up yesterday had not sheer weariness and seediness drove me to the country. I should cross over to you now had we not a Managers Meeting.[4]

Yours ever | John Tyndall.

IC HP 8:97
Typed Transcript Only

---

1. *[1872]*: dated by continuity with letter 3635 and Huxley's travels in the Mediterranean.
2. *back to your friends*: from his travels in Egypt and the Mediterranean; see letters 3608, 3616, 3623, 3631 and 3635.
3. *Your capital letter*: letter 3635.
4. *Managers Meeting*: of the RI.

From Joseph Dalton Hooker 8 April 1872 3645

Royal Gardens, | Kew, April 8th/72.

My dear Tyndall,

I am very sensible of Lord Derby's consideration and grateful too, more so than I can express.[1] I sincerely hope, that, should his lordship find opportunity to take any action in the matter,[2] it may be rather with a view to the position of this Establishment than to myself as its director.

We are all interested in our Gov[ernmen]t Scientific Institutions being placed upon a more satisfactory footing than they now are, in respect of their being made more useful and instructive, and more powerful instruments for good in the hands of the Government. This cannot be, until they are brought together under some intelligent control, their position better defined and their waste energies utilised.

I believe that were those of the Metropolis to be brought into harmonious action, an economy would be effected; and a body of active leading scientific men extemporized which would form a most efficient consulting body at the service of the Government in all cases where scientific aid or advice was required by it.

Such a body might consist of the Directors of the Institutions together with others (including some in Parliament) selected for their knowledge of the uses and requirements of scientific Institutions. This body might be, in all but financial matters, independent, as the British Museum[3] now is; all the directors might be placed under the Lords of the Committee of the council[4] (quasi Ministers of Education) to whom each Director should then be individually and solely responsible for the establishment he controls, but ready to join in advising in other matters, when called upon to do so by the Lords of the Committee.

As it is, the Lords have no scientific council that has the confidence of the public and Science is jumbled up with Art at South Kensington,[5] with the Parks under the Board of Works, with literature at the British Museum and so forth.

If the troubles of Kew were to eventuate in a better order of things I should never regret my share of them!

Ever yours sincerely | J. D. Hooker

RI MS JT/1/TYP/8/2617
LT Typescript Only

---

1. *I am . . . express*: referring to Derby's expression of interest in Hooker's conflict with Ayrton. See letters 3633, 3634 and 3643.

2. *should his lordship . . . the matter*: see letters 3654 and 3659 for Derby's actions.
3. *British Museum*: see letter 3576, n. 2.
4. *Lords of the Committee of the council*: probably the council on education, then headed by William Edward Forster (1818–86).
5. *South Kensington*: a complex of museums including the Natural History Museum, the Science Museum and the Victoria and Albert Museum (of design and decorative arts).

From Maria McKaye[1] 9 April 1872 3646

Exeter, New Hampshire U.S.A. | April 9th, 1872.

Dear Professor Tyndall,

My friend, Thomas Wentworth Higginson,[2] has kindly consented to be bearer of dispatches to you on the occasion of his first visit to England and I hope that nothing will occur to prevent your meeting; for I am sure that it will be a mutual pleasure for you to become acquainted with each other.

Mr Higginson's name has been so long prominent in American literature that it is possibly not unknown to you.

Your last letter[3] has been highly valued and carefully treasured. I have not acknowledged it before for fear of being troublesome. Please let this little volume of my cousin's travels[4] remind you of your name being an inspiration as well as a familiar and beloved household word across the Ocean: to us especially, who having met you once, in a way not to be forgotten, cherish the hope of showing you one day, in our own home, with what grateful regard we are your friends

Maria (and Henry) McKaye[5]

Prof. Tyndall | Introducing T.W. Higginson

RI MS JT/1/TYP/2/789
LT Typescript Only

1. *Maria McKaye*: Maria Ellery McKaye (sometimes MacKaye), an American woman who, with her son Henry, met Tyndall in 1867 in Zermatt. See letter 3190, *Tyndall Correspondence*, vol. 11.
2. *Thomas Wentworth Higginson*: Thomas Wentworth Higginson (1823–1911), a minister, author and social activist. Higginson was an abolitionist who resisted the Fugitive Slave Act, supported John Brown's rebellion and fought for the North in the American Civil War. After the war he turned his attention to literary work.
3. *Your last letter*: letter missing.
4. *this little volume of my cousin's travels*: McKaye's cousin was Clarence King. The book in question is almost certainly C. King, *Mountaineering in the Sierra Nevada* (Boston: Osgood, 1872).

5. *Maria (and Henry) McKaye*: Henry was Maria's son; Tyndall had also met him in Zermatt. See n. 1.

From Joseph Dalton Hooker 10 April 1872 3647

Royal Gardens | Kew | April 10th 1872.

Dear Tyndall

Do let us take our bags and some books to some quiet place at your selection next week. I want it as much as you do. New Forest?[1] Isle of Wight?[2] If you won't I will go myself somewhere.

The week is up and no answer to the questions which the Treasury profess to have considered.[3] My feeling is decided, and in God's name let Lord Derby go on.[4] Bentham has never wavered from his opinion that I should have no settlement short of Parliament, but he is too reticent and timid to recommend this. He says I must be wholly guided by yours and Huxley's and Lubbock's opinion, and he will stand by whatever you do.

I should not hesitate myself to ask Lord Derby to go on, but I greatly prefer that it should be said that my friends were not influenced by my wishes even. I should like to have it to say, that they did in this exactly what I would have wished and advised.

Ever aff[ectionatel]y yours | J. D. Hooker.

No time is to be lost and the ministers may consider their position at leisure in the Holidays,[5] after the papers are called for.[6] Could you not run down to St James's Square[7] and see Lord Derby? Telegraph if I am wanted.

RI MS JT/1/TYP/8/2618
LT Typescript Only

---

1. *New Forest*: in southern England.
2. *Isle of Wight*: island off England's southern coast.
3. *the questions which the Treasury profess to have considered*: probably a reference to a dispute between Hooker and Ayrton over the appointment of a clerk at Kew. Hooker was accustomed to filling such positions himself; Ayrton insisted on selecting a candidate through a civil service examination, with the result that the candidate chosen (Robert Smith) was, to Hooker's mind, not suited for the position. See Endersby, *Imperial Nature*, p. 283, and 'Correspondence between the Board of Works and Dr. Hooker relating to Changes proposed to be introduced into the Direction and Management of the Gardens at Kew', HL 213, Parliamentary Papers (1872), pp. 13–22.
4. *let Derby go on*: see letters 3633 and 3643 for Derby's offer of assistance, as well as letters 3654 and 3659.
5. *the Holidays*: Parliamentary holidays began in August and lasted through October.

6. *the papers are called for*: see letter 3633, n. 3.
7. *St James's Square*: a fashionable residential area in London and the location of Derby's London residence.

From Henrietta Huxley [between 8 and 10 April 1872][1] 3648

Dear Brother John

Thank you very much for the tickets[2] although I fear that I shall not be able to use them. My strength is only enough for doing home duties, and these seem to increase now that the children[3] grow older.

I am so sorry that you are not well—I wish you had a good wife to take care of you.

It is such a happiness to have Hal[4] back again[5]—He is really going to take care of himself. Can you come up on Sunday at 6—and I will give you some dinner—Let me know.

Ever | Yours affectionately | Nettie Huxley

RI MS JT/1/TYP/9/2951
LT Typescript Only

1. *[between 8 and 10 April 1872]*: Given the reference to Tyndall's lectures 'On Heat and Light' (see n. 2), we believe this letter was written before 11 April 1872. We also believe that, given the phrasing of Huxley's invitation ('Can you come up on Sunday'), it was written later than Sunday, 7 April.
2. *tickets*: Tyndall delivered a course of nine lectures, 'On Heat and Light', at the RI on Thursdays from 11 April to 6 June 1872 (*Roy. Inst. Proc.*, 6 (1870–2), p. 384). For Henrietta Huxley's attendance at other lectures of Tyndall's, see letters 3411 and 3608.
3. *children*: see letter 3411, n. 2.
4. *Hal*: her nickname for Thomas Huxley.
5. *back again*: Huxley returned to London from several months in Egypt, where he had travelled for his health, on 6 April 1872. For this trip, see letters 3608, 3616, 3623, 3631 and 3635. Tyndall wrote on 8 April 1872 to welcome Huxley back to England; see letter 3644.

From Gustav Wiedemann 10 April 1872 3649

Leipzig 10 April 1872.

Mein lieber Tyndall!

Ich habe, soeben von einer kleinen Erholungsreise zurückgekehrt, sogleich an Vieweg geschrieben, ob er die Absendung meines Galvanismus an

Sie verabsäumt hätte. Bitte, wenn Sie das Buch erhielten, schreiben Sie es mir wohl in ein Paar Worten, damit ich sicher bin, daß mein Auftrag befolgt ist

Der Titel „Stunden der Arbeit in den Alpen“ für Ihre hours etc . . . ist schon in den 3 ersten Bogen über den Seiten gedruckt; indeß können wir wenigstens auf dem Haupttitel „In den Alpen“ sehen; ein etwas zu bescheiden Titel, wie mir scheint. Sie werden damit einverstanden sein, daß wegen dieser nicht bedeutenden Sache die 3 ersten schon abgezogenen Bogen nicht völlig cassirt werden.

Zöllners Buch beklage ich auf das Tiefste als eine schwere Versündigung an der Wissenschaft, die frei von Persönlichkeiten sein sollte. Zöllner hat das Buch durchaus selbstständig verfaßt; die Ermahnungen und /Einwände ernstester/ Art, welche seine Freunde ihm bei gelegentlichen Aeußerungen vor der Publication gemacht haben, hat er nicht beachten wollen. Wer Zöllner näher kennt zweifelt nicht, daß er das Beste selbst gewollt hat und sich über etwaige persönliche Geneigtheiten selbst nicht klar geworden ist. Demnach liegt in dem ganzen Unternehmen eine solche (unbewußte) Selbstüberschätzung, und sind die Urtheile /darin/ so einseitig und hart, daß das Buch aufs höchste zu mißbilligen ist. Wie viele Leute haben nicht Freude am bloßen Scandal, wie /wenige/ freuen sich nicht aus *[2 words illeg]* und wenig edlen Motiven, wenn ein hochstehender Mann der Wissenschaft, der diesen oder jenen einmal unabsichtlich verletzt hat, einen Schlag bekommt? Selbst wenn Manches, was Zöllner tadelt, nicht ganz untadelhaft dasteht, so sollte es doch der gute Ton in der Wissenschaft mit sich bringen, nicht der Welt durch wissenschaftlichen Klatsch, wenn auch unabsichtlich, Vergnügen zu bereiten. Dies ist übrigens nicht nur mein Urtheil über Zöllners Buch, sondern das vieler unbefangener Freunde, so wie Clausius, Bentz, Helmholtz, der ja auch ganz ungerechtfertigt angegriffen ist, du Bois Reymond, Kronecker u. A.

Den Angriff auf Sie selbst, lieber Freund, werden Sie zu verschmerzen wissen. Wollte Zöllner Ihre Kometentheorie, die ich aber nur als eine „Theorie“ ansah, angreifen und critisiren, so hätte er es in einem besonderen Aufsatz über Cometen thun sollen. So ist der Angriff, der gerade durch die persönliche Behandlung den Leser /verlacht/, vollkommen am unrechten Ort. Uebrigens ist Zöllners Cometentheorie, ebenso wie seine /ebenso fromme/ Theorie des Erdemagnetismus, nach /dem Urtheil/ aller Physiker, die ich bisher gesprochen, völlig unhaltbar. Glauben Sie mir, daß Ihnen in der Meinung aller Einsichtigen Zöllners /Invectiven/ nichts schaden, und daß wir Alle, Ihre wissenschaftlichen Freunde, unvermindert Ihre rein wissenschaftlichen Verdienste, wie die großen Verdienste, welche Sie Sich um eine rationelle Popularisirung der Physik erworben haben, zu schätzen wissen. Ich selbst werde übrigens die Herausgabe Ihrer „hours“ zu einer /Vorrede/ benutzen, in der ich meine Ansicht über die betreffenden Punkte sehr ernst /äußern/ werde. Ich glaube, Helmholtz wird etwas ähnliches thun.

So wenig Ihnen der Angriff schadet, der Ihnen hoffentlich keine trüben Stunden bereitet, es ist dazu wirklich kein *[ernster]* Grund vorhanden—, so sehr beklage ich das Buch *[für]* Zöllner, der wirklich ein guter und begabter Mensch *[ist]*, aber offenbar auf falsche Bahnen durch seine speculative Methode der *[Forschung]* und sein etwas abgeschlossenes Leben gekommen ist und nun schadet, statt, wenn er es könnte, zu nützen. Wenn Sie es *[aber]* sich *[erwogen]*, bitte, repliciren Sie Selbst nicht. Ich glaube, die Reue wird bei Zöllner von selbst nicht ausbleiben.

Die Nationalität sollte in der Wissenschaft keine Rolle spielen & ich beklage deshalb mit Ihnen auch sehr, daß Maxwell & auch Thomson sich nicht lassen in der Literatur umsehen *[&]* die ihnen *[1 word illeg]* d.h. englischen Arbeiten zu sehr berücksichtigen. So schreibt u. A. Thomson die Erfindung der Ketten ohne Diaphragmen, die Meidinger ausgeführt, neuerdings Varley zu. Was nicht publicirt ist, gehört nicht der Wissenschaft; sonst *[könnte]* Jeder sagen, er habe vor so und so viel Jahren alles Mögliche gemacht. Diese Verhältniße müßen aufhören; *[&]* wir wollen dazu thun, was wir können. Möge doch mein Galvanismus, wenigstens in der Electricitätslehre, dazu beitragen, die Literaturkenntniße zu *[verbreiten]* und die *[traurigen]* Prioritätsstreitigkeiten zu vermindern.

Mit allerherzlichstem Gruß meiner Frau | in steter Freundschaft | Ihr | G. Wiedemann

Leipzig 10 April 1872.

My dear Tyndall!

I have just returned from a short holiday, and immediately written to Vieweg as to whether he might have neglected to send you my Galvanism.[1] Please, if you received the book, write so to me in a couple of words perhaps, so that I shall be sure that my instructions have been followed

The title „Stunden der Arbeit in den Alpen" for your hours etc . . . has already been printed at the top of the page in the initial 3 sheets; however, in the main title, at least, we can see "In den Alpen"—a somewhat too modest title, as it seems to me.[2] You will agree that because of this minor matter, the initial 3 sheets which have already been run off will not be charged at the full rate.

I deplore Zöllner's book[3] most profoundly as a grave sin against science, which should be free from personalities. Zöllner has written the book completely of his own accord; he *[did not want to heed]* the *[admonitions]* and *[objections of the most serious]* kind which his friends *[made]* to him in occasional comments prior to publication. Anyone who knows Zöllner well, does not doubt that he himself intended the best and was not himself aware of the possible personal tone of his remarks. Accordingly, there is such (unconscious) over-estimation of oneself in the whole undertaking, and the judgments *[in it]* are so one-sided and harsh, that the book is to be disapproved of to the highest degree. How many people do not find pleasure in sheer scandal, how */few/* do not enjoy, out of */2 words illeg/* and less noble motives, when a man of science of high standing, who on some occasion has unintentionally injured this person or that, receives a blow? Even if some of what Zöllner reproaches is not entirely beyond reproach, then good form in science should entail not giving the whole world amusement, even if unintentionally, through scientific gossip. This, incidentally, is not only my judgement of Zöllner's book, but that of many impartial friends, such as Clausius, Bentz,[4] Helmholtz, who of course is also quite unjustly attacked, du Bois Reymond, Kronecker[5] and others.

You will know, dear friend, how to get over the attack on you yourself. If Zöllner wanted to attack and criticise your theory of comets, which I regarded only as a "theory" though, then he should have done so in a special essay on comets. This way, the attack that */mocks/* the reader precisely through its personal treatment is completely in the wrong place. Incidentally, Zöllner's own theory of comets, just as his */just as pious/* theory on earth magnetism, is in the judgement of all the physicists I have spoken to up till now completely untenable. Believe me that Zöllner's */invectives/* will not do you any harm in the opinion of all reasonable men, and that we, your scientific friends, all appreciate your purely scientific services as much as before, as well as the great services which you have rendered to a rational popularising of physics. I myself, incidentally, shall take advantage of the publication of your "hours" for a */preface/* in which I shall */express/* my view about the points in question very seriously. I believe that Helmholtz will do something similar.

As little as the attack—which hopefully will not cause you any hours of gloom, there is really no */serious/* reason for this—will do you any harm, I deplore the book just as much */for/* the sake of Zöllner, who */is/* really a good and talented man, but has obviously got onto the wrong track through his speculative method of */research/* and his somewhat isolated life, and now causes harm, instead of, when he could do so, being useful. If you were */considering/* doing so, */however/*, please, do not reply yourself. I believe */remorse will/* not fail to appear */in Zöllner by/* itself.

Nationality should not play any role in science[6] */&/* I therefore deplore very much with you also that Maxwell */&/* even Thomson do not let themselves have look around in the literature */&/* take the works that */1 word illeg/* them, i.e. the English ones, too much into account. Thomson, among others, thus is lately ascribing the invention of circuits without diaphragms, which was done by Meidinger,[7] to Varley.[8] What is not published does not belong to science; otherwise anyone */could/* say he did everything possible so and so many years ago. These circumstances have to cease; */&/* we shall do what we can about it. May my Galvanism, however, at least within the theory of electricity, contribute to */widening/* knowledge of the literature and to lessening the sorry quarrels over priority.

With the very warmest of greetings from my wife[9] | in constant friendship | Your | G. Wiedemann

RI MS JT/1/W/52

1. *my Galvinism*: see letter 3615, n. 10; for discussion of the copy Wiedemann sent to Tyndall, see letters 3632 and 3640. Tyndall responded to the present letter in 3655.
2. *The title . . . to me*: The book under discussion is *Hours of Exercise in the Alps*; for how to translate the title, see letters 3602, 3615, 3618, 3632 and 3640.
3. *Zöllner's book*: Zöllner attacked Tyndall in his book *Über die Natur der Cometen*. See letter 3640, n. 8. For additional discussion of Zöllner, see letters 3642, 3652, 3655, 3695 and 3697.
4. *Bentz*: not identified.
5. *Kronecker*: Leopold Kronecker (1823–91), a Polish-born mathematician who lived and worked in Germany, though he also had significant international contacts. His research included number theory, the theory of elliptical functions and algebra. He is known for the remark, made in the course of a dispute with another mathematician, 'God Himself made the whole numbers—everything else is the work of men'.
6. *Nationality should not play any role in science*: see Tyndall's remarks on this subject in letter 3640.
7. *Meidinger*: Heinrich Meidinger (1831–1905), a German physicist who at the time was a professor of applied physics at the college of technology in Karlsruhe; he did work on galvanic batteries. He describes a type of galvanic battery that lacked a diaphragm in his paper 'Ueber eine völlig constante galvanische Batterie', *Annalen der Physik und Chemie*, 184:12 (1859), pp. 602–10.
8. *Varley*: Cromwell Fleetwood Varley (1828–83), an English engineer who was involved with the laying of the first transatlantic cables and who formed a partnership with William Thomson and Fleeming Jenkin. Varley was elected FRS in 1871. In 1868, Varley wrote to Tyndall regarding his belief in spiritualism (see letter 2822, *Tyndall Correspondence*, vol. 10). For Jenkin, see letter 3591, n. 6.
9. *my wife*: Clara Wiedemann.

From Jane Barnard 11 April 1872 3650

Barnsbury Villa, | 320 Liverpool Road, | N.

My dear Dr Tyndall

Many thanks to you for your kind remembrance & the tickets[1] which are very acceptable. I wish I could use one of them myself, but my Aunt[2] must be my first consideration & I cannot leave her. We have been speaking of you many times lately & wondering when we should have the pleasure of seeing you here & how we must not think of it, for I suppose, nine weeks to come. You have a heavy task before you & we fear from what you say that you are not feeling very equal to it.

We are in trouble, having had two losses among our near friends[3] this week; this & a heavy cold will prevent my being at the F. E.[4] as I had hoped.

My Aunt desires her very kind regards & with the same I am | yours very sincerely | Jane Barnard | 11 April 1872 | Dr Tyndall

RI MS JT/1/B/50

1. *the tickets*: for Tyndall's lectures on 'Heat and Light' at the RI; see letter 3639, n. 4.
2. *my Aunt*: Sarah Faraday.
3. *two losses among our near friends*: not identified.
4. *F.E.*: presumably an abbreviation for 'Friday Evening', that is, the Friday Evening Discourse at the RI. On Friday, 12 April 1872, John Morley spoke 'On the Influence of Rousseau'. See J. Morley, 'On the Influence of Rosseau', *Roy. Inst. Proc.*, 6 (1870–72), pp. 475–6.

From Jules Jamin[1] 12 April 1872 3651

Paris 12 avril 1872

Mon cher ami

Je vous envoie ci joint le reçu de la somme que vous m'avez autrefois envoyée pour les blessés de la dernière et malheureuse guerre. J'ai attendu jusqu'à présent ne voulant confier cette pièce qu'à des mains sûres.

La personne qui vous la remettra est un de nos amis les plus intimes, Mr Alluard Professeur à Clermont-Ferrand qui est chargé de construire un observatoire au sommet du Puy-de-Dôme. (Cette entreprise doit vous sourire.) Mr Alluard va en Angleterre pour *[étudier]* l'installation de vos grands établissemens. Je vous demande de vouloir bien l'accueillir comme un autre moi même et de lui faciliter les moyens de remplir ses missions.

Viendrez-vous bientôt à Paris ? Ma femme et moi nous serions bien heureux de vous y voir et vous nous ferez un bien grand plaisir en prenant votre logement à la maison. Faites-le je vous prie et croyez moi toujours

Votre ami dévoué | J. Jamin

Paris, 12 April 1872

My dear friend,

I am sending you with this letter the receipt for the sum that you sent me earlier for the wounded in the previous, unfortunate war.[2] I have waited until now, wishing to entrust this piece only to safe hands.

The person who will bring this to you is one of our closest friends, Mr Alluard,[3] Professor at Clermont-Ferrand,[4] who is responsible for the construction of an observatory at the summit of Puy-de-Dôme.[5] (This endeavour should delight you.) Mr Alluard is going to England to *[study]* the installation of your impressive facilities. I would ask you to kindly welcome him as you would welcome me, and to assist him with the means to fulfil his goals.

Will you be in Paris soon? My wife[6] and I would be very happy to see you, and you would give us great pleasure by staying at our home. Please do so, and believe me, always

Your devoted friend | J. Jamin

RI MS JT/1/J/16

---

1. *Jules Jamin*: Jules Jamin (1818–86), a French physicist and professor at the École Polytechnique and the Sorbonne; Jamin did important work in a range of subjects, including optics, magnetics and electricity.
2. *I am . . . war*: Tyndall contributed money to care for wounded soldiers on both sides of the Franco-Prussian War of 1870–1. See letter 3382, and *Ascent of John Tyndall*, pp. 267–8.
3. *Alluard*: Pierre-Jules-Émile Alluard (1815–1908), a French physicist and professor of physics and chemistry. Alluard's most notable work was the construction of a meteorological observatory on the Puy-de-Dôme; see n. 5.
4. *Clermont-Ferrand*: university in Central France.
5. *an observatory at the summit of Puy-de-Dôme*: Alluard began work on establishing the observatory in 1869 and it opened in 1876. Jamin himself may have been the source of the idea. See B. Brunhes, 'Notices sur les membres décédés', *Association amicale de secours des anciens élèves de l'École normale supérieure*, 10 January 1909, pp. 8–13, on pp. 10–11.
6. *My wife*: Thérèse Josephine Eudoxie Jamin (née Lebrun, 1832–80).

To Hermann Helmholtz 13 April 1872[1] 3652

London 13th. April 1872

My dear Helmholtz,

A few days ago our librarian[2] sent up two books to me to ascertain my opinion as to their being purchased for our library; the one was a work by Heer[3] of Zurich, the other a volume by Zöllner[4] of Leipzig. I had heard of this latter work from Hirst, who, during a recent visit to Paris, had found it on Bertrand's[5] table. Of Zöllner personally I know nothing, of his labours very little; but I had got the impression that he was a modest hardworking man who had done some good work in photometry and with the spectroscope.

When I turned over the leaves of his book I could hardly credit my senses; the discharge of animosity against me was so virulent, and the cause of it all so inadequate. From Germany I had small expectation of such an attack as this. I did not read all that Zöllner said regarding me, but handed the volume over to Bence Jones, who informs me that it is "savage and venomous" to the last degree.

Zöllner's indignation appears to have been aroused by a little hypothesis of mine regarding comets, which I threw out before the Cambridge Philosophical Society.[6] The history of the matter is this:—The secretary of the Society[7] had sent me an invitation pressing me earnestly to visit them and to make a communication to the Society. I replied that I had no time to prepare an original paper, but offered, if they wished, to repeat before them some of my experiments on the decomposition of vapours by light. The secretary assented, so I went to Cambridge, made the experiments, and spent five minute afterwards in the statement of the hypothesis which has borne such unexpected fruit in Zöllner's book. Adams,[8] Stokes, Cayley[9]—in fact all the scientific men of the University were present at the lecture, and I expressly invited them to tear the hypothesis to pieces if it contained any thing unworthy of enunciation. They did not do so, but received it with a favour greater than I had expected. Miller,[10] our foreign secretary, a most sagacious old fellow, as you know, came to me and said "I have been trying to pick a hole in your hypothesis but cannot do so". It was afterwards discussed by the Professors and the under graduates at the Secretary's rooms. Airy and Herschel both wrote to me about it, saying that it assuredly gave astronomers a great deal to think of. How it is that an idea that has run the gauntlet of so many men of the first eminence could have so maddened Zöllner passes my understanding.

But I dare-say, though I do not yet know it, that Zöllner objects to all that I have written with a view to public instruction. His wrath, I may remark, is a small matter to me when weighed against the fact that <u>you</u> have thought those popular books worthy of reproduction in Germany.[11] They took form in this way:—You know our audience at the Royal Institution, and how necessary it is to prepare both reasoning and experiment for people who though highly cultivated in a general way, are only partially scientific. Well, I gave myself some trouble in preparing my lectures, and for some years had been pressed by Longman[12] to publish them—I declined, and had it remained a publisher's question I should have continued to refuse to turn aside from my investigations to write such books. But at the Athenaeum Club[13] I had the opportunity of meeting and conversing with some of the most highly cultivated men in England, and hearing them speak about the state of science in this country. Among them I may signalise the late Sir Edmund Head,[14] once Governor General of Canada, for his words roused me more than any others. He spoke to me repeatedly of the deplorable ignorance and apathy of the so-called cultivated classes of England regarding science. He did not want to see these classes profoundly conversant with science, but he wished to see them competent to sympathise with it in an intelligent manner. These sentiments, often and earnestly repeated, coupled with the fact of my having already invested a good deal of labour in the subject, caused me to publish the Lectures on Heat.[15] No one knows their imperfections better than I do myself, but they did good and the knowledge of this caused me to add Sound[16] to Heat. This is the brief history of what I have done in the popular way. It was done to evoke sympathy for science among classes of persons whom our Universities had left wholly in the dark regarding science.

How this action could appear culpable to any sane mind passes my comprehension. I know as well as Zöllner, perhaps better, that it is to original thinkers, who work in the quiet of their own homes or cabinets, that science is to look for its growth and development, and I have tried to act practically on this conviction by throwing my private mite[17] into the fund of scientific knowledge. Of the proofs of Zöllner's fitness to sit in judgment on my work I know nothing. But it is probable that his self-conceit has caused him to commit an outrage upon men who have given proofs of devotion to science far greater and more varied than it has ever been in their assailant's power to offer.

Believe me dear Helmholtz | Yours ever faithfully | John Tyndall.

RI MS JT/1/T/489

---

1. *Helmholtz*: Tyndall sent this letter on 18 June 1872, accompanied by a short covering letter explaining that he had set it aside, 'not wishing to trouble you with a matter unworthy of your attention', but on second thoughts decided to send it after all (RI MS JT/1/TYP/2/506). This letter will be published in the forthcoming thirteenth volume of *The Correspondence of John Tyndall.*
2. *our librarian*: Benjamin Vincent (1818–99), librarian at the RI from 1849–89; before being appointed at the RI, Vincent worked as a translator and editor. He was, like Faraday, a Sandemanian.
3. *Heer*: Oswald Heer (1809–83), a Swiss geologist and naturalist whose work focused on paleobotany, plant geography and entomology. Heer was a professor of botany at the University of Zurich. The book may be O. Heer, *Fossile Flora der Bären Insel* (Stockholm: Norstedt, 1871).
4. *a volume by Zöllner*: see letter 3640, n. 8. For additional discussion of this issue, see letters 3642, 3649, 3655, 3695 and 3697.
5. *Bertrand's*: Joseph Louis François Bertrand (1822–1900), a French mathematician who held posts at the École Polytechnique and the Collège de France; he also became secretary at the Académie des Sciences and joined the Académie Française. Bertrand's mathematical work included the theory of curves and surfaces, differential equations and probability. He also worked in theoretical physics, and he published textbooks and popular articles. Hirst recorded seeing him in Paris in February 1872. See Brock and MacLeod, *Hirst Journals*, p. 1928.

6. *Cambridge Philosophical Society*: scientific society founded at Cambridge University in 1819 (see S. Gibson, *The Spirit of Inquiry: How One Extraordinary Society Shaped Modern Science* (Oxford: Oxford University Press, 2019)). The lecture was also published; see letter 3640, n. 9.
7. *The secretary of the Society*: not identified.
8. *Adams*: see letter 3407, n. 1.
9. *Cayley*: Arthur Cayley (1821–95), a mathematician and Sadleirian professor at Cambridge; Cayley also did work in mathematical astronomy and was elected FRS in 1852.
10. *Miller*: William Hallowes Miller (1801–80), a mineralogist who established an important system of crystallography and who also did significant work on standards of length and weight. Miller was elected FRS and served as the RS's foreign secretary from 1856 to 1873.
11. *you have . . . in Germany*: Helmholtz was involved with the publication of several of Tyndall's books in Germany, including *Heat Considered as a Mode of Motion* (London: Longmans, Green, and Co., 1863) and *Sound* (1867). See *Ascent of John Tyndall*, p. 207. For Helmholtz's involvement in the publication of *Fragments of Science*, see letter 3432.
12. *Longman*: London-based publisher for fourteen of Tyndall's sixteen books.
13. *Athenaeum*: see letter 3372, n. 14.
14. *Edmund Head*: Edmund Walker Head (1805–68), the governor-in-chief of British North America from 1854 until 1861. He had a wide range of intellectual interests, including art, political history, language, philology and ballads. In 1863 he was elected FRS.
15. *Lectures on Heat*: see n. 11. *Heat* was published in Germany as J. Tyndall, *Die Wärme, Betracht al seine Art der Bewegung* (Braunschweig: Vieweg, 1867).
16. *Sound*: *Sound* (1867).
17. *my private mite*: an allusion to the Biblical story of the widow's mite; see Mark 12:42–44, Luke 21:1–4.

From [Joseph Dalton Hooker][1] 15 April 1872 3653

Royal Institution | April 15th 1872.

Mr Helps is officially instructed by Lord Ripon to inform Dr Hooker that:—
"Mr Ayrton has been told that Dr Hooker should, in
"all respects, be treated as the Head of the local
"establishment at Kew; of course in subordination
"to the First Commissioner of Works"[2]

RI MS JT/1/TYP/8/2619
LT Typescript Only

1. *From [Joseph Dalton Hooker]*: A note by LT on this letter reads, 'This is copied by Tyndall on Royal Institution paper'. The original is a statement made to Hooker by Arthur Helps. We believe that Hooker may have written to Tyndall quoting from this letter. For the

conflict over Hooker's position at Kew, see letters 3630, 3633, 3634, 3638 and 3645. This message may also be referred to in letter 3658.

2. *the First Commissioner of Works*: the head of the Board of Works, the office of the British government concerned with public works; at this time, Ayrton.

From Edward Stanley 15 April 1872 3654

23 St James's Square | April 15 1872.

My dear Sir

I am again in town,[1] and will at once look into Dr Hooker's case.[2] I return his letter[3] enclosed in yours,[4] and remain,

Very faithfully yours | Derby. | Professor Tyndall.

RI MS JT/1/TYP/8/2618
Transcript Only

---

1. *I am again in town*: see letter 3643.
2. *will at once look into Dr Hooker's case*: for the results, see letter 3659.
3. *his letter*: probably letter 3647. Derby's letter appears as a handwritten transcript at the bottom of LT's transcription of 3647.
4. *in yours*: letter missing.

To Gustav Wiedemann 16 April 1872 3655

16th April 1872. | Tyndall

My dear Wiedemann

Your book[1] reached me, leaving evidence on every page of it of the labour and the learning with which it is executed. Surely, you have accomplished an invaluable work.

Ten years ago when my blood ran warmer than it now does, I might have replied to Zöllner[2] but not now. I have not read his entire book, but have seen enough of it to learn the nature of the attack which he has made upon me. I handed the book over to Bence Jones, who informs me that it contains things far worse than those that I had seen, and that it is "savage and venomous" to the last degree.

The only answer that I thought of writing was a private one to Zöllner himself; for he aroused a certain sympathy in me by the manner in which he spoke of Bunsen,[3] Wilhelm Weber[4] and others. But his conduct has been too outrageous to permit of my having any communication with him.

I can quite understand the frame of mind which prompted his assault. He has allowed his fancy to brood upon me until it has developed an image that exasperates him, and then he attacks the product of his own diseased imagination. He knows nothing of my life here; and that neither Bunsen in Heidelberg,[5] not Weber in Gottingen,[6] leads a simpler existence than I do in the heart of London with every possible opportunity of social enjoyment open to me.

Nor have I allowed the noisier doings of science itself to deflect me from my work. I decline habitually to attend public meetings—I have declined even the presidency of the British Association, though it has been twice offered to me by the Council.[7] For the sake of quietly pursuing my work I resigned my lectureship at the School of Mines, and my examinership at the University of London and at the various military colleges.[8] The remark of Zöllner's caught my eye regarding "lucrative public lectures." Foolish man! Three months work next autumn[9] would enable me to put Fifty thousand Thalers[10] in my purse, but I have declined to do the work on any such terms; and if I do it at all, it shall be done without a penny profit to me.

These popular books of mine were written at the prompting of men of the highest culture who deplored the utter absence of all sympathy with science on the part of powerful classes in this country.[11] And I am happy to believe they have fulfilled the need for which they were intended.

I have now said as much upon this subject as I intend to say. Zöllner has done his work—I mine; let the future judge between us.

Give my very best regards to your excellent wife.[12]

ever my dear Wiedemann | Yours | John Tyndall

EH FAM123

1. *your book*: Wiedemann's recent book on galvanism; see letter 3615, n. 10. For letters about the copy that Wiedemann sent to Tyndall, see letters 3632, 3640 and 3649. The present letter responds to letter 3649.
2. *Zöllner*: Zöllner attacked Tyndall in his book *Über die Natur der Cometen*. See letter 3640, n. 8. See also letters 3642, 3649, 3652, 3695 and 3697.
3. *Bunsen*: Robert Bunsen (1811–99), a German chemist whose work focused on experimentation and who invented or improved on various essential pieces of chemical equipment, including the Bunsen burner. Bunsen was a professor at the University of Marburg from 1838 until 1852, where Tyndall studied with him; see *Ascent of John Tyndall*, p. 44.
4. *Wilhelm Weber*: Wilhelm Eduard Weber (1804–91), a German physicist whose work on electricity and magnetism was especially significant.
5. *Heidelberg*: the University of Heidelberg, where Bunsen was a professor from 1852 until his death.

6. *Gottingen*: the University of Göttingen, where Weber held the physics chair from 1849 until his death.
7. *I have declined . . . by the Council*: In this volume, Tyndall declined the BAAS presidency for the meeting in Bradford in 1873. See letters 3627 and 3628.
8. *I resigned . . . Colleges*: Tyndall resigned from the School of Mines in the spring of 1868; we are not certain of the dates when he resigned the examinerships mentioned.
9. *Three months work next autumn*: presumably a reference to Tyndall's planned trip to America, for which he had insisted he receive nothing beyond his expenses; see, for example, letter 3639.
10. *Thalers*: 'a German silver coin; a dollar' (*OED*).
11. *These popular . . . country*: see letter 3652.
12. *your excellent wife*: Clara Wiedemann.

To Louisa Baring[1] 17 April [1872][2] 3656

17th April

My dear Friend.

Professor Frankland informs me that one of his students or assistants Mr. Ambrose Fleming[3] the son of a clergyman,[4] and he assures me "a nice fellow" would be willing to give your "May"[5] lessons. His address is

Ambrose Fleming Esq | Chemical Laboratory | Museum | South Kensington.[6]

But Mr. Fleming I fear will not be at liberty for 5 days of the week before 5 P.M.

There are no elementary lectures now going on at South Kensington—& I know not where such are given.

Always yours | John Tyndall

NLS, Ashburton Papers, Acc.11388 N0.101

---

1. *Louisa Baring*: see letter 3436, n. 1.
2. *[1872]*: We have assigned this year because Ambrose Fleming (see n. 3) began studying with Frankland in 1872.
3. *Ambrose Fleming*: John Ambrose Fleming (1849–1945), who became an electrical engineer, had been teaching science at the Rossall School in Lancashire and had recently moved to London to work with Frankland.
4. *a clergyman*: James Fleming (1816–79).
5. *your "May"*: Mary Baring; see letter 3436, n. 1.
6. *South Kensington*: center for scientific research and education in London.

From Joseph Dalton Hooker 17 April 1872 3657

Athenaeum[1] | April 17th 1872.

Dear Tyndall

I have consulted with Lubbock, Huxley, and Bentham with whom the general feeling is, that now it is the time that the affair[2] should be taken out of my hands; & these men are all willing and ready to consider what next is best done and to meet at the Athenaeum at 10.0 a.m. on Wednesday for the purpose.

Helps will get L[or]d Ripon's formal approval of his signature to the paper containing the message from the Cabinet to me,[3] and I will send it to you for communication to Lord Derby.

Meanwhile should you be able to see L[or]d Derby pray do, and learn his views.[4] Helps is most decidedly unfavourable to my addressing Mr Gladstone myself, or taking any further step.

Ever Sinc[ere]ly yours | J. D. Hooker.

I must confess that I feel a little sore at my own defeat as a scientific man on a scientific proposition and greatly regret having to hand my weapon to others; but I have fought for the position my father[5] made and left me to defend, and shall still always be to the fore.

RI MS JT/1/TYP/8/2620
LT Typescript Only

1. *Athenaeum*: see letter 3372, n. 14.
2. *the affair*: Hooker's conflict with Ayrton over the management of the botanical garden at Kew; see such letters as 3630, 3633, 3634, 3638, 3645, 3653, 3654, 3658 and 3659.
3. *message from the Cabinet to me*: possibly the message conveyed in letter 3653.
4. *should you . . . his views*: for a response to Hooker's request that Tyndall see Derby, see letter 3658.
5. *my father*: William Hooker.

To Joseph Dalton Hooker 18 April [1872][1] 3658

Royal Institution of Great Britain | 18th April.

My dear Hooker,

You will I am sure read the enclosed with pleasure[2]—Lord Derby is a brick.

We must enact some plan of action.[3]

Yours ever | John Tyndall

Keep your heart up my dear fellow. The nation would bear your ordeal if they only knew your case.[4]

RBG, Kew Papers Relating to Kew 1867–1872, Ayrton Controversy, volume 1, f. 149

---

1. *[1872]*: dated by reference to letters 3659 and 3661.
2. *You will . . . pleasure*: The enclosed is letter 3659. For Hooker's response, see letter 3661. The present letter may respond to Hooker's request, in letter 3657, that Tyndall consult with Derby.
3. *We . . . of action*: This comment may anticipate a meeting at the Athenaeum Club the following week; see letters 3665, 3666, 3667, 3668 and 3671.
4. *Keep . . . your case*: On the back of this letter appear the following sentences, very faint and difficult to read: 'I am unofficially informed that though Mr Helps was *[un]*officially | Mr Helps assumed it was official *[2–3 words illeg]* informed that it was not so | I am so informed that the *[*message*]* from *[*Ripon*]* was unofficial—'. It is unclear whether these are notes made by Tyndall or by Hooker, or if they were part of the original letter. They may have to do with the message in letter 3653.

From Edward Stanley 18 April 1872 3659

23 St James's Square | April 18th/72.

Private

Dear Professor Tyndall

I have gone carefully through the papers,[1] in Dr Hooker's case, and think he has a strong prima facie[2] ground of complaint. More of course cannot be said until one has heard the other side. I could move for papers,[3] which would bring the whole matter before the public: but inasmuch as to publish the quarrel in this way would be to make it impossible for Dr Hooker and Mr Ayrton to continue in their respective positions, and as it is quite probable that our rulers might not choose to remove Mr Ayrton (colleagues naturally stand by one another as long as they can) I think all peaceable means ought to be tried in the first instance. Finding that Lord Ripon was one of the committee of the Cabinet to whom the matter in dispute was referred, I took the opportunity of speaking to him; and he tells me that some communications, intended to be conciliatory,[4] have been addressed to Dr H[ooker]. (as I understand) from Mr Gladstone. This being so, I will take no step at present, awaiting the result of the overture which I understand to have been made. If it fails, I shall be ready and willing to move for the correspondence: which would be the foundation of any subsequent motion or debate that might take place. But if an amicable arrangement is practicable, I think that would be the wiser course.[5]

Believe me | Very faithfully yours | Derby.

Prof. Tyndall.

RI MS JT/1/TYP/8/2621
LT Typescript Only

---

1. *the papers*: perhaps Hooker's account of his conflict with Ayrton; see letter 3638, n. 2. See also letters 3643, 3647 and 3654.
2. *prima facie*: 'at first sight' (Latin).
3. *I could move for papers*: see letter 3633, n. 3.
4. *some communications, intended to be conciliatory*: probably a reference to the message that appears in letter 3653.
5. *if an amicable . . . wiser course*: Tyndall forwarded this letter to Hooker (see letter 3658); Hooker responded in letter 3661.

To Granville George Leveson-Gower[1] 19 April 1872 3660

ANGLO-AMERICAN COPYRIGHT.

The following Memorial of British authors on the subject of copyright in the United States of America was sent to the Right Hon[ourable]. Earl Granville[2] on Friday, April 19, by Mr. William Gilbert,[3] the hon[orary]. secretary:

"Looking forward with satisfaction to the prospect of harmonious relations being happily established between the United States and the United Kingdom, we, the undersigned, hope for a reconsideration of the policy in virtue of which authors, as authors, enjoy no rights which American citizens are bound to respect.[4] Letters from influential Americans—one of them a leading New York publisher—which have recently appeared here,[5] joined with the approval of them expressed in the journals of the United States, show the desire of the Americans for the conclusion of a Copyright Convention between their country and ours. We understand that the demands of publishers in this country have hitherto been the most formidable obstacles to the negotiation of a Copyright Convention. We are of opinion that the interests of our publishers in American copyrights are quite independent of the just claims of British authors, and that the latter may be fully admitted without recognition of the former. We think it would be a grave error if the settlement of this matter were retarded, or rendered impossible, in consequence of two classes of claims, which, in essence, are wholly distinct, if not antagonistic, being regarded by negotiators representing this country as identical and inseparable. Americans distinguish between the author, as producing the ideas, and the publisher, as producing the material vehicle by which these ideas are conveyed to readers. They admit the claim of the British author to be paid by them for his brain work. The claim of the British book manufacturer to a monopoly of their book market they do not admit. To give the British author a copyright is simply to agree that the American publisher shall pay him for work done. To give the British publisher a copyright is to open the American market to him on terms which prevent the American publisher from competing. Without dwelling on the argument of the Americans that such an arrangement would not be free trade, but the negation of free trade, and merely noticing their further argument, that while their protective system raises the prices of all the raw materials,[6] free competition with the British book manufacturer would be fatal to the American book manufacturer, it is clear that the Americans have strong reasons for refusing to permit the British publisher to share in the copyright which they are willing to grant to the British author. We venture to suggest, therefore, that, responding to the cordial feeling, recently expressed by Americans on the subject, and duly appreciating the force of their reasons for making the above distinction, future negotiations should be conducted with a view to secure a copyright on the conditions they specify. Without making it the foundation of a formal claim for reciprocity of treatment, we mention the fact that American authors may, if they please, secure all the advantages of copyright in the United Kingdom which are enjoyed by native authors.

(Signed) The Rev. Edward Henry Bickersteth,[7] Sir John Lubbock, Bart., M.P., Thomas H. Huxley, F.R.S., Thomas Hughes, M.P.,[8] the Rev. Dr. T. Guthrie,[9] the Rev. W[illia]m. Arthur,[10] Philip James Bailey,[11] J. E. Hilary Skinner,[12] J. B. Leicester Warren,[13] the Rev. Harry Jones,[14] the Rev. J. Martineau,[15] Andrew Edgar, LL.D.,[16] W. Durrant Cooper, F.S.A.,[17] Joseph Dalton Hooker, M.D., F.R.S., Harriet Martineau,[18] John Percy, M.D., F.R.S.,[19] George Scharf, F.S.A.,[20] Thomas Blizard Curling,[21] George Augustus Sala,[22] Robert Buchanan,[23] Henry Labouchere,[24] Augustus W. Franks, Vice-President Society of Antiquaries,[25] Richard A. Proctor,[26] Blanchard Jerrold,[27] Herbert Spencer, G. H. Lewis,[28] John Stuart Mill,[29] Helen Taylor,[30] J. A. Froude, T. Carlyle, John Ruskin,[31] John Morley,[32] W. F. Rae,[33] C. Tabor,[34] Eliza Tabor,[35] William Black,[36] Edward Dicey,[37] James Caird,[38] William Gilbert,[39] Sheldon Amos,[40] Edwin Pears,[41] James Paget,[42] Archibald Forbes,[43] J. C. Parkinson,[44] W. T. M'Cullagh Torrens, M.P.,[45] Charles Darwin, Erasmus Wilson,[46] Shirley Brooks,[47] Tom Hood,[48] Justin M'Carthy,[49] John Tyndall, F.R.S."

London Daily News, 6 May 1872, p. 6[50]

---

1. *Granville George Leveson-Gower*: Granville George Leveson-Gower (1815–91), second Earl Granville, a Whig politician. He was the eldest son of Granville Leveson-Gower (1773–1846), first Earl Granville, and Lady Henrietta Elizabeth Cavendish (1785–1862). Educated at Eton and Christ Church, Oxford (1832–6), he later became MP for Morpeth (1837–40). Under the Gladstone administration, Granville held several positions, including leader of the Liberal Party in the Lords (1866), colonial secretary (1868–70) and foreign secretary (1852–4, 1870–4, 1880–5).
2. *Right Hon[ourable]. Earl Granville*: see n. 1.
3. *Mr. William Gilbert*: possibly William Gilbert (1804–90), author of novels and nonfiction; Gilbert also wrote extensively for periodicals. His oldest son was the dramatist William S. Gilbert of Gilbert and Sullivan. We have not, however, been able to trace the senior Gilbert's involvement in international copyright issues, although a novel of his was plagiarized by a contemporary playwright in 1869 (J. W. Stedman, '"A Peculiar Sharp Flavour": The Contributions of Dr. William Gilbert', *Victorian Periodicals Review*, 19:2 (1986), pp. 43–50, on p. 44).
4. *a reconsideration . . . respect*: for Tyndall's involvement in the issue of international copyright, see letter 3484, especially n. 1, and letter 3545.
5. *Letters . . . appeared here*: The fall of 1871 saw a discussion in the British press of the copyright issue, including a series of letters in the *Times* under such headings as 'American Pirates' and 'International Copyright'. The 'leading New York publisher' is probably W. H. Appleton; see W. H. Appleton, letter to the editor, *Times*, 20 October 1871, p. 10.
6. *their protective system raises the prices of all the raw materials*: Protectionism dominated American trade policy in the postbellum era (M. W. Palen, 'Foreign Trade Policy from the Revolution to World War I', *Oxford Research Encyclopedias: American History* (22 November 2016), at https://doi.org/10.1093/acrefore/9780199329175.013.361 (accessed 24 May 2021)).
7. *Rev. Edward Henry Bickersteth*: Edward Henry Bickersteth (1825–1906), a bishop in the Church of England. He was Vicar of Christ Church, Hampstead from 1855–85, then Bishop of Exeter.
8. *Thomas Hughes, M.P.*: Thomas Hughes (1822–96), a British lawyer, social reformer and children's author. He served as MP for Lambeth from 1865–8 and for Frome from 1868–74.
9. *Rev. Dr. T. Guthrie*: Thomas Guthrie (1803–73), a Scottish minister and philanthropist.
10. *Rev. W[illia]m. Arthur*: William Arthur (1819–1901), an Irish minister, missionary and author.
11. *Philip James Bailey*: Philip James Bailey (1816–1902), a British poet.
12. *J. E. Hilary Skinner*: John Edwin Hilary Skinner (1839–94), a British author and journalist.
13. *J. B. Leicester Warren*: John Byrne Leicester Warren (1835–95), a British poet.

14. *Rev. Harry Jones*: possibly Harry Jones (1823–1900), author, broad churchman and vicar of St Luke's Berwick Street in London from 1852–72.
15. *Rev. J. Martineau*: James Martineau (1805–1900), a British Unitarian minister.
16. *Andrew Edgar, LL.D.*: Andrew Edgar (1831–90), a Scottish minister and author.
17. *W. Durrant Cooper, F.S.A.*: William Durrant Cooper (1812–75), a British antiquary and Fellow of the Society of Antiquaries of London.
18. *Harriet Martineau*: Harriet Martineau (1802–76), a British writer and journalist.
19. *John Percy, M.D., F.R.S.*: John Percy (1817–89), a British physician, metallurgist and professor at the Royal School of Mines.
20. *George Scharf, F.S.A.*: George Scharf (1820–95), a British artist, gallery director and Fellow of the Society of Antiquaries of London.
21. *Thomas Blizard Curling*: Thomas Blizard Curling (1811–88), a British surgeon.
22. *George Augustus Sala*: George Augustus Henry Fairfield Sala (1828–95), a British author and journalist. He wrote for the *Illustrated London News* and the *Daily Telegraph*.
23. *Robert Buchanan*: either Robert Buchanan (1785–1873), a Scottish minister, logician and playwright; or Robert Buchanan (1802–75), a Scottish minister and historian.
24. *Henry Labouchere*: Henry Du Pré Labouchere (1831–1912), a British journalist, publisher and politician.
25. *Augustus W. Franks, Vice-President Society of Antiquaries*: Augustus Wollaston Franks (1826–97), a British antiquary and museum administrator.
26. *Richard A. Proctor*: Richard Anthony Proctor (1837–88), a British astronomer and science writer.
27. *Blanchard Jerrold*: William Blanchard Jerrold (1826–84), a British journalist and playwright.
28. *G. H. Lewis*: possibly George Henry Lewis (1833–1911), a British lawyer, but we think this is more likely a misprint for G. H. Lewes; for the latter see letter 3448, n. 1.
29. *John Stuart Mill*: see letter 3617, n. 6.
30. *Helen Taylor*: Helen Taylor (1831–1907), a women's rights activist and Mill's stepdaughter; Taylor collaborated with Mill on issues including female suffrage.
31. *John Ruskin*: see letter 3605, n. 4.
32. *John Morley*: John Morley (1838–1923), a politician, journalist and author. He edited major Victorian periodicals including the *Fortnightly Review*, the *Pall Mall Gazette* and *Macmillan's Magazine*. Morley entered Parliament in 1883, where he worked on the question of Home Rule for Ireland with William Gladstone, whose biography he later wrote.
33. *W. F. Rae*: William Fraser Rae (1835–1905), a Scottish author.
34. *C. Tabor*: not identified, though possibly a relative of Eliza Tabor (see n. 35).
35. *Eliza Tabor*: Eliza Tabor, later Eliza Tabor Stephenson (1835–1914), a British novelist.
36. *William Black*: William Black (1841–98), a British journalist and novelist.
37. *Edward Dicey*: Edward James Stephen Dicey (1832–1911), a British author and journalist.
38. *James Caird*: James Caird (1816–92), a Scottish agriculturist, politician and writer.
39. *William Gilbert*: see n. 3.
40. *Sheldon Amos*: Sheldon Amos (1835–86), a British jurist and author.

41. *Edwin Pears*: Edwin Pears (1835–1919), a British lawyer and historian.
42. *James Paget*: James Paget (1814–99), a British surgeon and pathologist.
43. *Archibald Forbes*: Archibald Forbes (1838–1900), a Scottish journalist.
44. *J. C. Parkinson*: Joseph Charles Parkinson (1833–1908), a British journalist, civil servant and social reformer.
45. *W. T. M'Cullagh Torrens, M.P.*: William Torrens McCullagh Torrens (1813–94), an Irish politician and author. He was MP for Dundalk from 1848–52.
46. *Erasmus Wilson*: William James Erasmus Wilson (1809–84), a British surgeon, dermatologist and author.
47. *Shirley Brooks*: Charles William Shirley Brooks (1816–74), a British journalist and playwright.
48. *Tom Hood*: Thomas Hood (1835–74), a British humorist and journal editor.
49. *Justin M'Carthy*: Justin McCarthy (1830–1912), an Irish politician and historian.
50. This letter was also published in the *New York Times*, 20 May 1872, p. 8.

From Joseph Dalton Hooker [19 April 1872][1] 3661

Royal Gardens, Kew. | April 19th 1872.

My dear Tyndall

This is an excellent letter.[2] Perhaps the best plan would be that Lord Derby should see Mr Helps & and ask him whether he thinks that Dr Hooker has lost sight of any means of bringing this matter to an amicable issue. All I can say is, that I have refused offers from both sides of the House to bring it before the House of Commons—that I have refused repeated applications for information from the public journals, Horticultural & general,—that I have restrained deputations Scientific, Horticultural and Tradesmens (Nurserymen) from going to Mr Gladstone on the subject & that I have refrained from appealing to him through my near relations[3] who are amongst his most intimate friends.

Lubbock has copies of my correspondence with Mr Gladstone.—I am now asking him to send them to you for your perusal if you have time—& for you to forward to Lord Derby for his fuller information.[4] Lord D[erby]. can judge from them what chance there is of my ever getting on with Mr Ayrton & it would be well that he should also know how, through his mischief-making, my relations with officers of the Board[5] have become complicated.

I have been with Mr Helps this morning: he has seen Lord Ripon again who now declines to allow the communication he sent to me[6] to be regarded as official at all! Mr Helps sticks to his text; that he received it believing it to be an official communication, that he told me it was so—that he was further informed by L[or]d Ripon that it was final, & sent by Mr Gladstone himself. The fact is, that now knowing that it is unsatisfactory, the Cabinet endeavours to withdraw from it—& I shall get another kind of *[reply]* or letter from Mr G[ladstone]. Indeed Mr Helps informs me that Mr G[ladstone]. is to write but he may change his mind & I shall be told that no one is responsible *[for]* this information either. There is but one way out of it that is the removal of Kew from the Board of Works, where it should not remain.

Ever yours | J D Hooker

RBG, Kew Papers Relating to Kew 1867–1872, Ayrton Controversy, volume 1, f. 151
RI MS JT/1/TYP/8/2622

---

1. *[19 April 1872]*: date assigned by LT. The handwritten letter is very difficult to read, and the date is especially hard to make out, but given that this letter responds to letter 3658 and 3659, we believe that LT is correct.
2. *an excellent letter*: letter 3659, which Tyndall had forwarded to Hooker (see letter 3658).
3. *my near relations*: not identified.
4. *for you to forward to Lord Derby for his fuller information*: see letter 3665.
5. *the Board*: the Board of Works.
6. *the communication sent to me*: probably the message presented in letter 3653.

From Charles d'Almeida[1] 20 April 1872 3662

JOURNAL DE PHYSIQUE | THEORIQUE ET APPLIQUEE |
31 rue Bonaparte, Paris | Paris, le 20 Avril 1872

Cher Monsieur,

Je vous adresse le troisième numéro de notre Journal où vous trouverez la traduction de la lettre que vous avez bien voulu nous écrire pour nous souhaiter la bienvenue. Vos souhaits nous ont été des plus agréables et dès à présent le succès que vous nous avez prédit est, on peut dire, assuré.

Croyez bien que nous seront toujours heureux de publier *[les]* premiers vos ingénieuses expériences et agréez nos remerciements et pour le passé et pour l'avenir.

J'espère que nous aurons à vous les renouveler souvent et suis votre tout dévoué et respectueux serviteur.

Ch. d'Almeida

JOURNAL OF PHYSICS | THEORETICAL AND APPLIED |
31 Bonaparte Street, Paris | Paris, 20 April 1872

Dear Sir,

I am sending you the third issue of our journal[2] where you will find the translation of the letter[3] that you were willing to write us to welcome us.

Your wishes have been most appreciated and from now on the success that you have predicted is, we can say, assured.

Believe that we will always be happy to be [the] first to publish your ingenious experiments and accept our thanks for the past and for the future.

I hope that we will often have occasion to renew those thanks and I am your respectful and devoted servant.

Ch. d'Almeida

RI MS JT/1/D/6

1. *Charles d'Almeida*: see letter 3636, n. 1.
2. *our journal*: *Journal de Physique Théorique et Appliquée* (1872).
3. *the letter*: see letter 3636.

From Ferdinand Cohn[1] 22 April 1872 3663

Breslau 22 April 1872

Verehrter Herr

In dem Überbringer dieses Briefes beehre ich mich Ihnen meinen Collegen und Freund Herrn Prof. Dr. Zupitza vorzustellen, welcher bisher an hiesiger Universitaet wirkte, nunmehr aber als Professor der altgermanischen und altnordischen Sprachen an der Universitaet Wien angestellt ist, und mit Unterstützung der Oesterreichischen Regierung nach England reist, um daselbst Studien im Angelsaechsischen und Altenglischer Literatur zu machen. Ich empfehle diesen liebenswürdigen und strebsamen Mann Ihrer freundlichen Theilname.

Im Laufe des verflossenen Winters habe ich eine Reihe von Versuchen in Bezug auf die Beziehungen der Bacterien zur Fäulniß und zu Contagien gemacht, welche dieses dunkle Gebiet in manchen Punkten aufgeklärt zu haben scheinen. Interessant ist besonders die mir gelungene Erzeugung von rothen, blauen, grünen, gelben und bräunlichen Pigmenten aus klaren farblosen Lösungen (von weinsaurem und essigsaurem Ammoniak) durch Aussaat von Bacterien. Bis jetzt sind solche Pigmente nur zufällig auf Eiweißstoffen erhalten worden (Monas prodigiosa blaue und gelbe Milch, grüner Eiter). Ich erlaubte mir ein kurzes Referat Ihnen schon vor einigen Wochen zuzusenden und bin nun im Begriff, eine ausführliche Darstellung meiner Arbeiten zu publiciren.

In der Hoffnung, daß es dazu beitragen würde, mich bei Ihnen in freundlichem Andenken zu erhalten, erlaube ich mir, Ihnen mein Bild zuzusenden; es würde mir aber zu ganz besonderem Vergnügen gereichen, wenn ich als werthvolle Gegengabe dafür Ihre Photographie erhalten koennte.

Es würde mich sehr freuen, wenn uns das Glück der Reisenden wieder einmal mit Ihnen zusammenführte, ganz besonders, wenn Sie selbst einmal nach Breslau /kaemen/, und es mir und meiner Frau vergönnt waere, Sie in unserer Häuslichkeit zu empfangen.

Mit herzlichen Grüßen | ergebenst | Ferdinand Cohn

Breslau,[2] 22 April 1872

Dear Sir

In the bearer of this letter I have the honour to introduce to you my colleague and friend Prof. Zupitza,[3] who until now was working at the university here but is now employed as Professor of Old Germanic and Old Nordic languages at the University of Vienna, and is travelling to England with the support of the Austrian government in order to undertake studies there in Anglo-Saxon and in Old English literature. I entrust this amiable and industrious man to your kind concern.

Over the course of the past winter I have undertaken a series of experiments regarding the relationships of bacteria to putrescence and to contagia, which seem to have cleared up this obscure area in some points. Of particular interest is my successful production of red, blue, green, yellow and brownish pigments from clear, colourless solutions (of tartaric acid and acetic acid of ammonia) by seeding them with bacteria. Until now, such pigments have only been obtained accidentally on albumen (Monas prodigiosa blue and yellow milk, green pus). I permitted myself to send you a short paper[4] some weeks ago, and am now about to publish a detailed description of my work.[5]

In the hope that it would contribute to keeping me in your friendly remembrance, I permit myself to send you my picture; it would, however, be a quite special pleasure for me if I could receive your photograph as a valuable gift in return.

It would please me very much if the good fortune of travellers were to bring us together once again with you, most especially if you yourself /were to come/ to Breslau sometime, and my wife[6] and I would be granted the privilege of welcoming you in our home.

With warm regards | your devoted servant | Ferdinand Cohn

RI MS JT/1/C/39

1. *Ferdinand Cohn*: Ferdinand Julius Cohn (1828–98), a German botanist notable for his work in cellular biology and bacteriology; the journal he founded in 1872, *Beiträge zur Biologie der Pflanzen*, published the founding research of the latter field. Cohn was a professor at the University of Breslau.
2. *Breslau*: German city which became Wrocław, Poland after the Second World War.
3. *Zupitza*: Julius Zupitza (1844–95), a German philologist who specialized in the study of English, from Old English to the contemporary language.

4. *a short paper*: not identified.
5. *about to publish a detailed description of my work*: This may be F. Cohn, 'Untersuchungen über Bacterien', *Beiträge zur Biologie der Pflanzen*, 1:2 (1872), pp. 127–224.
6. *my wife*: Pauline Cohn (née Reichenbach, 1844–1907), whom Cohn married in 1867.

From Erik Edlund[1] 22 April 1872 3664

Stockholm d. 22. Apr. 1872

Hochgeehrtester Herr Professor

Ich nehme mir hiermit die Freiheit den Ueberbringer dieses Briefes, Herrn Doctor Sundell, Privat-Docenten der Physik an der Universität in Helsingfors, Ihnen zu empfehlen. Doctor Sundell hat in dem hiesigen physikalischen Laboratorium eine längere Zeit gearbeitet und darunter einige Untersuchungen ausgeführt, die in der Schwedischen Sprache publicirt sind. Er reiset nun im Auslande um sich weiters in seiner Wissenschaft auszubilden und wünschte sehr, daß es ihm möglich würde einige Zeit unter Ihrer Leitung arbeiten zu können. Wenn Gelegenheit dazu ihm bereitet werden könnte, würde ich, der ich mich sehr für ihm interessire, Ihnen sehr dankbar sein.

Mit grösster Hochachtung | Erik Edlund

Professor an der Akad. der | Wissenschaften in Stockholm

An Herrn Prof. D^r^. J. Tyndall.

Stockholm, 22. Apr. 1872

Most esteemed Professor

I herewith take the liberty of commending to you the bearer of this letter, Doctor Sundell,[2] adjunct lecturer in physics at the University of Helsinki.[3] Doctor Sundell has worked in the physics laboratory here for a considerable time and during which has carried out some investigations that are published in the Swedish language. He is now travelling abroad in order to further train in his discipline and was wishing very much that it would be possible for him to be able to work for some time under your direction. If the opportunity for this could be given to him, I, who take a great interest in him, would be very grateful to you.

With greatest esteem | Erik Edlund

Professor at the Acad. of | Sciences in Stockholm

To Prof. D^r^. J. Tyndall.

RI MS JT/1/E/2

1. *Erik Edlund*: Erik Edlund (1818–88), a Swedish physicist who worked primarily on the theory of electricity.
2. *Doctor Sundell*: August Fredrik Sundell (1843–1924), a Finnish physicist who was a pro-

fessor of mathematics, astronomy and physics at the University of Helsinki; his research focused on the theory of electricity.

3. *Helsinki*: In the German, Edlund writes 'Helsingfors', the Finnish name, rather than the German 'Helsinki'.

To Edward Stanley 23 April 1872 3665

Royal Institution | 23rd April 1872.

Dear Lord Derby

The accompanying letter to Mr Gladstone[1] will inform you how the case of Dr Hooker now stands.[2] With reference to "an amicable settlement," he writes to me thus:—"Perhaps the best plan would be that Lord Derby should see Mr Helps and ask him whether he thinks that Dr Hooker has lost sight of any means of bringing this matter to an amicable issue. All I can say is that I have refused offers from both sides of the House to bring it before the House of Commons, that I have refused repeated applications for information from the public Journals Horticultural and General, that I have restrained deputations, Scientific, Horticultural, and Tradesmen's (Nurserymen) from going to Mr Gladstone, and that I have refrained from appealing to him through my near relations, who are amongst his most intimate friends."[3]

Hooker some time ago entertained strong hopes that, by the intervention of Lord Ripon, the matter would be fairly arranged. That intervention, however, ended in a curt verbal statement,[4] which is copied verbatim in the letter which I now enclose. After he had received this intimation, Hooker came to me. He seemed much broken, and once or twice while commenting on his relations to Kew and the treatment he had received, it was with difficulty that he restrained himself from giving way altogether.

He seems resolved to withdraw from all further action himself, and to leave the conduct of his case to the care of his friends. Two or three of us[5] meet at the Athenaeum early on Friday morning to consider what is best to be done, but we should be immensely aided by your counsel;[6] and very grateful to you indeed for it. Indeed Hooker is disposed to regard your freely offered aid as his mainstay in this difficulty.

Most faithfully yours | J. Tyndall.

RI MS JT/1/TYP/8/2624
LT Typescript Only

---

1. *The accompanying letter to Mr Gladstone*: not identified, but see letter 3661, in which Hooker discussed 'the correspondence with Gladstone' and suggested that Tyndall forward it to Derby.

2. *how the case of Dr Hooker now stands*: This letter responds to 3659.
3. *Perhaps . . . friends*: quoted from letter 3661.
4. *a curt verbal statement*: probably the statement contained in letter 3653.
5. *two or three of us*: perhaps Lubbock, Huxley and Bentham; in the event, Hooker attended as well. See letter 3657, in which Hooker discussed plans for meeting at the Athenaeum Club; this meeting is also discussed in letter 3666, 3667, 3668 and 3671.
6. *we should be immensely aided by your counsel*: Derby did not attend this meeting; see letter 3667.

From Joseph Dalton Hooker 23 April 1872 3666

Royal Gardens, Kew | April 23rd/72.

My dear Tyndall

A thousand thanks.[1] I shall go to the Athenaeum to-morrow and bring the letters with me.[2] I shall not be with you, of course, but Bentham thinks that I should be within hail, to answer any questions and give information.

I am not surprised at the Duke's[3] reticence. I know that he, Bruce,[4] Cardwell,[5] and especially Lord Ripon, are heartily ashamed of the whole business. These men all know well that Kew is my father's[6] creation, and all have a thorough detestation of Ayrton.

An apology from Ayrton is an impossibility, It was talked of or rather hinted at 8 months ago, but I would not listen to it. "Quarter-deck"[7] apologies are blunders of the gravest description. The superior cannot apologise to the subordinate on compulsion without bitter hate, nor the subordinate receive the apology, but with exceeding scorn. I have no personal feeling towards Ayrton; I have no wish to see him humiliated, even if that were possible. I want to be placed in future in a better position in regard to the officials who surround me. I want to have the position of Director of Kew recognised as one of authority, trust and responsibility: given a status in short that cannot be attacked by an Ayrton. The position of the Director of Kew is no better now, than it was when my father took it as a garden of 9 acres, without Collections, Museum, Herbarium, Library publications, and a mundane correspondence. If he has raised it to a first-class scientific Establishment, it is time that the official responsibility of the Director were recognized accordingly.

Ever yours | J. D. Hooker.

RI MS JT/1/TYP/8/2623
LT Typescript Only

---

1. *A thousand thanks*: We believe that Tyndall may have shown Hooker letter 3665 and have sequenced the letters (written on the same day) accordingly.

2. *the letters*: This may refer to Hooker's idea of showing correspondence about his conflict with Ayrton to Derby; see letter 3661. For the meeting at the Athenaeum Club, see letter 3665, n. 5.
3. *the Duke*: possibly the Duke of Argyll, George Campbell. On Campbell's involvement in the Ayrton controversy, see MacLeod, "Ayrton Incident," p. 56.
4. *Bruce*: Henry Austin Bruce (1815–95), a politician and Gladstone's Home Secretary. Bruce first entered as a Liberal MP for Merthyr Tudful, in Wales, in 1852; during the 1860s he was particularly involved in issues relating to education. He became Baron Abedare in 1873.
5. *Cardwell*: Edward Cardwell (1813–86), a politician and Gladstone's Secretary of State for War. Cardwell entered Parliament in 1842 as a Conservative but switched allegiances to the Liberal party in 1852. As Secretary of State for War he undertook extensive reforms.
6. *my father's*: William Hooker.
7. *"Quarter-deck"*: the upper deck of a ship, reserved for the captain and officers; the term came to refer to behavior appropriate for a person in command (*OED*).

From Edward Stanley 24 April 1872 3667

26 St James's Square | April 24th/72

Dear Professor Tyndall

From the language held to me by Lord Ripon, I had hoped that something in the nature of explanation or apology was to be offered to Dr Hooker: but as that is not so, we must consider what can be done.[1]

I wish I could attend the consultation you speak of,[2] but an engagement of business takes me out of London for the whole day.

If parliamentary action is to be taken, the first step is to move for the correspondence.[3]

Tell me what your friends and you propose.[4]

I will see Mr Helps with pleasure if he can be of use.

Very faithfully yours | Derby.

RI MS JT/1/TYP/8/2625
LT Typescript Only

---

1. *From the . . . be done*: see letters 3659 and 3666. Tyndall had presumably forwarded 3666 or otherwise conveyed its contents.
2. *the consultation you speak of*: a meeting at the Athenaeum Club of Tyndall and others concerned with Hooker's case against Ayrton; see letter 3665, n. 5.
3. *move for the correspondence*: see letter 3633, n. 3.
4. *Tell me what your friends and you propose*: Letter 3674 may be the response to this letter.

From Joseph Dalton Hooker [25 April 1872][1] 3668

Linnean Society | Burlington H[ou]^{s}[e][2] London. W. |
Thursday 4 P.M.

Dear Tyndall

A thousand thanks for the correspondence.[3] Our guns are in position and we must now fight them. But Gladstone will be furious! I will bring the letters to the Athenaeum[4] and talk over the position.

Ever yours affect[tionately] | J. D. Hooker.

I left a slice of miall[5] (violet-wood) for your lady friend,[6] on your table.

RI MS JT/1/TYP/8/2626
LT Typescript Only

---

1. *[25 April 1872]*: dated by continuity with letters 3665, 3666 and 3667. 25 April was a Thursday in 1872.
2. *Linnean Society | Burlington H[ou]^{s}[e]*: The Linnean was a biological society founded in 1788. In 1857 it moved to Burlington House in London's Mayfair neighborhood, which became home also to the RS, Chemical Society, Royal Academy of Arts, and several other learned societies.
3. *the correspondence*: Tyndall may have forwarded his exchange with Derby; see letters 3665 and 3667.
4. *I will . . . Athenaeum*: see letter 3665, n. 5.
5. *miall*: myall, an Australian acacia with scented wood (*OED*).
6. *your lady friend*: not identified.

To Joseph Henry 26 April 1872 3669

26th April 1872.
ATHENAEUM CLUB | PALL MALL

My dear Professor Henry.

I am going to call into action a kind promise that you made to me when I had the pleasure of seeing you in Liverpool.[1] You were then good enough to say that if ever I intended to give lectures in the United States you would lend me your invaluable aid and counsel.[2]

Well the time is come for me to ask this aid and advice. I have received a great many applications relative to my visit to the United States, but I have answered them all in the same terms—namely, that I have placed myself wholly in the hands of my scientific friends. I want you to consent to be the foremost of these. Your position and pursuits naturally cause me to look up to you, and it would be an immense relief to me if you would consent to organize the plan of action which you consider most conducive to the interests of science and the gratification of the American people in this matter.

With regard to principal arrangements they are of the simplest possible kind. My desire would be to go to America and lecture there paying my own and my assistants' expenses and not accepting a dollar from anybody. For my visit is not a professional one, but a visit of goodwill. Still I am prepared to have my actual disbursements made good. I must engage two assistants, and purchase some apparatus. I shall also have travelling and hotel expenses to meet. But the proceeds over these items of actual outlay I should wish to devote to some scientific object in America, and nothing seems to me more suitable than to turn these to account in Chicago. The calamity there[3] has been overwhelming, and many needs will have to be looked to before scientific ones are considered. Therefore I think that good may be done by helping Chicago to get upon its scientific legs.

I earnestly hope that you will fall in with this idea. It simplifies the matter greatly, and altogether */diverts/* those questions about remuneration which I can see is a source of anxiety to many American institutions. Let the money question depend entirely on the success of the lectures, it being understood that what remains, after the expenses have been paid, shall be devoted to the helping of science in Chicago.

What towns I ought to lecture in you, I trust, will kindly decide. I should not be willing to lecture more than three days a week, the odd days being devoted to preparation, which implies the use of the lecture halls. This may be a difficulty, and if so I must contrive to meet it. With regard to New York there is a very good man there who would aid me in all possible ways—I mean D^{r}. Youmans, at present engaged by Appleton &C^{o}.

Would you also please to inform me what period of the year the lecture ought to begin? I intend to devote my vacation to the preparation of them. Of what character would you advise the lectures to be? I suppose they ought to be mainly experimental, weaving of course with the experiments various considerations regarding science & scientific culture in general. I thought of preparing say three lectures on some definite part of Light; three on some definite part of Heat; and three on some definite part of electricity—I would also, if it were deemed advisable, give two or three on sound. I wish you would let me know your views on these points: You know the ground that is least trodden, and this Knowledge communicated to me will be of immense service to me.

I hope I am not asking too much of you. If my request should appear to you unreasonable, pray forgive me. But as the leading physicist of the States,

I naturally seek the shelter of your shield & the encouragement of your countenance.

Pray present my best compliments to your daughter.[4]

& believe me | Yours ever faithfully | John Tyndall

John Tyndall to Joseph Henry, 26 April 1872, mssRH 3964, Box 56, William Jones Rhees Papers, The Huntington Library, San Marino, California

---

1. *seeing you in Liverpool*: at the meeting of the BAAS held at Liverpool in 1870.
2. *to give lectures in the United States*: Tyndall had been discussing a possible visit to the United States with Henry; see letters 3504, 3510 and 3542. For additional letters on this topic, see 3507, 3519, 3543, 3558, 3589, 3594, 3639, 3670, 3684, 3685, 3702 and 3703.
3. *the calamity there*: see letter 3639, n. 8.
4. *your daughter*: Henry had three daughters: Mary Henry (1834–1903), Helen Henry (1836–1912) and Caroline Henry (1839–1920). We do not know to which of them Tyndall sent greetings.

To Hector Tyndale 26 April 1872 3670

Royal Institution | 26th April 1872.

My dear Hector.

By this time you have received a letter which I wrote to you some weeks ago[1]—Since then I have received a copy of your remarks at the Banquet[2]—Trust me Hector all that you have done to illustrate the courage of your blood has been a source of pride to me.[3]

I will write immediately to Prof. Henry.[4] Indeed he was good enough to ask me to do so when I saw him in 1870.[5] Many applications and letters of enquiry have come to me lately but my reply to them, one & all, has been the same. Namely, that I have placed myself wholly in the hands of my scientific friends.

The terms on which I should like to lecture are stated in my last letter. It is not a professional call that I respond to in going to America, but a call of kindness and goodwill. Were I a millionaire I would go there in response to such a call and lecture without moving a dollar from any American purse. But though I am very wealthy! I am not wealthy enough to undertake so much, and I therefore consent that my actual disbursements shall be made good to me.

I do not however wish to set any limit to your generosity. You may give me a million dollars if you like, but then I claim the right of disposing of them in the way indicated in my last letter.[6]

I had an exceedingly pleasant note from Mr Emerson[7] some time ago introducing his daughter[8] and her husband (Lieut Col. Forbes).[9] I join them at dinner to-day at the house of a friend.[10] She is very charming and he is a most agreeable gentlemanly fellow. They have been very cordially received in London.

With regard to lodging I am willing to do whatever is agreeable to you. In England I usually indulge in the savage freedom of a hotel—But I daresay I shall be quieter under your roof than in a hotel. And I shall have no compunction in demanding from you every thing I need in the way of food & liquor. I hope Prof. Lesley will not think that I have played him false.[11]

John Tyndall is a fine young fellow—now the eldest son of our Gorey friend, George,[12] is dead.

Yours ever aff[ectionate]ly | John Tyndall

Kindest regards to Mrs Tyndall[13]

RI MS JT/1/T/1452
RI MS JT/1/TYP/4/1696–7

---

1. *a letter which I wrote to you some weeks ago*: see letter 3639.
2. *the Banquet*: not identified.
3. *all that . . . to me*: probably a reference to Tyndale's military service for the Union during the American Civil War, for which see *Ascent of John Tyndall*, p. 302.
4. *I will write immediately to Prof. Henry*: see letter 3669. Tyndall's syntax here suggests that he composed this letter to Tyndale before he wrote to Henry. These letters are part of a conversation about a planned trip to the United States; for additional letters on this topic, see 3504, 3507, 3510, 3519, 3542, 3543, 3558, 3589, 3594, 3684, 3685, 3702 and 3703.
5. *when I saw him in 1870*: see letter 3669, n. 1.
6. *in the way indicated in my last letter*: see letter 3639, where Tyndall wrote that he planned to donate any profits from his lectures to the cause of science in Chicago.
7. *an exceedingly pleasant note from Mr Emerson*: letter missing.
8. *his daughter*: Edith Emerson Forbes (1841–1929).
9. *her husband (Lieut Col. Forbes)*: William Hathaway Forbes (1840–97), the son of American businessman and railway developer John Murray Forbes (1813–98). William Forbes was involved in the development of the telephone system.
10. *the house of a friend*: unidentified.
11. *With regard . . . false*: see letter 3589, where Tyndall, writing that he preferred the 'barbarous liberty of a hotel', declined Lesley's invitation to stay at his house.
12. *the eldest son of our Gorey friend, George*: George Tyndall, but not otherwise identified.
13. *Mrs Tyndall*: Julia Tyndale; Tyndall's misspelling.

From Joseph Dalton Hooker [27 April 1872][1] 3671

Royal Gardens | Kew. | Saturday—7 a.m.

Dear old T.

I have taken a night's sleep—no, thought—over "the position," and feel more than ever determined to carry on the War. I was mightily fortified by you all yesterday,[2] and

"Have quitted not the harness bright
Neither by day nor yet by night:
I lay down to rest
With corselet laced,
Pillowed on buckler cold and hard.
I have carved my meal
With gloves of steel,
And drank the red wine through the helmet barred".[3]

Now for prose. No time must be lost. I will answer Treasury's insolent letter (signed by a Clerk!!!)[4] and bring the draft to the R[oyal]. S[ociety]. to-night. If I do not find you, I will send it to you. There never was such a chance as this for bringing the position of Science under Government into prominence. The horticulturalists and nurserymen have three times, within as many months, urged me to allow them to form a deputation, or draw up an address to Mr Gladstone on the subject of my first appeal (the Hothouses and Heating Apparatus)[5] and I promised to inform them of the result of my application to Mr Gladstone: this I must now do. They modestly disclaim all wish to interfere with, or take the position of scientific men or a scientific deputation, but do wish to call Mr G's attention to this matter from their own practical point of view, and to express their opinion regarding the importance of Kew to them. This will be the beginning of a row which I have earnestly striven to avert hitherto.

How would it do for Scientific men to attack him at the same time with an address pointing out the serious obstructions I have so long endured,[6] and the absolute necessity and urgency of an enquiry into the governance of Kew or of the relations between the Director and Government, and so put it, that he will be forced to refer the matter to the Science Commission[7] for a special Report. To-morrow afternoon I go and see a very experienced old official[8] at Coombe[9] and shall get his view of matters as they stand.

Yours ever | J. D. Hooker—

RI MS JT/1/TYP/8/2627
LT Typescript Only

1. *[27 April 1827]*: Hooker, Tyndall and a few others met on Friday, 26 April 1872, at the Athenaeum Club; the present letter follows up on that meeting. See letter 3665, n. 5.
2. *by you all yesterday*: those who met at the Athenaeum Club; see n. 1.
3. *Have quitted . . . barred*: W. Scott, *The Lay of the Last Minstrel* (London: Longman, Hurst, Reeds, and Orme; Edinburgh: A. Constable, 1805), Canto First, IV.
4. *Treasury's insolent letter (signed by a Clerk!!!)*: This letter confirmed the information conveyed to Hooker by Helps (in letter 3653). See MacLeod, "Ayrton Incident," p. 60.
5. *the Hothouses and Heating Apparatus*: the construction and heating of Kew's hothouses, one of the first conflicts between Hooker and Ayrton. See letter 3630, n. 1.
6. *How would it . . . so long endured*: This proposal evolved into a memorial letter written to Gladstone on behalf of scientific men in support of Hooker and published in *Nature* in July 1872. See C. Lyell, C. Darwin, G. Bentham, H. Holland, G. Burrows, G. Busk, H. C. Rawlinson, J. Paget, W. Spottiswoode, T. H. Huxley, J. Tyndall, 'Letter to W.E. Gladstone', *Nature*, 6:141 (11 July 1872), pp. 211–6. This letter and numerous letters to and from Tyndall concerning its drafting and distribution will be included in the forthcoming thirteenth volume of *The Correspondence of John Tyndall.*
7. *Science Commission*: the Royal Commission on Scientific Instruction. See letter 3468, n. 1.
8. *a very experienced old official*: not identified.
9. *Coombe*: Coombe Bank, William Spottiswoode's home.

To John Morley[1] 2 May 1872 3672

Royal Institution | 2nd. May 1872.

Dear Mr. Morley.

I can have no objection, since you desire it, to say to you what I have frequently said to others about you:—namely, that I consider your mind to possess the three qualities of Grasp, penetration, and logical power, and that I believe your store of Knowledge to be great.

I have moreover heard you lecture,[2] and can say with truth that the clearness and force of your exposition, the earnestness of your manner and the liberality of your views—technical 'liberals'[3] you know can be sometimes illiberal—gave me exceeding pleasure.

I offer no opinion about Political Economy but I can hardly imagine a man such as I suppose you to be applying for a chair of Political Economy[4] without being fully up to the mark as to the treatment of the subject.

Very faithfully yours | John Tyndall

Letter from John Tyndall to John Morley, 2 May 1872, Oxford, Bodleian Libraries, Archive of John Morley, MS. Eng. e. 3431 fols. 41–2

1. *John Morley*: see letter 3660, n. 32.
2. *heard you lecture*: For instance, Morley had lectured at the RI in April; see letter 3650, n. 4.
3. *technical 'liberals'*: presumably, belonging to the Liberal party in politics as Morley did.
4. *applying for a chair of Political Economy*: The university and exact position to which Morley was applying are unknown.

From Joseph Dalton Hooker 2 May 1872 3673

Royal Gardens, Kew. | May 2d 1872.

Dear Tyndall

I enclose herewith Treasury's letter, and answer thereto.[1] I believe the root of the whole evil lies in my undefined position under the Board,[2] and that a favourable answer to all my queries will go a very short way indeed towards setting me right, and will besides entail endless worry. What really is wanted is, that my appointment should be confirmed by a higher authority that the First Commissioner of Works[3] or that I or Kew should be removed altogether. I fear that to procure either will be a harder task that it should be.

I shall be most anxious to hear Lord Derby's views on the Treasury letter. I wish he would see fit to act upon it without reference to my answer. It would save a world of trouble.

Ever yours | J. D. Hooker.

The documents[4] go by book post.

RI MS JT/1/TYP/8/2628
LT Typescript Only

1. *Treasury's letter, and answer thereto*: These letters are missing, but their contents are discussed in letter 3671.
2. *the Board*: the Board of Works.
3. *First Commissioner of Works*: Ayrton.
4. *The documents*: letters relating to the Kew controversy, received by Derby on 3 May. See letter 3674.

To Edward Stanley 3 May 1872 3674

Royal Institution | 3rd May 1872.

Dear Lord Derby,

Would you permit me to place in your hands the whole of the Hooker correspondence?[1] The large parcel encloses it.

Hooker's last letter to Mr Gladstone[2] you will remember had reference to a verbal message received through Mr Helps from Lord Ripon.[3] The message was stated to be final.

This however did not turn out to be the case, it was afterwards characterised as a "private communication."

The official reply to Hooker's letter to Mr Gladstone[4] is forwarded herewith (in the smaller parcel). It bears date the 25th of April 1872. I would ask your Lordship to consider its terms, and decide whether any ordinary English brain, accustomed to straightforwardness in language can make out the meaning of him who dictated it.

Hooker's rejoinder, wherein he begs for a definition of the position, is also appended.[5] This I believe will be his last communication.

The scientific men of England will be unanimous in their support of Hooker; and I think Mr Gladstone will find them able and determined to state their case in language somewhat plainer than his reply.

The real and I believe only remedy for Kew would be to be removed altogether from under the authority of the First Commissioner.[6] The director of a great scientific establishment like Kew ought to be able to look to a higher authority than the First Commissioner of Works.

I wish and Hooker wishes[7] you could see your way to bringing the question forward in the house of Lords.[8]

I am, my Lord, | Very faithfully yours | John Tyndall.

[. . .][9]

RI MS JT/1/TYP/8/2629
LT Typescript Only

---

1. *the Hooker correspondence*: relating to Hooker's dispute with Ayrton. Hooker sent this material to Tyndall on 2 May 1872; see letter 3673. The present letter may be Tyndall's response to letter 3667.
2. *Hooker's last letter to Mr Gladstone*: see letter 3665, n. 1.
3. *a verbal message received through Mr Helps from Lord Ripon*: probably the message contained in letter 3653.
4. *The official reply to Hooker's letter to Mr Gladstone*: Enclosure missing, but this letter is discussed in letter 3671. See also MacLeod, "Ayrton Incident," p. 60.
5. *Hooker's rejoinder . . . appended*: Enclosure missing, but this letter is discussed in letter 3671.
6. *the First Commissioner*: First Commissioner of Works.
7. *Hooker wishes*: see letter 3673.
8. *I wish . . . the house of Lords*: for Derby's response, see letters 3675 and 3677.
9. *[. . .]*: LT copied letter 3675 here, in her own hand.

From Edward Stanley 3 May 1872 3675

23 St James's Sq[ua]^r[e] | May 3. 1872

Dear Professor Tyndall

I have your note & the papers.[1] I will read them tomorrow or Sunday.[2]

Very Truly Yours | Derby

RI MS JT/1/TYP/8/2629
Transcript Only

---

1. *I have your note & the papers*: letter 3674 and the correspondence concerning Hooker's conflict with Ayrton, which Tyndall had forwarded. This letter appears as a note, in LT's hand, at the bottom of letter 3674.
2. *I will . . . Sunday*: see letter 3677.

From Joseph Dalton Hooker 6 May 1872 3676

Royal Gardens, Kew, | May 6th 1872.

Dear Tyndall

After a vain search[1] for the terms upon which my father[2] undertook to organize Kew, I referred to the Treasury and find that he had neither warrant nor commission! That the Treasury simply sanctioned "the acceptance by the Commissioner of Woods and Forests[3] of Sir W. Hooker's offer to give his whole services to the management and improvement of the Botanic Gardens at Kew at a salary of £300 per annum and an allowance of £200 per annum to provide a house for his family and the deposit of his library and herbarium."

For this he left a Professorship in Scotland[4] yielding £1000 per annum (in 1840).

On my accession I received a warrant signed by the Secretary of the Board appointing me Director (just such another as a Park's superintendent receives) and my view is that I should now urge that the Director in future receives a more substantial recognition of his authority under Government, one that would give him inviolable control, under a minister of the Crown or under the Treasury.

If you agree with me, now would be the time to strike out for this, and for Huxley to urge it upon Lowe.[5] After what has passed it is not enough to restore me to the undefined and unsafe position I held.

Ever yours | J. D. Hooker

I am writing to Huxley—

How would it be for me to draw up an address to the Lords of the Treasury on the subject?

RI MS JT/1/TYP/8/2630
LT Typescript Only

---

1. *After a vain search*: This letter is part of the correspondence about Hooker's conflict with Ayrton over the management of Kew, and may be part of the work he did to prepare a memorial to Gladstone on the subject. On the memorial, see letter 3671, n. 6; for other letters relating to details of the history of Kew, see 3690, 3692, 3700, 3705 and 3706.
2. *my father*: William Hooker.
3. *First Commissioner of Woods and Forests*: at the time, John William Ponsonby (1781–1847), who became the earl of Bessborough in 1844. In 1851 the Board of Commissioners of Woods, Forests, Land Revenues, Works and Buildings was replaced by an Office of Woods and Forests and an Office of Works; see MacLeod, "Ayrton Incident," p. 50.
4. *a Professorship in Scotland*: at the University of Glasgow, where William Hooker was Regius Professor of Botany from 1820 to 1841.
5. *Lowe*: see letter 3634, n. 4.

From Edward Stanley 7 May 1872 3677

23 St James's Square | May 7th 1872.

Dear Professor Tyndall

I have been through the Hooker correspondence.[1] I agree that he has a good case, and if he concurs I should think that the best course was in the first instance to ask for the production of the papers.[2] I could ascertain whether they will be given or not, but I hardly think they can be refused.

We ought to have them in print before calling attention to them.

But I should like Dr Hooker's concurrence before taking this step.[3]

Very faithfully yours | Derby.

RI MS JT/1/TYP/8/2631
LT Typescript Only

---

1. *the Hooker correspondence*: sent by Tyndall to Derby; see letters 3674 and 3675.
2. *the production of the papers*: see letter 3633, n. 3.
3. *But I . . . this step*: see letters 3678 and 3691.

To Joseph Dalton Hooker 7 May [1872][1] 3678

7th. May.

My dear Hooker.

I send you Lord Derby's note[2] on the Hooker correspondence. What say you to his proposal?[3]

Yours ever | John Tyndall

RBG, Kew Papers Relating to Kew 1867–1872, Ayrton Controversy, volume 1, f. 186

---

1. *[1872]*: date assigned by the connection to letter 3677.
2. *Lord Derby's note*: letter 3677.
3. *his proposal*: to call for papers; see letter 3677. Hooker's response is missing, but see letter 3691.

From Joseph Dalton Hooker 8 May 1872 3679

Royal Gardens, Kew. | May 8th 1872.

Dear Tyndall,

The first-fruits of my letter to the Treasury[1] is another letter from the Board of Works requiring me to answer questions as to price etc. with a view to the sale of the copies of the Flora of Tropical Africa.[2]

The Board is utterly ignorant of the terms of the publication of a work published "under the authority of the First Commissioner of H.M. Works", as set forth on its title page, and are floundering accordingly.

I have a letter on the same subject from the Stationery Office[3] which is in correspondence with the Treasury about it. So here are literally four departments, Kew, B[oard]. of Works, Treasury, and Stationery O[ffice]., all wasting their time over a perfectly simple straight-forward matter, and Gregg[4] is the only one with brains enough to ask me what the terms of publication are.

Ever yours | J. D. Hooker.

Yours and Lord Derby's[5] just received. I will take it to Bentham.

RI MS JT/1/TYP/8/2632
LT Typescript Only

---

1. *my letter to the Treasury*: possibly the letter discussed in 3671.
2. *Flora of Tropical Africa*: D. Oliver, *Flora of Tropical Africa* (London: L. Reeve, 1868, 1871).

Kew commissioned a number of books on colonial floras; see C. Bonneuil, 'The Manufacture of Species: Kew Gardens, the Empire and the Standardisation of Taxonomic Practices in late 19th Century Botany', in M.-N. Bourguet, C. Licoppe, and O. Sibum (eds), *Instruments, Travel and Science: Itineraries of Precision from the Seventeenth to the Twentieth Century* (New York: Routledge, 2002), pp. 189–215, and Endersby, *Imperial Nature*, p. 163. Hooker and Ayrton came into conflict over the distribution of this volume; see, for instance, 'Correspondence between the Board of Works and Dr. Hooker relating to Changes proposed to be introduced into the Direction and Management of the Gardens at Kew', HL 213, Parliamentary Papers (1872), pp. 33–40. The issues discussed in the present letter are also taken up in letter 3680.

3. *Stationery Office*: department of the Treasury responsible for government printing and for supplying stationery to government departments.
4. *Gregg*: William Greg; see letter 3616, n. 11.
5. *Yours and Lord Derby's*: letters 3677 and 3678.

From Joseph Dalton Hooker [8 May 1872][1] 3680

I have sent a copy of this[2] to the Treasury with a letter saying that it is the first communication I have received regarding the matter[3] upon which their Lordships have given Mr Ayrton their views.

It shows that their Lordships have not been correctly informed as to the terms of publication, as sanctioned by the Treasury, the stock having all along been on sale at the Publishers.[4] The terms were, that the Government engaged to subscribe for 100 copies to encourage the publishers, guarantee a moderate priced work to the public and distribute copies to Government Institutions. Therefore to sell these copies would be to defeat their own intentions and injure the publisher.

I enclose copy of letter[5] showing this to be <u>their Lordship's</u> own arrangement.

If their Lordships still insist that the copies remain at Stationery Office[6] I bow to their decision; if that any be sold, I decline to superintend the preparation or publication of the remaining volumes. | J. H.

RI MS JT/1/TYP/8/2679–80
LT Typescript Only

---

1. *[8 May 1872]*: We assign this date on the belief that Hooker is following up on the letter from the Board of Works that he discussed in letter 3679.
2. *of this*: The enclosure is missing, but seems to have been a document concerning the publication of *The Flora of Tropical Africa* (see letter 3679, n. 2).
3. *the matter*: probably the sale of *The Flora of Tropical Africa*; see letter 3679.

4. *the Publishers*: L. Reeve, publishing house established by Lovell Augustus Reeve (1814–65) that specialized in natural history books.
5. *I enclose copy of letter*: enclosure missing; possibly the same document discussed in n. 2.
6. *Stationery Office*: see letter 3679, n. 3.

From Joseph Dalton Hooker 10 May 1872 3681

Royal Gardens, Kew. | May 10th 1872.

Dear Tyndall

I sent the correspondence[1] to Lord Russell at 8.0 this morning and got the enclosed[2] by 12.0. Should not Lord Derby see it?

Pray send it to him, if you think so.[3]

Ever yours | J. D. Hooker.

RI MS JT/1/TYP/8/2633
LT Typescript Only

---

1. *the correspondence*: concerning Hooker's conflict with Ayrton. It had previously been sent to Derby. See letters 3674 and 3677.
2. *the enclosed*: enclosure missing.
3. *Pray send . . . think so*: Tyndall did so; see letter 3688.

To Joseph Dalton Hooker [10 May 1872][1] 3682

Royal Institution of Great Britain

My dear Hooker.

I caught your letter[2] this morning post before starting for Birmingham[3] & sent it on to Huxley. The other two notes[4] have now reached me.

I am entirely of your mind that we ought to go to work[5] at once & I have told Huxley so.[6] I will see him tomorrow.[7]

Yours aff[ectionate]ly | John Tyndall

RBG, Kew Papers Relating to Kew 1867–1872, Ayrton Controversy, volume 1, f. 194

---

1. *[10 May 1872]*: This date appears at the top of the handwritten letter, in the following sequence: 'Rec[eive]d Saturday m[or]n[in]g | Friday night | 10 May 1872'. We surmise that these notes were added by Hooker. For similar issues in letters related to the Ayrton controversy, see letters 3634, 3683, 3690 and 3694. The date of 10 May 1872 makes sense in the timeline of this controversy.

2. *your letter*: possibly letter 3676, in which Hooker wrote, 'If you agree with me, now would be the time to strike out for this, and for Huxley to urge it upon Lowe'.
3. *before starting for Birmingham*: The purpose of Tyndall's trip is unknown. Birmingham is a major city located in the middle of England.
4. *other two notes*: probably 3679 and 3680.
5. *we ought to go to work at once*: probably on drafting the memorial to Gladstone; see letter 3671, n. 6.
6. *I have told Huxley so*: We do not have a letter that conveys this message.
7. *I will see him tomorrow*: Tyndall seems not to have seen Huxley; see letter 3683.

To Joseph Dalton Hooker [11 May 1872][1] 3683

Royal Institution of Great Britain | 10th May

My dear Hooker.

I went to see Huxley this afternoon but he was absent[2] & nobody knew anything about him.

I go to Oxford tomorrow,[3] but shall return on Monday.

If I can manage it I will get out with you for a couple of days towards the end of next week.[4]

I am fearfully low, and horribly entangled. Would to God I had a little more time I would this minute set about that memorial[5] and make a good thing of it.

To be a *[1 word illeg]* one of us[6] ought to draw it wholly up, and submit it to the others afterwards. I have no faith in the patchwork which results from pulling heads together.

Have you had any further communication with Lord Derby.[7]

I am still entirely of opinion that we ought to accept his offer.[8]

Yours aff[ectionate]ly | John Tyndall

RBG, Kew Papers Relating to Kew 1867–1872, Ayrton Controversy, volume 1, f. 196

---

1. *[11 May 1872]*: The following note appears at the top of this letter: 'Rec[eive]d Saturday night | 10th May 1872'. We believe that, as in letters 3634, 3682, 3690 and 3694, these notes were added by Hooker. We further surmise that he wrote '10th May' when he should have written '11th May', which was a Saturday; we think that this letter was written one day after letter 3682.
2. *I went . . . absent*: For Tyndall's intention of seeing Huxley, see letter 3682.
3. *I go to Oxford tomorrow*: We do not know the purpose of this trip.
4. *If I can . . . next week*: This excursion is also discussed in letters 3690, 3692, and 3693; it may refer back to Hooker's invitation in letter 3647.

5. *that memorial*: to Gladstone; see letter 3671, n. 6; preparing the memorial is further discussed in letters 3686 and 3690.
6. *one of us*: According to Endersby, Tyndall collaborated with Huxley, Spottiswoode and Lubbock on the memorial (Endersby, *Imperial Nature*, p. 285).
7. *Have you . . . Derby*: This may refer to Tyndall's query, in letter 3678, as to Hooker's response to Derby's proposal, in letter 3677, to move for the papers. Hooker responded in letter 3686.
8. *his offer*: to call for the production of papers. See letter 3677.

From Joseph Henry 11 May 1872 3684

Smithsonian Institution, | Washington May 11. 1872

My dear sir:

Your letter of April 26[1] has just been received and as I am on the point of starting for the Light House[2] *[session]* in New York to be absent a week I hasten to distill the following remarks.

I am very much gratified to learn that you have concluded to visit the United States and during my absence will endeavor to see Mr Lesley of Philadelphia and Prof Youmans of New York in regard to a definite arrangement as to the course of lectures that will be expected of you, viz,[3] at least one in Washington, another in Baltimore, a third in Philad[elphi][a], a fourth in New York, a fifth in Boston, with perhaps some others in the West, depending upon the time you can devote to lecturing.

The American Assoc[iatio][n] for the Advancement of Science[4] meets, this year in California on Aug 5. Doctor Gray, of Cambridge,[5] is the retiring President and arrangements are being made with the steam ship lines to bring free of expense a number of English savans to attend; among those to be invited are yourself, Professor Huxley and Dr. Hooker. The time of the voyage across the ocean is, on an average about ten days, while the trip to San Francisco, by rail, is seven. You should, therefore, leave England about the middle of July or, to give plenty of time, a week earlier. I can assure you, from personal experience last year, that you will find a journey across the continent, in a palace car,[6] very little fatiguing and abounding in objects of the highest interest and suggestive of trains of thought which could be elaborated by yourself into important deductions.

The lecturing season in the United States does not commence before October; the intermediate time will afford you an opportunity of becoming acquainted with the more salient peculiarities of the United States and Canada.

I know how valuable is your time to the advance of science but I hope you may conclude to visit the United States at the time I have mentioned and remain 'till the autumn to give the lectures which are so earnestly desired. Though by so doing you may interrupt for awhile your investigations, still the physical and even mental improvement which will result from your travels will, I think, much more than compensate for the time diverted from your usual pursuits.

I fully agree with you as to the subjects and character of the lectures mentioned in your letter as being well adapted to an American audience.

The proposition to give the proceeds of your lectures to the advance of Science in Chicago is a very generous one and will meet with high appreciation in the Country.

I am, yours, very truly, | Joseph Henry

Dr. John Tyndall

P.S. I shall write you again on return from New York[7]

Smithsonian Record Units 33, Vol 28, Page 769

---

1. *Your letter of April 26*: letter 3669, which, along with the present letter, is part of a conversation about Tyndall's plans to lecture in the United States. For Tyndall's response to the present letter, see letter 3702. See also letters 3504, 3507, 3510, 3519, 3542, 3543, 3558, 3589, 3594, 3639, 3670, 3685 and 3703.
2. *Light House*: Henry was chair of the American Light-House Board at the time.
3. *viz*: introducing more precise information (*OED*).
4. *American Association for the Advancement of Science*: association for the promotion of science in the United States, founded in 1848.
5. *Doctor Gray, of Cambridge*: Asa Gray (1810–88), an American botanist and professor at Harvard University, in Cambridge, Massachusetts, from 1842 until his death. In the United States Gray championed Darwin's theory of evolution, partly by arguing that evolution and Protestantism were compatible.
6. *palace car*: luxury sleeping car manufactured by the Pullman Palace Car Company. See J. Husband, *The Story of the Pullman Car* (Chicago: A.C. McClurg, 1917), p. 43.
7. *I shall write you again on return from New York*: letter missing, if written.

From Hector Tyndale 11 May 1872 3685

Philadelphia May 11th 1872

My dear John

Since my last,[1] of a week or so ago, I have yours of the 26th April.[2] I was absent the day it arrived, in New York by the sick bed of my friend T. Buchanan Read,[3] who died last night. You may remember him as one of the gentlemen who dined with us at Morley's Hotel in the Summer of 1860[4]—a small pleasant man, to whom you took a fancy as a poetical artist. He was both Poet and Painter: has been living in Rome for many years, took the fever there a year or so ago, has been in bad health since, came home a few days ago very feeble and died at a hotel in New York last night. He will be buried on Tuesday, just a fortnight after his arrival in the United States, to which he had hastened, seemingly, to die—and he had not time or strength to reach his Sister's house in Cincinnati.[5] I knew him from our infancy (I one year the eldest) I think about 46 or 47 years in all. That's a long time as things move here. A few years hence and his form shall appear to us, all unconscious, in trees and grasses and he, and you and I and all of us—where shall we be? Why, we cannot for one moment think or fancy ourselves as lying in dust, or even as organized anew. There remains for us, only to be as we are, consciously, and in memory as we are and were—or, we shall be nothing—to our present comprehensions. Around my soul, my elemental soul, what new forms may gather to build another life—another life than mine? What then of me? So then I may live, I indestructible, but my individuality the separate and identical element of Me, My Conscious Being—that is not to be? Then how can I live?—But there may be other lexicons to learn and our many words may be resolved into one, as colors into light! And then farewell to all of us! Other prisms may refract the self same single light, but Read's prism is broken and so good bye to him as Read? But he was no prism and we no screens to catch the colors of an unknown light behind him. If he is not an element, then all conscious thought is accident and individuality a bundle of chances, so that the very earnest words we use to demonstrate our souls, are wanton lies—elements are divisible and Truth—mighty Truth—a changeable reflection![6]

Last evening, with Prof:[7] Lesley, I called upon Prof: Henry,[8] who, by good fortune, was here a day or two—I had intended going down to Washington to see him to ask about the arrangements for your coming. As it is absolutely necessary that some single direction shall at once be made in the matter, that there may be no clashing or difficulties in the times and places of your lecturing &c. Prof: Henry agreed with me in this and very kindly and cordially consented to be the Director in the matter. In fact he had already begun and in the right way too. He had telegraphed Lesley to meet him and accidently I met the latter who told me and so saved me a journey. Henry said he had heard from you and had replied.[9] He had advised you to lecture in Washington, Baltimore, Philadelphia, New York and Boston and perhaps in other places. In Washington he said that a society (the Young Men's Christian Association[10] I think he said) would undertake the business of providing lecture-rooms, advertising &c., &c. In Baltimore, the Peabody Institute under Professor Morrison,[11] in Boston the Lowell Institute, under Mr Lowell,[12] would attend to the business. In New York he hoped that Prof Youmans (your friend) would undertake it and he had written to Youmans to meet him in New York to take action upon it. In Philadelphia he hoped that Lesley would do it. To speak of Philadelphia. Lesley will do all he can as will other scientific gentlemen here but no one of them can do all that will be necessary. But Lesley and myself will correspond with Prof: Henry and when the exact date of your coming to this city to lecture is known, we will get some one to take a place (such as the Academy of Music[13]) who will also do the advertising and other business. But it is absolutely necessary to arrange the exact days of lecture in each place at once, so that the days for each may be fixed without disagreement. I presume that the Lowell Institute will be the proper key, as to those days, as that is the most important of the Institutes holding lectures and is the most difficult, because of previous arrangements, to deal with. So that when the exact days for Boston have been fixed, the others, I think, can be named and arranged more easily. Prof. Henry said he would write to you again when he returned to Washington.[14] But in the meantime he wishes you would write him[15] (the sooner the better) and give him all the exact information you can. For fear of accident or of Prof: Henry's absence you had, perhaps, better write to me also upon the same matter.[16] I would advise you firstly to give the exact date of your departure from England. Secondly how long you can remain in this country. Thirdly how many assistants you will bring, and whether you will need any others here &c. Prof: Youmans can and no doubt will aid you in getting assistance or apparatus or work of any kind done in this Country if you should need it. Give Henry (and myself) all the other information you can, upon all points, so that he can better arrange as to times, places &c. Here in Philadelphia, there is one place so pre-eminently better than any other here for important lectures, viz the Academy of Music (a large and handsome theatre wherein all the best lectures are given) that it is important to secure that place as soon as possible. It is often taken up by different Associations &c. a whole year in advance it may be so now. So you see the necessity of at once arranging the dates by days.

Prof: Henry said he had advised you to attend the meeting of the American Scientific Association (I think that is the title) at San Francisco on the 5th August next.[17] To do this you will have to leave England about the middle of July, not later than the 15th. This would bring you in New York by about the 26th to 28th July—rest here a day or two. Start about the 28th to 30th and reach California in time. There (he said) in company with himself and others go to Van-Couver's Island,[18] to Yosemite Valley[19] and other places in California and on the backward railroad route—Pike's Peak[20] &c—All of which would take the month of August and thus to Niagara, to Canada & the most salient points of the Country and thus pass September, so that by October (in which month about the middle the lecture season commences) you would be ready to begin your lectures and to end your visit to this Country, if you liked, as soon as you were through lecturing.

If you can manage this, I think it a good programme. You can so go to the Pacific with some of the best of our men of Science who will no doubt take Cars (railway carriages) to themselves all the way through. In your letter to Prof Henry ask from him the exact day and place of his departure for the train to San Francisco—so that you may better regulate your day of sailing.

Should you adopt this suggestion of Henry's you can have your assistants leave England with their apparatus at any time after your departure. Say that they arrive in New York sometime in September so that you can have the apparatus shipped to any point you may desire, in time to have it opened, arranged, repaired &c. All of this will, of course, require some forethought. You will find plenty of help here if needed.

In relation to the money points. I think you had better tell, if not already, Prof Henry of your wishes. That you do not propose to lecture in aid of any Association or for any profit but that after deducting your expenses you wish the balance of receipts, if any, to be given by you to Science in Chicago &c. Of course Prof Henry will not consult me in anything about it, only that as one of your friends here I am to act in the business attending it in Philadelphia. I would give Henry to understand that you wish to give the surplus receipts (if any) yourself—Otherwise some Association may do it and in such a way as you would not like. This however is a delicate point. Can you find time to send a line to Lesley? I believe I have now touched on all points and fear at my tiresome length to you. Please tell me when George Tyndall (son of Dr: John) died & how?[21] Your rooms shall be ready for you and I think all your letters had better come to my house (No: 1021 Clinton St.) and all your business arranged there. It is a central point in this City which is itself geographically central for the five large Cities I have named. I wish Hirst and Debus, both, would come with you. We have room enough. You will have a private room for reading or writing or any other business you may require to do.

Mrs Tyndale sends her kindest remembrances to you and to Hirst—and give my love to Tom and Debus. Lesley sends respects and of course will hold you excused as our arrangement was made last year, besides the ownership you have of mine because of blood and friendship. Lesley's house and mine are in the same Street and within a dozen doors.

Affectionately your Cousin | Hector Tyndale
Prof: John Tyndall

RI MS JT/1/T/66
RI MS JT/1/TYP/5/1698–1702

---

1. *my last*: letter missing.
2. *yousr of the 26th April*: letter 3670.
3. *T. Buchanan Read*: Thomas Buchanan Read (1822–72), a painter, poet and sculptor. He spent a significant amount of time living in Europe (where he came in contact with the English Pre-Raphaelites), but returned to the United States to fight in the Civil War, which became a subject of his art. He was in a carriage accident in 1871, which may have been partly responsible for his death.
4. *one of the gentlemen . . . 1860*: There is a Morley's Hotel in London, but we have not further identified this dinner.
5. *his Sister's house in Cincinnati*: Cincinnati, a major city in the American state of Ohio. At the age of fifteen Read moved there to live with a married sister (name unknown).
6. *a changeable reflection*: Tyndale's letter includes a break at this point marked by a line centered on the page.
7. *Prof:*: for this use of a colon, see letter 3507, n. 11.
8. *with Prof: Lesley, I called upon Prof: Henry*: Tyndall had been in touch with Tyndale, Lesley and Henry about arranging a course of lectures in the United States. See letters 3504, 3507, 3510, 3519, 3542, 3543, 3558, 3589, 3594, 3639, 3669, 3670, 3684, 3702 and 3703.
9. *Henry . . . replied*: letters 3669 and 3684.
10. *the Young Men's Christian Association*: Founded in London in 1844, the Young Men's Christian Organization (YMCA) aimed to provide services to improve the spiritual conditions of young men, especially in cities.
11. *the Peabody Institute under Professor Morrison*: see letter 3510, n. 4; the name is Morison.
12. *the Lowell Institute, under Mr Lowell*: see letter 3589, n. 4 and n. 6.
13. *Academy of Music*: concert hall in Philadelphia.
14. *Prof. Henry . . . Washington*: see letter 3684, n. 7.
15. *he wishes you would write him*: Tyndall did so in letter 3702.
16. *you had . . . same matter*: Tyndall may have done so in a letter of 13 June 1872. This letter will be published in the forthcoming thirteenth volume of *The Correspondence of John Tyndall*.
17. *Prof: Henry . . . next*: see letter 3684.
18. *Van-Couver's Island*: large island off the west coast of Canada, just above the border with the United States.
19. *Yosemite Valley*: valley in California's Sierra Nevada mountain range, home to the Yosemite people. After being seized by white settlers, the valley became a National Park in 1864; the Yosemite people were driven out. See M. Dowie, *Conservation Refugees: The Hundred-Year Conflict between Global Conservation and Native Peoples* (Cambridge: MIT Press, 2011), pp. 1–14.
20. *Pike's Peak*: in the Rocky Mountains in the western United States.
21. *Please tell me when George Tyndall (son of Dr: John) died & how?*: Tyndall had told Tyndale of this death in letter 3670, see n. 12.

From Joseph Dalton Hooker 13 May 1872 3686

Royal Gardens, Kew. | 13th May.

Dear Tyndall,

I much fear that in your overwrought condition[1] I was wrong in letting you (and other overwrought friends) take my case[2] in hand.

You are right:—it should be the work of one hand and brain, which should submit its propositions to the rest.[3] All I can do is to offer my services as unrecognised Secretary, to be used as best you can.

What I fear is that my friends underrate the difficulty of the task they have taken in hand. I am as it were pledged to resign if the Gov[ernmen]t do not satisfy me.

I assume that Lord Derby is instructed to call for the correspondence (i.e. mine with Gladstone) and that you sent him Lord Russell's letter.[4]

The Memorial[5] is the next essential matter now, and should I suppose be sent in on the production of the correspondence.

Telegraph for me and I will run in at any time this evening and dine with you at the Athenaeum[6] if you like.

Ever yours | J. D. Hooker.

RI MS JT/1/TYP/8/2634
LT Typescript Only

1. *overwrought condition*: see letter 3683, to which this letter responds.
2. *my case*: his case against Ayrton; see, for example, letters 3671 and 3673.
3. *You . . . right*: Tyndall makes this point in letter 3683.
4. *I assume . . . letter*: see letters 3677 and 3681.
5. *The Memorial*: see letter 3671, n. 6.
6. *Athenaeum*: see letter 3372, n. 14.

From Thomas Henry Huxley 13 May 1872 3687

Jermyn Street | May 13th 1872.

My dear Tyndall,

Routing among my papers yesterday—I came upon the inclosed cinders of an old fire[1] which I always told you, you should see some day. They will be better in your keeping than mine.[2]

I hope you are better. We were all very sorry to miss you on the 7th.[3]

Ever yours faithfully | T. H. Huxley.

P.S. I reopen this to say, in reply to your note just received,[4] that I know of no reason why Lord Derby should not move for the papers.[5]

I wrote to Hooker to that effect immediately on receipt of your last.[6] | T. H. H.

IC HP 8.97
Typed Transcript Only

---

1. *the inclosed cinders of an old fire*: LT appended the following note to this letter: 'This refers to a movement made by Thomson, Murchison, Whewell, Lyell, Tait, and others to give Forbes the Copley medal [Nov. 3. 1859]. Huxley, who was not then on the Council of the Roy. Soc. wrote a long paper which Frankland read at the meeting, the result of which was that the Copley medal was otherwise awarded. Tyndall knew nothing about the matter at the time'. LT's dating is, however, incorrect; the year of the medal in question was 1854, when the RS chose Hooker over Forbes (Desmond, *Huxley*, p. 205). On Forbes, see letter 3462, n. 3. Huxley was forwarding the paper he wrote to Frankland on this occasion.
2. *They will . . . mine*: Tyndall responded in letter 3689.
3. *the 7th*: possibly a birthday celebration for Huxley (born 4 May 1825). See letter 3704.
4. *your note just received*: letter missing, but may follow up on letters 3682 and 3683, in which Tyndall discusses consulting with Huxley about the Kew affair.
5. *move for the papers*: see letter 3677.
6. *receipt of your last*: letter missing.

To Edward Stanley 13 May 1872 3688

13th May 1872

Dear Lord Derby

I beg to forward you a note from Lord Russell to Dr Hooker.[1] It contains its own explanation.

Hooker gratefully accepts your offer to move for the correspondence,[2] and others besides Hooker congratulate themselves that the matter has fallen into your hands.[3]

The next step of his friends will be to draw up a memorial[4] setting forth the facts, and signed by the principal scientific men of England, to be forwarded to Mr Gladstone.

Most faithfully yours | John Tyndall

RI MS JT/1/TYP/8/2635
Transcript Only

---

1. *a note from Lord Russell to Dr Hooker*: enclosure missing; see letter 3681.
2. *your offer to move for the correspondence*: see letter 3677. Hooker's acceptance was not explicitly articulated in any of the letters here, but letter 3686 conveys his belief that Derby

will proceed. Because letter 3686 also contains a reminder to Tyndall to send the note from Russell to Derby, we have placed it before the present letter. Derby responded to the present letter in letter 3691.

3. *others . . . your hands*: Tyndall was surely referring to himself here, but was perhaps also making to reference to other friends of Hooker; see especially Huxley's opinion that Derby should call for the correspondence in letter 3687.
4. *a memorial*: see letter 3671, n. 6. For discussion of drawing up the memorial, see letters 3683, 3686 and 3690.

To [Thomas Henry Huxley][1] 13 May 1872 3689

Royal Institution. | 13th. May, 1872.

What a capital simile "Cinders of an old Fire."[2]

I have duly labeled the document as received from you this day.

J. T.

Your every day illustrations are such as I might reach by hard discipline.

IC HP 8.98

Typed Transcript Only

---

1. *[Thomas Henry Huxley]*: identified by continuity with letter 3687.
2. *Cinders of an old Fire*: see letter 3687, to which this letter responds, particularly n. 1. Tyndall seems to have labeled the letter thus: 'HUXLEY'S LETTER TO FRANKLAND &c. | On the occasion of Prof. Forbes being proposed for the Copley Medal. | Sent to me by Huxley May 13 1872. | I had not previously seen this document. | John Tyndall | 13th May 1872' (RI MS JT/1/TYP/9/2988).

To Joseph Dalton Hooker [14 May 1872][1] 3690

Royal Institution of Great Britain | 14th May

My dear Hooker

Lord Derby has everything in his possession.[2] I expressed to him your grateful acceptance of his offer to move for the correspondence; and also said that many besides yourself had reason for congratulation that the matter had fallen into his hands.[3]

You will enormously facilitate matters by pulling the materials in a rough way together.[4] The circumstances attending the establishment of Kew and of your fathers[5] acceptance of it.[6] Its development to its present wonderful pitch. Its relations to the colonies:[7] Its scientific relations: The good it has done throughout the world to Botanical Science.[8] I think the Correspondence will suffice for the development of the present difficulties.[9] A great case may certainly be made of it.

If Huxley is not able to undertake it I am willing to do so. I have twenty five pages of my International book to write.[10] This I hope to accomplish by Sunday evening next. And then I should be willing to attack the memorial.[11]

I think going away this week is out of the question.[12] This confounded book[13] will hold us back. But I should willingly make a rush for three days of the week following.

Yours ever | John Tyndall

RBG, Kew Papers Relating to Kew 1867–1872, Ayrton Controversy, volume 1, f. 204

---

1. *[14 May 1872]*: The date '14 May' appears at the top of this letter in a different ink; we believe it was added by Hooker, but it makes sense in the sequence of letters. For similar issues, see letter 3633, 3682, 3683 and 3694.
2. *Lord Derby . . . possession*: papers relating to the Kew controversy; see letter 3674, 3675 and 3677.
3. *I expressed . . . into his hands*: in letter 3688.
4. *You will . . . way together*: Hooker expressed his intention to do so in letter 3692, the response to this letter. The list of topics that Tyndall provided is covered in the memorial to Gladstone, for which see letter 3671, n. 6.
5. *your fathers*: William Hooker.
6. *acceptance of it*: William Hooker became director of Kew in 1841.
7. *Its relations to the colonies*: For an introduction to the complicated ways in which science at Kew was related to the British Empire, see Endersby, *Imperial Nature*, pp. 17–20.
8. *Its scientific . . . Botanical Science*: Kew's scientific significance, and the respect therefore due to it and its director, was a major point of the controversy. See MacLeod, "Ayrton Incident," p. 52, and Endersby, *Imperial Nature*, pp. 277–8.
9. *I think . . . present difficulties*: Evidence suggests that in fact Hooker put together notes about the 'present difficulties'; see letters 3700 and 3705.
10. *I have . . . write*: probably *Forms of Water*, which was part of the International Scientific Series; see letter 3611, n. 3.
11. *the memorial*: to Gladstone; see n. 4.
12. *I think . . . question*: Hooker and Tyndall seem to have planned a trip out of London; possibly this is the trip referred to in letter 3647. Hooker's response to this letter (letter 3692) appears to overlook Tyndall's comment here.
13. *This confounded book*: *Forms of Water*; see n. 10.

From Edward Stanley 14 May 1872 3691

23 St James's Square, | May 14th 1872.

Dear Professor Tyndall

I will move for the papers[1] as soon as the House meets again. Your note of yesterday[2] conveying Dr Hooker's sanction for my doing so did not arrive in time to allow me to give the necessary notice. Meanwhile I can ascertain if there will be any difficulty about giving them. I have the M.S.S. here[3] and will return them to you if desired, as they may be wanted for the memorial.

Very faithfully yours | Derby.

RI MS JT/1/TYP/8/2636
LT Typescript Only

1. *I will move for the papers*: Derby was responding to letter 3688.
2. *Your note of yesterday*: letter 3688.
3. *the M.S.S.*: probably the correspondence that Tyndall had sent to Derby on Hooker's behalf, and that had also been sent to Russell. See letters 3674, 3675, 3677 and 3681.

From Joseph Dalton Hooker 15 May 1872 3692

Royal Gardens, Kew. | May 15th 1872.

Dear T.

Stick to your book[1] by all means and dismiss all else that you can from your mind.

Meanwhile I will put the memoranda together.[2] Nothing can now be done till the 27th when the House meets.[3]

Lord J. Manners[4] has written for copies of the correspondence (which I am supplying) "with a view to the course it may be advisable to take in the interests of the public."

A hot fight is raging between the Treasury and Ayrton a propos of the Clerk,[5] who threatens me with an action, I hear! Ayrton throws the blame on me, the Treasury on Ayrton. Neither party has troubled me as yet: no doubt I soon shall be in the thick of it.

Now, dear fellow, compose yourself to your book and I will to my work till we meet at the railway-station[6] at your time and bidding.

Yours | J. D. Hooker.

RI MS JT/1/TYP/8/2637
LT Typescript Only

1. *your book*: probably *Forms of Water*, which Tyndall discusses working on in letter 3690, to which this letter responds.
2. *Meanwhile . . . together*: as Tyndall had asked him to do in letter 3690.
3. *Nothing . . . meets*: see letter 3691; Hooker is referring to Derby's intention of moving for papers.
4. *Lord J. Manners*: John James Robert Manners (1818–1906), MP and adviser to Disraeli, the leader of the Conservative party; Manners held various cabinet posts in Disraeli's governments. He became the seventh duke of Rutland in 1888.
5. *a propos of the Clerk*: Robert Smith; see letter 3647, n. 3.
6. *meet at the railway-station*: This may refer to the trip that Tyndall cancelled in letter 3690 and that may also be the subject of discussion in letter 3647. See also letter 3693.

From Joseph Dalton Hooker 16 May 1872 3693

May 16/72

Dear Tyndall

I quite forgot that we dine with the Flowers[1] on Whitsunday,[2] so I fear we must put off[3] till the Saturday following our run.

I have nothing to trouble you with meanwhile. I entirely understand your note.[4]

Ever y[ou]rs | J. D. Hooker

RI MS JT/1/TYP/8/2638
Transcript Only

1. *the Flowers*: probably the family of William Henry Flower (1831–99), a zoologist who served as curator of the Hunterian museum at the Royal College of Surgeons and as Hunterian chair of comparative anatomy and physiology.
2. *Whitsunday*: 'the seventh Sunday after Easter, observed as a Christian festival in commemoration of the events described in Acts 2, when the Holy Spirit descended on the disciples' (*OED*). In 1872, Whitsunday fell on 19 May.
3. *I fear we must put off*: may refer to the excursion mentioned in letter 3692.
4. *your note*: letter missing.

To Joseph Dalton Hooker [18 May 1872][1] 3694

Royal Institution of Great Britain | 18th May.

My dear Hooker

I am in the thick of my work & I hope to break its heart by tomorrow.[2]

We ought to have the decks cleared for action by the 31st. It will be a great thing to be able to distribute correspondence[3]—therefore I propose acting at once on Lord Derby's suggestion to have it printed.[4]

If you do not know a printer whom you would wish to have it I shall place it at once in the hands of George Spottiswoode.[5] or *[Clowes]*.[6]

By Wednesday you could have the proofs & you would be best able to correct them.

Yours ever | John Tyndall

Keep up your heart old fellow. They shall know that men of science can use a sledge hammer—

RBG, Kew Papers Relating to Kew 1867–1872, Ayrton Controversy, volume 1, f. 210

---

1. *[18 May 1872]*: The date '18 May' appears at the top of this letter in a different ink; we believe it was added by Hooker, but it makes sense in the sequence of letters, as the present letter seems to respond to letter 3692. For similar issues in dating, see letters 3634, 3682, 3683 and 3690.
2. *I am . . . tomorrow*: probably on the book *Forms of Water*; see letter 3690, n. 10.
3. *distribute correspondence*: for discussion of the correspondence, see letters 3677 and 3688.
4. *Lord Derby's . . . printed*: in letter 3677.
5. *George Spottiswoode*: William Spottiswoode was the Queen's printer and seems to have handled this correspondence (see letter 3706, n. 8). As we cannot locate a George Spottiswoode, we think Tyndall may have written George when he meant William; possibly he was thinking of Spottiswoode's partner, George Edward Eyre (1818–87).
6. *[Clowes]*: possibly the printing firm William Clowes and Sons, first established in 1803.

From Peter Guthrie Tait 18 May 1872 3695

17 Drummond Place, | Edinburgh, 18/5/72.

My dear Tyndall,

I have only today succeeded in getting hold of Zöllner's Treatise on Comets:[1] and have read rapidly through it. I write now merely to call your attention to it in case you may not have seen it—for you, in the excellent company of Thomson (not to say Helmholtz and Hofmann[2]) are pitched into in most unmitigated style. I suffer, but mainly by being employed as a sort of missile to fling at you and the others. However, there will be a splendid row, which is some little consolation.

Yours truly | P. G. Tait.

RI MS JT/2/11/120

1. *Zöllner's Treatise on Comets*: see letter 3640, n. 8. Zöllner's attack on Tyndall is also discussed in letters 3642, 3649, 3652 and 3655. For Tyndall's response to the present letter, see 3697.
2. *Hofman*: Auguste Wilhelm von Hofmann (1818–92), a German chemist who specialized in organic chemistry. Hofmann was the first director of the Royal College of Chemistry, a position he held from 1845 until 1864.

To Frances Colvile 19 May [1872][1] 3696

Royal Institution of Great Britain | 19th May

My dear Lady Colville[2]

Take your choice & throw the others in the fire.

I send you two of the prints[3] which you will probably prefer.

Yours ever | John Tyndall

I cannot feel natural when the camera is pointed at me.

BL Add MS 60634, f. 130

1. *[1872]*: date proposed by LT.
2. *Colville*: Tyndall spelled her name incorrectly here.
3. *two of the prints*: enclosures missing. In letter 3465 Tyndall discussed a photograph taken of himself by the firm Elliot and Fry. He was also photographed by Barraud and Jerrard and by Whitlock in this period.

To Peter Guthrie Tait 19 May 1872 3697

Royal Institution of Great Britain | 19th May 1872.

My dear Tait

Oddly enough the thought of you had crossed my mind two minutes before your note[1] came. Hirst brought me intelligence about Zöllners book[2] from Paris. He saw it on Bertrand's table;[3] and I have since glanced over it—not read it, myself. I can see that he means to mangle me—Kill me first and chop me to mincemeat afterwards. But whether it is that the fire of my life has fallen to a cinder, the book has produced singularly little disturbance in my feelings. I suppose I am supported by the growing conviction that though the average estimate of what little a man can do in this world may be based on exaggeration on the one side and depreciation on the other, it is sure to be pretty fair at last.

Ten years ago I should have been at the throat of Zöllner, but not now.

I would rather see you and Clausius friends than Zöllner and myself.[4] Trust me C. is through & through an honest high minded man.

Yours ever | John Tyndall

RI MS JT 2/13c/1399–1400

1. *your note*: see letter 3695.
2. *Zöllners book*: see letter 3640, n. 8, and for additional discussion of this topic, letters 3642, 3649, 3652 and 3655.
3. *Bertrand's table*: see letter 3652, n. 5.
4. *I would rather see you and Clausius friends than Zöllner and myself*: see letters 3591 and 3642.

From Edward Stanley 22 May 1872 3698

Knowsley, Prescot. | May 22d 1872.

Dear Professor Tyndall,

I wrote some days ago to Lord Granville,[1] as leader of the House of Lords, to ask whether he would object to lay the papers.[2] He said he would consult the departments concerned, and let me know: and I am awaiting his answer. If the papers are agreed to be given, as I hope and rather expect, I shall move for them and they will be printed as public documents, without cost or trouble to any one. If they are refused, we will then see about the expediency of circulating them on our own account. But I hope there may be no need of that.

Very faithfully yours | Derby.

RI MS JT/1/TYP/8/2640
LT Typescript Only

1. *Lord Granville*: see letter 3660, n. 1.
2. *lay the papers*: see letter 3691, which this letter follows.

From Joseph Dalton Hooker 22 May 1872 3699

Royal Gardens, Kew. | May 22d 1872.

Dear Tyndall,

I really do not know what to think of the enclosed.[1] I wish it was more to the purpose, but you must judge.

I have nothing more from the Board.[2]

Please return me the documents I put into your hands at Colvile's.[3] I have no copy of Board's memo. If Lord Derby has returned you the papers,[4] it would be well to have regard to my two first letters to Gladstone.

Tyler and wife[5] lunch here on Friday. If you want a walk do trot out.

I shall be at Phil[osophical]. Club[6] to-morrow.

Ever yours | J. D. Hooker.

RI MS JT/1/TYP/8/2639
LT Typescript Only

---

1. *the enclosed*: We surmise that this was letter 3698, forwarded by Tyndall to Hooker (though any covering letter is missing). Perhaps Hooker felt that Derby was not moving matters along quickly enough.
2. *I have nothing more from the Board*: the Board of Works; Hooker may be making reference to the matter discussed in letter 3679.
3. *Colvile's*: James William Colvile (1810–80), a lawyer and husband of Frances Colvile. Colvile served as a judge in India from 1845 until 1859, the last four years of which he was the chief justice of Bengal. He was elected FRS in 1875.
4. *If Lord Derby has returned you the papers*: probably those referred to in letters 3674, 3675, and 3677.
5. *Tyler and wife*: possibly LT's misspelling for Tylor, in which case Hooker's guests may be the anthropologist Edward Burnett Tylor (1832–1917) and Anna Tylor (née Fox, n.d.).
6. *Phil[osophical]. Club*: the RS's Philosophical Club, founded in 1847 (see letter 3380, n. 2). Tyndall was elected to it on 10 May 1855 (Journal, 12 May 1855, RI MS JT/2/13c/745).

From Joseph Dalton Hooker 25 May 1872 3700

Royal Gardens, Kew. | May 25th 1872.

Dear Tyndall,

I don't know where to hunt for Ayrton's onslaught on the Queen,[1] but will get it for you somehow.

I am putting concisely the cases of the reprimand and of the clerk[2] and will forward them this afternoon to you.

Lord J. Manners[3] writes that he approves the correspondence being called for,[4] and thinks its "probable" effect will be the removal of Ayrton. I do not. Lord Russell was as eager as ever, but so feeble bodily, that I shall be astonished if he carries out his resolution of attending at the House of Lords. He wholly approves of Lord Derby's action, but thinks nothing direct will come of the movement, but much indirect good.

He has immense faith in the appeal to Gladstone from scientific men.[5]

He thinks the Directorship of Kew should be a Treasury appointment and have a seat at the Board of Works at any rate when Kew matters are discussed.

But he is so terribly deaf and frail that I doubt his being able to back these views; except perhaps by a letter to Gladstone when the memorial goes in.

Lady R[ussell]. is anxious that Lord Rollo,[6] now at Oxford, should stick to science of which he is very fond, but has gone off to History and law, not seeing his way to anything in science.

Charlie[7] goes to lunch there to-day and birds' nests[8] with Lady Agatha.[9]

Ever yours | J. D. Hooker.

Lady Agatha wants an autograph of Carlyle.

RI MS JT/1/TYP/8/2641
LT Typescript Only

1. *Ayrton's onslaught on the Queen*: In 1866, Ayrton criticized Queen Victoria for withdrawing from public life following the death of her husband, Prince Albert (*ODNB*). Hooker seems to be responding to a missing letter from Tyndall. This correspondence probably has to do with the memorial to Gladstone, for which see letter 3671, n. 6.
2. *the cases of the reprimand and of the clerk*: on the clerk (Robert Smith), see letter 3647, n. 3. By 'the reprimand', Hooker probably has in mind an incident that occurred soon after Ayrton became commissioner of works: Ayrton reprimanded Hooker over Kew's accounts, then realized that the reprimand was based on a mistaken reading of the accounts. See Endersby, *Imperial Nature*, p. 283, and MacLeod, "Ayrton Incident," p. 53.
3. *Lord J. Manners*: see letter 3692, n. 4.
4. *the correspondence being called for*: see letters 3691 and 3692.
5. *the appeal to Gladstone from scientific men*: see letter 3686, n. 5.
6. *Lord Rollo*: see letter 3490, n. 3.
7. *Charlie*: Charles Paget Hooker (1855–1933), Hooker's son and third child; he became a physician.
8. *birds' nests*: 'to search for bird's nests' (*OED*).
9. *Lady Agatha*: Mary Agatha Russell.

To Frances Colvile 28 May [1872][1] 3701

28th May

My dear Lady Colvile,

I hope I do not misinterpret your note regarding the photograph.[2] I understood you to desire me to sign one which you purposed to send to me, and to save you the trouble of sending it I forestalled matters by sending you such as I possessed. If I have blundered excuse me—I shall be only too willing to correct my blunder.

Yours ever | John Tyndall

BL Add MS 60634, ff. 131–2

1. *[1872]*: dated by continuity with letter 3696.
2. *your note . . . photograph*: Colvile's letter is missing.

To Joseph Henry 28 May 1872 3702

28th May 1872

My dear Professor Henry

I lose not a minute in thanking you for your excellent letter.[1] Trust me that not the least honouring and gratifying part of this transaction is the fact that you have consented to act in it as my friend & adviser.

There is but one drawback to my happiness in going to the United States.—I cannot go to California.[2] I have been so oppressed with work that no time has been at my disposal to put together the lectures that I intend to give in the United States. I am also weary and must have some Alpine air so that I shall not be able to reach America before October.

I am perfectly willing to lecture in the cities you mention giving in each of them a course of six lectures. This would occupy a fortnight in each city, for I should not like to give more than three a week.

I have never tried the experiment of repeating the same lecture at different places, and may find it easier than I anticipate: but it is always safe to have a day beforehand for the preparation of experiments & the arrangement of apparatus. Possibly after a first trial I may be able to lecture oftener—but this is doubtful for the night after a lecture is only too likely to be a sleepless one.

As the time approaches I feel more and more pleasure in the thought of going to America. And I trust that this will not be my last visit and that some time not far distant I may go over expressly to see that glorious country[3] which has been so vividly described by Prof. Whitney,[4] and by my excellent friend M^r^. Clarence King.

I ought to be in London on the last day of this year, and I am willing to be in America as early as you like in October. I am told that your lecturing season does not commence till about the middle of October.

I intend to take two assistants with me, one a thoroughly skilled one whom I have trained myself. The other a young fellow who has besides helping in the lectures & experiments acted as a kind of secretary for some years.[5]

With renewed thanks believe me | yours ever faithfully | John Tyndall

John Tyndall to Joseph Henry, 28 May 1872, mssRH 3965, Box 56, William Jones Rhees Papers, The Huntington Library, San Marino, California

---

1. *your excellent letter*: letter 3684, which, along with the present letter, is part of a conversation about Tyndall's plans to lecture in the United States. See letters 3504, 3507, 3510, 3519, 3542, 3543, 3558, 3589, 3594, 3639, 3669, 3670, 3685 and 3703.
2. *to California*: to attend the meeting of the American Association for the Advancement of Science; see letter 3684.
3. *that glorious country*: possibly the American West.
4. *Prof. Whitney*: see letter 3510, n. 16.
5. *two assistants . . . some years*: The first of these is John Cottrell. The second, a young man named Miller, died of typhoid in Philadelphia (*Ascent of John Tyndall*, p. 309).

To Edward Livingston Youmans 28 May 1872 3703

May 28, 1872.

MY DEAR YOUMANS: You will have your kindness toward me tested by Prof. Henry in regard to the coming lectures. I wrote to him saying that I knew you would help me, and he has written to me to say he would call upon you.[1]

He proposes five cities (and perhaps others) in which to lecture—I have expressed my willingness to give a course of six lectures in each at the rate of three a week.[2] Two things render it desirable that the number should not exceed three a week. Firstly, I must keep up my physical vigor, and the night subsequent to a lecture is only too likely to be a sleepless one. Secondly, it is above all things desirable to make sure of the experimental arrangements the day before the lecture. . . .

Yours ever, | JOHN TYNDALL.

E. A. Youmans, 'Tyndall and His American Visit', *Popular Science Monthly*, 44 (February 1894), pp. 502–14, on p. 508

1. *I wrote . . . you*: see letters 3669 and 3684. These, and the present letter, concern Tyndall's plans to lecture in the United States. See also letters 3504, 3507, 3510, 3519, 3542, 3543, 3558, 3589, 3594, 3639, 3670, 3685 and 3702.
2. *He proposes . . . a week*: see letters 3684 and 3702.

From Thomas Henry Huxley 28 May 1872 3704

May 28th 1872.

"Will have a great deal of pleasure in seeing everybody," etc. &c. &c Oh yes!! (Aside, Damn[1] that bothering fellow)—I shouldn't risk it anyhow.[2]

Perhaps you forget that you owe me a dinner by reason of shameful desertion on my birthday.[3] Now pay your debts like a man. Sir J. D. Hay (who was so civil to me at Tangiers) and his wife[4] are coming to dine with us on Friday May 31st at 7 p.m. and I want you to meet them. We have asked the Hookers and a couple of Yankees[5] who are over here. Now be an amiable Pagan and write yes. The wife[6] shall send you a reminder in form!!

Ever yours faithfully | T. H. Huxley.

RI MS JT/1/TYP/9/2959
LT Typescript Only

1. *Damn*: This word appears as 'D-n' in the transcript, but the censorship was probably LT's.
2. *"Will . . . anyhow*: The context of this remark is not clear; the letter may be responding to a missing letter from Tyndall, or may be a fragment of which we have only the second half.
3. *my birthday*: 4 May; see letter 3687.
4. *Sir J. D. Hay and his wife*: see letter 3608, n. 8; John Hay Drummond Hay (1816–93), a British diplomat to Morocco, was married to Annette Cazytensen, the daughter of the Danish privy councilor.
5. *a couple of Yankees*: not identified; Americans.
6. *The wife*: Henrietta Huxley.

From Joseph Dalton Hooker [between 15 and 29 May 1872][1] 3705

CONFIDENTIAL | To Dr Tyndall.

(1) On Sunday Dec. 18th 1870, Mr Ayrton, unknown to me, visited my subordinate, the Curator of the Gardens,[2] and proposed his acceptance of a position in London as Secretary of the Parks (involving authority over me at Kew) and that he should supply me in his place with an officer of inferior

qualifications, pay and position; and he desired the Curator not to inform me of the proposal. He subsequently officially applied for the Treasury's sanction for this measure, and informed the Curator that he had done so, keeping me still in ignorance of it. The treasury (as I was privately informed) refused in a letter of sharp rebuke. The Curator had communicated to me every step of this transaction; which formed the subject of an official remonstrance on my part with Mr Ayrton.[3]

(2) At about the same time Mr Ayrton, without communicating with me, sent for the Curator to London, and instructed him to superintend the grounds, works, and planting round the memorial[4] in Hyde Park[5] (a work not under my authority, and of months' duration), and to inform me that I was to make arrangements for his doing so, and thus make my requirements subservient to my Curator's convenience. This was also the subject of a long correspondence which terminated in Mr Ayrton's sending the Curator back to Kew,[6] but only when I threatened to bring the matter before a higher tribunal.

(3) Mr Ayrton, without my knowledge, had plans and estimates made and an estimate submitted to the Treasury with the annual estimates, for extensive alterations in the principal Museum at Kew,[7] which would not only have been perfectly useless, but would have cost a very large sum. They would have necessitated the closing of the building for months, the rearrangement of its entire contents (it contains upwards of 20,000 articles) the making of new paths, cabinet work, heating apparatus, furniture, etc. for all which no provision was made in the estimate submitted to the Treasury, and they would have resulted in the disfigurement of the grounds and building too. I accidentally heard of this and communicated with Mr Stansfeld,[8] when the estimates were sent back to the Board for revision. Upon this my Curator (not I, the Director) was sent for by Mr Ayrton, to discuss the subject; and he, acting under my instructions, succeeded in getting Mr Ayrton to withdraw the estimate. In June, 1871, I found myself, without enquiry or notice of any sort superseded in the control of the construction and repairs of the Plant Houses and heating apparatus, which I had myself built and arranged, and for the supreme control of which I had a special warrant from the Board.[9] I respectfully requested an explanation, and referred Mr Ayrton to the warrant. After a month's delay I was officially informed that I had been thus suspended months previously, and was instructed to "govern myself accordingly".[10] I remonstrated officially, stating that I considered this act as "another of the conspicuous proofs of the disregard of the office of Director of the Royal Gardens, or of want of confidence in myself as Director, that I had experienced at the hands of the present First Commissioner, and that I claimed the privilege of forwarding a copy of the correspondence to the First Lord of the Treasury." This I did officially, Aug. 19th 1871[12] (receipt acknowledged Aug. 21st)

Aug. 22d Mr Ayrton asked me to furnish him with exact dates and particulars of the "proofs of disregard" etc.[13]

Aug. 31st. I answered, citing five conspicuous proofs, including the above transactions; and transmitted a copy of them, with another official letter to the First Lord of the Treasury.[14] (receipt acknowledged Sept. 4th)

Oct. 4th. Mr Gladstone informed me, in a private note, that he had communicated with Mr Ayrton, who "had sent him certain explanations", a copy of which he forwarded to me; adding that he had Mr Ayrton's assurance, that on the occasion of his visit to Kew (when he intrigued with my subordinate) he went "for the very purpose of personal and friendly communication with me."[15]

(After delay, from the illness of a relative[16]), on Oct. 16th, I officially addressed the First Lord of the Treasury, thanking him for his note, and pointing out that Mr Ayrton's "explanations" referred to but two subjects out of the five I cited;—that he not only evaded the first three of them, but had so expressed himself as if it was his object to lead Mr Gladstone to suppose that I had not brought them forward;—that his explanation of the 4th was at variance with the facts as I have stated them;—and that his excuses for the 5th were wholly unsatisfactory to me. I added that with regard to the visit to Kew, his assurances were at variance with the fact that though I was in my house when he came, he never asked for me there or at the garden gates, or of the Curator whom he went to at once:—that he desired the latter to conceal the proposal from me, and that he further concealed from me his application to the Treasury, which followed at once.

This was followed by a private communication from Mr Gladstone's secretary, that "a plan was under the consideration of the Government, by which my position in regard to the First Commissioner of Works would be materially altered."[17]

After three months of hope consequent on this, I was privately informed that the plan could not be carried out, but that the subject should not be dropped.

March 13th. I was summoned to meet some of the members of the Cabinet at Lord Ripon's, where I found Lord Ripon, Lord Halifax and Mr Cardwell, to whom I explained my official position and left the enclosed document.[18]

Since my appeal to the First Lord of the Treasury Mr Ayrton has proceeded unchecked with his arbitrary measures. In regard to scientific and other assistants, he refuses me the power of selection that Clause VII of the Civil Service rules allows to Heads of Departments. He is now trying to force an incompetent man (formerly in service here) upon this establishment, as a Clerk, <u>to be trained at Kew for subsequent transference to the Board in London</u>.[19] He has, without enquiry, withdrawn from my charge at Kew the distribution copies of the "Tropical African Flora", a work subscribed for by the Government and published at Kew under my direction, and he has sent them to the Stationery Office for sale.[20] He has, without communication with me, struck my name (Director) out of the official Blue Book of Estimates for the Parks, Gardens etc. and substituted that of the Secretary of the Board, thus reducing me to the position of a "Park Superintendent" in the matter of the control of the expenditure, as he already had of the control of the works.

Mr Ayrton's object being to deprive the Director of all authority, /his action/[21] has seriously compromised my position in respect of my subordinates. His intrigue with my Curator, had it not been met by that officer's integrity, would have proved subversive of all discipline; his arbitrary acts in regard to the control of the Plant-houses, Museums and appointments, are well known to the clerks in the office in London and to the Foremen and Gardeners at Kew, and some of them were in fact known to the latter for long before I became cognizant of them.

Meanwhile I have avoided any act that could embarrass the Government; I have refused offers to bring the subject before Parliament; I have withheld deputations both of scientific men and of Horticulturists from troubling Mr Gladstone, and I have refrained from asking mutual friends to interfere in the matter. I have relied on my own standing in the service (now 32 years); on the fact of the Garden, Museum, Herbarium, and Library of Kew being all creations of the late Director[22] and myself, acting throughout under the instructions, and with the encouragement of successive governments; and on the recognized value of the Institution to the public no less that to Botanists and Horticulturists in an instructional and scientific point of view, and to India and the Colonies for many of their sources of industry and prosperity.

I should add that I have never given Mr Ayrton any cause of offence. Previous to taking office in his address to his constituents he paraded his contempt of my profession and its duties,[23] and in the very week that he took office, he administered to me a wholly unmerited reprimand,[24] the first I ever received. His object I believe to be to force me to retire, to turn Kew into a park and to alter its whole character; and to this end he has so far succeeded as to deprive me of all authority as Director, and subject me to that of the Secretary to the Board[25] (who has, however, taken no advantage of this) and the Director of Works.

I believe that the removal of this establishment altogether from the category of the Public Parks is the only remedy now available.

I assume that Lord Ripon has now reported to Mr Gladstone the result of his and his colleagues' enquiry, but I do not know, and until I do, I am of course precluded from taking any further step.

Sir John Lubbock has copies of my correspondence with Mr Ayrton and Mr Gladstone.

RI MS JT/1/TYP/8/2612–5
LT Typescript Only

1. *between 15 and 29 May 1872*: We surmise that this letter was composed before 29 May 1872 (the date of letter 3706, in which Hooker mentioned a draft of the memorial to Gladstone, for which see letter 3671, n. 6) and after 15 May 1872 (the date of letter 3692, in which Hooker mentioned 'putting the memoranda together'; the present letter may be part of that memoranda).
2. *the Curator of the Gardens*: John Smith (n.d.); see Endersby, *Imperial Nature*, p. 283 and p. 377, n. 33, and MacLeod, "Ayrton Incident," p. 53.
3. *the subject . . . Mr Ayrton*: J. Hooker to the First Commissioner, 30 December 1870, in 'Correspondence between the Board of Works and Dr. Hooker relating to Changes proposed to be introduced into the Direction and Management of the Gardens at Kew', HL 213, Parliamentary Papers (1872), p. 120.
4. *the memorial*: The Prince Consort National Memorial, otherwise known as the Albert Memorial, which commemorates Queen Victoria's husband Prince Albert; the memorial was opened in 1872. See R. J. Callendar to J. Hooker, 20 December 1870, in 'Correspondence between the Board of Works and Dr. Hooker relating to Changes proposed to be introduced into the Direction and Management of the Gardens at Kew', HL 213, p. 120.
5. *Hyde Park*: a major park in central London. Originally a hunting park of King Henry VIII, Hyde Park was redesigned by Queen Caroline in 1728. The Albert Memorial stands in Kensington Gardens, part of the park.
6. *a long correspondence . . . to Kew*: See 'Correspondence between the Board of Works and Dr. Hooker relating to Changes proposed to be introduced into the Direction and Management of the Gardens at Kew', HL 213, pp. 119–25.
7. *Mr Ayrton . . . Museum at Kew*: These alterations seem to have involved replacing a staircase in one of the museums at Kew; see MacLeod, "Ayrton Incident," p. 53.
8. *Mr Stansfeld*: James Stansfeld (1820–98), a politician and social reformer. Gladstone made Stansfeld third Lord of the Treasury and financial secretary in 1869.
9. *In June . . . from the Board*: on the Plant Houses, see letter 3630, n. 1.
10. *I was officially . . . accordingly*: R. Callendar to J. Hooker, 17 August 1871, in 'Correspondence between the Board of Works and Dr. Hooker relating to Changes proposed to be introduced into the Direction and Management of the Gardens at Kew', HL 213, p. 28.
11. *First Lord of the Treasury*: William Gladstone; the prime minister traditionally holds this position.
12. *I remonstrated . . . August 19th, 1871*: J. Hooker to First Commissioner of Works, 19 August 1871, in 'Correspondence between the Board of Works and Dr. Hooker relating to Changes proposed to be introduced into the Direction and Management of the Gardens at Kew', HL 213, p. 107.
13. *"proofs of disregard" etc.*: See R. Callendar to J. Hooker, 22 August 1871, in 'Correspondence between the Board of Works and Dr. Hooker relating to Changes proposed to be introduced into the Direction and Management of the Gardens at Kew', HL 213, p. 108.

14. *I answered . . . First Lord of the Treasury*: J. Hooker to First Commissioner of Works, 31 August 1871, in 'Correspondence between the Board of Works and Dr. Hooker relating to Changes proposed to be introduced into the Direction and Management of the Gardens at Kew', HL 213, pp. 108–10.
15. *Mr Gladstone . . . with me*": This letter, and the correspondence referred to in subsequent paragraphs, is discussed in the memorial to Gladstone, for which see letter 3686, n. 5.
16. *illness of a relative*: Hooker's mother, Maria Sarah Hooker (1797–1872) became ill in the fall of 1871 and Hooker traveled to see her in September. See J. Hooker to E. Darwin, 15 September 1871, in F. Burkhardt, et al. (eds), *The Correspondence of Charles Darwin*, 28 vols (New York: Cambridge University Press, 2005), vol. 19 (1871), pp. 574–5; this is letter 7945 in the online edition of the Darwin correspondence.
17. *a private communication . . . materially altered*": See 'Correspondence between the Board of Works and Dr. Hooker relating to Changes proposed to be introduced into the Direction and Management of the Gardens at Kew', HL 213, p. 47.
18. *March 13th . . . enclosed document*: for this meeting, see letter 3630. The enclosure is missing.
19. *He is now . . . London*: for the affair of the clerk, see letter 3647, n. 3.
20. *He has . . . for sale*: see letter 3679, n. 2.
21. *his action*: in the typescript letter, these words are superscripted with a question mark.
22. *late Director*: William Hooker.
23. *Previous to taking . . . its duties*: In 1869, as he assumed the position of Commissioner of Public Works, Ayrton had given a speech denouncing gardeners (*Times*, 9 November 1869, pp. 6–7, on p. 7).
24. *wholly unmerited reprimand*: see letter 3700, n. 2.
25. *Secretary to the Board*: George Russell (1830–1911), a civil servant who became assistant secretary to the Board of Works in 1856 and secretary in the 1870s.

From Joseph Dalton Hooker 29 May 1872 3706

Royal Gardens, Kew. | May 29th 1872.

Dear Tyndall

Your two notes[1] and draft[2] came last night. I will see to the corrections the first thing after my return from Edinburgh, whither I go to-night, to enquire into this squabble concerning the Museum.[3] I hope to be back on Saturday. Meanwhile I am overwhelmed with preparations.

The draft is uncommonly well done, whatever may be the opinion of the others as to its form and substance. I have been hunting up verifications of the payments and estates,[4] hitherto with very partial success. Sometimes I had a few shillings more pay than I have stated; at others <u>none at all</u>, I find. It may be better to avoid details. We had all <u>double pay</u> on the expedition to the S[outh]. Pole,[5] as all similar expeditions had; per contra, I got nothing for my zoological and meteorological work any more than for botanical, and, as I stated, no pay whatever except for my duties as Ass[istan]t Surgeon, which was the pay of my rank. Again, I find I had my 1/2 pay when I returned from India[6] and was employed on the publications;[7] and on their conclusion I received a lump sum, calculated on a commutation of Ass[istan]t Surgeon's half-pay and Surgeon's full pay.

I cannot profess to be easy about printing the correspondence with Gladstone. I know it is against official etiquette, till called for in the House, or leave obtained, which, if refused, almost compels resignation, previous to publication. No doubt that Spottiswoode's office[8] is as safe as the Bank.

I am sorry about the Telegraph[9] which was only to ask if I should go in on Monday to Athenaeum.[10] I could not tell that you were not to dine elsewhere.

Ever yours | J. D. Hooker

RI MS JT/1/TYP/8/2642
LT Typescript Only

---

1. *Your two notes*: notes missing.
2. *draft*: presumably of the memorial to Gladstone; see letter 3671, n. 6.
3. *this squabble concerning the Museum*: not identified.
4. *verifications of the payments and estates*: The memorial (see n. 2) contains details of Hooker's pay in his various positions of government service.
5. *the expedition to the S[outh]. Pole*: Hooker traveled as assistant surgeon on HMS *Erebus* from 1839 to 1843.
6. *India*: where Hooker traveled from 1847 until 1851.
7. *the publications*: probably including J. D. Hooker, *Rhododendrons of Sikkim-Himalaya* (London: Reeve, Bentham, and Reeve, 1849–51), and J. D. Hooker, *Himalayan Journals; or, Notes of a Naturalist in Bengal, the Sikkim and Nepal Himalayas, the Khasia Mountains, &c.*, 2 vols (London: John Murray, 1854).
8. *Spottiswoode's office*: William Spottiswoode was Queen's printer.
9. *Telegraph*: LT adds a note: 'i.e. telegrams'. This correspondence is missing.
10. *Athenaeum*: see letter 3372, n. 14.

BIOGRAPHICAL REGISTER

This register contains the names and biographical details of people who are mentioned three or more times in the letters included in this volume. Biographical information about people mentioned once or twice is contained in the notes appended to the appropriate letter(s).

Conventions

The sources for the biographical entries are standard biographical dictionaries, such as the *Complete Dictionary of Scientific Biography*, the *Oxford Dictionary of National Biography*, and *American National Biography*. These have not been specified in the entries unless necessary. Additionally, the editors have found material from previous volumes of the Tyndall correspondence very useful. Where additional sources add important material, they are given in parentheses at the end of the entry.

Acheson, Millicent (1812–87), was the daughter of Archibald Acheson, 2nd Earl of Gosford (1776–1849) and the second cousin and wife of Henry Bence Jones. The pair married in May 1842 and had seven children.

Agassiz, Louis (1807–73), was a Swiss-born geologist who moved to the United States in 1847 to accept a position at Harvard University. He earned his PhD in Germany in 1829, and in 1830 earned his MD from Munich. He studied with a number of well-known naturalists. While at Harvard, he began to make his mark on geology, ichthyology, and paleontology. He earned the Wollaston Medal from the Geological Society in London, and the Copley Medal from the RS. He was notably opposed to Darwinian evolution, but continued to make strides in natural history throughout his career.

Airy, George Biddell (1801–92), was an English astronomer. He graduated from Cambridge in 1823 as senior wrangler (mathematics) and first Smith's Prizeman. His *Mathematical Tracts on Physical Astronomy* was published in 1826, the same year he was elected Lucasian Professor of Mathematics at Cambridge. He won the RS's Copley Medal in 1831 and became Astronomer

Royal in 1835, but was not elected FRS until 1836. He supported Greenwich becoming the meridian in 1884. He was president of the RS from 1871 to 1873.

Allen, Robin (1820–99), was a British poet who was appointed as a clerk in the secretary's office at Trinity House in 1837. From 1867 to 1881 he was the secretary of Trinity House. Some of his correspondence is held at the Cambridge University Library (MS Add.9225; "The Trinity House," *Times*, 25 April 1881, 9).

Argyll, 8th Duke of: see Campbell, George Douglass.

Ayrton, Acton Smee (1816–86), was a Liberal party politician and barrister. He practiced as a solicitor in India until 1851 and then served as an MP from 1857 to 1874. He was the first commissioner of works from 1869 to 1873 during the first Gladstone administration. Ayrton was transferred away from the office due to professional conflicts, one of which was the Ayrton scandal at the Royal Botanic Gardens at Kew involving Joseph Dalton Hooker. He is also known for the Ayrton light installed at the top of the clock tower of the Palace of Westminster.

Ball, John (1818–89), was an Irish politician and glaciologist. He was admitted to Cambridge University in 1835, but, being a Catholic unwilling to subscribe to the Protestant Articles, he did not have his degree conferred. During the 1840s he began to study the glaciers of the Alps, returning to Ireland to work for the Poor Law Commission during the Irish famine (1845–52). In Ireland he noted that the geological formations of the Dingle peninsula resembled those of the Alps and proposed that the landscape must have been formed by ancient glaciers. He became MP for Carlow County in 1852 and in 1855 was named assistant undersecretary of state in the colonial department by Lord Palmerston. He was the first president of the Alpine Club (1858–60) and edited the club's first publication, *Peaks, Passes, and Glaciers* (London: Longman, Green, Longman, and Roberts, 1859). His multivolume work *Alpine Guide*, published beginning in 1863, was a respected resource on mountain travel. From April to June of 1871, Ball traveled to the Atlas Mountains with Joseph Dalton Hooker, and a joint work about the trip was published in 1878.

Barnard, Jane (1832–1911), was one of more than eighty nieces and nephews of Michael and Sarah Faraday. She was the daughter of Michael's favorite younger sister, Margaret, and Sarah's brother, John Barnard. Jane Barnard lived with Sarah and Michael from the mid-1860s, as Michael's health declined.

Together, the women took over most of Faraday's administrative roles at the RI. After his death in 1867, the women had to leave the RI, but they settled in Islington, where they began work on his papers and his biography. They made his manuscripts more available to Faraday's earliest biographers, including Tyndall and Bence Jones. Like her aunt and uncle, Barnard was a member of the Sandemanian Church.

Bence Jones, Henry (1813–73), was a wealthy English physician and medical chemist. He was educated at Harrow School and Trinity College, Cambridge, before studying medicine at Saint George's Hospital, London. Attracted by the potential applications of chemistry to clinical medicine, he studied organic chemistry with Justus Liebig at Giessen (1841–43) before assuming a career as a successful physician in private practice and at Saint George's Hospital. In May 1842 he married a cousin, Lady Millicent Acheson, daughter of the 2nd Earl of Gosford, with whom he had seven children. Jones was elected FRS in 1846. He published important research on protein chemistry and made an abstract in English of du Bois-Reymond's work on electrophysiology (1852) that annoyed Carlo Matteucci (1811–68), the Italian pioneer of electrophysiology. In 1853 he became a manager (trustee) of the RI and served as its secretary from 1860 to 1872. Jones helped get Tyndall appointed a professor at the RI in 1853 and frequently entertained Tyndall at both his London home and Folkestone, where he kept a summer house.

Bentham, George (1800–84), was a lawyer and botanist. Independently wealthy after the death of his father and his uncle, philosopher Jeremy Bentham (1748–1832), George Bentham devoted himself to botany, working at Kew and collaborating with Joseph Dalton Hooker, with whom he published the three-volume *Genera plantaram* (1862–83). Bentham became a fellow of the Linnean Society of London in 1828, and served as president from 1862 to 1874. He was elected to the Athenaeum Club and became honorary secretary of the Horticultural Society in 1830. In 1854 he donated a sizable horticultural collection to the Royal Botanic Gardens at Kew. Bentham was elected FRS in 1862, after receiving the a Royal Medal from the RS in 1859.

Bodmer, Frederick (1853–71), was a student who was lost in the Alps during the summer of 1871. Tyndall and his companions searched for him for days, but never found him. Tyndall wrote the only information known about him: "He was 18 years and two months old, and was the only child of his father, who is a teacher in a secondary school at Neumünster, near Zürich. His father describes him as having but one defect—'a breakneck rashness'" (see letter 3508).

Budd, William (1811–80), was a British epidemiologist known for recognizing that infectious diseases were contagious. He studied in Paris, London, and Edinburgh from 1828–37. In 1842 he was appointed to Saint Peter's in Bristol, a poor-law hospital. He soon moved to the Bristol Royal Infirmary, and also was lecturer in medicine at the Bristol medical school from 1845 to 1855. In 1849, because of a cholera outbreak, he and others in Bristol argued that the disease was caused by a fungus, likely in contaminated air and water. He also worked on typhoid fever and tuberculosis, as well as animal diseases such as anthrax, pig typhoid fever, and cattle plague. He was elected FRS in June 1871. Tyndall and Budd became friends in 1866.

Busk, George (1807–86), was a surgeon and naturalist, with interests in paleontology and microscopy. After finishing his medical education at Saint Thomas's and Saint Bartholomew's Hospitals in London, he served on hospital ships at Greenwich from 1830 to 1855. He became well known for using the microscope in studying cholera and scurvy, and founded the Microscopical Society in 1839. He started studying paleontology in the mid-1850s with a focus on microscopic studies of lower life forms, but is well known for his work on Neanderthal skulls. He was elected FRS in 1850 and received the Royal Medal from the RS in 1871. He was a member of the Geological Society and the Anthropological Institute, as well as a member of the X Club.

Campbell, George Douglas, 8th Duke of Argyll (1823–1900), was a politician and a writer on science and society. In politics, he was noted for his strong opinions and combativeness; he held posts in various governments and served as Gladstone's secretary of state for India from 1868 until 1874. Campbell's scientific interests focused particularly on geology and ornithology, as well as on arguing against evolution; he insisted that living things exhibited evidence of design, a position that brought him into direct conflict with Darwin and Huxley.

Carlyle, Thomas (1795–1881), was a Scottish essayist, novelist, and historian. He was a major and highly influential figure of the period, the author of well-known works including *Sartor Resartus* (1838), *On Heroes, Hero-Worship, and the Heroic in History* (1841), and *Past and Present* (1843). After reading the latter, Tyndall commented that Carlyle "must be a true hero. My feelings towards him are those of worship 'transcendental wonder' as he defines it" (Journal, 18 July 1847, RI MS JT/2/13a/231). Carlyle was a formative influence for Tyndall as well as other scientific practitioners of the time. Tyndall met Carlyle and his wife Jane Baillie Carlyle (née Welsh) in the 1850s, and became close friends with Carlyle in the 1860s.

Carpenter, William Benjamin (1813–85), was an English physiologist, psychologist, and biologist. As a young man Carpenter was apprenticed to a doctor before attending medical school. After receiving his MD in 1839, he focused on teaching and writing about medicine. During his career, Carpenter advanced ideas about unconscious cognition and published *The Principles of Mental Physiology* in 1874. He also used microscopes to study *Foraminifera* (microscopic marine organisms) and theorized about oceanic movements. Carpenter was appointed Fullerian Professor of Physiology at the RI in 1844, the same year he became FRS. He was an administrator at UCL from 1856 to 1879 where he was instrumental in the unification of the University of London.

Clausius, Adelheid Rimpau (d. 1875), was the wife of Rudolf Clausius, and mother of John Clausius. Rudolf and Adelheid had six children together.

Clausius, Rudolf Julius Emanuel (1822–88), was a German physicist who was well known for being one of the founders of the science of thermodynamics. He was educated at the University of Berlin (BA 1844) and the University of Halle (PhD 1847). His famous paper on the theory of heat "Ueber die bewegende Kraft der Warme" was published in the *Annalen der Physik* in 1850. He taught physics at the Royal Artillery and Engineering School in Berlin in 1850 before moving to a professorship in mathematical physics at the Polytechnicum in Zurich in 1855. Tyndall translated many of his major works into English for publication in the *Phil. Mag.* and *Scientific Memoirs* in the early 1850s. In 1858 Clausius nominated Tyndall to the Naturforschende Gesellschaft in Zurich for his work on Swiss glaciers. Although Clausius would later move on to the University of Würzburg in 1867, and the University of Bonn in 1869, his time at Zurich proved to be the most academically fertile period in his life. It was here that he further developed thermodynamics and contributed to the kinetic theory of gases. In the 1870s his work laid more emphasis on electrodynamics.

Colvile [Colville], Frances Elinor (née Grant, 1838–1919), was the wife of Sir James William Colvile. The couple married in Calcutta Cathedral in 1857. Her father was Sir John Peter Grant (1807–93), lieutenant-governor of Bengal and then governor of Jamaica. After the Colviles returned to London in 1859, Lady Colvile established a salon at their Hyde Park residence, which was patronized by Lawrence Alma-Tadema, Robert Browning, George Eliot (Marian Evans), James Anthony Froude, Harriet Grote, Frederick Leighton, John Millais, Alfredo Piatti, and George Frederic Watts. Lady Frances and Sir James had one child, a son named Andrew John Wedderburn (1859–76).

Cottrell, John (n.d.), was the assistant in the Physical Laboratory at the RI from 1866 to 1885. He went with Tyndall to the United States in 1872–73 as his experimental assistant, and the audience liked his work. While in Philadelphia, Cottrell trained a man to help with the experiments and Tyndall brought him along when another assistant fell ill and ultimately died from typhoid. Cottrell also joined Tyndall as an assistant in his lighthouse observations, and in 1875 in his glacier observations.

Croullebois, Marcel (1843–86), was a lecturer in physics at the University of Aix-Marseille. In 1868 he presented public lectures on experiments related to an induction-coil machine developed by Heinrich Ruhmkorff (1803–77). In April of 1873 he published a work in *Annales de Chimie et de Physique* on the optical characteristics of quartz prisms. He was a member of the Marseille Academy, elected in 1874. In 1885 he translated Fleeming Jenkin's (1833–85) seventh edition of *Electricity and Magnetism* into French.

d'Alcantara, Pedro, Emperor Dom Pedro II (1825–91), also known as the emperor of Brazil, Dom Pedro II "the Magnanimous," was the last monarch of Brazil. He began his reign under regents at the age of five, and in 1840, at the age of fourteen, assumed the throne. He became FRS in 1871 and was made an honorary member of the RI during a visit to England in 1871. He was very popular in Brazil for most of his forty-nine-year reign, and ruled until he was ousted by a military coup in 1889. Two days after the coup he and his family were exiled to Paris, where he died two years later.

Darwin, Charles Robert (1809–82), was an English naturalist and geologist. Having received a BA at Christ's College, Cambridge in 1831, Darwin spent the following five years on HMS *Beagle*, making geological observations and collecting specimens for those interested in natural history. Darwin is best known for his book *On the Origin of Species* (London: John Murray, 1859), which he had meant to be an abstract of a planned larger work. *Origin* served as a starting point for other books that expanded elements of his work such as *The Variation of Animals and Plants under Domestication* (1868) that would advance his ideas on variation. Though he suffered from episodes of poor health, Darwin had two major works published in 1871–72 and continued his botanical observations with the assistance of his son Francis. In *The Descent of Man, and Selection in Relation to Sex* (London: John Murray, 1871) Darwin outlined his theories about human evolution and sexual selection. Darwin also identified evolutionary continuity in the facial expressions of humans in *The Expression of the Emotions in Man and Animals* (London: John Murray,

1872). While these works were met with some criticism, they were widely read, and Darwin's evolutionary theory was a major element of the scientific worldview promoted by Tyndall, Huxley, and other scientific naturalists.

Debus, Heinrich (1824–1915), was a German chemist. At the polytechnic in Kassel, he was taught chemistry by Robert Bunsen, until Bunsen left for the University of Marburg in 1839. Eventually Debus joined him there, and worked as Bunsen's personal assistant. During his time at Marburg, he finished a PhD in 1848, and met Tyndall, Frankland, and Hirst, who were being taught by Bunsen as well. Debus moved to England in 1851, where thanks to Frankland he was given a post as a chemistry teacher at Queenwood College and was for a time a colleague of Tyndall's. After leaving Queenwood in 1867, he spent three years at Clifton College, Bristol, then moved to London as lecturer in chemistry at Guy's Hospital Medical School (1870–88), adding to this the position of professor of chemistry at the Royal Naval College, Greenwich (1873–88) working under Hirst. Debus was the vice-president of the Chemical Society from 1871 to 1874 and a member of the RS's council in 1870–72. Most of his research was in organic chemistry, including the oxidization of ethyl alcohol to form glyoxal ($C_2H_2O_2$) and glyoxylic acid ($C_2H_2O_3$).

de la Rive, Auguste-Arthur (1801–73), was a Swiss physicist and one of the founders of the electrochemical theory of batteries. He served as professor of natural philosophy at the Academy of Geneva and worked principally on electricity, although he also investigated the specific heats of gases and calculated the temperature of the Earth's crust. His first publication, in 1822, focused on the Earth's magnetic field. He also published on electrical discharges through rarefied gases and on the aurora borealis. His principal book was *Traité d'électricité théoretique et appliquée*, 3 vols. (Paris: J.-B. Baillière et H. Baillière, 1854–58). Charles Vincent Walker (1812–82) was translating volume 3 into English when de la Rive asked Tyndall to read the proofs and correct what de la Rive called Walker's "Gallicisms." De la Rive was a member of the Society of Physics and Natural History in Geneva, of which Tyndall was elected as an honorary member in 1859.

Derby, Lord: see Stanley, Edward Henry.

Doty, Henry Harrison, was an American ship captain who lived in Norfolk, Virginia. In the 1860s he developed a lighthouse burner that could safely use kerosene; this was a significant invention because lighthouses were seeking new sources of fuel as whale oil had become more expensive. Though he offered his burner first to Trinity House in 1868, their lack of interest led him to take it to the French, who adopted a modified version for their national lighthouses. By 1874 versions of Doty's lamp had been adopted internationally, though not in England or the United States, where Doty came into conflict with the lighthouse authorities over patent issues (Thomas A. Tag, "The Doty Dilemma: Technological Advancements in Lighthouse Lamps and Subsequent Patent Infringement," U.S. Lighthouse Society *Keeper's Log* 16, no. 4 [Summer 2000]: 28–33).

Douglass, James Nicholas (1826–98), was a civil engineer in England, a member of the Institution of Civil Engineers (1861), and FRS (1887). He was appointed the chief engineer for Trinity House in 1862, and held the position until he resigned in 1892 due to an onset of paralysis. He worked on lighthouses for much of his career, collaborating with Tyndall and Faraday on experiments for lighting them efficiently. He was knighted in 1882 for his redesign of the Eddystone lighthouse, and in 1884 was appointed to a Trinity House committee for investigating the effectiveness of various lighthouse lamps.

du Bois-Reymond, Emil Heinrich (1818–96), was a German physiologist whose work helped to develop modern neuroscience. He is credited with discovering that nerve signals are transmitted electrically in the brain and muscular tissues, and developed equipment capable of measuring weak bioelectric currents. Many of his experimental investigations focused on the shocks generated by electric fish. Du Bois-Reymond was working in the laboratory of Heinrich Gustav Magnus (1802–70) in Berlin when Tyndall visited in the summer of 1851. Du Bois-Reymond made several trips to England during the early 1850s and delivered a course of lectures on electrophysiology at the RI in the spring of 1855. An abstract of his work on electrophysiology was published by Bence Jones in 1852. He was elected to the Berlin Academy of Sciences in 1851 and from 1858 until his death held a professorship at the University of Berlin, where he had also trained as a student.

Egerton, Mary Alice "May" (1848–1924), was the eldest daughter of Edward and Mary Egerton. She married Beauchamp Tower in 1902 ("Births," *Times*, 19 February 1848, 8; "Wills and Bequests," *Times*, 11 June 1924, 15).

Egerton, Mary Frances (née Pierrepont, 1819–1905), was the daughter of Charles H. Pierrepont (1778–1860), 2nd Earl Manvers. She and Edward Christopher Egerton (1816–69) were married in August of 1845. They had five children, including Hugh Edward Egerton (1855–1927), Charles "Charley"

Augustus Egerton (1846–1912), Mary Alice "May" Egerton (1848–1924), and Emily Margaret Egerton (d. 1894). She became one of Tyndall's close friends and admirers after they began exchanging letters in 1867.

Emerson, Ralph Waldo (1803–82), was an American poet, essayist, and a leader of the transcendentalist movement. He attended Harvard Divinity School and afterward moved to Concord, Massachusetts. He is best known for his essays, including "Nature" (1836), "The American Scholar" (1837), and "Self-Reliance" (1841). He toured widely, giving lectures and spreading his philosophical ideas. He became known as the "Sage of Concord," and worked alongside other American thinkers and poets. Tyndall was deeply influenced by Emerson's writing, which he urged upon his protégés Thomas Hirst and James Craven.

Emperor Dom Pedro II: see d'Alcantara, Pedro.

Faraday, Michael (1791–1867), was a natural philosopher, Fullerian Professor of Chemistry at the RI, and a member of the Sandemanian Church. His extensive research focused on electricity and encompassed electrochemistry, electromagnetism, diamagnetism, and magneto-optics (the Faraday effect). Faraday also held the positions of director of the laboratory (1825–67) and superintendent of the house (1852–67) at the RI. He was Tyndall's patron and helped him obtain the position of professor of natural philosophy at the RI in 1853. Despite significant differences between their views on religion and on the nature of magnetism, they held the highest mutual regard for each other. Tyndall's *Faraday as a Discoverer* (London: Longmans, Green, 1868) was his tribute to his mentor. In 1867 Tyndall succeeded Faraday as superintendent of the house and director of the laboratory at the RI.

Faraday, Sarah (née Barnard, 1800–79), was the daughter of Mary Boosey and Edward Barnard, a successful London silversmith. She was a Sandemanian and likely came to know Michael Faraday at the Sandemanian meeting house in Paul's Alley, London. They married in 1821. She was a staunch supporter of her husband. When the couple lived in the RI and in Hampton Court, Sarah managed most of the domestic work of the RI. Along with her niece, Jane Barnard (1832–1911), she took over Faraday's administrative duties in the 1860s, when his health was in decline. Tyndall frequently dined and socialized with Michael, Sarah, and Jane. Tyndall wrote of Sarah Faraday: "She is very motherly to me, and Faraday says that were it not for her own feebleness she must become quite a mother to me, and take care of me" (Journal, 2 March 1855, RI MS JT/2/13c/730).

Frankland, Edward (1825–99), was a chemist and one of Tyndall's closest friends. He worked initially as assistant to the chemist Lyon Playfair and then taught at Queenwood College, where he met Tyndall and taught laboratory-based science courses with him. He and Tyndall then commenced studies at the University of Marburg, but in 1848 Frankland proceeded to Justus Liebig's laboratory at the University of Giessen. Returning to England, he taught briefly at the Putney College of Engineering before being appointed to the chair of chemistry at Owens College, Manchester in 1851; later he became professor at the Royal College of Chemistry. One of Tyndall's closest friends, he was elected FRS in 1853 (with Tyndall's support) and was awarded the Royal Medal from the RS in 1857. Frankland held a tenured post at the RI, following the retirement of Michael Faraday, from 1863–68. Frankland held a temporary position at the Royal College of Chemistry from 1865 to 1868, which became a permanent post in 1868. During his time at the college Frankland began courses in practical laboratory skills for teachers, a subject most were inexperienced with at the time. In his early researches Frankland isolated a number of new organometallic compounds, thus helping to establish the field of organic chemistry. During the years 1868–76 Frankland was active in research on water quality, campaigning for higher standards for potable water. From 1871 to 1872 he was president of the Chemical Society.

Froude, James Anthony (1818–94), was a historian, journal editor, and man of letters. Froude's *Nemesis of Faith* (1849) caused a scandal by revealing his loss of traditional faith. His *History of England*, published between 1856 and 1870, was popular but came in for criticism by historians. Over the summers of 1869–70 Froude resided temporarily in Ireland while working on *The English in Ireland in the Eighteenth Century*. In that work he presented ultraunionist views about the English governance of Ireland, on the basis of what he perceived to be the racial inferiority of the Irish. He was a friend of Tyndall's and also a member of the Athenaeum Club, the Breakfast Club, and other organizations.

Galton, Douglas (1822–99), was a British engineer, best known for his work in sanitary engineering. He was educated at the Royal Military Academy, Woolwich and was commissioned in the Royal Engineers in December 1840. He was elected FRS in 1859. He was a member of the Royal Commission on Railways in 1866 and the director of Public Works and Buildings from 1869 to 1875. Starting in 1871, Galton served as a general secretary for the BAAS, and was elected president of the organization in 1895. He advocated for improving education for women and for children with learning disabilities. He also received recognition for his work with the sick and wounded on both sides of the Franco-Prussian War.

Ginty, Margaret (née Roberts, 1824–92), was the wife of William Gilbert Ginty (ca. 1820–66), Tyndall's friend and colleague from the Ordnance Survey of Ireland, for which Tyndall worked from 1839 to 1842. Margaret and William married on 18 May 1846, in Killucan, Westmeath, Ireland. Together they had five children: Kathleen, William Roberts, Fanny, George, and Fred. In 1855 Margaret and the children traveled to Rio de Janeiro to join her husband, who had gone there the previous year to manage Rio's gas works (*Ascent of John Tyndall*, 91, 100). The sons remained in England for their education. When William died in 1866, Tyndall became Margaret's trustee and "looked out for the education and employment of her children" (*Ascent of John Tyndall*, 198). Margaret Ginty returned to England within a few years of her husband's death.

Gladstone, William Ewart (1809–98), was a Liberal politician, author, and prime minister. Gladstone was born into a wealthy family. His father, Sir John Gladstone (1764–1851), was a successful merchant and politician who owned slave-labor sugar plantations in the West Indies and traded agricultural goods across the Atlantic. William Gladstone was educated at Eton College and Christ Church, Oxford as part of his father's desire to see his son a successful politician. Gladstone was elected to the House of Commons in 1832, was appointed president of the Board of Trade in 1843, served as chancellor of the exchequer starting in 1852, and became prime minister (the first of four times he would hold the position) in 1868. Gladstone also repaired relations with the United States, which had been tense since the American Civil War (1861–65), and enacted military reforms. In this last, as in other matters, his aims were hampered by an increasingly tense relationship with Queen Victoria. Gladstone's government had good relations with the scientific community (despite Gladstone's lack of personal engagement with science) until Joseph Dalton Hooker came into conflict with Ayrton; Gladstone attempted to remain detached from the matter, but in essence sided with Ayrton.

Goethe, Johann Wolfgang von (1749–1832), was a preeminent German Romantic writer, best known for his *Götz von Berlichingen* (1773), *The Sorrows of Young Werther* (1774), and *Faust* (1808/1832). His extensive writings included his theory of colors and a study of the structure and function of plants. During his time in Marburg (1848–50) Tyndall became familiar with some of Goethe's works. Tyndall and Hirst frequently included quotes and poetry from Goethe in their correspondence. Tyndall later wrote an essay on Goethe's theory of color, "Goethe's Farbenlehre: Theory of Colors II," *Popular Science Monthly* 17 (July 1880): 312–21, which he first delivered as a discourse at the RI on 19 March 1880.

Hamilton, Archibald Henry (1835–1914), was a son of mathematician William Rowan Hamilton (1805–65). He was educated at Trinity College, Dublin and did well. He went to the BAAS meeting in Liverpool in 1854 with his father, expressing an interest in science. In 1860 he was ordained and began his duties as a clergyman in Castleknock, eventually getting his own parish in 1864 in Clogher. After 1867, he held a series of curacies, one of them being Omagh in 1871–72. Reportedly, by 1873, he had intellect "of a high order" but "eccentricity in him seem[ed] fast ripening into insanity" (Thomas L. Hankins, *Sir William Rowan Hamilton* [Baltimore, MD: Johns Hopkins University Press, 1980], 382).

Helmholtz, Anna (née von Mohl, 1834–99), was a prominent salonnière in Berlin and a translator of a number of scientific works, including Tyndall's *Sound* and *Faraday as a Discoverer* (London: Longmans, Green, 1868). She was the second wife of Hermann Helmholtz. The couple married on 16 May 1861, a year after the death of Hermann's first wife. Anna was an intelligent, well-educated woman. Fluent in English, French, and German, she was a well-known translator and largely did the translations of works from English into German when her husband had been tasked to do them.

Helmholtz, Hermann Ludwig Ferdinand (1821–94), was a German physicist and physician. Educated in military medicine at the Friedrich-Wilhelms Institute in Berlin, he became the first to hold a regular post as teacher of anatomy at the Berlin Academy of Arts in 1848. His interest in physics was already developed by 1847, when he published his paper *Über die Erhaltung der Kraft*, which discussed the mathematical principles of the conservation of energy. This work was based on his study of muscle metabolism, and he argued that there was no such thing as a vital "force." He became professor of physiology at the University of Königsberg, and became professor of anatomy and physiology at the University of Bonn in 1855. It was at Bonn that he increasingly turned his attention to physics, a process that continued during his time at Heidelberg from 1858 onward. In 1871 he became a professor in physics at Humboldt University in Berlin, working on acoustics and electromagnetics. In addition to his own scientific work, Helmholtz also translated works of science into German, but this was usually the work of his wife, Anna Helmholtz. Called "an intellectual giant" by James Clerk Maxwell, Hermann Helmholtz was involved in the fields of hydrodynamics, thermodynamics, and electrodynamics.

Helps, Arthur (1813–75), was a career public servant, author, and social reformer. He was educated at Eton College and at Trinity College, Cambridge, and was elected to the Cambridge Apostles, also known as the Cambridge Conversazione Society, in 1833. Helps served as a private secretary to the chancellor of the Melbourne administration and to the chief secretary for Ireland until 1841. Afterward he directed his attention toward literary aspirations and social causes. In 1860 Helps was appointed to the position of clerk of the Privy Council, which he held for the rest of his life. As clerk, he acted as an intermediary in the communications between Hooker and Gladstone's administration during the dispute over the Royal Botanic Gardens at Kew.

Henry, Joseph (1797–1878), was an American physicist and first director of the Smithsonian Institution from 1846 to 1878. He was a professor at the Albany Academy in New York from 1825 to 1832, when he moved to New Jersey College (now Princeton University). He was an inventor of the electromagnetic motor in 1829 and also studied astronomy, sunspots, and solar radiation. His work in meteorology at the Smithsonian helped to establish the United States Weather Bureau. Henry was chair of the American Lighthouse Board from 1871 to 1878.

Herschel, Alexander Stewart (1836–1907), was an astronomer and professor of natural philosophy. He is known primarily for his work on meteor spectroscopy and for his identification of comets. He was the fifth of twelve children of Sir John Frederick William Herschel (1792–1871) and Margaret Brodie (1810–84). Herschel completed his BA in mathematics at Trinity College, Cambridge in 1855, and eventually his MA in 1877. While an undergraduate student at Trinity College, he worked with James Clerk Maxwell. In 1861 he began his career of observing meteors at the Royal School of Mines, London. Herschel then became a professor of natural philosophy at Anderson's University, Glasgow from 1866 to 1871. In 1871 he received a new appointment as first professor of physics and experimental philosophy at the University of Durham College of Science, Newcastle upon Tyne. He was elected FRS in 1884.

Herschel, John Frederick William, first baronet (1792–1871), was an English astronomer and mathematician. He was the only child of astronomer William Herschel and Mary Pitt. He was educated at Cambridge, where he was senior wrangler in mathematics and a founding member of the Analytical Society. He took over his father's observations in 1822, and focused on double stars and nebulae. In 1831 his scientific and intellectual prowess had earned him a knighthood in the Royal Guelphic Order. He was often compared to Newton, and was master of the mint from 1850 to 1855. He was buried in Westminster Abbey, next to Newton.

Hirst, Thomas Archer (1830–92), was a mathematician and one of Tyndall's closest friends. His father, Thomas, died when Hirst was twelve, and he was denied much of the inheritance. Tyndall and Hirst met when they were working on the West Yorkshire Railway. Hirst studied mathematics and chemistry, following Tyndall's path to Marburg, Germany. He was elected FRS in 1861 and joined the council in 1864. He was a founding member of the X Club and helped form the London Mathematical Society in 1865. In August of that year he was appointed professor of mathematical physics at University College, London; by 1873 he was the director of studies at the Royal Naval College in Greenwich. In 1883 he was awarded a Royal Medal from the RS. He kept a journal from the age of fifteen.

Holland, Henry, first baronet (1788–1873), was an English physician and president of the RI from 1865 to 1873, seeing it through a major construction project that added new laboratories for chemistry and physics. He was also an alpinist and explorer who made eight visits to the American continent. He was physician to Caroline (1768–1821), Princess of Wales (1795–1820) and later queen (1820–21). Upon the accession of Queen Victoria in 1837, he was appointed physician-extraordinary and in 1840 became physician-in-ordinary to Prince Albert. He was a firm supporter of the RI and other groups that spread scientific knowledge.

Hooker, Joseph Dalton (1817–1911), was a British botanist who specialized in taxonomy and the geographical distribution of plants; he was the younger son of William Hooker. He arguably became the most famous British botanist of the Victorian era. After studying medicine at the University of Glasgow he became a naval surgeon and spent the years 1839–43 on James Clark Ross's geomagnetic survey of the southern oceans, where he collected plants and served as surgeon. Subsequently he explored and collected plants in India (1847–51), achieving considerable mountaineering success, before joining his father as deputy director of the Royal Botanic Gardens at Kew in 1855. Hooker's research focused in part on plant distribution, a question that had theoretical implications but also practical ones, as plants were essential to Britain's empire. Hooker assumed the directorship of Kew upon his father's death in 1865. In this position, which he held until 1885, Hooker pursued the gardens' involvement in empire first established by his father. His tenure was also marked by tensions over Kew's public role: in many ways, Hooker resisted the public's interest in the gardens, which he tried to preserve as a center of scientific research. Hooker served as president of the BAAS (1868), and was knighted several times and received many scientific awards, including one of the RS's Royal Medals for 1854 and its Copley Medal for 1887.

Hooker, William Jackson (1785–1865), was a British botanist. He was appointed regius professor of botany at Glasgow University in 1820, and greatly expanded the university's gardens and the number of botany students. In 1841 he relocated to London to become the first full-time director of the Royal Botanic Gardens at Kew. Hooker expanded the gardens, which became enormously popular with the public and made Kew into a center for botanical researchers, through his collection of herbaria. He also amplified the gardens' role in the British empire; for example, under Hooker's oversight, the *Cinchona* plant (the source of quinine, and native to South America) was cultivated in India. Hooker's son Joseph Dalton Hooker succeeded him as director of Kew.

Huxley, Henrietta Anne (née Heathorn, 1825–1915), was born in the West Indies to Henry Heathorn and Sarah Henrietta Richardson (née Harris), and was schooled in Neuwied, Germany, for two years as a young woman. In 1843 she, her mother, and her half sister moved to Australia to join her father. In 1847 she was keeping house for her sister Oriana and brother-in-law William Fanning in Sydney, Australia, when she met Thomas Huxley, who was then a surgeon and zoologist on HMS *Rattlesnake*. After an eight-year engagement she arrived in London in 1855 and married Huxley. She assisted her husband in his work by translating German and drawing diagrams for his lectures. She was largely responsible for arranging his trip to Egypt in 1872, and during his time away cared for their seven children (ages six to fourteen) as well as the house they were building. She was also a poet, and at the age of eighty-six published a volume of her collected poems, *Poems of Henrietta A. Huxley with Three of Thomas Henry Huxley* (London: Duckworth, 1913).

Huxley, Thomas Henry (1825–95), was a British zoologist, educator, and science popularizer, and one of Tyndall's closest friends. Following medical training at Charing Cross Hospital in London, he was a ship's surgeon aboard the HMS *Rattlesnake* expedition to Australia (1846–50) under Captain Owen Stanley (1811–51). While in Australia he met Henrietta Anne Heathorn, whom he married in July 1855 after an eight-year engagement. He became professor of natural history at the Royal School of Mines, London (1854–71) and moved with the school to South Kensington, where he taught in a new laboratory until 1885. He joined Tyndall twice at the RI as Fullerian Professor of Natural History (1855–58, 1866–69). In 1864 Huxley, Tyndall, Busk, Frankland, Hirst, Hooker, Lubbock, Spencer, and Spottiswoode formed the X Club, a monthly dining club that collaborated to promote science in British society. Huxley was the secretary of the RS from 1871 to 1880. In 1870 he was the president of the BAAS in Liverpool and served on a number of royal commissions. Owing to overwork from lectures, reviewing, numerous administrative positions, and a seat on the first London School Board, in January 1872 he had a breakdown, as a result of which he took a three-month recuperative holiday in Egypt and the Mediterranean.

Jenni, Peter (n.d.), was one of Tyndall's Alpine guides on a number of his climbing trips. In 1864 they had an accident in an avalanche on the Piz Morteratsch, which Tyndall retold in *Hours of Exercise in the Alps* (206–18). Tyndall had a lot of respect for Jenni, who, Tyndall said, "behaved with the greatest coolness and courage" even though Tyndall also argued that Jenni "ought not to have taken us down the ice-slope so late in the day" (*Hours of Exercise in the Alps*, 218; see also *Ascent of John Tyndall*, 233–34, 276).

Joule, James Prescott (1818–89), was the son of a brewer, and studied for a brief time under John Dalton (1766–1844), a chemist and natural philosopher, before focusing his own work on electromagnetism. In 1838 Joule published his first paper on electromagnetic engines, and spent the next decade studying their efficiency. This research led to his formulation of the mechanical equivalent of heat, which he presented to the BAAS in 1843, and for which the RS awarded him a Royal Medal in 1852. Tyndall, however, promoted Mayer's work on the same subject, arguing that the latter deserved credit for the discovery; this brought Tyndall and Mayer into conflict with Joule and the Scottish physicists (including Thomson and Tait) who championed Joule's work as establishing the conservation of energy through experiment. By the 1870s Joule, though in poor health, was still pursuing experimental work.

King, Clarence (1842–1902), was a geologist and first director of the US Geological Survey. Educated in chemistry at Yale, King became interested in geological sciences and in 1863 joined Josiah D. Whitney's geological survey of California. In 1867 he convinced Congress to support a geological survey of the Fortieth Parallel; he was assigned to map, survey, and assess the natural resources of the area between the Sierra Nevadas and the Great Plains. This fieldwork was described in King's book *Mountaineering in the Sierra Nevada* (1872), first published as a series of essays in the *Atlantic Monthly*. He became the director of the US Geological Survey in 1879, a post he held until 1881. Though already married, in 1888 King secretly married a Black woman, Ada Copeland. He lived an elaborate double life in which he concealed his marriage from the public, and his true name and identity from Copeland, with whom he had five children, until his death.

Ladd, William (1815–85), was a British optician and maker of scientific instruments with a London shop at 11–12 Beak Street (see G. Clifton, *Directory of British Scientific Instrument Makers 1550–1851* [London: Zwemmer, 1995], 125–26, 161). He and his partner filed a number of patents for their instruments, and some of them have ended up in the collection of the Science Museum, London (https://collection.sciencemuseumgroup.org.uk/people/cp52344/william-ladd).

Lake, William Charles (1817–97), was the dean of Durham from 1869 to 1894. He was educated at Rugby and Oxford, becoming a fellow of Balliol in 1838. In 1869 he was appointed to the deanship at Durham, which also meant he was the warden of Durham, at a time when the university had only fifty students. He emphasized the study of theology at Durham, and in 1871 the Durham College of Science at Newcastle was founded. Under his tenure, the cathedral at Durham was also renovated and restored. He was the dean until he resigned due to failing health in 1894.

Lesley [Leslie], Joseph Peter (1810–1903), was an American author and geologist. He worked with the first Pennsylvania Geological Survey from 1836 to 1842, then became a minister. In 1852 he left the ministry to devote all of his time to geology. He was a professor at the University of Pennsylvania for decades. Elected to the American Philosophical Society in 1856, he served as the society's librarian (1858–85), secretary (1859–87), and vice-president (1887–98). He was also a member of the National Academy of Sciences in the United States, and in 1884 he was president of the American Association for the Advancement of Science.

Lockyer, Joseph Norman (1836–1920), was an English astronomer and a journal editor. He made preliminary observations of the moon's surface, then of Mars. He published his first scientific paper on these observations in 1863. He founded the journal *Nature* in 1869 and became its first editor, retiring in 1919. He was elected FRS in 1869, and in 1874 he was awarded the RS's Rumford Medal. In 1875 he was elected a corresponding member of the Académie des Sciences in Paris and was awarded its Janssen Medal. He was the secretary of the Royal Commission on Scientific Instruction and the Advancement of Science, which sat from 1871 to 1874. He attended a number of solar eclipses and was concerned with observations of the corona.

Longman family (*per.* 1724–1972), was a London-based publishing family that established itself in the trade in the eighteenth century. Brothers Thomas Longman (1804–79) and William Longman (1813–77) became head of the firm in 1842. A major force in the world of Victorian print, Longmans published authors including John Stuart Mill, Herbert Spencer, and Anthony Trollope, as well as the *Edinburgh Review*. Tyndall published fourteen of his sixteen books with them.

Lubbock, John (1834–1913), was an English banker and a science writer. He was the head of the family bank by 1865, but had a lifelong interest in natural history fueled by his close friendship with Darwin. He was a member of the RI, the Geological Society of London, and the X Club; he was elected FRS in 1858. His first major work in science was the 1865 *Pre-historic Times as Illustrated by Ancient Remains*. He served as vice-chancellor of London University from around 1872, and as its MP from 1880.

Lyell, Charles, first baronet (1797–1875), was a Scottish geologist and lawyer, whose *Principles of Geology*, 3 vols. (London: John Murray, 1830–33) advocated a "uniformitarian" view of the processes of geological change. Educated at Exeter College, Oxford, he studied law at Lincoln's Inn, London. He was briefly professor of geology at King's College, London (1831–33), after which he lived by private means. In July 1832 he married Mary Elizabeth Horner (1808–73), who collaborated in her husband's scientific work. A prominent figure in the Geological Society of London, Lyell was twice elected as its president (1835–37; 1849–51). He was knighted in 1848. Lyell and Hooker were instrumental in bringing Darwin's theory of natural selection to public attention. The discovery in 1859 that humans had lived alongside the mammoths of the last ice age led to Lyell's last major publication, *Geological Evidences of the Antiquity of Man* (London: John Murray, 1863).

Maxwell, James Clerk (1831–79), was a Scottish physicist and mathematician. He graduated in the Cambridge mathematics Tripos as second wrangler in 1854, after which he held professorships at Marischal College, Aberdeen (1856–60), King's College, London (1860–65), and, from 1871, the Cavendish professorship at the University of Cambridge. Maxwell is best known for his contributions to the study of electromagnetic radiation. He developed Faraday's field theory and demonstrated that electricity, magnetism, and light were instances of the same phenomenon. Maxwell's work culminated in his publication *A Treatise on Electricity and Magnetism* (Oxford: Clarendon Press, 1873).

Mayer, Alfred Marshall (1836–97), was an American physicist whose research included work on magnetism and sound, and who popularized physics, particularly for children. He was a self-taught chemist, and his first paper brought him to the attention of Joseph Henry. He went on to be a professor of chemistry in Maryland and of physical science in Missouri, moving to Gettysburg College, then Lehigh University, and finally setting up the physics department at the Stephens Institute of Technology in New Jersey in 1871. He became a member of the National Academy of Sciences in 1872 and was also a member of the American Philosophical Society and the American Academy of Arts and Sciences.

Mayer, Julius Robert von (1814–78), was a German doctor and theoretical physicist, noted for his work on heat. He practiced medicine in Heilbronn, and his work in physics emerged out of physiological observations. In a paper published in 1842 he formulated one of the first statements on the conservation of energy, a discovery for which Joule is generally credited. Partly because of Mayer's lack of institutional affiliation, his work remained obscure until the 1850s, when it was discovered by Helmholtz, Clausius, and Tyndall, the last of whom helped to introduce his name in Britain, most famously in an 1862 lecture at the RI. Mayer suffered from depression and spent time at asylums in the 1850s.

Murray, John, III (1808–92), was a publisher and the third generation of the Murray family publishing dynasty, which his grandfather had established in the late eighteenth century. Educated at Edinburgh and interested in natural history and travel, he filled a number of notebooks about the topics, which became the Murray's Handbooks for Travellers about countries across Europe and the world. He took over the company upon his father's death in 1843, inheriting a prestigious backlist of books and high status in the publishing world. Under John Murray III the firm was notable for publishing travel literature and works of science, including Charles Lyell's *Principles of Geology* (1830–33) and Darwin's *On the Origin of Species* (1859). Murray maintained the firm's conservatism, a position expressed in the house periodical, the *Quarterly Review*. He was elected a member of the Alpine Club in 1858 and likely became Tyndall's friend then. Although Murray published *Glaciers of the Alps* (1860), most of Tyndall's subsequent books were published with Longmans, whom he preferred for financial reasons.

Newton, Isaac (1642–1727), was an English natural philosopher and mathematician. Educated at Trinity College, Cambridge from 1661 to 1665, he became a fellow at Trinity upon his graduation. He became Lucasian Professor of Mathematics in 1667 and held this post until 1695. He was elected FRS in 1672, and became president of the RS in 1703, a position he held until his death. He was also master of the mint from 1696 until his death. He was best known for his *Principia* (1687), in which he presented his law of universal gravitation.

Ogle, William (1827–1912), was an Oxford physician and naturalist. He was a lecturer on physiology at Saint George's Hospital from 1858 to 1869 and assistant physician from 1869 to 1872. He also published on flower structure and mechanisms for fertilization. He was a correspondent and acquaintance of Darwin.

Owen, Richard (1804–92), was an English comparative anatomist and paleontologist who was known for his work on fossils and who coined the word *dinosaur*. He was educated in Lancaster, Edinburgh, and London, and held positions at the Hunterian Museum and the British Museum; he founded the Natural History Museum in South Kensington, which opened in 1881. He was elected FRS in 1834, and he won the Royal Medal of the RS in 1846, then their Copley Medal in 1851. He earned a number of honors due to his work in comparative anatomy and medicine, and was knighted in 1884. He was often at odds with Tyndall and other members of the X Club.

Pasteur, Louis (1822–95), was a French microbiologist who is best known for his work on microorganisms. His work helped to advance the prevention of diseases, especially rabies and anthrax, through vaccination. He also influenced the manufacture of beer and wine through his research on fermentation. Tyndall and Pasteur were mutually supportive of each other's work.

Peel, Emily (née Hay, 1836–1924), was the daughter of George Hay (1787–1876), 8th Marquess of Tweeddale, and Lady Susan Montagu (1797–1870). In 1856 she married Robert Peel, and the couple had three daughters and one son. Peel was a regular attendee of Tyndall's lectures. She was clearly interested in science and was an intelligent and well-traveled woman. Later in life, she left her husband and moved to Italy to live in Geneva and later Florence, where she died in 1924.

Peel, Robert, third baronet (1822–95), was a politician and Liberal. The son of Prime Minister Robert Peel (1788–1850), he took over his father's seat in the Commons in 1850 and held it until 1880. He was involved in politics for the rest of his life, and also enjoyed breeding and racing horses, under the name Mr. F. Robinson. He made poor financial decisions—in 1871 had to sell some of his family's fine art collection to the National Gallery, and in 1884 the Peels had to sell their Drayton Estate. He died of a brain hemorrhage in 1895.

Petitjean, Tony (1823/4–73), was a French chemist who patented a process for silvering glass in July 1855. In 1871 he was granted £25 from the RS's Scientific Relief Fund, which Tyndall helped administer. He was in a convalescent home recovering from an illness in April of 1871.

Poggendorff, Johann Christian (1796–1877), was a German physicist at the University of Berlin. His major contribution to science was as editor for fifty-two years (1824–76) of *Annalen der Physik und Chemie*, which became known as *Poggendorff's Annalen* during his tenure. His wife, Charlotte E. Kneser (1805–65), aided in his editorial work. He used his broad knowledge of science and scientific men in producing the *Biographisch-literarisches Handworterbuch zur Geschicte der exacten Wissenschaften*, the first two volumes of which were published in 1863.

Pollock, Juliet (née Creed, 1819–99), was a daughter of Catherine Herries and the Reverend Henry Creed, vicar of Corse, Gloucestershire. She married William Frederick Pollock, the barrister and author, in 1844, and had three sons with him. She and her husband were loyal supporters of the RI, attended Tyndall's lectures, frequently invited Tyndall to dinner, and were among his closest friends. Pollock and Tyndall had nicknames for each other: she called him Boreas, he called her Eolia. In classical mythology, Eolia is the master of the wind and Boreas is the north wind. Juliet Pollock was the author of several books, including *Julian and His Playfellows* (London: Grant and Griffith, 1852), *New Friends* (London: John W. Parker and Son, 1858), and *Macready as I Knew Him* (London: Remington, 1885).

Pollock, William Frederick, second baronet (1815–88), was a British lawyer and author. The eldest son of Jonathan Frederick Pollock, first baronet (1783–1870), and Frances Rivers (d. 1827), he married Juliet Creed in 1844. He was admitted to Trinity College, Cambridge, in 1832, obtaining his BA (1836) and MA (1840). Admitted to the Inner Temple in 1833 and called to the bar in 1838, he then became a master of the court of exchequer in 1846 and was appointed Queen's remembrancer in 1874. Pollock's *Personal Remembrances*, 2 vols. (London: Macmillan, 1887), illustrates how the family circulated among London's elite and entertained at their Montagu Square home. Both William and Juliet were loyal supporters of the RI, and Tyndall was frequently among their guests.

Rankine, William John Macquorn (1820–72), was a Scottish civil engineer and physicist. He studied natural philosophy at the University of Edinburgh, and also studied chemistry outside the university. He became secretary of section A (Mathematical and Physical Science) of the BAAS in 1850 and became very well known in the field. He was elected FRS in 1853. He became the chair of Civil Engineering and Mechanics at the University of Glasgow in 1855. He also began work related to the architecture of ships for the Royal Navy. He worked with Thomson on the new science of thermodynamics and developed his own Rankine scale for temperature; it was similar to the Kelvin scale but in degrees Fahrenheit.

Ripon, Lord: see Robinson, George Frederick Samuel.

Robinson, George Frederick Samuel, 1st Marquess of Ripon (1827–1909), was a politician with radical, Christian socialist views. He was first elected to Parliament in 1852, where he led a small group of radicals but was also friendly with members on all sides of the House of Commons. Robinson became the Earl of Ripon in 1859. In the 1860s he held the posts of secretary of state for war and secretary of state for India. He joined Gladstone's first administration in 1868 as lord president of the council. Robinson negotiated with the United States over conflicts that emerged from the American Civil War (1861–65); his success in this matter earned him a marquessate.

Rothschild, Albert Salomon Anselm Freiherr von (1844–1911), was an Austrian banker, and member of the banking family Rothschild. He held the title of baron. Rothschild had a number of interests outside of banking and used his money to fund these. He owned the first car in Vienna, built an observatory, and collected art.

Russell, Frances Anna Maria Elliott (née Kynynmound, 1815–98), was the daughter of Gilbert Elliott Murray Kynynmound (1782–1859), 2nd Earl of Minto, and Mary Brydone (1786–1853). In 1841 she married John Russell, twenty years her senior, and was drawn into the center of London political life. Although she disliked the social aspects of this position, she was deeply interested in politics and committed to liberal political causes. With her husband, she supported the RI and regularly attended Tyndall's lectures.

Russell, John, 1st Earl Russell (1792–1878), was a politician and first entered Parliament in 1814. He was the son of John Russell (1766–1839), the 6th Duke of Bedford, and Georgiana Elizabeth (née Byng, ca. 1768–1801). Russell became a leader of the Whigs and was a crucial figure in the passage of the First Reform Act of 1832. He was prime minister from 1846 to 1852 and from 1865 to 1866. In 1856 he became the first Earl Russell. Russell and his wife, Frances Russell, supported the RI and regularly attended Tyndall's lectures.

Russell, Mary Agatha (1853–1933), was an author and the daughter of John and Frances Russell. She coedited her mother's memoirs, *Lady John Russell: A Memoir with Selections from Her Diaries and Correspondence*, and published them in 1910. There are two portraits of Mary Russell in the National Portrait Gallery in London. She became friends with Tyndall by a connection with her cousin Arthur Russell, a Liberal party politician.

Sabine, Edward (1788–1883), was an Irish astronomer and geophysicist, as well as a general in the British Army from 1803 to 1870. He was introduced to London scientific circles by his older brother, Joseph Sabine, after the Napoleonic Wars were over in 1815. He was a member of a number of scientific expeditions, which focused efforts on measuring latitude and longitude and measuring gravitational forces at different latitudes. He was elected FRS in 1818 and was president of the RS from 1861 to 1871. He was awarded the RS's Copley Medal in 1821 for his work in the Arctic, and a Royal Medal in 1849 for his writings on geomagnetism.

Sharpey, William (1802–80), was a Scottish anatomist and physiologist. He was known as the "father of British physiology." Educated at Edinburgh, he took his MD in 1823 with a thesis on stomach cancer. After practicing with his stepfather, he became the new chair of anatomy and physiology at UCL in 1836, a position he held until 1874. He became FRS in 1839 and a member of the RS's council in 1844. He was also the secretary of the RS from 1854 to 1872. Some of his pupils at UCL included Joseph Lister, John Marshall, and J. B. Hayes.

Spencer, Herbert (1820–1903), was an English biologist, philosopher, journalist, and sociologist. Through family connections, at the age of twenty-eight he secured a post as a subeditor at the *Economist*, which put him at the heart of literary London. He published his first book, *Social Statics*, in 1850 and his second book, *Principles of Psychology*, in 1855. Spencer was a respected philosopher, but one of his most lasting contributions was his *Principles of Biology* (1864), in which he coined the phrase "survival of the fittest." Along with Tyndall and seven others, Spencer was a member of the X Club, a monthly dining club in which they promoted science in British society. Spencer was the only member who was not an FRS.

Spottiswoode, William (1825–83), was an English mathematician and physicist. He was also a partner of Eyre & Spottiswoode, the Queen's printer. He was educated at Harrow, then went to Oxford where he earned a first-class degree in mathematics in 1846 and an MA in 1848. Working with William Rowan Hamilton's theories on quaternions, he began publishing papers in continental journals. In 1851 he published a mathematics textbook, *Elementary Theorems Relating to Determinants*. He began working on experimental physical science in 1871 and gave lectures at the RI and the BAAS. He was a founding member of the X Club and was elected FRS in 1853. He was treasurer of the RI from 1865 to 1873, and of the RS from 1871 to 1878. He became president of the BAAS in 1878 and was elected president of the RS that same year.

Stanley, Edward Henry, 15th Earl of Derby (1826–93), was a politician and diarist. He earned his BA in 1848 at Trinity College, Cambridge, where he was a member of the Cambridge Apostles, an undergraduate society that influenced many British intellectuals. He entered Parliament in 1848 as the Conservative member for King's Lynn, although he also had sympathies with aspects of the Liberal party, which he would join several decades later. His father, Edward George Geoffrey Smith Stanley, the 14th Earl of Derby, served as prime minister in the 1850s and 1860s and helped advance his son's political career; the younger Stanley also became a friend and political ally of Benjamin Disraeli. In the early 1870s he was engaged in pursuing reconciliation with the United States over conflicts that emerged out of the American Civil War (1861–65). He went to the House of Lords in 1869 after being an MP for King's Lynn for twenty years.

Steuart, Elizabeth Dawson (née Duckett, 1802–93), was the wife of Tyndall's friend William Richard Steuart (1792–1852), and a friend of Tyndall's from Leighlinbridge. She helped Tyndall manage affairs in Ireland, and gather intelligence about family members (and others) who continued to ask for his aid. She also took a keen interest in Tyndall's career and was probably his most consistent and long-lived patron.

Stevenson, David (1815–86), was a Scottish civil engineer and the brother of Thomas Stevenson. He was educated in Edinburgh, and in 1838 he joined his father and brother Alan in his father's engineering firm, Robert Stevenson & Sons. The firm was largely responsible for harbor construction and bridges over major rivers. He was elected FRSE in 1844. In 1851 he started working with Thomas for the fishery board. In 1853 he became an engineer for the Northern Lighthouse Board, and in 1855 he and Thomas started working together to build lighthouses. They built twenty-eight beacons and thirty lighthouses. He retired from the company D. and T. Stevenson in 1884.

Stevenson, Thomas (1818–87), was a Scottish civil engineer, meteorologist, and the brother of David Stevenson. He was educated in Edinburgh and apprenticed with his father's engineering firm in 1839–41. By 1846, on his father's retirement, he became a junior partner in the company that became D. and T. Stevenson. The company specialized in maritime engineering and lighthouses. Together David and Thomas built twenty-eight beacons and thirty lighthouses. He was elected FRSE in 1848, and became president of the Royal Society of Edinburgh in 1884.

Stokes, George Gabriel (1819–1903), was an Irish physicist and mathematician. He was Lucasian Professor of Mathematics at Cambridge University from 1849 to 1903. He became FRS in 1851 and served as secretary of the RS from 1853–85. As secretary, part of his duties involved understanding and explaining experiments to RS members. He was president of the RS from 1885 until 1890, and president of the BAAS in 1869.

Sumner, Charles (1811–74), was an American senator from Massachusetts from 1851 until his death. He graduated from Harvard University in 1830 and from its law school in 1833, and then moved to Boston to begin a law practice. In 1856 he was attacked for his abolitionist views while at his desk in the Senate by the representative from the slave state of South Carolina, Preston Brooks, and was absent from service for over three years due to his injuries. Sumner traveled widely and had influential friends all over the world.

Tait, Peter Guthrie (1831–1901), was a Scottish physicist and mathematician. Educated at Cambridge, he was a professor of mathematics at Queen's College Belfast from 1854 until 1860 before being appointed chair of natural philosophy at the University of Edinburgh. There, he joined Thomson, Joule, and other North British physicists in establishing a new science of energy. This North British group, which had strong Presbyterian commitments, came into conflict with Tyndall and Huxley, and Tait in particular clashed with Tyndall when the latter championed Julius Mayer's work.

Thomson, William (1824–1907), was a Scottish mathematician and physicist. He was educated at Cambridge where he graduated senior wrangler. After his graduation Thomson returned to Glasgow, where he worked at the university, studying electricity and the new science of thermodynamics. He was elected FRS in 1851 and awarded the RS's Copley Medal in 1883. Thomson was president of the RS after George Gabriel Stokes, from 1890 to 1895. He presided over a number of scientific meetings, including the BAAS in Edinburgh in 1871. As a leader of the North British group of physicists, Thomson came into conflict with Tyndall and Huxley. In 1892 he was made Baron Kelvin. He was unhappy with the seeming arbitrariness of temperature scales, so he devised a scale for temperature change that includes an absolute zero; the Kelvin scale bears his name today.

Trevor, Charles Cecil (1830–1921), was the assistant secretary in the Harbour Department at the Board of Trade from 1867 to 1895. He was educated at Rugby and then Saint Catharine's College, Cambridge, where he earned his BA in 1852 and his MA in 1855. He was called to the bar at Lincoln's Inn in that same year. While at the Board of Trade, he worked with Tyndall on Irish lighthouses. He was also a member of the Athenaeum Club. He was knighted in 1896. After his retirement from the Board of Trade, he was a conservator of the River Thames, until 1902.

Tyndale, Hector (1821–80), was Tyndall's distant American cousin who had been a Union general during the American Civil War (1861–65). He joined his father's business of importing china and glassware from Europe in 1845. He made several tours of Europe, including a visit to London in March and April of 1871. With Joseph Henry and Edward L. Youmans, Tyndale helped to organize Tyndall's American lecture tour from October 1872 to February 1873, and the three acted as trustees for the Tyndall Fund to support pure science research in the United States.

Tyndale, Julia (née Nowlen, 1823–97), was Hector Tyndale's wife.

Tyndall, Caleb (ca. 1798–1879), was Tyndall's uncle who lived in Leighlinbridge, and a drain on Tyndall's accounts. Over time, Tyndall gave money to Caleb's family for a marriage, a cow, and the purchase of a farm.

Tyndall, Emily (1817–96), sometimes called Emma by her family, was Tyndall's sister. Upon their mother's death in 1867, she moved in with Dr. John Tyndall, a Gorey relative.

Tyndall, John (ca. 1815–75), was a relative of Tyndall's in Gorey, County Wexford, Ireland. A doctor and member of the Royal Historical and Archaeological Society of Ireland, he died in 1875 of acute bronchitis.

Tyndall, Sarah (née Macassey [Macasey/McAssey], 1793–1867), was Tyndall's mother, who was widowed in March 1847 and lived in Leighlinbridge. He continued to give her money each year up until her death. Elizabeth Steuart was the liquidator of her estate.

Valentin, William George (1829–79), was a German chemist. He arrived in England to study at the Royal College of Chemistry, where he quickly became senior assistant in the laboratory. He also taught at the Science Schools, South Kensington. He held the post of gas examiner for the Great Western Railway, and finally as the chemical advisor to Trinity House. He died suddenly of apoplexy in 1879.

Victoria, Queen of the United Kingdom of Great Britain and Ireland, and empress of India, (1819–1901), reigned from 1837 until her death. Victoria largely isolated herself from public appearances after the death of her husband, Albert, the prince consort (1819–61), in 1861, though she remained involved in politics. When Gladstone became prime minister in 1868 he urged her to resume public duties, as her seclusion had made the monarchy unpopular and had contributed to growing republican sentiment. But this tendency reversed in 1872 after the illness and recovery of Victoria's son Albert Edward, the Prince of Wales (1841–1910); a public thanksgiving for which Victoria drove in state through London; and a failed assassination attempt on her two days later.

Vieweg, or Friedrich Vieweg und Sohn, was a publishing house in Germany, founded by Johann Friedrich Vieweg (1761–1835), and run by his son, Eduard Vieweg (1797–1869). The publishing house remains in operation today as Springer Vieweg Verlag.

Washington, Catherine (n.d.), was the daughter of a friend from Leighlinbridge, Tyndall's birthplace and hometown. He continued to send her money.

Whymper, Edward (1840–1911), was a British mountaineer and wood engraver. He was apprenticed to his father's engraving firm at the age of fourteen. By the time he was twenty, the publisher Longmans had commissioned him to make illustrations of the Alps, so he went to visit the Alps to gain knowledge for those engravings. He was elected to the Alpine Club in 1861 after he climbed Mont Pelvoux. He is known for a number of first ascents, including the Matterhorn in 1865. On the descent, there was an accident that killed four people. His *Scrambles amongst the Alps in the Years 1860–69* (London: John Murray, 1871) was one of the most popular mountaineering books ever written. Tyndall viewed Whymper as a main competitor in his mountaineering pursuits.

Wiedemann, Clara (née Mitscherlich, 1827–1914), was a translator and the wife of Gustav Wiedemann. She collaborated with Anna Helmholtz on translations of Tyndall's and others' works, from English into German. Neither woman received credit for their translation of Tyndall, which instead went to their husbands (*Historisches Lexikon der Schweiz*; B. Meijer et. al., eds., *Nordisk familjebok*, 2nd ed. [Stockholm: Nordisk familjeboks förlags, 1904–26]).

Wiedemann, Gustav Heinrich (1826–99), was a German physicist and author. He was well known for his work *Lehre von Galvanismus und Elektromagnetismus*, which was first published in 1861 and which, in 1871, was in the process of being revised for a new edition. Wiedemann was trained at the University of Berlin by Heinrich Gustav Magnus and came to know Hermann Helmholtz there. He held professorships of physics at Basel (1854–63), Brunswick (1854–66), and Karlsruhe (1866–71) before taking the chair of physical chemistry at Leipzig in 1871. After the death of Johann Poggendorff in 1877, Wiedemann took over as editor of the *Annalen der Physik und Chemie*, which subsequently took on the name *Wiedemann's Annalen*.

Youmans, Edward Livingston (1821–87), was an American writer, lecturer, and founder in 1871 of the "International Scientific Series," with the publisher Appleton. In this series, Youmans published works of scientists (such as Tyndall and Spencer) in a number of modern languages. He founded and edited *Popular Science Monthly* in 1872. With Joseph Henry and Hector Tyndale, Youmans helped to organize Tyndall's American lecture tour from October 1872 to February 1873, and the three acted as trustees for the Tyndall Fund to support pure science research in the United States.

Young, James (1811–83), was a Scottish chemist and philanthropist who patented the process of making paraffin oil and the oil itself, called Young's paraffin. He was educated in Glasgow and worked at University College, London with Thomas Graham. He began working at Tennant, Clow & Co. in 1844 in Manchester. He improved methods of production at their plants, including a cheaper indigo dye. In 1848 he patented a method to produce sodium and potassium stannate from tinstone. He also experimented with the dry distillation of coal and patented the process for doing so from shale. He established Young's Paraffin Light and Mineral Oil Company in 1864 and by his retirement in 1870 had become known as Paraffin Young. He was president of Anderson's University, Glasgow from 1868 to 1877, and elected FRS in 1873.

Zöllner, Johann Karl Friedrich (1834–82), was a German astrophysicist. He studied with Heinrich Magnus and Wiedemann, receiving his PhD in 1857 with a thesis on photometry. He became a professor at the University of Leipzig in 1866. His research included the development of several instruments for astrophysical research as well as more theoretical work, including a theory of comets. In the 1870s and 1880s he began using his published work to attack other scientific practitioners, including Tyndall. This, along with his growing commitment to spiritualism, led to his increasing isolation from the scientific community; some considered that he suffered from mental illness.

INDEX

The index refers to the letters using their serial numbers. The letter 'n' denotes a footnote reference and appears if an item is not otherwise mentioned in the letter to which the footnote pertains. A number in **bold** signifies a letter to or from a correspondent.